Everything You Always Wanted to Know About…

Physics

5th edition

STERLING
Education

Customer Satisfaction Guarantee

Your feedback is important because we strive to provide the highest quality educational materials. Email us comments or suggestions.

info@sterling-prep.com

We reply to emails – check your spam folder

5 4 3 2 1

ISBN-13: 979-8-8855730-3-0

Sterling Education materials are available at quantity discounts.

Contact info@sterling–prep.com

Sterling Education
6 Liberty Square #11
Boston, MA 02109

Published by Sterling Education

Printed in the U.S.A.

STERLING
Education

From the foundations of Newtonian physics to the complex mechanisms of atomic and nuclear theories, this comprehensive text is an excellent guide for learning physics. This book provides an essential overview of physical laws, principles, and approaches elucidating multidimensional concepts of mechanics, statics and dynamics, as well as particle physics.

This book is for those who want to develop a better understanding of our physical universe, as well as the relationships between different laws of physics. It explores how physical phenomena continuously intersect with our daily life.

The content of the book covers essential laws and mechanisms affecting physical processes. You will learn about kinematics and dynamics, statics and equilibrium, foundations of gravity, energy, work, sound and light, electricity and magnetism, basic principles of atomic physics, as well as heat and thermodynamics.

Created by highly qualified physics instructors and researchers, this book takes readers on a journey through the fascinating, complex, and sometimes incomprehensible world of physics. This book educates and empowers readers, regardless of whether they took physics courses or are unfamiliar with the basics.

The editors sincerely hope that this guide is a valuable resource for learning fundamental concepts of physics.

250724akp

If you benefited from this book, please leave a review on Amazon so others can learn from your input. Reviews help us understand our customers' needs and experiences while keeping our commitment to quality.

Page intentionally left blank

Table of Contents

Table of Contents (*continued*)

Table of Contents (*continued*)

Table of Contents (*continued*)

Table of Contents (*continued*)

Table of Contents (*continued*)

CHAPTER 3: Force, Motion, Gravitation (*continued*)

Table of Contents (*continued*)

Table of Contents (*continued*)

Table of Contents *(continued)*

Table of Contents (*continued*)

Table of Contents (*continued*)

Table of Contents (*continued*)

Table of Contents (*continued*)

Table of Contents *(continued)*

Table of Contents (*continued*)

CHAPTER 7: Light & Optics (*continued*)

Table of Contents (*continued*)

Table of Contents (*continued*)

Table of Contents (*continued*)

Table of Contents (*continued*)

CHAPTER 8: Fluids & Solids (*continued*)

Table of Contents (*continued*)

Table of Contents *(continued)*

Table of Contents (*continued*)

Table of Contents (*continued*)

CHAPTER 9: Thermal Physics (*continued*)

Table of Contents (*continued*)

Table of Contents (*continued*)

Table of Contents (*continued*)

Table of Contents (*continued*)

Table of Contents (*continued*)

Table of Contents (*continued*)

CHAPTER 11: Electric Circuits (*continued*)

Table of Contents (*continued*)

Table of Contents (*continued*)

Table of Contents (*continued*)

Table of Contents (*continued*)

CHAPTER 12: Atomic & Nuclear Structure (*continued*)

Table of Contents (*continued*)

Table of Contents (*continued*)

Page intentionally left blank

Chapter 1

Kinematics & Dynamics

The Nature of Science

Units and Dimensions

Vectors and Vector Components

Vector Addition & Subtraction

Speed and Velocity

Acceleration

Freely Falling Bodies

Chapter Summary

Page intentionally left blank

The Nature of Science

Scientific discovery

Observation is the first step toward scientific discovery. Observations recognize patterns and anomalies in unexplained phenomena.

Hypothesis is advanced as a proposed explanation for the observation.

Evaluating a hypothesis involves experiments and additional observations (i.e., recording and collecting data).

Experimental observations are made while manipulating *independent variables* (e.g., how long the plant is exposed to sunlight per twenty-four hours) with *dependent variables* (data collected regarding height measurements).

Observations generate *data* that:

1) *support the hypothesis* (i.e., different from proves) or

2) *refute the hypothesis*.

Hypothesis are modified by narrowing or expanding their application to explain additional data. More experiments are performed, and the results either *support* (not prove) or *refute* the hypothesis during this iterative process.

Principles are the initially proposed explanations that are specific and apply to a narrow range of phenomena.

Theories build upon valid hypotheses

Theories are proposed to account for the observations when a hypothesis is valid over a range (e.g., sunlight, temperature, or humidity).

Theories are advanced to explain the phenomena and to predict what will happen under other conditions with different variables.

Further observations determine whether the predictions were accurate.

Research (re-search or *again search*) continues as these many observations support hypotheses and theories.

When theories are accurate over a broad range of scenarios and experiments, *Laws* are developed based on these repeatedly replicated results.

Laws as established theories

Laws are brief descriptions of how nature behaves in a broad set of circumstances. They are the consolidation of several theories. Laws are the products of multiple theories tested and proven, culminating in a broad claim regarding those predictions.

Laws are widely applicable explanations. Laws with broad and straightforward claims are more robust than those that make claims on a narrower, more specific set of variables and parameters.

The number of Laws describing physical phenomena is minuscule compared to the number of theories and magnitudes, which are less than the number of supported hypotheses.

The number of refuted (unsupported) hypotheses may be several magnitudes larger.

Failed experiments do not typically refer to the physical process where the researcher made a human error during the experiment. A failed experiment means that the *data does not support the hypothesis.* Therefore, after additional trials to validate the "failed" results, the hypothesis must be modified to predict the results of either future experiments or objective observations in the physical world.

Theories, models and laws

Theories emerge from a hypothesis that experimental data has repeatedly supported. Theories are detailed statements that provide testable predictions of the behavior of natural phenomena.

Theories are established to explain observations and then tested (supported or refuted) based on their predictions. Tested theories can articulate the boundaries of the tested hypothesis.

Theories are more specific than *hypotheses* but vaguer and more encompassing than *models*.

Models are constructed to explain the observed behavior if a hypothesis accurately predicts a phenomenon. They are more specific in their application than theories.

Data collected to develop the theories provide the conceptual foundation for constructing models. The model creates mental pictures (e.g., complicated weather patterns displayed as visual pictures for the viewer) of the physical phenomena.

Models should be consistent with scientific theories' predictions and support a comprehensive understanding of a phenomenon. Models allow for predictive outcomes under specific theories (e.g., when the wind speed reaches x, then y will occur).

Multiple models are integrated to explain a portion, or the entire scope, of a *theory*.

Understand the limitations of a model and do not apply it dogmatically unless its applicability is evaluated. A model is an underlying representation of a part of a theory; it is not intended to provide a complete picture of every phenomenon under that theory.

Instead, models provide the fundamental aspects of the theory.

Units and Dimensions

Measurement and uncertainty

No measurement is exact; there will always be some uncertainty due to limited instrument accuracy and precision.

Accuracy depends on how well-calibrated the actual quantitative values of measurements are. For example, an empty scale holding no objects initially calibrated to 0 grams is more accurate than the same empty scale calibrated to 1 gram of mass

Precision is based upon the scale of the measurement, with measurements on a smaller scale being more precise than measurements on a larger scale.

For example, a meter stick with millimeter markings (0.001 m) is more precise than one with centimeter markings (0.01 m).

Estimated uncertainty of a measurement is written with a ± sign.

8.8 ± 0.1 cm reads as "8.8 centimeters, *give or take* 0.1 centimeters."

Percent uncertainty is the ratio of the uncertainty to the measured value multiplied by 100.

For example, the percent uncertainty in the measurement is:

$$\frac{0.1}{8.8} \times 100 \approx 1\%$$

Significant figures

Significant figures are the number of reliably known digits in an expressed value.

For example,

- 22.41 cm has *four* significant figures.
- 0.058 cm has *two* significant figures because the initial zeroes before the significant digits are ignored. For three significant figures, write 0.0580 cm.
- 70 km has *one* significant figure (zeroes that serve as placeholders without a decimal point are not significant).

For one significant figure, write 70 km.

For two significant figures, write 70. km (i.e., decimal point changes the significance of the trailing zero).

For three significant figures, write 70.0 km.

Significant figures equal the fewest figures in the calculation when multiplying or dividing.

12.5 cm × 5.5 cm = 68.75 cm, but because 5.5 cm has two significant figures, the value 68.75 cm must be rounded to 69 cm.

The answer is no more precise than the least precise value used when adding or subtracting.

Round the quantity with higher precision to the same decimal place as the quantity with lower precision. For example:

68.75 cm - 32.1 cm = 36.65 cm; 32.1 cm has three significant figures.

Therefore, 36.65 cm must be rounded to 36.7 cm with three significant figures.

Calculators may not display the correct number of significant figures or too many digits depending on the instrument's setting.

Order of magnitude for rapid estimating

Estimate a calculated quantity by rounding numbers to one significant figure, resulting in the proper *order of magnitude*.

Round to the nearest power of 10.

$3{,}321 \times 401 = 1{,}331{,}721$

$1{,}331{,}721 \approx 13.3 \times 10^5$

Estimate by rounding:

3,321 is rounded to 3×10^3

401 is rounded to 4×10^2

Calculate:

$(3.0 \times 10^3) \times (4.0 \times 10^2) = 12 \times 10^5$

Compare:

13×10^5 (calculated) $\approx 12 \times 10^5$ (estimated)

This estimated value is close to the accurate solution (e.g., the difference in the value from estimating compared to the calculated value) and the same order of magnitude.

Rounding is invaluable because 1) estimating *saves time* when solving questions, and 2) it ensures that the proposed solution is the *proper approach* to producing a valid answer.

Percent error from estimating:

$(1.0 \times 10^5) / (12 \times 10^5) \times 100 = 8.3 \times 10^{-5}$ % (or 0.000083%)

Dimensions and dimensional analysis

Dimensions of a quantity are the base units and are expressed using square brackets around the base units (e.g., [m/t] or meters/second).

speed = distance / time

dimensions of speed: [m/t]

Quantities added or subtracted must have the *same dimensions*.

Dimensional analysis helps check the approach for solving calculations.

Quantity calculated should have the correct dimensions, so:

One dimension only contains the magnitude of quantities.

Two dimensions contain quantities on a 2D plane (x- and y-coordinates).

Three dimensions contain quantities in 3D space (x-, y-, and z-coordinates).

Four dimensions contain quantities in 3D space at a given time (x-, y-, and z-spatial coordinates, plus time coordinate t).

Units, standards, and the SI System

Quantity	**Unit**	**Standard**
Length	Meter (m)	Length of the path traveled by light (in a vacuum) in 1/299,792,458 seconds
Time	Second (s)	Time required for 9,192,631,770 periods of radiation emitted by cesium atoms
Mass	Kilogram (kg)	A platinum cylinder in the International Bureau of Weights and Measures, Paris

Units in different dimensions

Quantity units have comparable properties:

unit + unit = unit

unit− unit = unit

unit × unit = unit^2

unless multiplying different units (e.g., meter × second = m·s)

unit / unit = no unit

unless dividing different units (e.g., meter / second = m/s)

Powers of units represent dimensions:

unit = one dimension

unit^2 = two dimensions

unit^3 = three dimensions

International System of Units (SI) units are meter (m), kilogram (kg), and second (s).

Complex units (e.g., velocity, force, power) derive from SI units.

SI units for translational motion

Basic Translational Motion SI Units		
Quantity	SI Unit	Name
Length	m	meter
Mass	kg	kilogram
Time	s	second

Complex Translational Motion SI Units		
Speed or Velocity	m/s	meter per second
Acceleration	m/s^2	meters per second squared
Area	m^2	square meter
Volume	m^3	cubic meter
Density	kg/m^3	kilogram per cubic meter
Force	$N\ (kg \cdot m/s^2)$	Newton
Energy	$J\ (N \cdot m)$	Joule
Power	W (J/s)	Watt

$$Volume = \frac{kg}{1} \times \frac{m^3}{kg}$$

Kilograms cancel on opposite sides of the division bar.

Units for volume:

$$Volume = m^3$$

As problems become more complex, check units to ensure the solution represents the proper units and that no errors have been introduced.

Metric prefixes for powers of 10

Metric (SI) Prefixes		
Prefix	**Abbreviation**	**Value**
peta	P	10^{15}
tera	T	10^{12}
giga	G	10^{9}
mega	M	10^{6}
kilo	k	10^{3}
deci	d	10^{-1}
centi	c	10^{-2}
milli	m	10^{-3}
micro	μ	10^{-6}
nano	n	10^{-9}
pico	p	10^{-12}

Many of these prefixes are familiar, while some are rarely used.

Converting units

Converting between metric units involves powers of 10.

$1 \text{ kg} = 10^3 \text{ g}$

$1 \text{ mm} = 10^{-6} \text{ km}$

$5 \text{ s} = 5 \times 10^3 \text{ ms}$

Converting to and from imperial (SI) units is more involved.

1 m = 3.28084 ft

1 kg = 2.204 lbs.

1 km = 0.621 mile

Conversion examples

Use dimension analysis to ensure the conversion is evaluated correctly.

Write the conversion factor and cancel units during the calculations.

1. Convert 2.6 kg to grams. (use conversion of 1 kg = 10^3 g)

   2.6 kg × (10^3 g / kg)

   2.6 ~~kg~~ × (10^3 g / ~~kg~~)

   2.6 × (10^3 g) = 2.6 × 10^3 g

2. Convert 3.0 miles to km. (use conversion of 3.28 ft = 1.00 m)

   3.0 miles × (1 km / 0.621 mile)

   3.0 ~~miles~~ × (1 km / 0.621 ~~miles)~~

   3.0 × (1 km / 0.621) = 4.83 km

3. Convert 3.0 km to ft. (use conversion of 3.28 ft = 1 m and 1 km = 10^3 m)

   3 km × (1000 m / 1 km) × (3.28 ft / 1 m)

   3 ~~km~~ × (1000 m / 1 ~~km~~) × (3.28 ft / 1 m)

   3000 m × (3.28 ft / 1 m)

   3000 ~~m~~ × (3.28 ft / 1 ~~m~~)

   3000 × (3.28 ft) = 9,840 ft

Notes for active learning

Vectors and Vector Components

Kinematics in one dimension

Measurements of *position*, *distance*, or *speed* must be made with respect to a reference frame.

For example, if a person walks down the aisle of a moving train, the person's speed with respect to the train is a few miles per hour. Speed with respect to the ground outside is much higher; it is the combined speed of the train and the person walking on the train.

Kinematics describes how objects move concerning a reference frame.

Within a reference frame, *displacement* (movement in position) signifies a change in position.

Displacement and *distance* are different.

Displacement *vs.* distance

Displacement (a solid line shown as 40 m on the graph below) is how far the object is from its starting point, regardless of how it got there.

Distance traveled (a dashed line shown as 70 m) is measured along the path (*magnitude* of displacement and the *direction* of travel) of the object.

Distance is 100 meters, but the displacement is 40 meters:

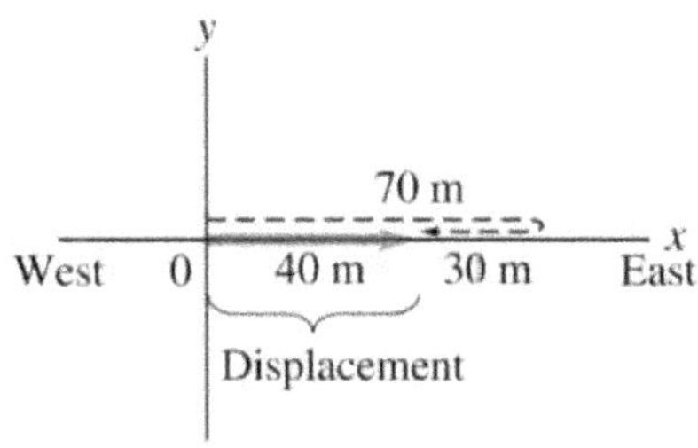

Displacement compared to distance traveled

Displacement is expressed as $\Delta x = x_2 - x_1$.

Displacement involves *direction*; it can be *positive* or *negative* about the coordinate directions.

$\Delta x = x_2 - x_1$

$\Delta x = 30\text{ m} - 10\text{ m}$

$\Delta x = 20\text{ m}$

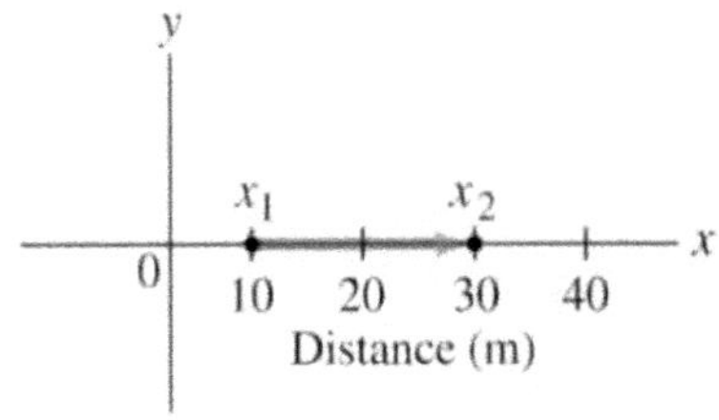

Example of negative displacement:

$\Delta x = x_2 - x_1$

$\Delta x = 10\text{ m} - 30\text{ m}$

$\Delta x = -20\text{ m}$

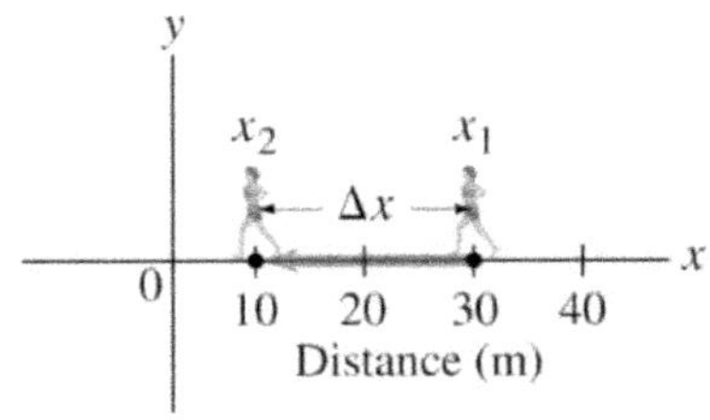

The choice of direction is arbitrary and is selected when choosing a coordinate system.

Computed answers will be correct if the vector components are consistently labeled for the chosen coordinate system direction.

Vectors and vector components

Scalar quantities represent a magnitude without a direction.

Length, time, and mass are quantities without direction.

Vector quantities include both magnitude and direction.

Displacement, acceleration, and force inherently contain a direction applied.

Trigonometric rules for vectors

Three trigonometric rules apply for calculating vector components.

A mnemonic is SOH CAH TOA

- **SOH**: *sin θ* = **o**pposite / **h**ypotenuse
- **CAH**: *cos θ* = **a**djacent / **h**ypotenuse
- **TOA**: *tan θ* = **o**pposite / **a**djacent

"Opposite," "adjacent," and "hypotenuse" for vectors about angle *θ*.

Resolving vector components

For example (diagram below):

vector adjacent to θ_x is v_x

vector opposite θ_x is the vertical dotted line (equivalent to v_y)

hypotenuse is v.

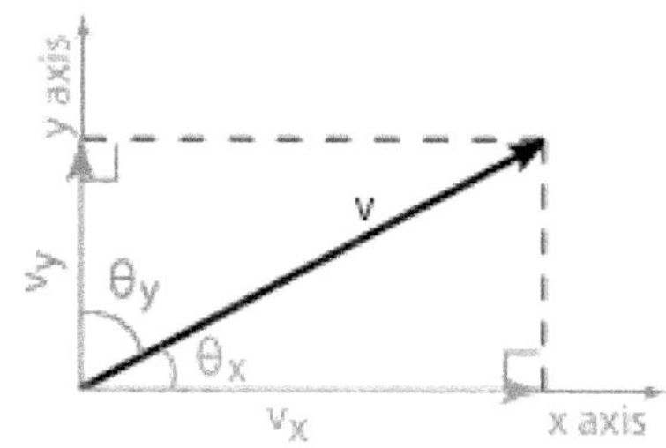

Axis-vector components along x- and y-axis

Another method for the relationship expressed in "SOH CAH TOA" is from the phrase "**S**ome **O**ld **H**airy **C**amels **A**re **H**airier **T**han **O**thers **A**re."

Vector components are portions of the vector in each direction.

Coordinate system for component vectors

Typically, the components are calculated along the axis directions in a coordinate system.

A vector may have a horizontal component and a vertical component in a two-dimensional (2D) coordinate system (i.e., *x*- and *y*-component):

$$v_x = \vec{v}\cos\theta_x = \vec{v}\sin\theta_y$$

$$v_y = \vec{v}\cos\theta_y = \vec{v}\sin\theta_x$$

Pythagorean Theorem calculates the magnitude of the resulting vector:

$$\vec{v}^2 = v_x^2 + v_y^2$$

Vector components in a 2D coordinate system

Examples of vectors and vector components in a 2D coordinate system:

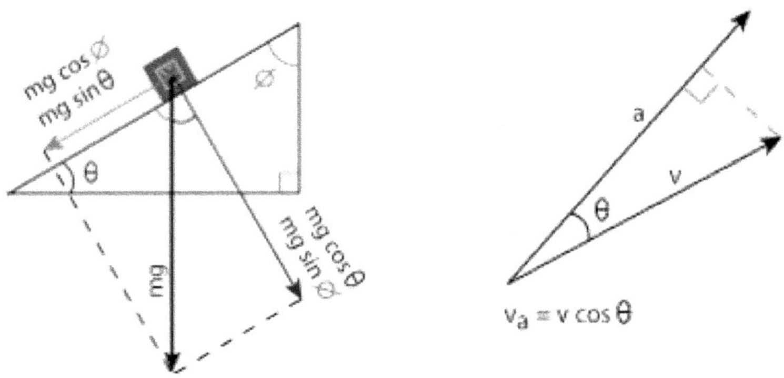

Vector components in a 2D coordinate system with gravity component on slope (shown on left above) compared to vector-vector components (shown on right above)

Vector components in a 3D coordinate system

In a 3D coordinate system (i.e., *x*-, *y*- and *z*-axis), a third *z*-component of the vector is included:

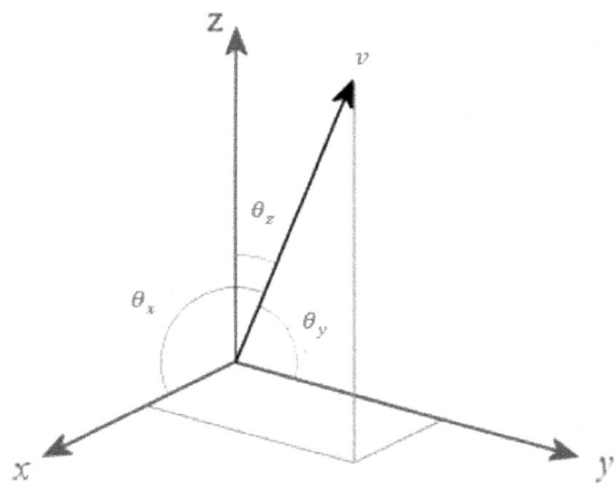

Vector components in a 3D coordinate system

$$v_x = \vec{v} \cos \theta_x$$

$$v_y = \vec{v} \cos \theta_y$$

$$v_z = \vec{v} \cos \theta_z$$

$$\vec{v}^2 = v_x^2 + v_y^2 + v_z^2$$

Vector Addition & Subtraction

Vector addition yields a vector or scaler

Vector addition is performed *graphically* or by using *vector components.*

Vectors can be added in the same dimension (direction). Otherwise, the separate addition of each vector's x-, y- and z-components must be calculated.

Sum of all components of a vector equals the vector itself. Therefore, the added components are the *resultant vector.*

Operations with two vectors may or may not result in a vector (i.e., magnitude and directions).

For example, kinetic energy derived from the square of two velocity vectors is a scalar quantity.

Operations using a *vector* and a *scalar* result in a *vector.*

Operations using a *scalar* and a *scalar* result in a *scalar.*

Adding and subtracting vectors in one dimension

For vectors in one dimension, merely add and subtract.

Vector addition in one dimension (same direction):

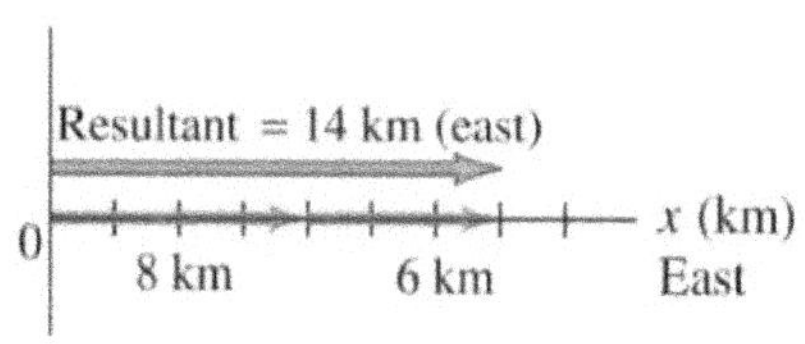

$$8\ km_{East} + 6\ km_{East} = 14\ km_{East}$$

Vector addition in same direction

Vector subtraction in one dimension (opposite directions):

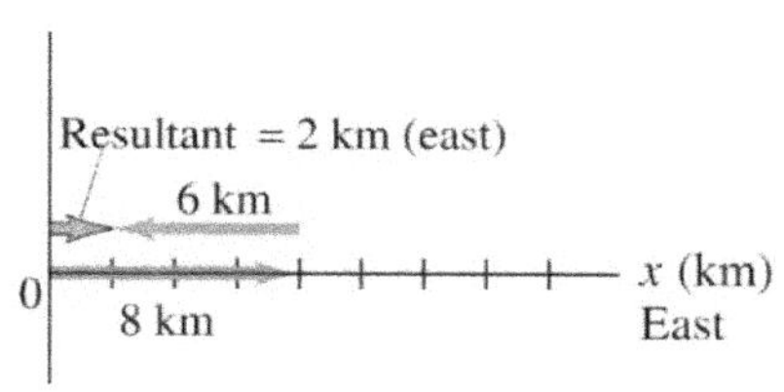

$$8\ km_{East} - 6\ km_{East} = 2\ km_{East}$$

Vector subtraction in same direction

Graphical methods for adding right-angle vectors

If there is two-dimensional motion, the analysis is more complicated.

If travel paths are at right angles, the displacement (*d*) is calculated by Pythagorean Theorem:

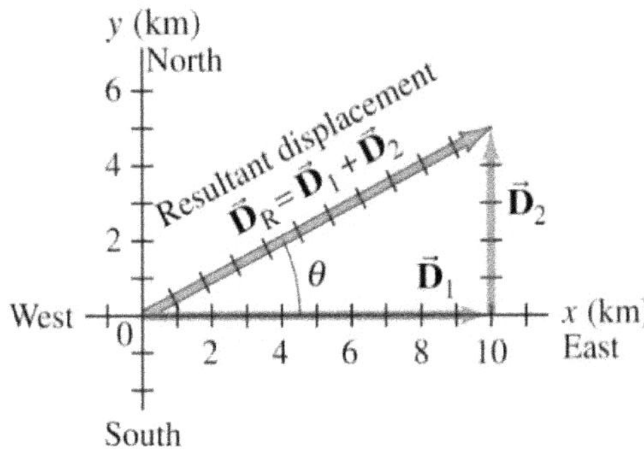

Pythagorean Theorem calculates resultant vector

Adding the vectors in the other sequence gives the same result:

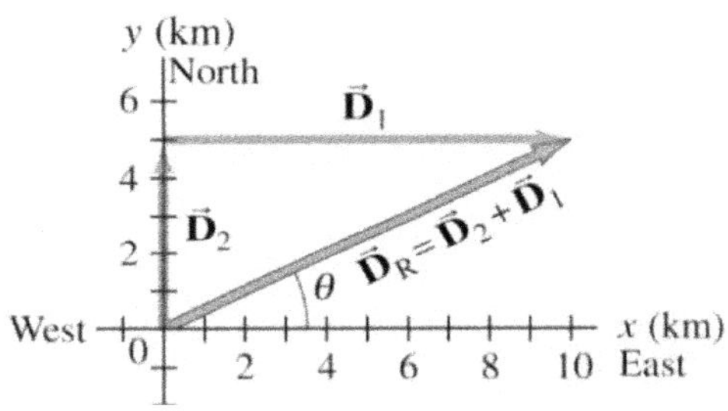

$$|\vec{D}_R| = \sqrt{\vec{D}_1^{\,2} + \vec{D}_2^{\,2}}$$

Tip-to-toe method for vectors

For vectors not at right angles, add using the "*tail-to-tip*" method.

Tail-to-tip graphically places each vector's tail at the tip of the arrow of the vector before it.

After vectors are arranged tail-to-tip, draw the resultant vector from the original vector's tail to the tip of the last vector.

For example, *tail-to-tip* method for resolving three vectors:

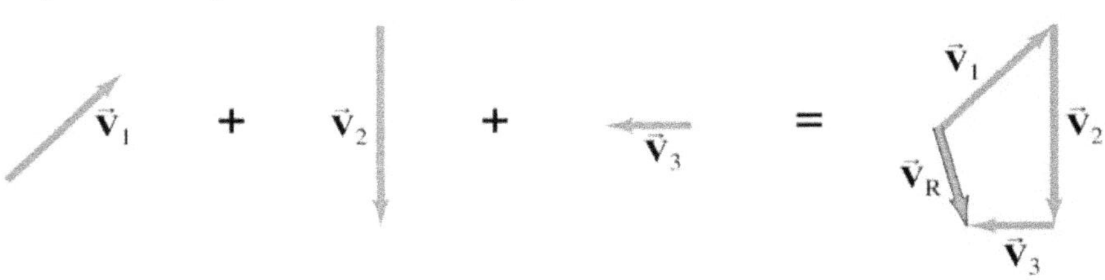

Tail-to-tip method for resolving three vectors

Parallelogram method

Parallelogram method arranges two vectors such that their tails are connected.

Starting from the tips of the two vectors, draw two dashed lines that complete the parallelogram.

Resultant vector is drawn from the tails of the original to the corner of the drawn parallelogram.

Using *tail-to-tip* (a), *parallelogram* method (b) for resolving vectors:

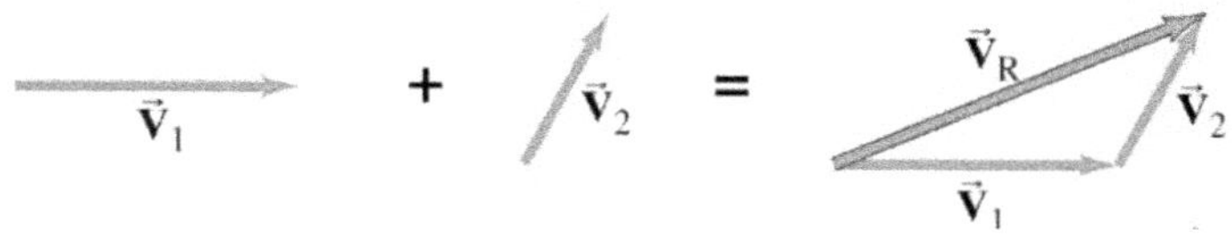

Resolving two vectors using the tail-to-tip and parallelogram method

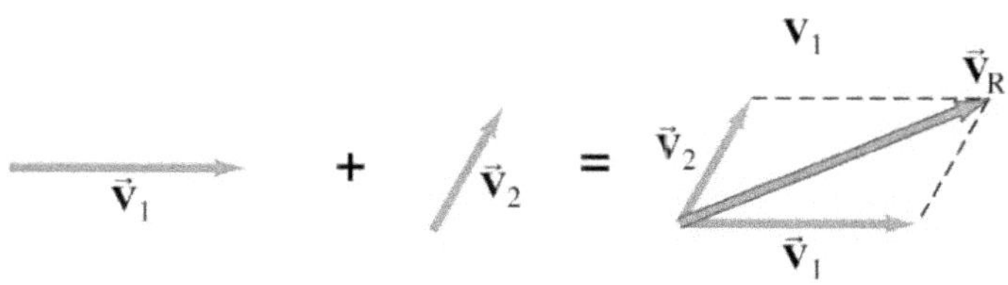

Resolving two vectors using the tail-to-tip and parallelogram method

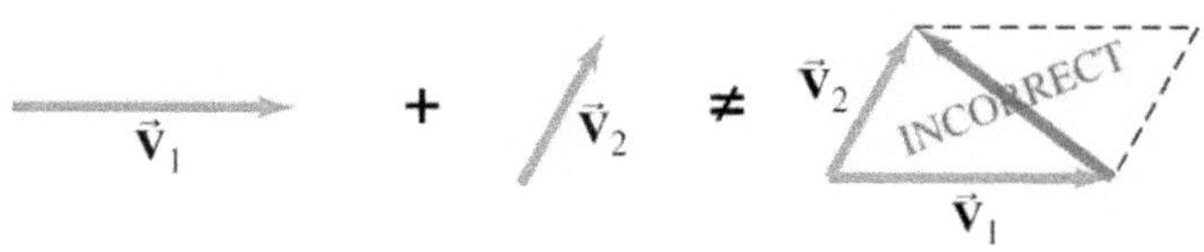

Incorrect method for resolving two vectors

Negative vectors

To subtract vectors, define the negative direction with the same magnitude but pointing in the opposite direction.

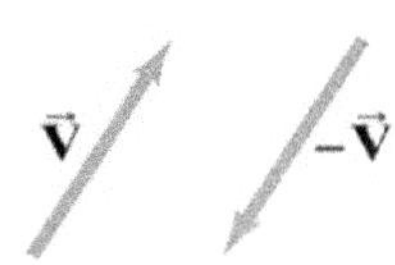

Defining coordinate system

Then add the negative vectors:

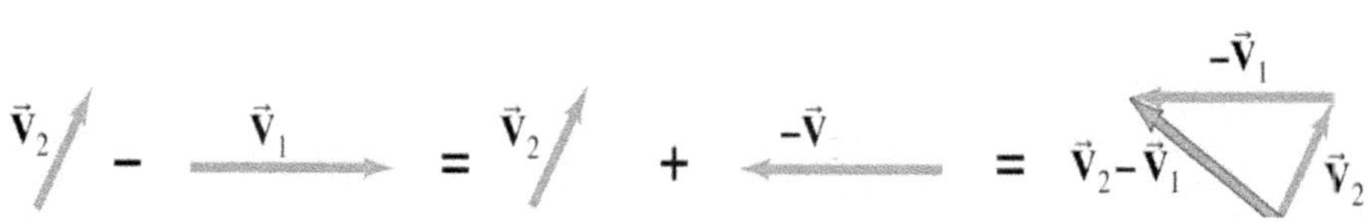

Adding negative vectors

Multiplying vectors by a scaler

A vector, $\vec{V}$ can be multiplied by a scalar c.

The result is a vector that has the same direction but magnitude $c\vec{V}$.

If c is negative, the resultant vector points in the opposite direction.

An example of a vector, $\vec{V}$ being multiplied by different scalars:

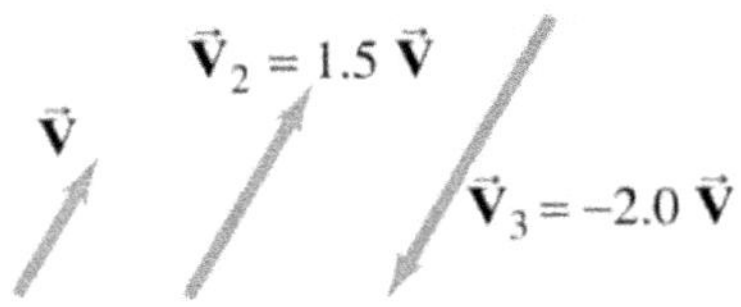

Vectors with different scalers

Adding vectors by components

A vector can be expressed as the sum of its component vectors.

Component vectors are chosen so that they are perpendicular:

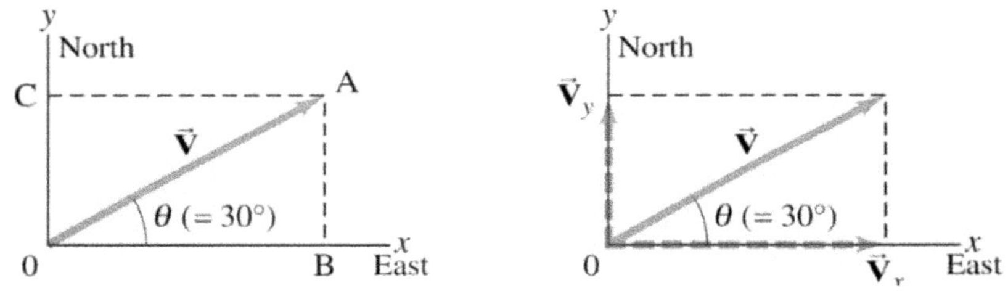

Vectors expressed as the sum of perpendicular component vectors

Trigonometric functions to calculate components

For perpendicular components, calculate the values using trigonometric functions.

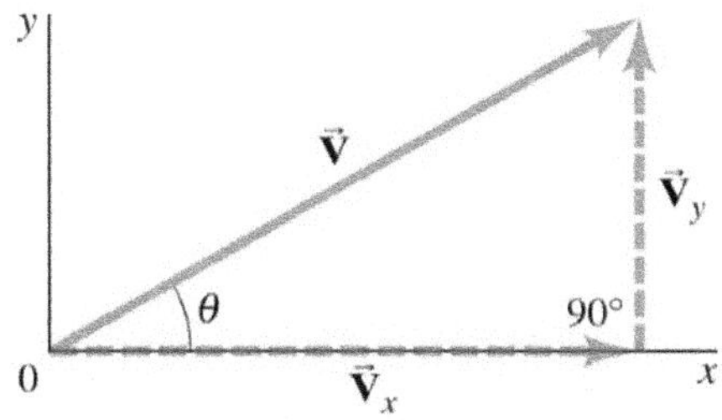

Trigonometric functions for perpendicular vector components

$$\sin\theta = \frac{V_y}{V}$$

$$\cos\theta = \frac{V_x}{V}$$

$$\tan\theta = \frac{V_y}{V_x}$$

$$V^2 = V_x^2 + V_y^2$$

Resolving vectors algebraically

Components are effectively one-dimensional, so add algebraically:

$$V_{Ry} = V_{1y} + V_{2y}$$

$$V_{Rx} = V_{1x} + V_{2x}$$

$$|\vec{V}_R| = \sqrt{\vec{V}_{Rx}^{\,2} + \vec{V}_{Ry}^{\,2}}$$

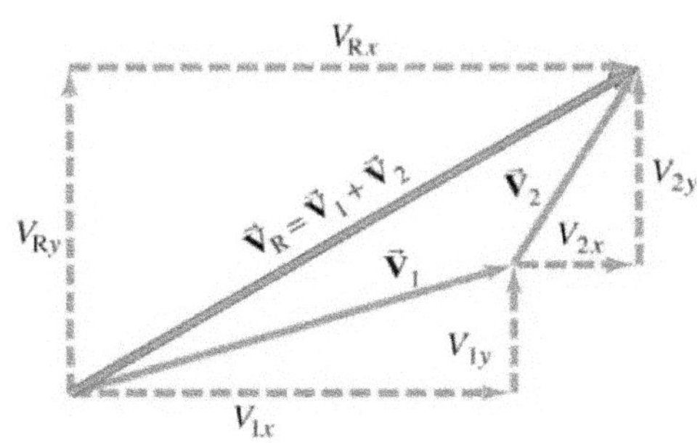

Resolving vectors from component vectors

Five rules for adding vectors

1. Draw a diagram of the vectors on a coordinate system.
2. Establish the x and y-axes.
3. Resolve each vector into x and y components.
4. Calculate each component using sines and cosines.
5. Add components in each direction to determine the resultant vector.

   For magnitude and direction of the resultant vector, use:

$$\vec{v} = \sqrt{v_x^2 + v_y^2} \qquad \tan\theta = \frac{v_y}{v_x}$$

Speed and Velocity

Speed as a scalar quantity

Speed is the quantity of how far an object travels in a time interval.

Speed is the rate of change in the *distance*; it does not include a direction and is a scalar quantity.

Average speed is the distance (scalar) traveled divided by the time elapsed:

$$speed_{average} = v_{avg}$$

$$v_{avg} = \frac{d}{t}$$

$$\frac{d}{t} = \frac{distance\ traveled}{time\ elapsed}$$

Velocity as a vector quantity

Velocity quantifies the speed of an object in a direction.

Velocity is the *rate of change* in *displacement.* Since velocity includes direction, it is a vector quantity (unlike speed, a scalar quantity).

Average velocity is displacement (vector) divided by elapsed time:

$$velocity_{average} = \vec{v}_{avg}$$

$$\vec{v}_{avg} = \frac{s}{t}$$

$$\frac{s}{t} = \frac{displacement}{time\ elapsed}$$

$$\frac{displacement}{time\ elapsed} = \frac{final\ position - initial\ position}{time\ elapsed}$$

Instantaneous velocity as a vector

Instantaneous speed is the speed at a point in time (infinitesimal time interval).

Instantaneous velocity is the velocity at a point in time (i.e., infinitesimal time interval). It is the mathematical limit (i.e., time becomes infinitesimally small) of displacement over imperceptible time intervals:

$$\vec{v}_{inst} = \lim_{\Delta t \to 0} \frac{\Delta x}{\Delta t}$$

Instantaneous velocity *is* a vector and has a direction.

Instantaneous speed is *not* a vector and *does not* indicate a direction.

Direction of instantaneous velocity is tangent to the path at that point. Its magnitude is equal to that of the instantaneous speed at that point.

Graphs for *constant velocity* (*a* on left) and *varying velocity* (*b* on right):

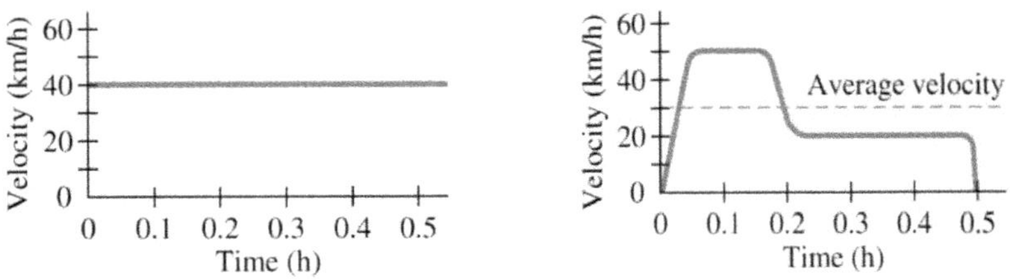

Constant velocity (on left) vs varying velocity (on right)

Relative velocity

Relative speed moves in two dimensions and calculated by adding (or subtracting) magnitudes.

Velocities are vectors that must be added and subtracted.

Component velocity is:

1) labeled with the object and

2) with the reference frame in which it has this velocity.

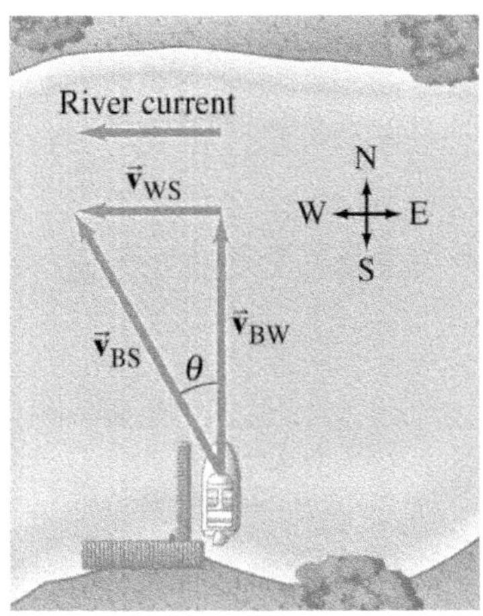

Relative velocity in two dimensions

$\vec{v}_{BS}$ is the velocity of the boat relative to the shore

$\vec{v}_{BW}$ is the velocity of the boat relative to the water

$\vec{v}_{WS}$ is the velocity of the water relative to the shore

Relationship between the three velocities is:

$$\vec{v}_{BS} = \vec{v}_{BW} + \vec{v}_{WS}$$

Acceleration

Acceleration as a change in velocity

Acceleration is the rate of change in velocity and expressed as:

$$average\ acceleration = \vec{a}_{avg}$$

$$\vec{a}_{avg} = \frac{\Delta\vec{v}}{\Delta t}$$

$$\frac{\Delta\vec{v}}{\Delta t} = \frac{\vec{v}_f - \vec{v}_i}{t}$$

$$\frac{\vec{v}_f - \vec{v}_i}{t} = \frac{change\ in\ velocity}{time\ elapsed}$$

From the calculated average acceleration, instantaneous acceleration can be determined.

Instantaneous acceleration

Instantaneous acceleration is the average velocity over the limit of the time interval as time becomes infinitesimally small:

$$\vec{a}_{inst} = \lim_{\Delta t \to 0} \frac{\Delta\vec{v}}{\Delta t}$$

Like velocity, acceleration is a vector (i.e., magnitude and direction).

In one-dimensional motion (e.g., speed), only a positive or negative sign is needed to calculate the resulting motion (e.g., net speed).

Constant acceleration

If the magnitude of the acceleration is *constant,* and there is *no change in direction*, the magnitude of speed (a scalar quantity) and velocity (a vector quantity) are interchangeable.

If the magnitude of the acceleration is *constant,* and there is *no change in direction*, the magnitude of distance (scalar) and displacement (vector) are interchangeable.

Reference the direction of the acceleration.

Negative acceleration *vs.* deceleration

Negative acceleration and deceleration describe different phenomena.

Negative acceleration is an acceleration in the coordinate system's negative direction (e.g., a booster rocket dislodged from the satellite). The direction of the acceleration is independent of velocity.

Deceleration is relative to an object's velocity; it occurs when acceleration is opposite in direction to the velocity (e.g., a car skidding to a stop).

If the object travels in a positive direction (i.e., velocity is positive), then deceleration and negative acceleration are the same.

Positive acceleration (i.e.., car advances) is shown:

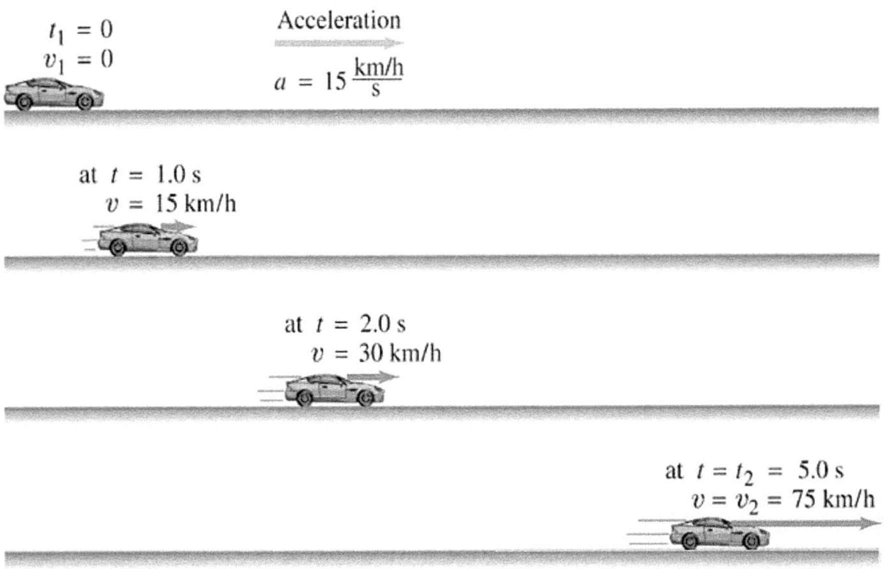

Positive acceleration at time = 0, 1.0s, 2.0s and 5.0s

Negative acceleration (i.e., acceleration in the negative direction) is shown:

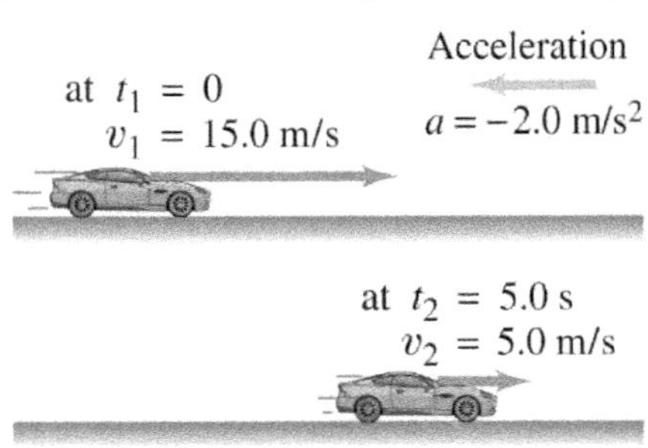

Positive acceleration at time = 0, 1.0s, 2.0s and 5.0s

Motion at constant acceleration

Average velocity of an object during a time interval t is:

$$\bar{v} = \frac{\Delta x}{\Delta t} = \frac{x - x_0}{t - t_0} = \frac{x - x_0}{t}$$

Constant velocity expression is:

$$\bar{v} = \frac{v_0 + v}{2}$$

Constant acceleration expression is:

$$\vec{a} = \frac{v - v_0}{t}$$

Combine the expressions for *average velocity, constant acceleration,* and *constant velocity*; where x is the position, x_0 is the initial position, v_0 is the initial velocity, t is the time:

$$x = x_0 + \bar{v}t$$

$$x_0 + \bar{v}t = x_0 + \left(\frac{v_0 + v}{2}\right)t$$

$$x_0 + \left(\frac{v_0 + v}{2}\right)t = x_0 + \left(\frac{v_0 + v_0 + at}{2}\right)t$$

or simplified as:

$$x = x_0 + v_0 t + \frac{1}{2}at^2$$

Solve to eliminate t:

$$v^2 = v_0^2 + 2a(x - x_0)$$

Equations for constant acceleration problems

- $v = v_0 + at$
- $x = x_0 + v_0 t + \frac{1}{2}at^2$
- $v^2 = v_0^2 + 2a(x - x_0)$
- $\bar{v} = \frac{v_0 + v}{2}$
- $s = v_{avg}t$
- $v_{avg} = \frac{v_f + v_i}{2}$
- $a = \frac{\Delta v}{\Delta t} = \frac{v_f - v_i}{t}$
- $v_f^2 = v_i^2 + 2as$
- $s = \frac{1}{2}at^2 + v_i t$
- $l_f = l_i + s$

Learn to rearrange, combine, and apply the equations appropriately.

Using these equations, assign one direction as positive and the opposite as negative; use this orientation for all subsequent calculations.

Cartesian coordinates

Cartesian coordinates (i.e., horizontal x and vertical y-axes) take upward and rightward motions as positive and downward and leftward motions as negative.

For free falls, take downward (gravitational acceleration) as positive.

Direction for motion can be assigned arbitrarily, provided that the opposite direction has the opposite sign.

Graphical analysis of linear motion

x *vs.* t (i.e., distance *vs.* time) graph for an object moving with constant velocity is shown below.

Slope is the rise/run (x-axis / y-axis) on the graph.

Velocity is the slope of the x *vs.* t (i.e., distance *vs.* time) graph.

If the slope is positive, the velocity is positive.

If the slope is negative, the velocity is negative.

Slope of the line is linear (straight line) when velocity is constant:

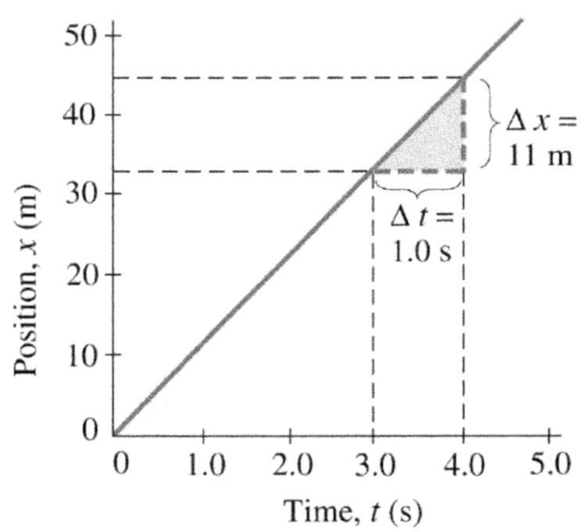

Position vs. time graph shows a positive linear slope for constant velocity

Freely Falling Bodies

Motion with constant acceleration

Motion with constant acceleration is *free fall* (e.g., an apple falling to the ground from a tree).

Near the Earth's surface, objects experience acceleration due to gravity ($g = 9.8 \text{ m/s}^2$ or use 10 m/s^2 for calculations).

The differences in acceleration due to gravity between various objects (e.g., feather and marble) are related to air resistance. A larger surface area (e.g., a flat sheet of paper compared to a crumpled sheet of the same paper) means higher air resistance.

In a vacuum (i.e., no air resistance), gravity causes objects to fall with the *same acceleration* ($g = 9.8 \text{ m/s}^2$). Acceleration is *constant* because of *no air resistance.*

Bodies in free fall

Whenever an object is in the air, it is in free fall, regardless of whether it is tossed upwards, downwards, or at an angle.

Label the downward direction for dropped objects as positive, so g (acceleration) is positive.

Objects traveling downward have initial velocity, and g as positive.

Objects traveling upward have initial velocity and acceleration (g) with opposite signs.

Choose *either* the upward or downward direction as positive.

Projectiles move in two directions

Projectiles are objects moving in two dimensions under the influence of the Earth's gravity. The projectile's path of travel is a *parabola.*

Motion along this path is *projectile motion* described by vertical and horizontal components.

Speed in the horizontal (x-direction) is constant.

Speed in the vertical (y-direction) is constant acceleration g.

Projectiles are freely falling bodies, so the vertical component of the projectile motion is accelerating toward the Earth at a rate of g.

When the projectile travels upward, the downward acceleration of gravity causes the projectile to decelerate when traveling upward.

There is no acceleration due to gravity in the horizontal component.

Horizontal component is constant

Horizontal component of velocity is constant.

Time elapsed while the projectile is in motion is dependent on only the vertical (not horizontal) component of the projectile motion.

Horizontal distance is the horizontal speed multiplied by elapsed time (speed × time).

An object tossed straight up comes down to where it started, and the total displacement s for the trip is zero.

Initial velocity and final velocity are equal and opposite:

$$(\vec{v}_{\text{final}} + \vec{v}_{\text{initial}} = 0)$$

Acceleration is the opposite sign to the initial velocity.

Falling objects with different x-direction motion

Two balls that fall at the same time are illustrated below.

Ball on the right in figure has an initial velocity in the x-direction, while the ball on the left was dropped straight downward (i.e., zero initial velocity in the x-direction).

The two balls' vertical positions are identical at identical times.

The ball's horizontal position on the right increases linearly.

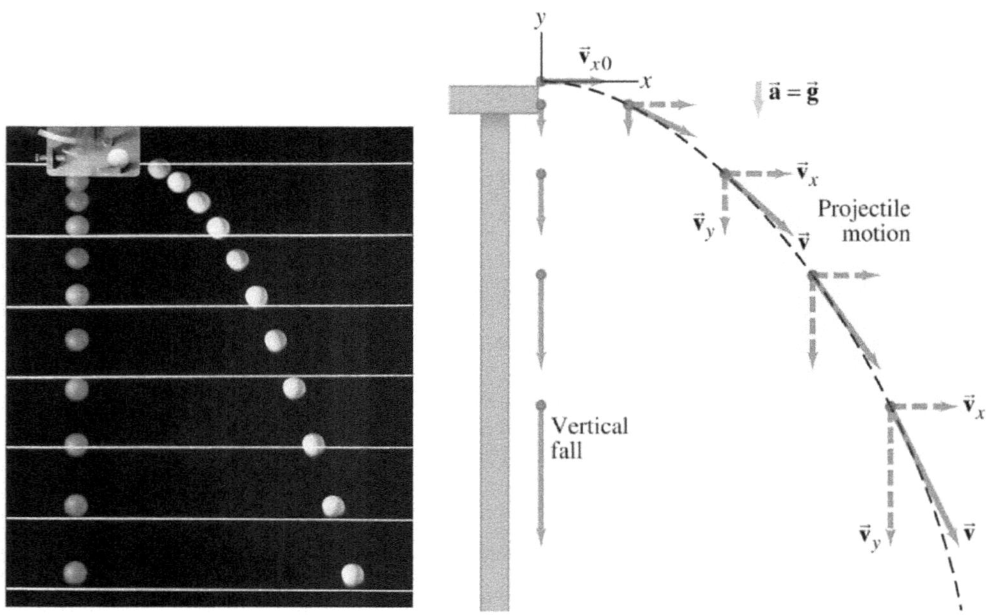

Comparison of a ball dropped vs. projected with a horizontal velocity

Object is launched at an initial angle of θ_0 with the horizontal.

Analysis is similar, except that the initial velocity has *horizontal* and *vertical* components.

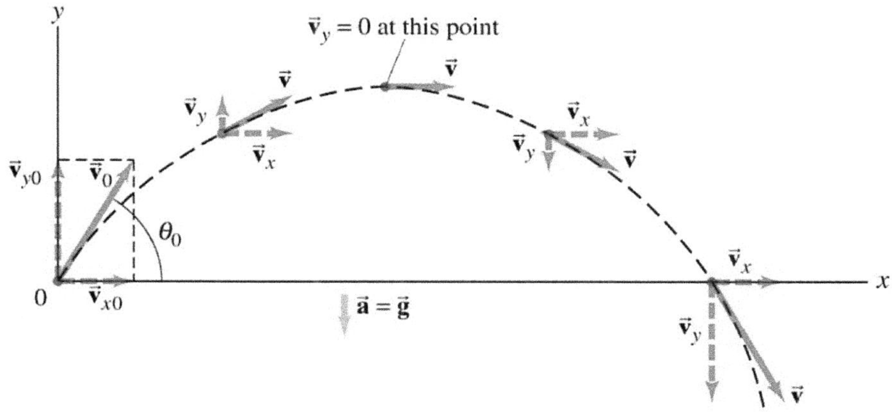

Projectile with both x- and y-component for velocity

Projectile motion is parabolic

To demonstrate that projectile motion is parabolic, write y as a function of x.

Parabola is expressed by the equation:

$$y = Ax^2 - Bx + C$$

where A, B, and C are constants.

A is a scalar (magnitude) multiplier of the *object's acceleration.*

B as a scalar multiplier of the *velocity.*

C as a scalar multiplier of the object's *initial position.*

Direction of the object's acceleration is determined by the signs of the constants for velocity and initial displacement from the origin.

Solving projectile motion problems

Projectile motion has constant acceleration in one dimension.

For acceleration due to gravity, the direction is downward.

Kinematic Equations for Projectile Motion (y positive downward; $a_x = 0$, $a_y = g = 9.8$ m/s^2)	
Horizontal Motion ($a_x = 0$, v_x = constant)	Vertical Motion ($a_y = g$ = constant)
$v_x = v_{x0}$	$v_y = v_{y0} - gt$
$x = x_0 + v_{x0}\, t$	$y = y_0 + v_{y0}\, t - (½)gt^2$
	$v_y^2 = v_{y0}{}^2 - 2g(y - y_0)$

Seven steps for solving projectile motion problems

1. Read the problem carefully and choose the object(s) to be analyzed.
2. Draw a diagram of the described actions.
3. Choose a point of origin and a coordinate system.
4. Determine the time interval;

   same in both directions and

   only the time object is moving, with constant acceleration g.
5. Evaluate x and y-components of motion separately.
6. List known and unknown quantities.

   v_x never changes, and

   $v_y = 0$ at the highest point.
7. Use appropriate equations, combining some equations.

Free fall *vs.* non-free fall

In free fall, gravity is the only force (ignoring air resistance) acting on the object.

Assuming no air resistance, the only resistance to change is the object's inertia.

Objects fall at a *constant acceleration* of 9.8 m/s^2 downward.

Since Earth is not a vacuum, air resistance (force opposing the motion downward) is generally accounted for when objects fall.

Non-free fall is when the upward force due to air resistance counteracts the downward force due to gravity.

Net force of this interaction is the difference between force due to gravity and air resistance.

Therefore, the total force on an object in a non-free fall is *less* than that of the same object in free fall, where the force from air resistance is zero.

As an object falls, the force from *air resistance increases* with the object's increased speed, and the force downward decreases until the two opposing forces are equal. Acceleration is zero.

Terminal velocity, weight and gravity

Terminal velocity is when the object travels at a constant velocity downward due to the *maximum force of air resistance.*

Terminal velocity occurs when air resistance *equals* the weight of the object.

Weight is the mass under the influence of gravity.

Mass of an object on Earth and the Moon are the same.

However, weight (not mass) measured is different on the Earth and Moon due to gravity differences.

Gravity is an attractive force between two bodies.

Massive objects exert a tremendous attractive (gravitational) force.

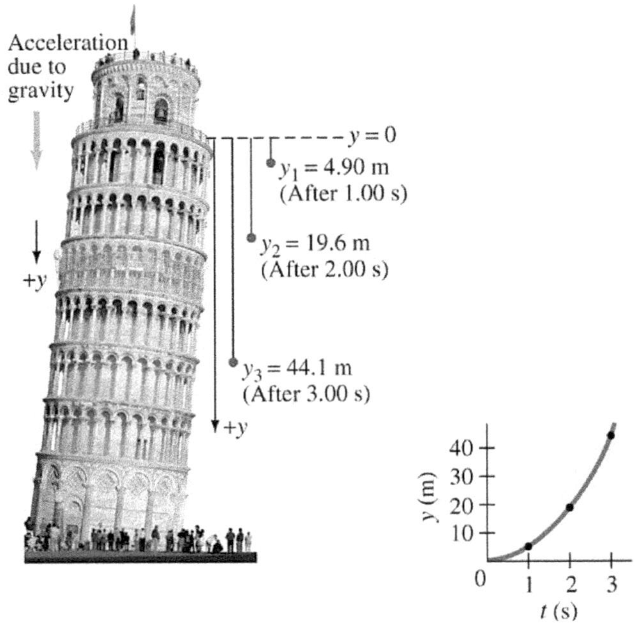

Graph plots distance vs. time of the falling object

Satellites and weightlessness

Satellites are routinely put into orbit around the Earth.

Tangential speed of the satellite must be enough so that the satellite does not fall to Earth.

However, the tangential speed cannot be so high that the satellite escapes Earth's gravity.

Satellites orbiting the Earth are in continuous free fall, and their *centripetal acceleration* equals the acceleration from the Earth's gravity.

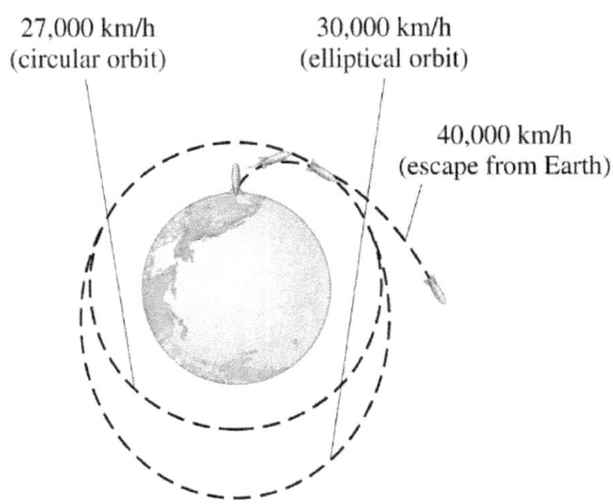

Tangential speed of the satellite determines its orbit

Satellites are kept in orbit around the Earth by its speed.

Satellites are continually falling and accelerating toward Earth. They do not crash into Earth's surface because Earth curves away underneath at the same rate as the satellite falls toward Earth.

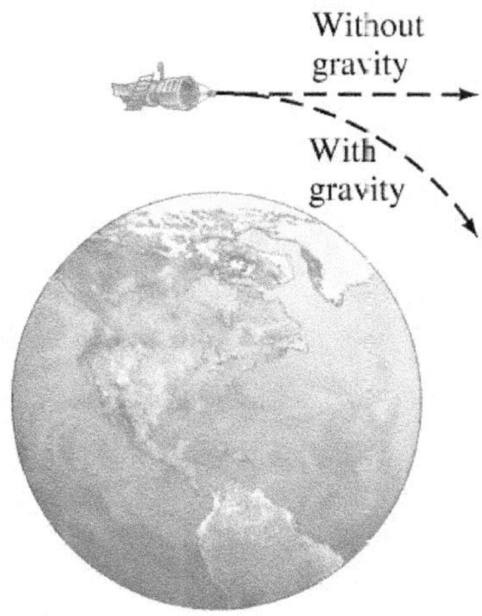

Satellites orbiting the Earth are in continuous free fall

Objects in orbit are said to experience weightlessness, even though they have a gravitational force acting on them.

Satellite and its contents are in free fall, so there is no normal force, leading to the experience of weightlessness.

Apparent *vs.* real weightlessness

Apparent weightlessness as the gravitational force still exists.

Apparent weightlessness is experienced on Earth, but only briefly.

For example, when a roller coaster rushes down a hill, the normal force temporarily goes to zero, and the riders experience apparent weightlessness.

Real weightlessness occurs when no net gravitational force is acting on an object.

1) void of the attractive force of gravity when massive objects are light-years away or

2) the object does not experience a net gravitational force because it is between two other objects with equal gravitational forces that cancel.

Notes for active learning

Chapter Summary

Graphical analysis

Graphs are helpful to visualize the motion of an object.

Graphs solve problems when graphical information describes a physical phenomenon.

For graphing, consider the following:

- First, look at the axes of the graph.

  A common mistake for a velocity *vs.* time graph is confusing it with a position *vs.* time graph.

- Do not assume one interval is one unit; inspect the axis numbers.

- Align the position *vs.* time graphs *directly above* the velocity *vs.* time graphs and acceleration *vs.* time graphs so key points among the graphs can be matched.

Slope and area

Slope of a distance *vs.* time (x *vs.* t) graph gives *velocity.*

Slope of a velocity *vs.* time (v vs. t) gives *acceleration.*

Area under an acceleration *vs.* time (a *vs.* t) graph gives the *change in velocity.*

Area under a velocity *vs.* time (v vs. t) gives the *displacement.*

Horizontal and vertical motion

Five Equations describe the motion of an object in one dimension.

Identify the given, determine what must be calculated, and use an equation with those variables.

Note hidden (assumed) information (e.g., acceleration due to gravity $g = -10$ m/s^2)

		Missing variable
#1*	$\Delta x = vt$	a
#2	$v = v_0 + at$	x
#3	$x = x_0 + v_0 t + \frac{1}{2}at^2$	v
#4	$x = x_0 + vt - \frac{1}{2}at^2$	v_0
#5	$v^2 = v_0^2 + 2a(x - x_0)$	T

* Because the acceleration is uniform: $\Delta x = \frac{1}{2}(v_0 + v)t$

Horizontal and vertical components for projectiles

For projectiles, separate the horizontal and vertical components.

Horizontal Motion	Vertical Motion
$x = v_x t$	$y = y_0 + v_0 t + \frac{1}{2}gt^2$
$v_x = v_{0x} =$ constant	$v_y = v_{0y} + gt$
$a_x = 0$	$a = g = -10\frac{m}{s^2}$

At any given time, the relationship between v, v_x and v_y is given by:

$v^2 = v_x^2 + v_y^2$

$v_x = v\ cos\ \theta$

$v_y = sin\ \theta$

$\theta = tan^{-1}(\frac{v_y}{v_x})$

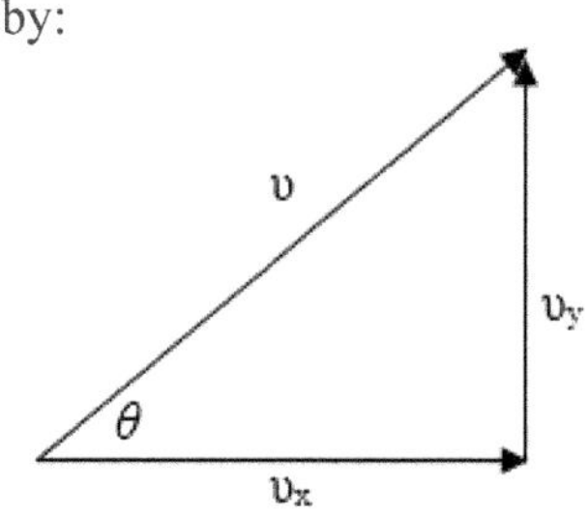

Resolving vectors from the known x- and y-component of the separate vectors

Solving motion problems

Problems involving motion:

1. Read the entire problem carefully.

   Evaluate what is provided and what is asked to be solved.

2. Decide which objects are under study and what the time interval is.

3. Draw a diagram and choose coordinate axes.

4. Evaluate known quantities and identify unknowns needed.

5. Determine which equations relate to the known and unknown quantities.

   Solve algebraically for unknowns and evaluate results (e.g., magnitude and dimensions).

6. Calculate the solution and round to the appropriate number of significant figures.

7. Evaluate if the result is reasonable by comparing it with approximations (e.g., positive, negative, order of magnitude).

8. Check that the units on one side of the equation match the other.

Notes for active learning

Practice Problems: Kinematics & Dynamics

1. Which graph represents a constant non-zero velocity?

I. 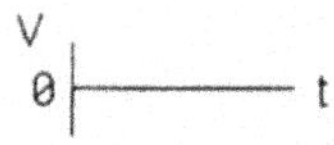

II. 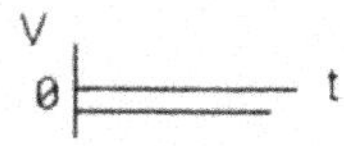

III. 

A. I only
B. II only
C. III only
D. I and II only
E. II and III only

2. A train starts from rest and accelerates uniformly until it has traveled 5.6 km and has acquired a velocity of 42 m/s. The train then moves at a constant velocity of 42 m/s for 420 s. The train then slows down uniformly at 0.065 m/s^2 until it stops moving. What is the acceleration during the first 5.6 km of travel?

A. 0.29 m/s^2
B. 0.23 m/s^2
C. 0.16 m/s^2
D. 0.12 m/s^2
E. 0.20 m/s^2

3. A cannonball is fired straight up at 50 m/s. Ignoring air resistance, what is the velocity at the highest point it reaches before starting to return towards Earth?

A. 0 m/s
B. 25 m/s
C. $\sqrt{50}$ m/s
D. 50 m/s
E. 100 m/s

4. The tendency of a moving object to remain in unchanging motion in the absence of an unbalanced force is:

A. impulse
B. acceleration
C. free fall
D. inertia
E. momentum

5. Which statement is correct when an object moving in the $+x$ direction undergoes an acceleration of 2 m/s^2?

A. It travels at 2 m/s
B. It travels at 2 m/s^2
C. It decreases its velocity by 2 m/s every second
D. It decreases its velocity by 2 m/s^2 every second
E. It increases its velocity by 2 m/s every second

6. What is the increase in speed each second for a freely falling object?

A. 0 m/s
B. 9.8 m/s^2
C. 9.8 m/s
D. 19.6 m/s
E. 4.8 m/s

7. A marble initially rolls up a slight incline at 0.2 m/s and, starting at $t = 0$ s, decelerates uniformly at 0.05 m/s². At what time does the marble come to a stop?

A. 2 s
B. 4 s
C. 8 s
D. 12 s
E. 6 s

8. An 8.7-hour trip is made at an average speed of 73 km/h. If the first third of the journey was driven at 96.5 km/h, what was the average speed for the rest?

A. 54 km/hr
B. 46 km/hr
C. 28 km/hr
D. 62 km/hr
E. 37 km/hr

9. The lightning flash and the thunder are not observed simultaneously because light travels much faster than sound. Therefore, it can be assumed as instantaneous when the lightning occurs. What is the distance from the lightning bolt to the observer if the delay between the sound and the lightning flash is 6 s? (Use speed of sound in air $v = 340$ m/s)

A. 2,040 m
B. 880 m
C. 2,360 m
D. 2,820 m
E. 1,460 m

10. If a cat jumps at a 60° angle off the ground with an initial velocity of 2.74 m/s, what is the highest point of the cat's trajectory? (Use acceleration due to gravity $g = 9.8$ m/s²)

A. 9.46 m
B. 0.69 m
C. 5.75 m
D. 1.55 m
E. 0.29 m

11. What is the resultant vector AB when vector A = 6 m and points 30° North of East, while vector B = 4 m and points 30° South of West?

A. 8 m at an angle 45° East of North
B. 8 m at an angle 30° North of East
C. 2 m at an angle 30° North of East
D. 2 m at an angle 45° North of East
E. 8 m at an angle 45° North of East

12. How much farther would an intoxicated driver's car travel before he hits the brakes than a sober driver's car if both cars are initially traveling at 49 mi/h, and the sober driver takes 0.33 s to hit the brakes while the intoxicated driver takes 1 s to hit the brakes?

A. 38 ft
B. 52 ft
C. 32 ft
D. 48 ft
E. 60 ft

13. A projectile is fired at time $t = 0$ s from point 0 at the edge of a cliff with initial velocity components of $v_{0x} = 80$ m/s and $v_{0y} = 800$ m/s. The projectile rises, then falls into the sea at point P. The time of flight of the projectile is 200 s. What is the height of the cliff? (Use the acceleration due to gravity $g = 10$ m/s^2)

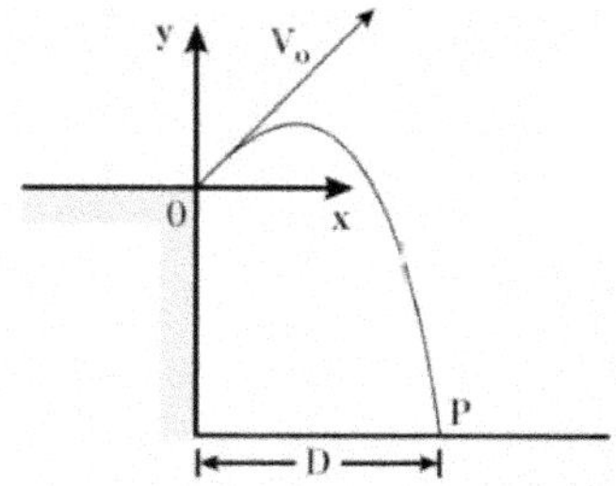

A. 30,000 m
B. 45,000 m
C. 47,500 m
D. 35,000 m
E. 40,000 m

14. Which of the following is a unit that can be used to measure weight?

A. kilogram
B. kg·m/s
C. Newton
D. milligram
E. liter

15. Which of the statements about the direction of the velocity and acceleration is correct for a ball that is thrown straight up, reaches a maximum height, and falls to its initial height?

A. Its velocity changes from upward to downward, and its acceleration points downward
B. Its velocity points downward, and its acceleration points upward
C. Both its velocity and its acceleration point downward
D. Both its velocity and its acceleration point upward
E. Its velocity points downward, and its acceleration changes from upward to downward

Solutions: Kinematics & Dynamics

1. B is correct.

Graph I depicts a constant zero velocity process.

Graph II depicts a constant non-zero velocity.

Graph III depicts non- constant velocity.

2. C is correct.

$$v_f^2 = v_i^2 + 2ad$$

$$a = (v_f^2 - v_i^2) / 2d$$

$$a = [(42\text{ m/s})^2 - (0\text{ m/s})^2] / [2(5{,}600\text{ m})]$$

$$a = (1{,}764\text{ m}^2/\text{s}^2) / 11{,}200\text{ m}$$

$$a = 0.16\text{ m/s}^2$$

3. A is correct.

At the maximum height, the velocity = 0, and the cannonball stops moving up and begins to come back.

4. D is correct.

An object's inertia is its resistance to change in motion.

5. E is correct.

$$v = v_0 + at$$

If $a = 2\text{ m/s}^2$ then the object has a 2 m/s increase in velocity every second.

Example:

$$v_0 = 0\text{ m/s},\ a = 2\text{ m/s}^2,\ t = 1\text{ s}$$

$$v = 0\text{ m/s} + (2\text{ m/s}^2)\cdot(1\text{ s})$$

$$v = 2\text{ m/s}$$

6. C is correct.

Acceleration due to gravity is 9.8 m/s^2.

If an object is in freefall, then every second, it increases velocity by 9.8 m/s.

$v = v_0 + at$

$v = v_0 + (9.8\text{ m/s}^2)\cdot(1\text{ s})$

$v = v_0 + 9.8\text{ m/s}$

7. B is correct.

To determine the time at which the marble stops:

$v_2 - v_1 = a\Delta t$

where $v_1 = 0.2$ m/s and $a = -0.05$ m/s^2

$\Delta t = (v_2 - v_1) / a$

$\Delta t = (0\text{ m/s} - 0.2\text{ m/s}) / (-0.05\text{ m/s}^2)$

$\Delta t = (-0.2\text{ m/s}) / (-0.05\text{ m/s}^2)$

$\Delta t = 4\text{ s}$

8. D is correct.

$v_{avg} = (1/3)v_1 + (2/3)v_2$

$73\text{ km/h} = (1/3)\cdot(96.5\text{ km/h}) + (2/3)v_2$

$73\text{ km/h} = (32\text{ km/h}) + (2/3)v_2$

$(2/3)v_2 = 41\text{ km/h}$

$v_2 = (41\text{ km/h})\cdot(3/2)$

$v_2 = 62\text{ km/h}$

9. A is correct.

Distance to the lightning bolt is calculated by determining the distance traveled by sound.

Distance = velocity × time

$d = vt$

$d = (340\text{ m/s})\cdot(6\text{ s})$

$d = 2{,}040\text{ m}$

10. E is correct.

Calculate the vertical component of the initial velocity:

$v_{up} = v(\sin \theta)$

$v_{up} = (2.74 \text{ m/s}) \sin 60°$

$v_{up} = 2.37 \text{ m/s}$

Then, solve for the upward displacement given the initial upward velocity:

$d = (v_f^2 - v_i^2) / 2a$

$d = [(0 \text{ m/s})^2 - (2.37 \text{ m/s})^2] / 2(-9.8 \text{ m/s}^2)$

$d = (-5.62 \text{ m}^2/\text{s}^2) / (-19.6 \text{ m/s}^2)$

$d = 0.29 \text{ m}$

11. C is correct.

30° North of East is the opposite direction of 30° South of West, so set one vector as negative, and the resultant vector will be in the direction of the larger vector between A and B.

Since the magnitude of A is greater than the magnitude of B and vectors were added "tip to tail," the resultant vector is:

AB = A + B

AB = 6 m + (–4 m)

AB = 2 m at 30° North of East

12. D is correct.

The solution is measured in feet, so first convert the car velocity into feet per second:

$v = (49 \text{ mi/h})\cdot(5{,}280 \text{ ft/mi})\cdot(1 \text{ h}/3{,}600 \text{ s})$

$v = 72 \text{ ft/s}$

Sober driver's distance:

$d = vt$

$d_{sober} = (72 \text{ ft/s})\cdot(0.33 \text{ s})$

$d_{sober} = 24 \text{ ft}$

continued...

Intoxicated driver's distance:

$d = vt$

$d_{drunk} = (72 \text{ ft/s})\cdot(1 \text{ s})$

$d_{drunk} = 72 \text{ ft}$

The differences between the distances:

$\Delta d = 72 \text{ ft} - 24 \text{ ft}$

$\Delta d = 48 \text{ ft}$

13. E is correct.

$d = v_0t + \frac{1}{2}gt^2$

Set g as negative because it is in the opposite direction as +800 m/s:

$d = (800 \text{ m/s})\cdot(200 \text{ s}) + \frac{1}{2}(-10 \text{ m/s}^2)\cdot(200 \text{ s})^2$

$d = (160{,}000 \text{ m}) - (200{,}000 \text{ m})$

$d = -40{,}000 \text{ m}$

The projectile traveled –40,000 m from the top of the cliff to the sea at point P, which means the cliff is 40,000 m from the base to the top.

14. C is correct.

1 Newton is the force needed to accelerate 1 kg of mass at a rate of 1 m/s^2.

$F = ma$

$\text{W} = mg$

$\text{N} = \text{kg}\cdot\text{m/s}^2$

15. A is correct.

When the ball is thrown upwards, its velocity is initially upwards, but as it falls back down, the velocity direction points downward.

Acceleration due to gravity always points down and does not change.

Notes for active learning

CHAPTER 2

Equilibrium & Momentum

Page intentionally left blank

Concept of Equilibrium

Newton's First Law of Motion

Equilibrium is related to Newton's First Law of Motion.

Newton's *First Law of Motion* (Law of Inertia) states that:

> *An object at rest tends to stay at rest, and an object in motion will remain in motion at a constant velocity unless acted upon by an external force.*

Object is in equilibrium until an outside force is applied and the object begins to *accelerate*.

Acceleration (i.e., a vector with magnitude and direction) is the change in the velocity (vector); change may be just for the direction of motion.

Translational and rotational equilibrium

Equilibrium is linked to *acceleration*; objects undergoing acceleration are not in equilibrium.

Objects with zero acceleration are in equilibrium.

Equilibrium is *translational* or *rotational,* which may exist concurrently or separately.

Objects in *translational equilibrium* have no *linear* acceleration.

Objects in *rotational equilibrium* have no *angular* acceleration.

Zero linear and zero angular acceleration means translational and rotational equilibrium.

However, an object with zero linear acceleration but non-zero angular acceleration is only in translational equilibrium.

Static and dynamic equilibrium

Equilibrium can be separated into static and dynamic equilibrium.

Static equilibrium refers to objects with zero velocity (linear or angular velocity).

Dynamic equilibrium for objects with constant non-zero velocity (linear or angular velocity).

Force and Linear Acceleration

Newton's Second Law of Motion

Force causes objects to accelerate (i.e., change speed), change direction, or both.

Newton's *Second Law of Motion* states that:

> *The acceleration of an object is directly proportional to the applied force and inversely proportional to the mass of the object*:
>
> $$\vec{a} = \frac{F}{m}$$

or

$$F = m\vec{a}$$

Labeling forces

An arrow above a symbol indicates a force, with the arrowhead pointing toward the force.

Magnitude of the force is labeled near the arrow and measured in units of Newtons (N).

Typically labeled force vector:

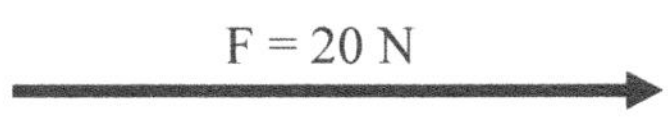

Linear and angular acceleration

The concept of force is essential for *equilibrium*, as force imparts linear or angular acceleration to an object.

Linear acceleration is measured in units of m/s^2 and is the *rate of change* of linear velocity of an object over time:

$$\vec{a} = \frac{\Delta\vec{v}}{\Delta t}$$

Angular acceleration is measured in units of rad/s^2 and is the *rate of change* of angular velocity of an object over time:

$$\vec{\alpha} = \frac{\Delta\vec{\omega}}{\Delta t}$$

Translational Equilibrium

Forces acting on objects

Translational equilibrium refers to an object undergoing zero linear acceleration.

In translational equilibrium, either:

1) no forces act upon the object, and linear acceleration is zero, or

2) forces acting upon the object cancel, and the object experiences zero linear acceleration in any direction.

Typical situations are when forces acting upon the object cancel; thus, translational equilibrium is achieved when forces along each coordinate axis *sum to zero.*

Force is expressed by:

$$\Sigma\vec{F} = 0 \text{ N}$$

Analyzing forces on incline planes:

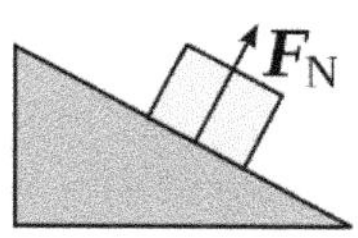

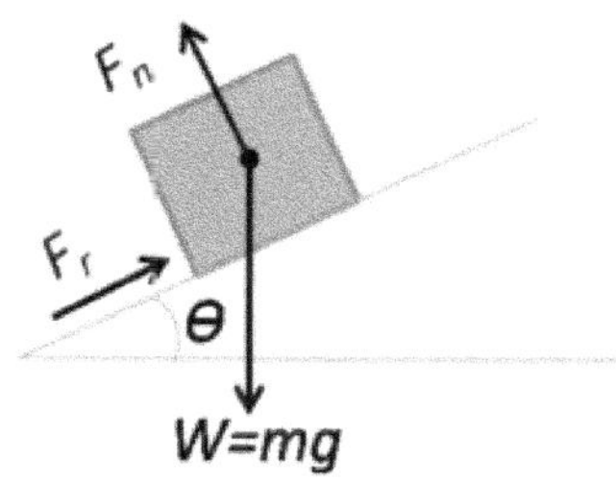

Block rests on an incline plane in translational equilibrium

Normal force

Normal force is the component of the contact force perpendicular to an object's contacts.

Book has *zero net linear acceleration* and remains *stationary.*

Book has *zero velocity* and is an example of a *static system.*

Equal and opposite normal upward force of the table cancels the force due to gravity.

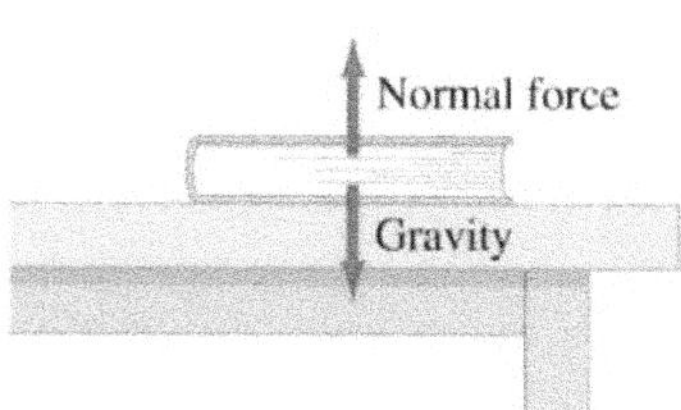

$$\Sigma F_y = F_N + F_G$$

$$\Sigma F_y = 0 \text{ N}$$

$$\vec{v} = 0 \text{ m/s}$$

Static translational equilibrium

Below is another system in *static translational equilibrium.*

A two-coordinate axis (x- and y-*axis*) is presented, and the forces along each axis sum to zero.

Like the previous example, no velocity vector is labeled (no movement of the object).

Therefore, the system is *static* and has no velocity.

$F_{Ax} = F_A \cos 60°$

$F_{Ay} = F_A \sin 60°$

$\Sigma F_x = F_{Ax} + F_B = 0 \text{ N}$

$\Sigma Fy = F_{Ay} + 1{,}960 \text{ N} = 0 \text{ N}$

$\vec{v} = 0 \text{ m/s}$

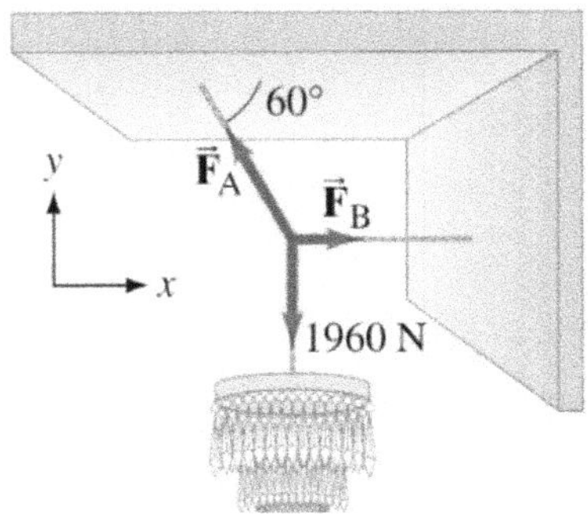

Forces sum to zero for objects in static equilibrium

Dynamic translational equilibrium

Skydiver is in *dynamic translational equilibrium.*

Skydiver achieved *terminal velocity*, and equal and opposite forces balance the force of gravity (weight) from air resistance.

Skydiver has a *constant non-zero velocity* and is a *dynamic translational* system.

$\Sigma F_y = F_G + F_{Air\ Resistance} = 0 \text{ N}$

$\vec{v} =$ constant

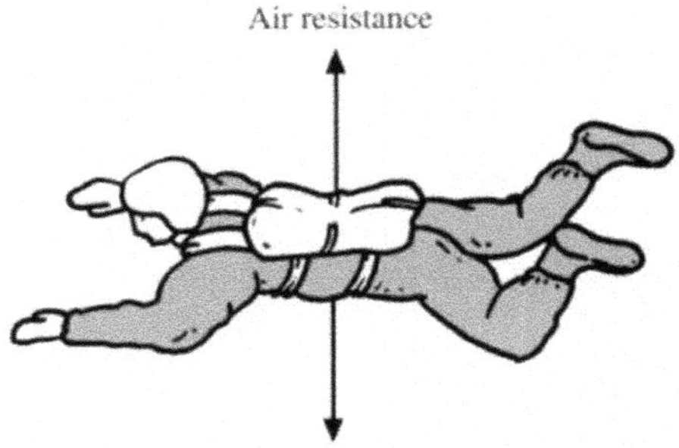

Dynamic translational equilibrium with constant non-zero velocity

Translational equilibrium *vs.* non-equilibrium

Scenario	Type of equilibrium	Translational non-equilibrium
Apple at rest	Static translational	Apple falling toward Earth with an acceleration of g
Car moving at constant velocity	Dynamic translational	Car either accelerating or decelerating
Skydiver falling at terminal velocity	Dynamic translational	Skydiver before reaching terminal velocity

Stable equilibrium

Stable equilibrium is when the forces on an object return the object to its equilibrium position. For example, a hanging ball connected to a string is in stable equilibrium. If the ball were momentarily disturbed, the net force would return it to its original equilibrium position.

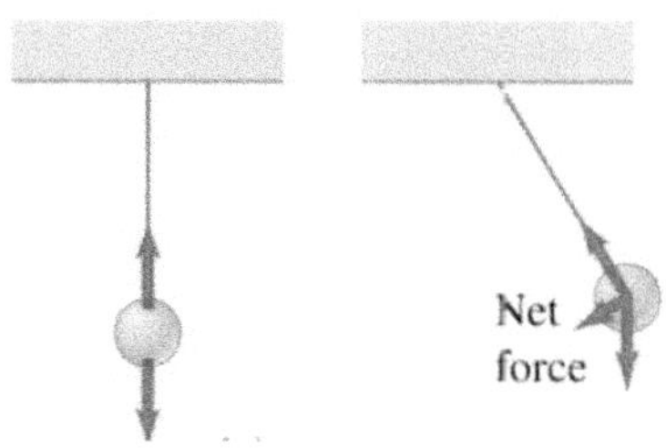

Object returns to its equilibrium position when the net force is removed

Unstable equilibrium

Unstable equilibrium is when the forces tend to move an object away from its equilibrium point.

A pencil balanced on its tip is in unstable equilibrium. When disturbed from its equilibrium position, the net force will not return the pencil to its original equilibrium position (i.e., a pencil resting on the tip) but cause it to settle in a new equilibrium position (e.g., lying on its side).

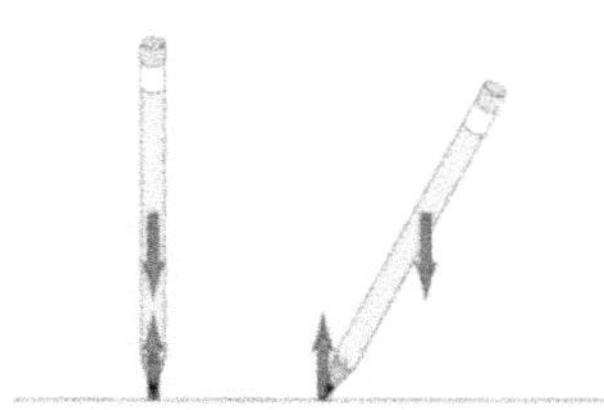

Center of gravity outside the pivot point for unstable equilibrium

When disturbed, an object in stable equilibrium transitions to an unstable equilibrium, so its center of gravity (CG) is outside the pivot point.

A resting refrigerator exhibits stable equilibrium while its center of gravity (CG) is vertically positioned over the pivot point.

Suppose the refrigerator's center of gravity deviates vertically above the pivot point. It transitions to an unstable equilibrium, and the net force causes it to settle in a new equilibrium position (e.g., on its side).

Object exhibits unstable equilibrium when its center of gravity deviates from the pivot point

Torques and Lever Arms

Torque force applied at an angle

Force applied to an object or system that induces *rotation* is torque.

Torque is the angular equivalent of force; it causes objects to rotate, has angular acceleration, and changes angular velocity.

Position and direction of the force applied to the rotating body about the rotation axis determine the torque's magnitude.

Torque is the product of the applied force and lever arm.

Torque is directly proportional to the applied force's magnitude and the length of the lever arm, expressed in Newton-meter units (N·m).

$$\vec{\tau} = r_{\perp}\vec{F} = r\ sin\ \theta\vec{F}$$

The lever arm is a distance from the axis of rotation to the line along which the perpendicular component of the force acts.

Force of lever arm from the angle between the applied force and distance to the axis of rotation:

$r_{\perp} = r\ sin\ \theta$

$\theta \leq 90$

$r_{\perp} \leq r$

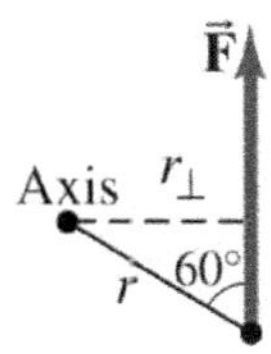

Torque and lever arm

Maximum and minimum torque

Maximum torque is when the force is applied perpendicular to the radius.

For maximum torque, the angle $\theta = 90°$ between the force vector and radius:

$$\vec{\tau}_{max} = r\ sin\ (90)\overrightarrow{F} = r_{\perp}\vec{F})$$

where sin 90° = 1

Minimum torque is when force applied to rotational body is directed toward the center of body:

$$\vec{\tau}_{min} = r\sin(0)$$

$\vec{F} = 0$ N·m where $\sin\theta = 0°$

Lever arms

For a door with two forces (A and B) applied at different points along its length:

$\vec{F}_A = \vec{F}_B$

$\theta_A = \theta_B = 90°$

$\vec{\tau} = r_{\perp}\vec{F}$

$r_{\perp A} > r_{\perp B}$

$\vec{\tau}_A > \vec{\tau}_B$

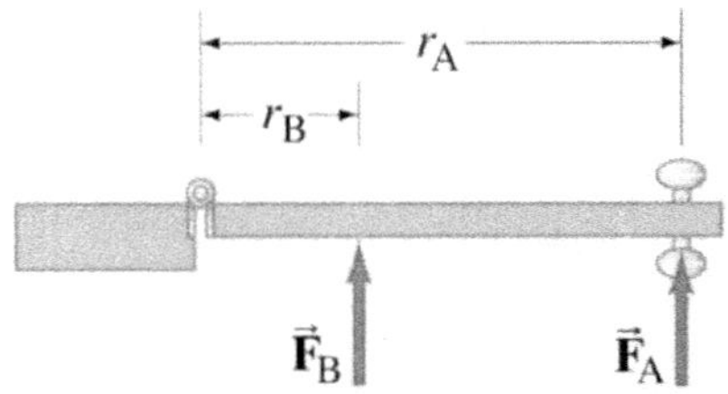

In the diagram above, F_A and F_B are equal in magnitude, and each is applied perpendicularly to the door (i.e., $\theta = 90°$).

Moving the door by F_A is easier than by using F_B because the lever arm for F_A is longer than the lever arm for F_B (torque = r *sin* θ *F*).

Torque produced by F_A is larger than that by F_B, and the door can be opened more easily with this larger torque.

Imagine a door with force applied at angles relative to the doorknob:

$\vec{F}_A = \vec{F}_C = \vec{F}_D$

$90° = \theta_A > \theta_C > \theta_D$

$\vec{\tau} = r_{\perp}\vec{F} = r \sin\theta\vec{F}$

$r_{\perp A} > r_{\perp C} > r_{\perp D}$

$\vec{\tau}_A > \vec{\tau}_C > \vec{\tau}_D$

Forces A, C, and D are equal in magnitude and applied at the same point from the axis of rotation (length from the doorknob to the hinge).

Torques by forces A, C, and D are unequal because forces are applied at different angles.

Force A produces the maximum torque because it is applied perpendicular (i.e., sin (90°) = 1) to the door and has the largest lever arm.

Force C is applied at a non-perpendicular angle to the door; it has a smaller lever arm and produces a smaller torque.

Force D is applied parallel to the door and produces no torque because the force and its lever arm are equal to zero (*sin* (0°) = 0).

Rotational Equilibrium

Conditions for rotational equilibrium

Objects in rotational equilibrium have *zero angular acceleration.*

Rotational equilibrium has no torques applied or around every rotational axis equal to zero.

Rotational equilibrium:

$$\sum \vec{\tau} = 0 \text{ N} \cdot \text{m}$$

Objects either do not rotate (static rotational equilibrium) or rotate at a constant rate (dynamic rotational equilibrium).

Positive and negative torques

Positive torques act *counterclockwise.*

Negative torques act *clockwise.*

Positive Convention +

Negative Convention -

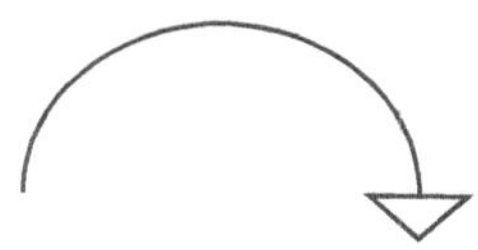

Pulley systems

Pulley systems demonstrate static rotational equilibrium.

Force 1 and 2 are equal and applied at opposite ends tangent to the pulley. Resulting torques 1 and 2 are equal and opposite.

Sum of torques along the pulley's rotational axis is zero. Pulley experiences no angular velocity and is a static system.

$$\vec{F}_1 = \vec{F}_2 \text{ and } \theta_1 = \theta_2 = 90°$$

$$r_{\perp} = d$$

$$\vec{\tau}_1 = \vec{F}_1(d \sin \theta_1)$$

$$\vec{\tau}_2 = \vec{F}_2(d \sin \theta_2)$$

$$\sum \vec{\tau} = \vec{\tau}_1 + \vec{\tau}_2 = 0 \text{ N} \cdot \text{m}$$

$$\vec{\omega} = 0 \text{ rad/s}$$

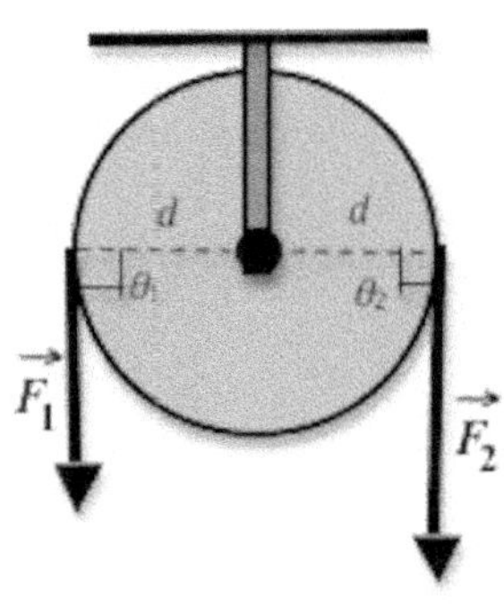

Torque and tension

Static rotational equilibrium system involves a mass attached to a cable-supported beam.

Mass hangs perpendicularly from a beam at distance *L*, creating a clockwise torque from the gravitational force.

Cable attached at a distance *d* produces its torque from the tension force and cancels the torque from the mass.

Sum of torques about the rotation axis is zero, and the system is *static with no angular velocity.*

$$\vec{F}_G \neq \vec{F}_T$$

$$\vec{\tau}_G = \vec{F}_1(L\,sin\,90°)$$

$$\vec{\tau}_T = \vec{F}_T(d\,sin\,\theta)$$

$$\sum\vec{\tau} = \vec{\tau}_G + \vec{\tau}_T = 0\ \mathrm{N\cdot m}$$

$$\vec{\omega} = 0\ \mathrm{rad/s}$$

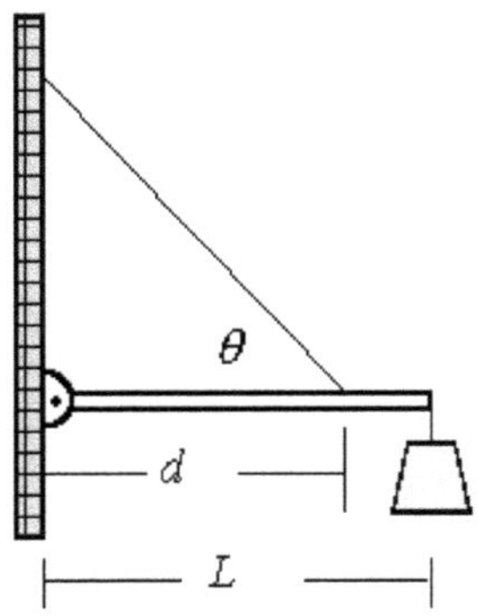

Static rotational equilibrium system with mass attached to a cable-supported beam

Static and dynamic rotational equilibrium

Scenario	Equilibrium type	Rotational non-equilibrium
Equal masses on a balance	Static rotational	Unequal masses on balance cause angular acceleration
Propeller spinning at a fixed frequency	Dynamic rotational	Propeller spins faster and faster
Asteroid rotating at a constant velocity	Dynamic rotational	Asteroid rotation slows

Solving Equilibrium Problems

Steps for solving equilibrium problems

1. Identify objects and draw *free-body diagrams* of forces acting and where.
2. Choose a coordinate system and set axis's negative and positive directions.
3. Resolve forces into components (the normal and parallel components for inclined planes).
4. Add the force components; the resulting x-, y- and z-components determine the net force acting on the object.
5. Choose an axis perpendicular to the plane of the forces and use equilibrium equations for the forces and torques.

   Proper choice of equations simplifies the problem.
6. Determine the type of equilibrium (refer to the table below).
7. Pythagorean Theorem for the magnitude of net force from components.

Use trigonometry to solve for the angles.

Angles are measured in *radians*, a whole circle = 2π radians.

Acceleration, velocity and equilibrium type

Acceleration	Velocity	Equilibrium type
$a = 0$ m/s^2	$v = 0$ m/s	Static translational
	$v \neq 0$ m/s	Dynamic translational
$\alpha = 0$ rad/s^2	$\omega = 0$ rad/s	Static rotational
	$\omega \neq 0$ rad/s	Dynamic rotational

Example 1: static equilibrium system

A 15,000 kg printing press is set upon a 1,500 kg table. Assuming the table's center of gravity contains all its mass, what are the reaction forces from the table legs A and B?

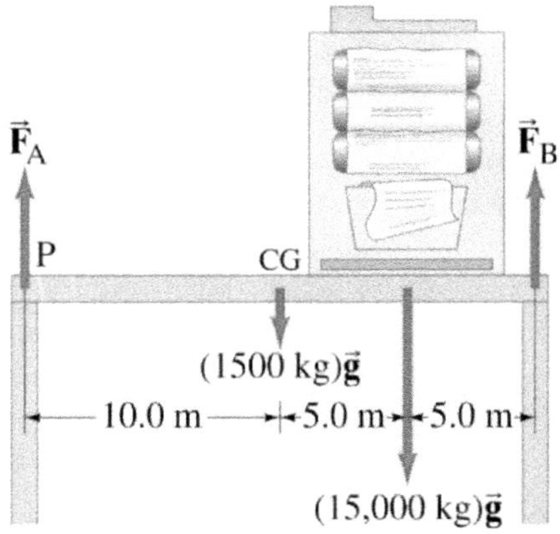

Static equilibrium system

Sum forces in *y*-axis:

$\Sigma F_y = F_A + F_B - (1{,}500 \text{ kg})\cdot(9.8\text{m/s}^2) - (15{,}000 \text{ kg})\cdot(9.8\text{m/s}^2) = 0$

$F_A + F_B = 161{,}700 \text{ N}$

Sum torques about point P:

$\Sigma\tau_P = [-(1{,}500 \text{ kg})\cdot(9.8\text{m/s}^2)\cdot(10 \text{ m}) - (15{,}000 \text{ kg})\cdot(9.8 \text{ m/s}^2)\cdot(15 \text{ m})] + F_B(20 \text{ m}) = 0$

$F_B(20 \text{ m}) = 2{,}352{,}000 \text{ N}\cdot\text{m}$

$F_B = 117{,}600 \text{ N}$

Solve for the remaining forces:

$F_A + 117{,}600 \text{ N} = 161{,}700 \text{ N}$

$F_A = 161{,}700 \text{ N} - 117{,}600 \text{ N}$

$F_A = 44{,}100 \text{ N}$

If calculated force is negative, the opposite direction of the initially chosen positive direction.

There is no "correct" way to choose which direction is positive.

The requirement is that it is kept consistent throughout the problem.

Example 2: static equilibrium system with torque

A beam with a mass of 500 kg located at the center of gravity is attached to two supporting legs. According to the diagram, what are the reaction forces from leg A and leg B? (Use the acceleration due to gravity, g = 9.8 m/s²)

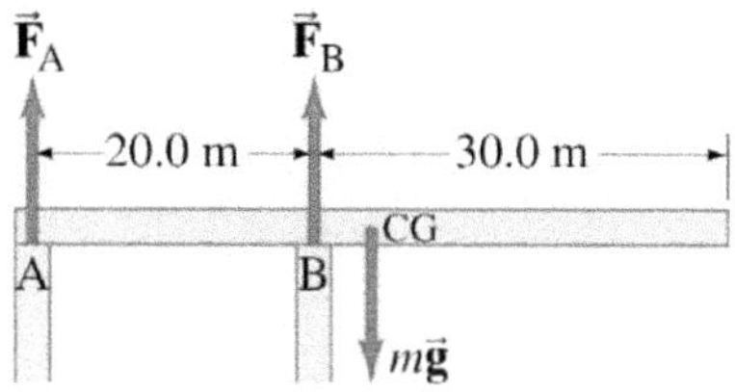

Static equilibrium system with torque

Sum forces in y-axis:

$$\Sigma F_y = F_A + F_B - (500\text{ kg})\cdot(9.8\text{ m/s}^2)$$

$$\Sigma F_y = 0$$

$$F_A + F_B = 4{,}900\text{ N}$$

Sum torques about point A (the center of gravity is the midpoint of the beam):

$$\Sigma\tau_A = -(500\text{ kg})\cdot(9.8\text{m/s}^2)\cdot(25\text{ m}) + F_B(20\text{ m})$$

$$\Sigma\tau_A = 0$$

$$F_B(20\text{ m}) = 122{,}500\text{ N}\cdot\text{m}$$

$$F_B = 6{,}125\text{ N}$$

Solve for the remaining forces:

$$F_A + 6{,}125\text{ N} = 4{,}900\text{ N}$$

$$F_A = -1{,}225\text{ N}$$

The cable attached supports tension forces (F_T) only along its length.

Force applied perpendicular to F_T would cause it to bend.

Example 3: static equilibrium system with cables

A sign support consists of a beam attached to a hinge, a sign attached at the opposite end, and a cable supporting the load. The beam has a length of 10 m, a mass of 500 kg, and the center of gravity is at the 5 m point. The sign has a mass of 1,000 kg 10 m from the hinge. If the support cable forms an angle of 45° with the beam, what are the reaction forces labeled in the diagram?

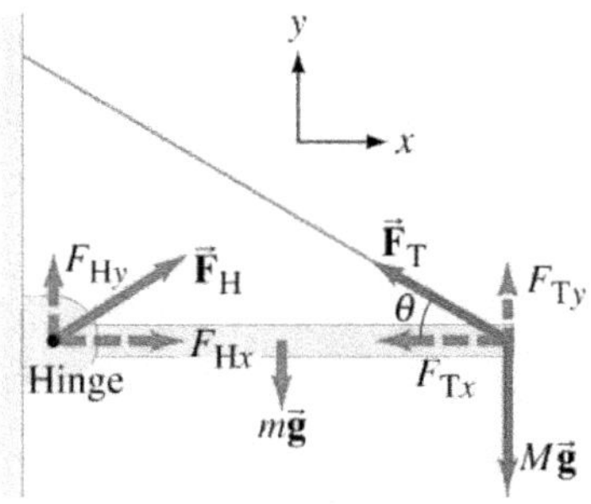

Static equilibrium system involving cables

Sum forces in y-axis:

$$\Sigma F_y = F_{Hy} + F_{Ty} - (500\text{ kg})\cdot(9.8\text{ m/s}^2) - (1{,}000\text{ kg})\cdot(9.8\text{ m/s}^2)$$

$$\Sigma F_y = 0$$

$$F_{Hy} + F_{Ty} = 14{,}700\text{ N}$$

Sum forces in x-axis:

$$\Sigma F_x = F_{Hx} - F_{Tx} = 0$$

$$F_{Hx} = F_{Tx}$$

Sum torques about hinge H (center of gravity is at the beam's midpoint):

$$\Sigma\tau_H = -(1{,}000\text{ kg})\cdot(9.8\text{ m/s}^2)\cdot(10\text{ m}) - (500\text{ kg})\cdot(9.8\text{ m/s}^2)\cdot(5\text{ m}) + F_T(10\text{ m})\,sin\,(45°)$$

$$\Sigma\tau_H = 0$$

$$F_T(10\text{ m})\,sin\,(45°) = 122{,}500\text{ N}\cdot\text{m}$$

$$F_T = 17{,}324\text{ N}$$

Solve for the remaining forces:

$F_{Ty} = F_T\,sin\,(45°)\text{ N}$ and $F_{Hx} = F_{Tx} = F_T\,cos\,(45°)\text{ N}$

$F_{Ty} = (17{,}327\text{ N})\,sin\,(45°)$ and $F_{Hx} = F_{Tx} = (17{,}327\text{ N})\,cos\,(45°)$

$F_{Ty} = 12{,}250\text{ N}$ and $F_{Hx} = F_{Tx} = 12{,}250\text{ N}$

$$F_{Hy} + 12{,}250\text{ N} = 14{,}700\text{ N}$$

$$F_{Hy} = 2{,}450\text{ N}$$

Example 4: static equilibrium system with lever arm

Principles of force and torque can be applied to the human body.

In an arm, the elbow acts as a fulcrum, and the forearm acts as a lever arm. Biceps are modeled as a cable containing tension.

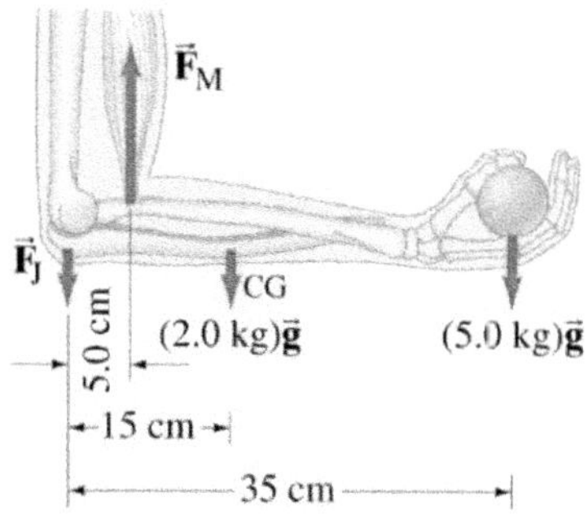

Lever arm biceps model a static equilibrium system

Sum forces in y-axis:

$$\Sigma F_y = F_M - F_J - (2\text{ kg})\cdot(9.8\text{ m/s}^2) - (5\text{ kg})\cdot(9.8\text{ m/s}^2)$$

$$\Sigma F_y = 0$$

$$F_M - F_J = 69\text{ N}$$

Sum torques about point J:

$$\Sigma \tau_J = F_M(0.05\text{ m}) - (2\text{ kg})\cdot(9.8\text{ m/s}^2)\cdot(0.15\text{ m}) - (5\text{ kg})\cdot(9.8\text{ m/s}^2)\cdot(0.35\text{ m})$$

$$\Sigma \tau_J = 0$$

$$F_M = 402\text{ N}$$

Solve for the remaining forces:

$$402\text{ N} - F_J = 69\text{ N}$$

$$F_J = 333\text{ N}$$

Notes for active learning

Momentum

Inertia is related to mass

Newton's *First Law of Motion* (Law of Inertia) states that:

> *An object at rest tends to stay at rest, and an object in motion will remain in motion at a constant velocity unless acted upon by an external force.*

This is due to the object's *inertia* (i.e., resistance to change).

Inertia is directly related to an object's mass; therefore, massive objects have more resistance changes in motion.

Momentum

When an object is in motion (i.e., it has a non-zero velocity), inertia translates into momentum.

Momentum (p) is the vector product of mass and velocity:

$$\vec{p} = m\vec{v}$$

where m is mass measured in kg, $\vec{v}$ is the velocity measured in m/s, and momentum $\vec{p}$ is measured in units of kg·m/s or N·s (Newton·seconds)

Total momentum is the vector sum of individual momenta (note velocity directions of individual momenta).

Conservation of momentum

Law of Conservation of Momentum states that *in an isolated system of objects, the total momentum of the objects is conserved.*

Conservation of momentum has essential implications for understanding and predicting the behavior of natural phenomena.

A classic example of the Law of Conservation of Momentum is the predictable behavior of colliding objects (e.g., bumper cars or pool balls).

Conservation of Linear Momentum

Isolated systems

In an isolated system, the total momentum for any collision is conserved.

Isolated system has objects free from net external forces.

Forces present in the isolated system are supplied by the objects.

If two balls collide on a frictionless surface, it is considered an isolated system since no external forces are present, and the conservation of momentum applies.

Friction as a force opposing motion

If non-negligible friction is present, the balls experience an external force (force of friction).

Momentum changes with force supplied by friction.

Force of friction is in the direction *opposite to the motion.*

Law of Conservation of Momentum does *not* apply if:

1) an external force is supplied, and

2) the motion is not within an isolated system.

Conservation of momentum in a collision (within isolated systems) is when total momentum (p) before a collision equals the total momentum after the collision.

$$\vec{p}_{initial} = \vec{p}_{final}$$

Momentum has direction and magnitude

Momentum (p) is a vector, so assign one direction as positive and the opposite as negative when adding the component momenta vectors.

Momentum of a bomb at rest equals the vector sum of the momenta of all shrapnel from the explosion. Because shrapnel tends to travel in all directions, the momenta of individual pieces cancel, and total momentum is conserved.

Conservation of momentum applies to a rocket in flight if the rocket and its fuel are considered one system to account for the mass loss of the rocket's fuel.

When a rocket fires in space, it does not need air to push against it because the momentum of the combustion gases must be conserved. Thus, the rocket moves in the opposite direction.

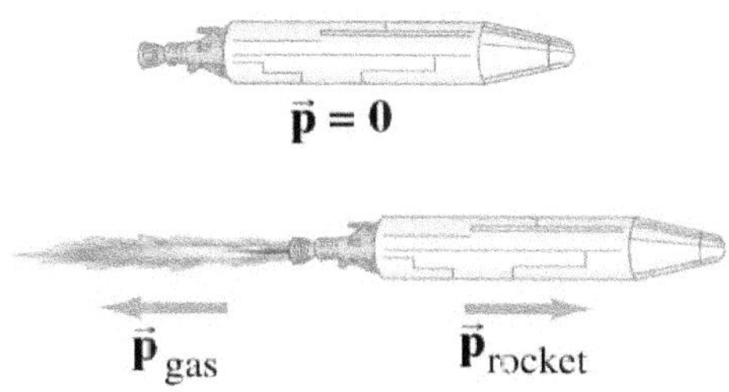

Conservation of momentum observed with the rocket as a system

Momentum before = Momentum after = 0 N·s

$$m_{Gas}\vec{v}_{Gas} = m_{Rocket}\vec{v}_{Rocket}$$

Elastic and inelastic collisions

In elastic and inelastic collisions *momentum is conserved.*

In elastic collisions *kinetic energy* (KE) *is conserved* (e g., pool balls bounce after collision)

In inelastic collisions *kinetic energy* (KE) *is not conserved* (e.g., bullet embedded in wooden block).

Notes for active learning

Elastic Collisions

Conservation of kinetic energy

Elastic collisions conserve kinetic energy so that the initial total kinetic energy equals the final total kinetic energy.

$$KE_{Before} = KE_{After}$$

Kinetic energy is a scalar quantity (not a vector):

$$KE = \frac{1}{2}mv^2$$

The unit for energy is Joule (J), and since energy is a scalar (no direction specified), there are no associated positive or negative signs for KE.

In a perfectly elastic collision, a dropped ball bounces back to its original height (e.g., ignoring gravity and air resistance).

For example, a ball strikes a wall and bounces at the same speed it struck it.

Conservation of momentum

Elastic collisions conserve momentum.

For elastic collisions, use the conservation of kinetic energy and conservation of momentum.

The example below has two objects with known masses and initial speeds colliding elastically.

Top diagram below is before impact, and the bottom is after collision.

Since *momentum and kinetic energy* are conserved, both equations can be written to solve for the unknown final speeds via substitution:

$$\vec{p}_A + \vec{p}_B = \vec{p}_B' + \vec{p}_A'$$

$$m_A\vec{v}_A + m_B\vec{v}_B = m_A\vec{v}_A' + m_B\vec{v}_B'$$

$$KE_{Before} = KE_{After}$$

$$KE = ½mv^2$$

$$\frac{1}{2}m_A\vec{v}_A^{\,2} + \frac{1}{2}m_B\vec{v}_B^{\,2} = \frac{1}{2}m_A\vec{v}_A'^{\,2} + \frac{1}{2}m_B\vec{v}_B'^{\,2}$$

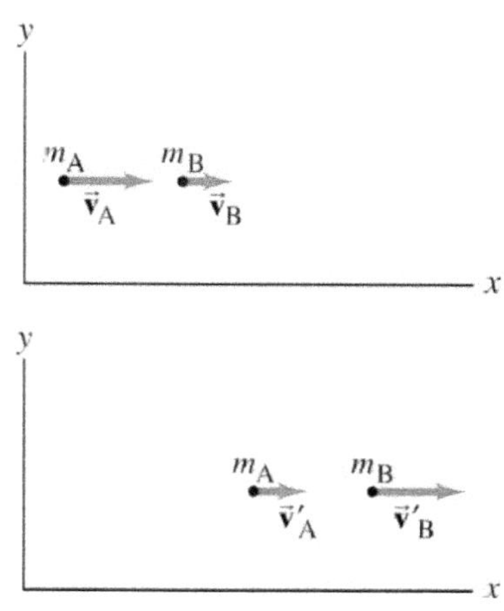

Elastic collisions conserve kinetic energy and momentum

Elastic collisions in opposite directions

For *elastic collisions* of objects traveling in opposite directions:

initial momentum equals final momentum.

initial kinetic energy equals final kinetic energy;

Sign of velocity changes depending on the coordinate axis.

$$\vec{p}_A + \vec{p}_B = \vec{p}_B' + \vec{p}_A'$$

$$m_A\vec{v}_A + m_B\vec{v}_B = m_A\vec{v}_A' + m_B\vec{v}_B'$$

$$KE_{Before} = KE_{After}$$

$$KE = \frac{1}{2}mv^2$$

$$\frac{1}{2}m_A\vec{v}_A^{\,2} + \frac{1}{2}m_B\vec{v}_B^{\,2} = \frac{1}{2}m_A\vec{v}_A'^{\,2} + \frac{1}{2}m_B\vec{v}_B'^{\,2}$$

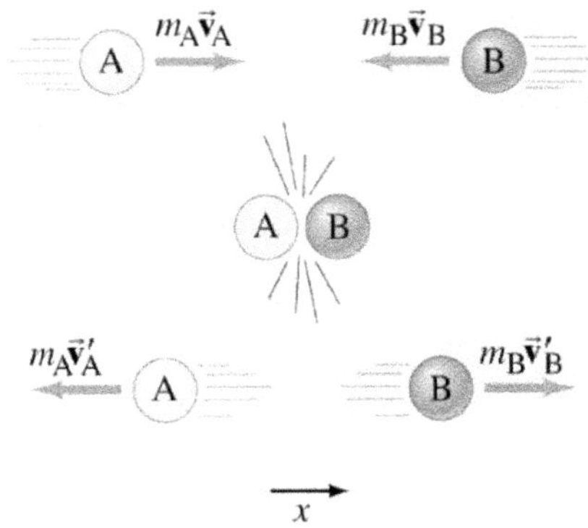

Elastic collision of objects traveling in opposite directions

Inelastic Collisions

Inelastic collisions do not conserve kinetic energy

Inelastic collisions *conserve momentum* but do not conserve kinetic energy ($KE_{initial} > KE_{final}$).

This does not violate the Law of Conservation of Energy because kinetic energy is converted into another form during the collision.

Collisions in everyday life are often inelastic.

For example, if a ball is dropped, it does not reach the same height as it was released after bouncing off the ground. The collision is inelastic, with momentum conserved but not KE. Some of the ball's KE was converted to other forms (e.g., acoustic, thermal, vibrational).

Elastic *vs.* inelastic collisions

Diagram illustrates two objects that experience elastic collision (i.e., conserves both momentum and KE) and inelastic collision (i.e., conserves momentum only, not KE).

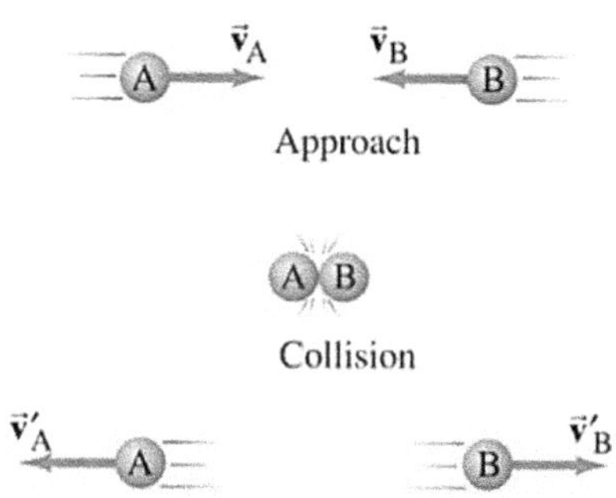

Approaching objects experience elastic collision with conservation of momentum and KE

In inelastic collisions objects stick together after a collision.

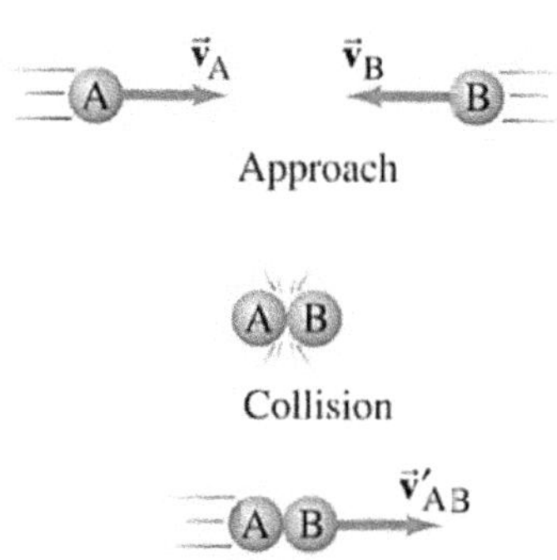

Approaching objects experience inelastic collision with conservation of momentum

Inelastic collision of a bullet conserves momentum

Below, a bullet with known velocity and mass is shot into a hanging block of known mass.

To solve the height attained by the block + bullet (inelastic collision), apply the conservation of momentum to find the velocity after the collision.

After the collision, the KE of the system is converted to potential energy.

Conservation of energy is applied, and the height can be calculated.

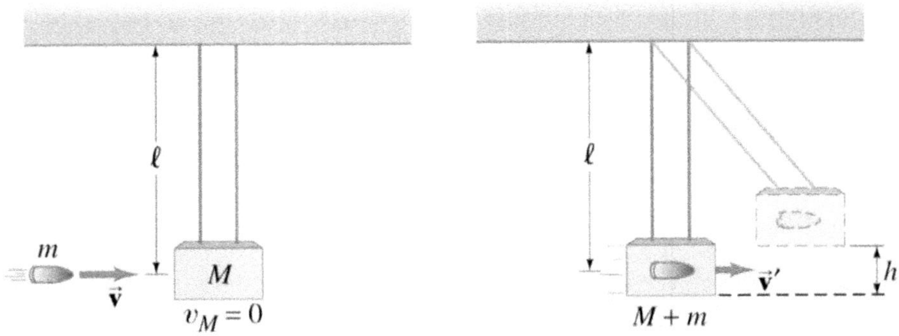

Inelastic collision with conservation of momentum

$$\vec{p}_{Bullet} + \vec{p}_{Block} = \vec{p}'_{Bullet+Block}$$

$$m\vec{v} + M\vec{v}_M = (M + m)\vec{v}'$$

$$KE_{Bullet\ Before} \neq KE_{System\ After}$$

$$KE_{System\ After} = PE_{System\ After}$$

$$KE = \frac{1}{2}mv^2$$

$$PE = mgh$$

$$\frac{1}{2}(M + m)\cdot(\vec{v}')^2 = (M + m)gh$$

Solving Collision Problems

Collisions in two or three dimensions

Conservation of kinetic energy and momentum analyzes collisions in two or three dimensions.

A moving object m_A collides with an object m_B initially at rest.

Determining final velocities requires:

1) masses,

2) initial velocities, and

3) angles of the object's post-collision trajectory.

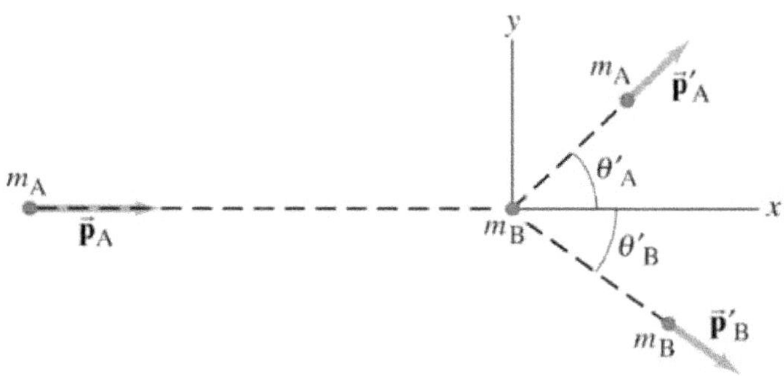

Final velocities of collision with conservation of momentum and kinetic energy

$$\Sigma\vec{p}_x = \vec{p}_{Ax} + \vec{p}_{Bx} = \vec{p}'_{Bx} + \vec{p}'_{Ax}$$

$$m_A\vec{v}_A = m_A\vec{v}'_A \cos\theta'_A + m_B\vec{v}'_B \cos\theta'_B$$

$$\Sigma\vec{p}_y = \vec{p}_{Ay} + \vec{p}_{By} = \vec{p}'_{By} + \vec{p}'_{Ay}$$

$$0 = m_A\vec{v}'_A \sin\theta'_A + m_B\vec{v}'_B \sin\theta'_A$$

Eight steps for solving collision problems

1. Choose the system.

   If it is a complex system (e.g., involves both elastic and inelastic collisions), subsystems should be chosen to apply one or more conservation laws.

2. Choose a system so that no external forces are present.

3. Draw diagrams of the initial and final, with the momentum vectors labeled.

4. Choose a coordinate system. If an object's momentum is at an angle to the axis, separate motion into x- and y- components:

   $$p_x = \vec{p}\cos\theta \qquad \text{and} \qquad p_y = \vec{p}\sin\theta$$

5. Apply conservation of momentum with separate equations for motion in each dimension.

   Total initial momentum in x-direction equals the total final momentum in x-direction ($\Sigma p_{xi} = \Sigma p_{xf}$).

   Total initial momentum in the y-direction equals the total final momentum in the y-direction ($\Sigma p_{yi} = \Sigma p_{yf}$).

6. Do not assume the collision type (elastic or inelastic); either the type is given or evaluate if kinetic energy is conserved.

   If KE conserved, collision is elastic.

   If KE not conserved, collision is inelastic.

7. For elastic collisions, use substitution between the conservation of kinetic energy and momentum; inelastic collisions use momentum conservation.

   $$\vec{p} = \sqrt{p_x^2 + p_y^2}\text{, and the angle is given by } \theta = tan^{-1}\left(\frac{p_y}{p_x}\right)$$

8. Check units and magnitudes to evaluate if they make sense.

Example of an inelastic collision

Train car A (m =1,000 kg) collides inelastically with train car B of equal mass. After the collision, the train cars stick (inelastic collision that conserves momentum but not kinetic energy). What is the final velocity of the attached train cars?

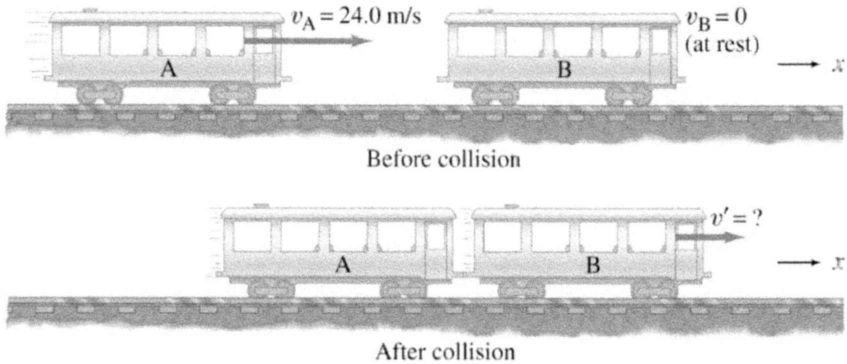

Inelastic collisions conserve momentum

Apply Law of Conservation of Momentum:

$$\vec{p}_A + \vec{p}_B = \vec{p}'_{A+B}$$

$$m_A\vec{v}_A + m_B\vec{v}_B = (m_A + m_B)\vec{v}'$$

Solve for final velocity:

$$(1{,}000 \text{ kg}) \cdot (24 \text{ m/s}) + (1{,}000 \text{ kg}) \cdot (0 \text{ m/s}) = (1{,}000 \text{ kg} + 1{,}000 \text{ kg})\vec{v}'$$

$$\vec{v}' = 12 \text{ m/s}$$

Determine the magnitude of the difference in kinetic energy:

$$KE = \frac{1}{2}mv^2$$

$$KE_{\text{Before}} = \frac{1}{2}(1{,}000 \text{ kg}) \cdot (24 \text{ m/s})^2 = 288 \text{ kJ}$$

$$\text{KE}_{\text{After}} = \frac{1}{2}(2{,}000 \text{ kg}) \cdot (12 \text{ m/s})^2 = 144 \text{ kJ}$$

$$\text{KE}_{\text{Before}} \neq \text{KE}_{\text{After}}$$

Notes for active learning

Impulse

Impulse as a change in momentum

Objects (e.g., cars) are deformed during collisions due to the large forces involved.

Softer objects (e.g., pillows) are deformed more dramatically over a larger time frame than objects made of stiffer materials (e.g., metal).

Impulse (J) relates the extent to which an object deforms due to an impact, the time the object contacts another mass.

Impulse (i.e., the change in momentum, p) is represented as $F\Delta t$, where F is a force, and Δt is the time interval during which the force acts:

$$\Sigma\vec{F} = \frac{\Delta\vec{p}}{\Delta t}$$

$$\vec{F}\Delta t = \Delta\vec{p}$$

Calculating impulse

Total force equals the change in momentum divided by the time interval:

$$\Sigma\vec{F} = \frac{\Delta\vec{p}}{\Delta t}$$

Impulse ($\vec{J} = \vec{F}\Delta t$) can be derived using the equation $\vec{F} = m\vec{a}$.

$$\vec{F} = m\vec{a} = m\frac{\Delta\vec{v}}{\Delta t} \quad \rightarrow \quad \vec{F}\Delta t = m\Delta\vec{v} = \Delta\vec{p} = \vec{J}$$

Impulse determines if a change in momentum occurs by *either* a:

1) large force acting for a *short time,* or

2) small force acting for a *long time*.

Impulse is affected by airbags

Automobile airbags are helpful because they increase the time between the initial impact of the two objects (the airbag and person) and the moment both objects come to rest.

Without an airbag, this time interval would be shorter (smaller impulse), and therefore, the force of impact is greater and more harmful (i.e., the impulse is inversely proportional to force).

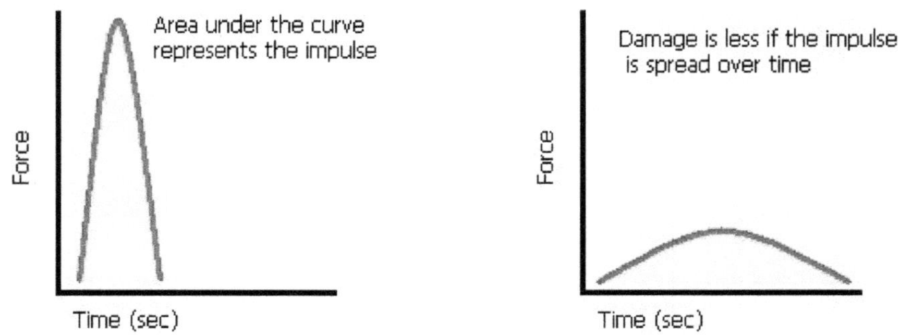

Airbags increase the time interval for contact and reduce the force applied

Force *vs.* time diagrams

For force *vs.* time diagrams, the magnitude of force changes with time.

Area under the force *vs.* time diagrams curve represents the impulse.

For the above diagram on the right, the magnitude of the force is less because the impulse is extended over a greater range of time.

Impulse is inversely proportional to force

In practice, the collision time is brief, and the time dependence of the force is less relevant.

Since the time interval is relatively short, the average force, rather than the time-dependent force, can be used when calculating impulse (diagrams below).

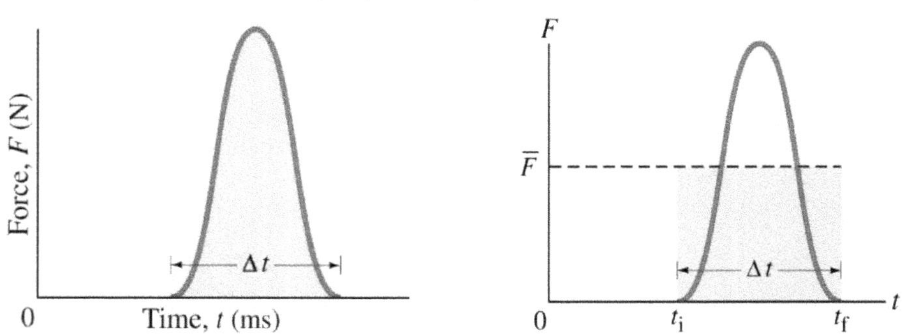

Area under the curve represents the impulse

Chapter Summary

Translational equilibrium has no linear acceleration.

Rotational equilibrium has no angular acceleration.

Two forms of equilibrium may exist concurrently or separately.

Static equilibrium refers to objects with zero velocity (linear or angular)

Dynamic equilibrium refers to objects with constant non-zero velocity (linear or angular).

In translational equilibrium, either:

1) no forces act upon the object, and linear acceleration is zero or

2) all forces acting upon the object cancel, and object experiences no linear acceleration in any direction.

Force applied to an object or system induces rotation as *torque*.

Torque can be expressed as:

$$\vec{\tau} = r_{\perp}\vec{F} = r \sin\theta \vec{F}$$

When objects are in rotational equilibrium, the angular acceleration is zero.

Therefore, either

1) no torques are applied, or

2) sum of torques around each rotational axis equals zero.

Acceleration	Velocity	Equilibrium type
$a = 0$ m/s^2	$v = 0$ m/s	Static translational
	$v \neq 0$ m/s	Dynamic translational
$\alpha = 0$ rad/s^2	$\omega = 0$ rad/s	Static rotational
	$\omega \neq 0$ rad/s	Dynamic rotational

Momentum (p) is the vector product of mass × velocity, represented by the equation:

$$\vec{p} = m\vec{v}$$

Law of Conservation of Momentum: total momentum is conserved in an isolated system.

$$\vec{p}_{initial} = \vec{p}_{final}$$

Elastic collisions, momentum and total kinetic energy are conserved:

$$\frac{1}{2}m_A\vec{v}_A^{\,2} + \frac{1}{2}m_B\vec{v}_B^{\,2} = \frac{1}{2}m_A\vec{v}_A'^{\,2} + \frac{1}{2}m_B\vec{v}_B'^{\,2}$$

Inelastic collisions, momentum is conserved (not the kinetic energy).

$(KE_{initial} > KE_{final})$

This does not violate the Law of Conservation of Energy because kinetic energy is converted into another form of energy (e.g., light, heat, sound) during the collision.

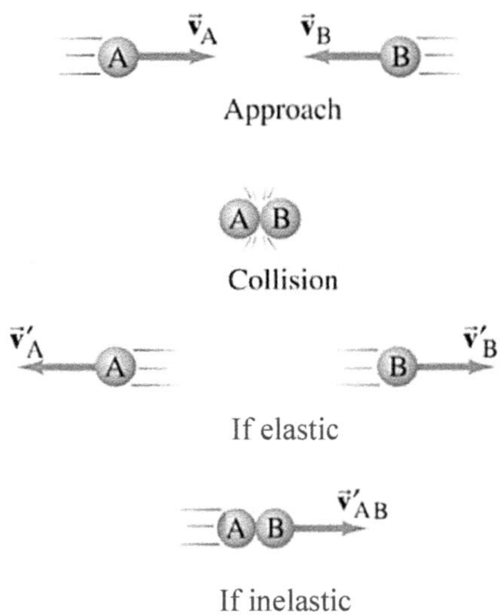

Impulse (J), the time the object contacts another mass, determines the extent to which an object deforms (e.g., airbags) due to an impact.

Impulse, the change in momentum, is represented as $F\Delta t$, where F is a force, and Δt is the time interval during which the force acts:

$$\sum\vec{F} = \frac{\Delta\vec{p}}{\Delta t}$$

Practice Problems: Equilibrium & Momentum

1. A 1.2 kg asteroid travels toward the Orion Nebula at a speed of 2.8 m/s. Another 4.1 kg asteroid is traveling at 2.3 m/s in a perpendicular direction. The two asteroids collide and stick. What is the change in momentum from before to after the collision?

A. 0 kg·m/s
B. 2.1 kg·m/s
C. 9.4 kg·m/s
D. 12 kg·m/s
E. 19.7 kg·m/s

2. A 1,120 kg car experiences an impulse of 30,000 N·s during a collision with a wall. If the collision takes 0.43 s, what was the car's speed just before the collision?

A. 12 m/s
B. 64 m/s
C. 42 m/s
D. 27 m/s
E. 18 m/s

3. Does the centripetal force acting on an object do work on the object?

A. No, because the force and the displacement of the object are perpendicular
B. Yes, since a force acts and the object moves, and work is force times distance
C. Yes, since it takes energy to turn an object
D. No, because the object has a constant speed
E. Yes, because the force and the displacement of the object are perpendicular

4. Cars with padded dashboards are safer in an accident than cars without padded dashboards because a passenger hitting the dashboard has:

I. increased the time of impact
II. decreased impulse
III. decreased impact force

A. I only
B. II only
C. III only
D. I and III only
E. I and II only

5. The acceleration due to gravity on the Moon is only one-sixth of that on Earth, and the Moon has no atmosphere. If a person hit a baseball on the Moon with the same effort (and therefore at the same speed and angle) as on Earth, how far would the ball travel on the Moon compared to on Earth? (Ignore air resistance on Earth)

A. The same distance as on Earth
B. 6 times as far as on Earth
C. 1/6 as far as on Earth
D. 36 times as far as on Earth
E. $\sqrt{6}$ times as far as on Earth

6. A uniform meter stick weighing 20 N has a weight of 50 N attached to its left end and a weight of 30 N attached to its right end. The meter stick is hung from a rope. What is the tension in the rope, and how far from the left end of the meter stick should the rope be attached so that the meter stick remains level?

A. 100 N placed 37.5 cm from the left end of the meter stick
B. 50 N placed 40 cm from the left end of the meter stick
C. 80 N placed 37.5 cm from the left end of the meter stick
D. 80 N placed 40 cm from the left end of the meter stick
E. 100 N placed 40 cm from the left end of the meter stick

7. To catch a ball, a baseball player extends her hand forward before impact with the ball and then lets it ride backward in the direction of the ball's motion upon impact. Doing this reduces the force of impact on the player's hand principally because the:

A. time of impact is decreased
B. time of impact is increased
C. relative velocity is less
D. force of impact is reduced by $\sqrt{2}$
E. none of the above

8. A 1,200 kg car, moving at 15.6 m/s, collides with a stationary 1,500 kg car. What is their combined velocity immediately after the collision if the two vehicles lock together?

A. 12.4 m/s
B. 5.4 m/s
C. 6.9 m/s
D. 7.6 m/s
E. 11.8 m/s

9. A 0.05 kg golf ball, initially at rest, has a velocity of 100 m/s immediately after being struck by a golf club. If the club and ball were in contact for 0.8 ms, what is the average force exerted on the ball?

A. 5.5 kN
B. 4.9 kN
C. 11.8 kN
D. 7.1 kN
E. 6.3 kN

10. Which of the following is an accurate statement for a rotating rigid body?

A. All points on the body are moving with the same angular velocity
B. Its center of rotation is its center of gravity
C. Its center of rotation is at rest and, therefore, not moving
D. Its center of rotation must be moving with a constant velocity
E. All points on the body are moving with the same linear velocity

11. Cart 1 (2 kg) and Cart 2 (2.5 kg) run along a frictionless, level, one-dimensional track. Cart 2 is initially at rest, and Cart 1 travels 0.6 m/s toward the right when it encounters Cart 2. After the collision, Cart 1 is at rest. Which of the following is true concerning the collision?

A. The collision is completely elastic
B. Kinetic energy is conserved
C. Momentum is conserved
D. Total momentum is decreased
E. Potential energy is conserved

12. What is the reason for using a long barrel in a gun?

A. Allows the force of the expanding gases from the gunpowder to act for a longer time
B. Increases the force exerted on the bullet due to the expanding gases from the gunpowder
C. Exerts a larger force on the shells
D. Reduces frictional losses
E. Reduces the force exerted on the bullet due to the expanding gases from the gunpowder

Questions **13-15** are based on the following:

A 5-gram bullet is fired horizontally into a 2 kg block of wood suspended from the ceiling by 1.5 m strings. The bullet becomes embedded within the block of wood, and they move together at a speed of 1.5 m/s. Gravity can be ignored during the bullet's impact on the block. Then, the woodblock with the bullet swings upward by height h.

13. What best describes the energy flow as the bullet impacts the wood block?

A. Kinetic to heat and kinetic
B. Potential and kinetic to heat
C. Kinetic to potential
D. Potential to heat
E. Kinetic to heat and potential

14. What is the bullet's velocity just before it enters the block?

A. 30 m/s
B. 60 m/s
C. 600 m/s
D. 90 m/s
E. 180 m/s

15. Immediately after the bullet embeds itself in the wood, what is the kinetic energy of the block and bullet?

A. 2.26 J
B. 4.5 J
C. 9 J
D. 18 J
E. 36 J

Solutions: Equilibrium & Momentum

1. A is correct.

Since the momentum is conserved during the collision, the change is $p = 0$.

$$p_{initial} = p_{final}$$

$$m_1v_1 + m_2v_2 = (m_1 + m_2)v_3$$

2. D is correct.

$$J = \Delta p = m\Delta v$$

$$\Delta v = J / m$$

$$\Delta v = (30{,}000\ \text{N·s}) / (1{,}120\ \text{kg})$$

$$\Delta v = 27\ \text{m/s}$$

$$\Delta v = v_f - v_i$$

$$27\ \text{m/s} = 0 - v_i$$

$$v_i = 27\ \text{m/s}$$

3. A is correct.

In a uniform circular motion, the centripetal force does *no work* because the force and displacement vectors are at *right angles*.

$$W = Fd$$

Work equation applies only if force and displacement direction are the same.

4. D is correct.

$$J = F\Delta t$$

Impulse remains constant, so the time of impact increases, and the impact force decreases.

5. B is correct.

Consider a dropped baseball on Earth:

$$d_E = -\tfrac{1}{2}gt^2$$

On Moon:

$$d_M = -\tfrac{1}{2}(g / 6)t^2$$

$$d_E / d_M = (-\tfrac{1}{2}gt^2) / [(-\tfrac{1}{2}(g / 6)t^2)]$$

$d_E / d_M = 1/6$, so the distance is 6 times greater on the Moon.

6. E is correct.

Total downward force on the meter stick is:

20 N + 50 N + 30 N = 100 N

Total upward force on the meter stick –provided by the tension in the supporting rope – must also be 100 N to keep the meter stick in static equilibrium.

Let x be the distance from the left end of the meter stick to the suspension point.

From the pivot point, balance the torques.

Counterclockwise (CCW) torque due to the 50 N weight at the left end is 50 x.

Total clockwise (CW) torque due to weight of the meter stick and the 30 N weight at the right end is:

$(50\text{ N})x = (20\text{ N})\cdot(50\text{ cm} - x) + (30\text{ N})\cdot(100\text{ cm} - x)$

$(50\text{ N})x = [1{,}000\text{ cm} - (20\text{ N})x] + [3{,}000\text{ cm} - (30\text{ N})x]$

$(50\text{ N})x = 4{,}000\text{ cm} - (50\text{ N})x$

$(100\text{ N})x = 4{,}000\text{ cm}$

$x = 40\text{ cm}$

7. B is correct.

$J = F\Delta t$ is constant

increased t = decreased F

8. C is correct.

$m_1v_1 = (m_1 + m_2)v_2$

$v_2 = (m_1v_1) / (m_1 + m_2)$

$v_2 = (1{,}200\text{ kg})\cdot(15.6\text{ m/s}) / (1{,}200\text{ kg} + 1{,}500\text{ kg})$

$v_2 = (18{,}720\text{ kg·m/s}) / (2{,}700\text{ kg})$

$v_2 = 6.9\text{ m/s}$

9. E is correct.

$F\Delta t = m\Delta v$

$F = (m\Delta v) / \Delta t$

$F = [(0.05 \text{ kg})\cdot(100 \text{ m/s} - 0 \text{ m/s})] / (0.0008 \text{ s})$

$F = (5 \text{ kg·m/s}) / (0.0008 \text{ s})$

$F = 6{,}250 \text{ N} = 6.3 \text{ kN}$

10. A is correct.

In a circular motion, all points along the rotational body have the same angular velocity regardless of their radial distance from the center of rotation.

11. C is correct.

In both elastic and inelastic collisions, momentum is conserved.

Compare the KE before and after the collision to determine if the collision is elastic (KE is conserved) or inelastic (KE is not conserved).

Initial KE:

$KE_i = ½m_1v_1^2$

$KE_i = ½(2 \text{ kg})\cdot(0.6 \text{ m/s})^2$

$KE_i = 0.36 \text{ J}$

Conservation of momentum:

$m_1v_1 + m_2v_2 = m_1u_1 + m_2u_2$

$m_1v_1 = m_2u_2$

$(m_1 / m_2)v_1 = u_2$

$u_2 = (2 \text{ kg} / 2.5 \text{ kg})\cdot(0.6 \text{ m/s})$

$u_2 = 0.48 \text{ m/s}$

Final KE:

$KE_f = ½m_2u_2^2$

$KE_f = ½(2.5 \text{ kg})\cdot(0.48 \text{ m/s})^2$

$KE_f = 0.29 \text{ J}$

KE_i (0.36 J) does not equal KE_f (0.29 J); thus, KE is *not conserved.*

Therefore, the collision is inelastic, and only momentum is *conserved.*

12. A is correct.

A longer barrel gives the expanding gas more time to impart a force upon the bullet and thus increase the impulse upon the bullet.

$$J = F\Delta t$$

13. A is correct.

Energy starts as kinetic. Initially, there is no change in potential energy.

Upon impact, most of the bullet's kinetic energy is converted into heat, some is transferred into kinetic energy of the block, and a small amount remains as kinetic energy for the bullet.

Since gravity is ignored, the increase in the height of the block does not lead to potential energy.

14. C is correct.

Conservation of momentum:

$$m_1v_1 = (m_1 + m_2)v_2$$

$$(0.005\text{ kg})v_1 = (0.005\text{ kg} + 2\text{ kg})\cdot(1.5\text{ m/s})$$

$$v_1 = 600\text{ m/s}$$

15. A is correct.

$$\text{KE} = \tfrac{1}{2}mv^2$$

$$\text{KE} = \tfrac{1}{2}(m_1 + m_2)v^2$$

$$\text{KE} = \tfrac{1}{2}(2.005\text{ kg})\cdot(1.5\text{ m/s})^2$$

$$\text{KE} = 2.2556\text{ J} \approx 2.26\text{ J}$$

Notes for active learning

CHAPTER 3

Force, Motion, Gravitation

Page intentionally left blank

Newton's First Law: Inertia

Newton's First Law of Motion

Newton's First Law of Motion states:

> *an object at rest remains at rest and an object in motion (with constant velocity) remains in motion (with that same constant velocity) unless acted on by an external force.*

Nothing changes about an object's motion regarding speed and direction unless an outside force acts on it.

Inertia

All objects resist changes in their state of motion (even if that motion is zero) until an outside force manipulates them into a new state of motion. This is because all objects have *inertia.*

Inertia is an inherent property of objects and tends to resist changes in the state of motion.

Inertia cannot be calculated, but it can be measured by an object's *mass*; the more massive the object, the more inertia.

In the SI system, mass is measured in kilograms (kg).

Mass *vs.* weight

Mass is different from weight (i.e., $w = mg$).

Weight (w) depends on gravity (and can change depending on location), the mass (m) of an object is always the same, and gravity (g) depends on the attractive force between two objects.

Newton's First Law of Motion is only valid for an inertial reference frame, any frame of reference that moves with constant velocity (i.e., no acceleration) relative to the inertial system under observation.

Newton's Second Law: *F = ma*

Force measured in Newtons

Newton's Second Law states that force is the *vector product of mass and acceleration*:

$$\Sigma\vec{F} = m\vec{a}$$

Net force equals mass times acceleration.

Newton is the SI unit for force, the product of the units of mass and acceleration ($N = kg{\cdot}m/s^2$).

When considering force, it is essential to note the units.

Pound is a unit of force, not mass. Therefore, it can be equated to the Newton, not kilograms.

Conventionally, kilograms are used to measure weight (e.g., a person weighs 186 lbs. or 80 kg).

Measuring weight in kilograms is *not* technically correct.

Kilogram is a unit of *mass*.

Weight of an object is a *measurement of force*. This everyday use of kilograms is incorrect, and do not mistake kilograms for weight when solving problems.

Mass and force units

System	Mass	Force
SI	kilogram (kg)	Newton ($N = kg{\cdot}m/s^2$)
CGS	gram (g)	dyne ($= g{\cdot}cm/s^2$)
Imperial	slug	pound (lb)
Conversion factors: 1 dyne = 10^{-5} N; 1 lb ≈ 4.45 N; 1 slug ≈ 14.6 kg		

Force as a vector

Newton's Second Law of Motion uses x-, y- and z-components because force is a vector:

$$\Sigma F_x = ma_x$$

$$\Sigma F_y = ma_y$$

$$\Sigma F_z = ma_z$$

From the above equations, force is directly proportional to the mass of the object. This further explains Newton's First Law of Motion (or *Law of Inertia*).

More force is required to push a boulder up a hill than a pebble because the mass of the boulder is higher than that of the pebble.

Applying Newton's First and Second Laws

Applying Newton's First Law, the boulder has higher inertia than the pebble and more resistant to changes in motion.

Applying Newton's Second Law, the force needed to move (accelerate) the boulder is directly proportional to its mass and greater than the force needed to move (accelerate) the pebble.

Acceleration, force and mass

Rearranging Newton's Second Law ($F = ma$) shows that acceleration is directly proportional to force and inversely proportional to mass:

$$a = F / m$$

Force on an object with a mass of 2 kg is 4 N, therefore:

$$a = \frac{F}{m}$$

$$a = \frac{4\text{ N}}{2\text{ kg}}$$

$$a = 2\text{ m/s}^2$$

If the force is 8 N and mass of the object is 2 kg, then:

$$a = \frac{F}{m}$$

$$a = \frac{8\text{ N}}{2\text{ kg}}$$

$$a = 4\text{ m/s}^2$$

Acceleration and mass relationship to force

Acceleration is *directly proportional* to force.

By doubling the force, acceleration of the object is *doubled.*

The opposite holds when doubling mass. Since mass is *inversely proportional* to acceleration, doubling the mass will halve the acceleration.

It makes sense, considering the mass of the boulder and pebble in the previous example.

Exerting the same force on both will not give the same acceleration because each object's mass (representing inertia or the resistance to change) is vastly different.

Problem example

If the pebble has a mass of 5 kg and its acceleration is 20 m/s^2, what force was exerted on it?

$$a = \frac{F}{m}$$

$$20\ \mathrm{m/s^2} = \frac{F}{5\ \mathrm{kg}}$$

$$F = 100\ \mathrm{N}$$

What is the acceleration if the boulder has a mass of 20 kg, and the same 100 N is exerted?

Solution: the boulder has four times the mass of the pebble, so the acceleration is one-fourth that of the pebble: 5 m/s^2 compared to 20 m/s^2 for the pebble.

Force acts in a direction

Force and acceleration are *vectors* because they have magnitude and direction.

Many physics questions omit direction because it is assumed. For example, when an apple falls to the ground, gravity acts downwards, and the apple falls downwards.

Questions may involve substituting values into appropriate formulas.

However, more complex questions have directional attributes associated with them.

For example, when a bar of soap slides down an inclined plane, the force of gravity acts downwards, but the acceleration is not entirely downwards; instead, it is slanted.

Therefore, *vector analysis* is needed.

Solving Newton's Second Law problems

- Visualize (draw a sketch) and always use a free-body diagram.
- Clearly define an appropriate coordinate system.

  This is important when solving inclined planes because the angle between the *normal* and *gravitational forces* acting on the object is altered.
- Apply Newton's Second Law to each mass separately.
- If there are forces in two or more directions, apply Newton's Second Law to each direction separately.
- "Translational equilibrium" implies two equations:

  $$\Sigma F_x = 0$$

  and

  $$\Sigma F_y = 0$$
- When $\Sigma F = 0$, $a = 0$ and vice versa.
- Solve the component equations.

  If the acceleration of x or y is known, or the object is moving at constant velocity ($a = 0$), the component equations can solve unknown forces.
- Use the unit circle to calculate trigonometric functions.

  Use degrees if the angles are given in degrees.

  Use radians if angles are given in radians.

Notes for active learning

Newton's Third Law: Equal and Opposite Forces

Newton's Third Law and force

For Newton's Third Law, it is essential to understand how forces work.

A force must have a source object (i.e., any force exerted on one object caused by another).

When a person interacts with an object by exerting a force, that object exerts the same amount, or magnitude, of force back on it.

This is often simplified into the dogma for Newton's Third Law:

every action has an equal and opposite reaction.

If body A exerts a force F_A on body B, then body B exerts a force F_B on body A:

$$F_B = -F_A$$

Action and reaction pairs

Applying Newton's Third Law, it is essential to understand that the forces are exerted on different objects and should not be treated as acting on the same object.

These matched forces are *action and reaction pairs*.

Notation: the first subscript is the object on which the force is being exerted, and the second is the source object.

For example, a person (P) walking forward exerts a force on the ground (G), which exerts an equal and opposite force to move them forward.

These forces can be expressed as:

$$\vec{F}_{GP} = -\vec{F}_{PG}$$

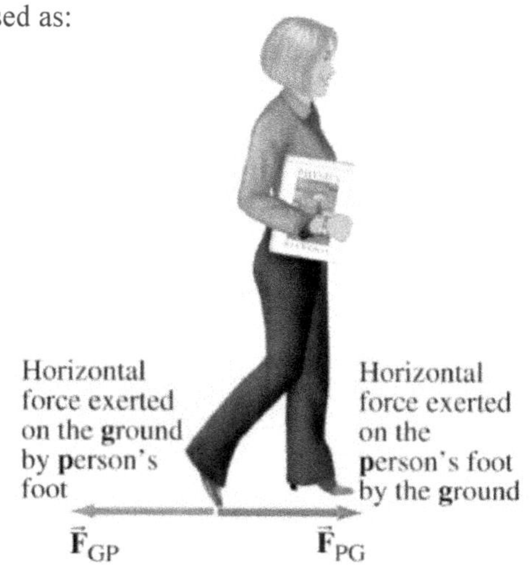

$\vec{F}_{GP}$ is the horizontal force exerted *on* the ground by the woman's foot and $\vec{F}_{PG}$ is the horizontal force exerted on the woman's foot *by* the ground.

Rocket propulsion

Using Newton's Third Law for rocket propulsion, when a rocket is in flight, hot gases from combustion spew out of the tail of the rocket at high speeds (*action force*).

Reaction force propels the rocket forward; rocket does not need anything to "push" against.

Whether traveling within Earth's atmosphere or in the vacuum of space, the rocket can propel itself forward because of the action-reaction forces from combustion.

All three of Newton's Laws of Motion can be extrapolated and observed in most situations.

Cannon and cannonball

A cannon and cannonball provide an example:

- A cannon is initially at rest.

  The cannon has a large mass and, therefore, large inertia.

  The cannon remains at rest until a force is applied (Newton's First Law).

- When a cannon fires, a large force is applied by the cannon to the cannonball to overcome inertia of the cannonball and cause acceleration (Newton's Second Law).

- The tremendous force applied by the cannon is reciprocated by the cannonball back onto the cannon, and the cannon recoils (Newton's Third Law).

Why does the cannonball accelerate faster than the recoiling cannon?

Consider all three laws, but the crucial component is inertia.

Third Law: the forces exerted on the cannon and cannonball must be the same.

Second Law states that force and mass (inertia) determine acceleration; since the forces are equal, but the masses are considerably different, the *accelerations are considerably different.*

Cannonball accelerates faster because its mass (i.e., inertia) is less than the cannon's mass:

$$a = \frac{F}{m}$$

Once the cannonball is airborne, it remains in flight (First Law) unless a force (air resistance and gravity) acts upon it to change its motion.

Air resistance and gravity eventually cause the cannonball to slow and fall to the ground.

Solving free-body diagram problems

1. Draw a sketch.
2. Draw a free-body diagram showing the forces acting *on* the object.

   Make the magnitudes and directions accurate.

   Label each force.

   Draw a separate diagram for each object.

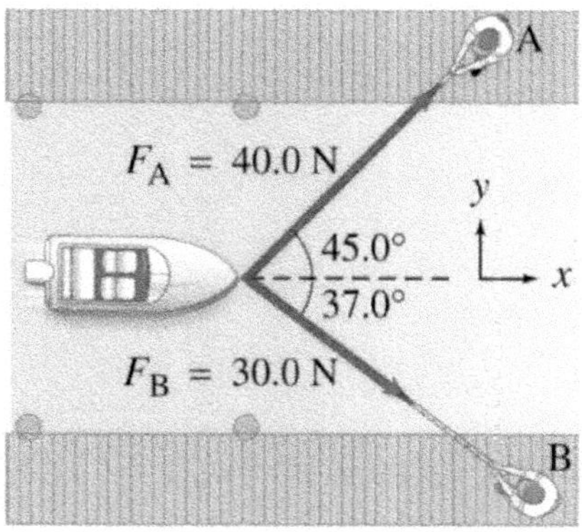

3. Resolve the vectors into components.
4. Apply Newton's Second Law of Motion ($F = ma$) to each component.
5. Solve for the net force in each component direction.
6. Combine components to determine the total net force.

Notes for active learning

Weight

Weight as a force

Weight is the gravitational force (i.e., the force of gravity) that acts on a mass when an object is near the Earth's surface.

Gravitational force changes with substantial increases or decreases in altitude.

Weight is a force, therefore, a vector quantity.

Gravitational force is:

$$\vec{F}_g = m\vec{g}$$

where g is the acceleration due to gravity at 9.8 m/s^2 (rounded to 10 m/s^2).

Movement affects weight

An object's weight changes if it moves. Since weight is calculated by multiplying object's mass by the acceleration due to gravity, if the acceleration due to gravity changes, the object's weight will also change.

For example, when one drives a car over a bump or rides a roller coaster, their weight changes because they are experiencing acceleration change.

An object weighs more on an elevator accelerating up because:

$$F = m_g + m_a$$

where a is the acceleration of the elevator.

An elevator accelerating up has the same force as an elevator decelerating down; only the direction of the acceleration changes.

Weight in space, on Earth and Moon

An object weighs less when it is further away from the Earth because force of gravity decreases with distance.

When an object orbits in space (e.g., an astronaut in a spacecraft), it is not genuinely weightless.

An astronaut falls toward Earth at the same rate as the spacecraft.

If the astronaut's spacecraft falls from space to Earth's surface, its weight will increase during the fall due to the increase of gravitational acceleration as it gets closer to the Earth's surface.

Its weight on Earth is different for a given mass than on the Moon.

If a person stands on the Moon with a gravitational acceleration of about 1/6 g, that person is 1/6 of their weight on Earth.

Their mass, however, is the same in both locations.

From Newton's Second Law ($F = ma$), an object at rest must have no net force on it.

Objects have a gravitational force acting upon them.

Normal force acts perpendicular to surface

For both Newton's Second and Third Laws, there must be an equal but opposite force.

For an object resting on a flat surface, the force of the surface pushing up on the object is the *normal force.*

The normal force is *perpendicular* to the plane of the object.

When an object is still on a horizontal surface, the normal force is equal and opposite to the weight (i.e., the force of gravity).

Normal force balances the object's downward force (if force gets too big, something breaks).

If a statue is on a table, the normal force from the table is equal and opposite to force of gravity.

As a result, the statue experiences no net force and does not move.

$$-\vec{F}_G = \vec{F}_N$$

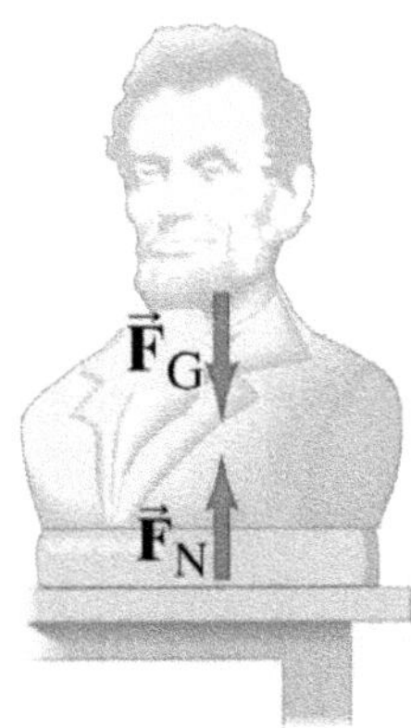

Gravitational force equals the normal force for objects at rest

Center of Mass

Defining the center of mass

Every object has a center of mass.

Center of mass is a single, approximate point at which all of an object's mass is concentrated.

External forces can be considered to act on the object at the center of mass.

If an object is balanced on its center of mass, it spins uniformly about this axis.

Locating the center of mass

If an object is uniform along its length, the center of mass is the center of the object.

If the object is not uniform, the center of mass is the point obtained by taking an average of all the positions weighted by their respective masses:

$$x_{cm} = \frac{\sum x_i m_i}{\sum m_i} \qquad y_{cm} = \frac{\sum y_i m_i}{\sum m_i} \qquad z_{cm} = \frac{\sum z_i m_i}{\sum m_i}$$

It is unnecessary to have absolute coordinates when calculating the center of mass.

Set the point of reference for convenience and use relative coordinates.

The center of mass *need not* be within the object.

For example, a doughnut's center of mass is in the center of its hole.

Center of mass for two-body system

For two objects, the center of mass is closer to the more massive object.

Imagine two balls of an unequal mass attached to a massless stick. The equation and image below describe the situation and the center of mass calculation:

$m_A > m_B$

$$x_{CM} = \frac{m_A x_A + m_B x_B}{m_A + m_B} = \frac{m_A x_A + m_B x_B}{\sum M}$$

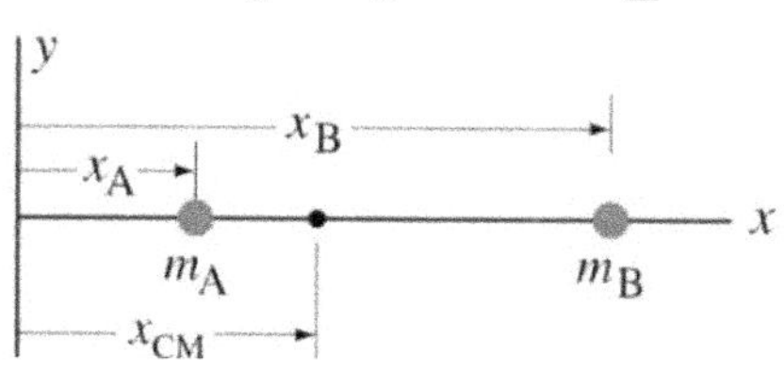

Center of mass is closer to the more massive object

If the difference in mass between two objects is substantial, the center of mass can be assumed to be that of the more massive object.

For example, the center of mass of Earth and a man in space is almost at Earth's center because the man is comparatively tiny. Therefore, his coordinate, weighted for his mass, is negligible.

Translational motion using center of mass

When problems involve directional forces on an object and the subsequent path that the object takes, these directions and paths are based on the object's center of mass.

The object is effectively treated as a point (or particle) with the same mass as the object but is located at the object's center of mass.

Sum of forces acting on a system is equal to the total mass of the system multiplied by the acceleration of the center of mass:

$$ma_{CM} = F_{net}$$

This simplifies calculations because calculating the motion of an irregularly shaped object involves complex mathematics.

Center of gravity

Center of gravity (CG) is the average location of the weight of an object; it is the point at which the gravitational force is considered to act.

Center of gravity is the same as the center of mass *only if* the gravitational force does not vary among parts of the object.

Center of gravity is found *experimentally* (image below) by suspending an object from points.

A string about the object's length is securely attached to a point near the object's edge. The string near the connection point holds the object, and a line is traced along the hanging string.

The string hangs perpendicular to the ground because gravitational force acts in the y-direction. The object is rotated, and the steps repeated. The point at which the drawn lines cross is the center of gravity. If hundreds of lines are drawn onto the object, lines intersect at this point.

Like the *center of mass*, an object rotates perfectly around its *center of gravity*.

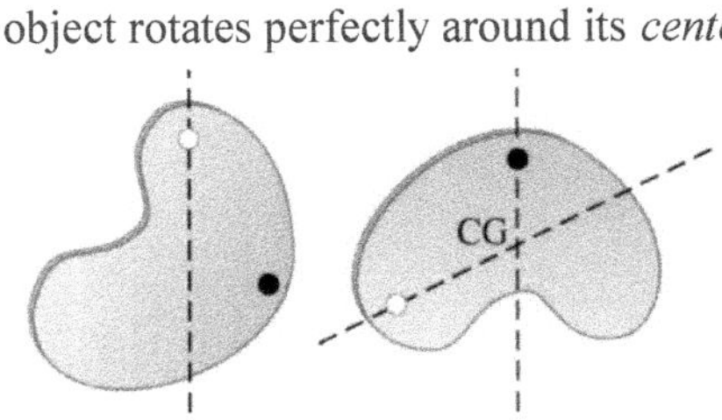

Drawing imaginary lines to determine the center of gravity (CG)

Friction: Static and Kinetic

Friction opposes motion

Friction is a force that opposes the direction of motion.

Like other forces, friction is a *vector* (i.e., it has both magnitude and direction).

For example, a dog can walk, and cars can drive because of friction.

Lubricants (e.g., oil, grease) reduce friction because they change surface properties and decrease the coefficient of friction.

Heat is produced as a by-product of the force of friction.

Friction depends on contact surfaces

Friction involves multiple factors, and changes depending on surface contact. On a microscopic scale, most surfaces are rough.

As one surface moves over another, the crevices of the two surfaces catch and release. This is where the force resisting motion (i.e., friction) originates.

For example, in the figure below, the block moves to the right, and as its rough surface moves over the table's rough surface, the force of friction acts to the *left*.

Friction always acts in the *direction opposing the motion*.

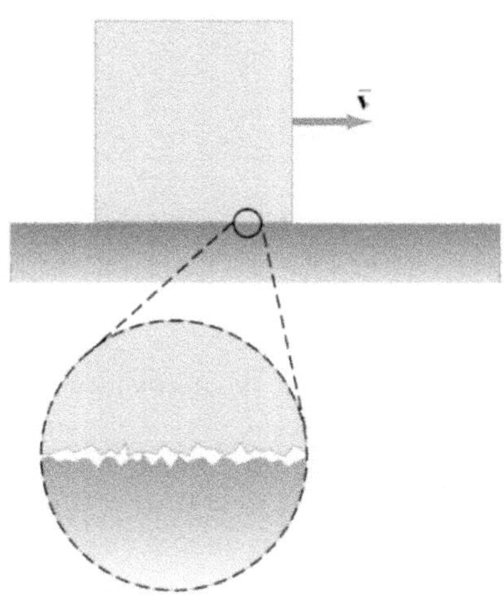

Friction exists at the contact between two surfaces with the force of friction opposing motion

In some situations, more frictional forces act on an object than in others.

For example, a puck gliding across an air table experiences almost no friction or pushing force.

A block pulled up a rough incline experiences a frictional force (opposing motion) and a force of tension. Both objects experience a normal force and a gravitational force.

Coefficient of friction

Frictional force is calculated as the product of the normal force of an object and a dimensionless value for the *coefficient of friction.*

Coefficient of friction is determined empirically and is intrinsic to the material properties of the surface and the object.

Static and kinetic friction

Friction occurs in two types: static and kinetic.

Static friction pertains to objects sitting still (an object can remain stationary on an inclined plane because of static friction).

Kinetic friction pertains to objects in motion (a vase sliding across a table has its motion resisted by a force of kinetic friction).

Static friction is greater than kinetic friction

Force and coefficient of *static friction* are *greater* than force and coefficient of *kinetic friction.*

It takes a greater force to start an object moving on a rough surface than keeping the object moving on the same surface.

Equations to determine friction:

Static friction:

$$F_s = \mu_s F_N$$

Kinetic friction:

$$F_k = \mu_k F_N$$

where μ is the coefficient of friction, and F_N is the normal force.

Kinetic and static friction relationships:

$$F_s > F_k$$

$$\mu_s > \mu_k$$

Coefficients of static and kinetic friction

Friction coefficients depend on the characteristics of both surfaces.

Surface	Coefficient of static friction μ_s	Coefficient of kinetic friction μ_k
Wood on wood	0.4	0.2
Ice on ice	0.1	0.03
Steel on steel (unlubricated)	0.7	0.6
Rubber on dry concrete	1.0	0.8
Rubber on wet concrete	0.7	0.5
Rubber on other solid surfaces	1 – 4	1
Teflon® on Teflon® in air	0.04	0.04
Teflon® on steel in air	0.04	0.04
Lubricated ball bearings	< 0.01	< 0.01
Synovial joints (in human limbs)	0.01	0.01
Values are approximate and intended only as an illustration of the concepts.		

Calculating static friction force

An object at rest on a frictional surface has the force of *static friction* acting on it.

Force of static friction is:

$$F_s \leq \mu_s F_N$$

It is only equal when the object is about to move, meaning that the static friction is on the verge of changing to kinetic friction.

For an object that starts at rest and then has a force applied to it, the force of static friction increases as the applied force increases *until* the applied force overcomes the maximum value of static friction.

Exceeding static friction

Objects start moving when the force of *kinetic friction* exceeds *static friction.*

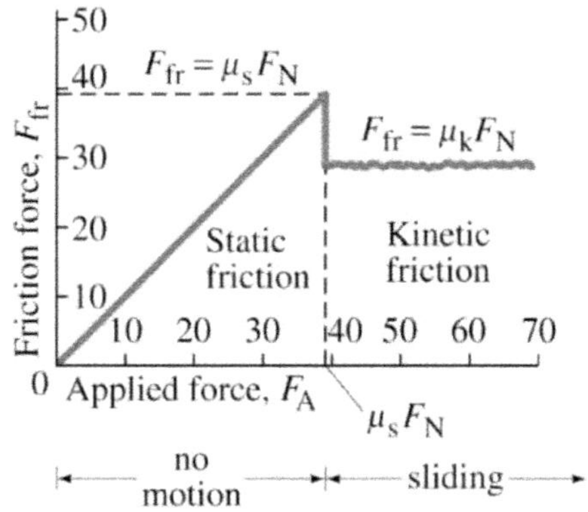

Applied force is greater than force of friction when object is in motion

Normal force of an object on a horizontal surface equals weight ($w = mg$) of the object.

However, on an inclined plane, the normal force is perpendicular to the inclined plane, and its magnitude is equal to the weight of the object times the cosine of the incline angle:

$$F_N = w \cos \theta$$

(*see inclined planes*)

Motion on an Inclined Plane

Force of gravity for inclined planes

When an object is on an inclined plane, the force of gravity is divided into two components: one component is *normal* (i.e., perpendicular) to the plane surface, and the other component is *parallel* to the plane surface.

The image and equations describe forces on the block:

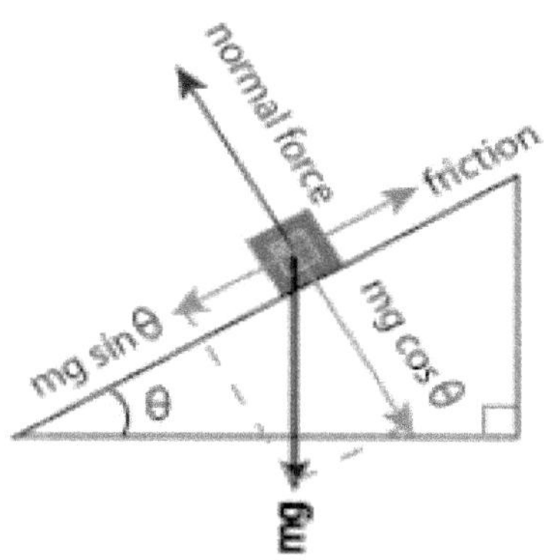

Calculating force of gravity for inclined planes

Parallel Force: $F_{\parallel} = mg \sin \theta$

Perpendicular Force: $F_{\perp} = F_N$

$F_N = mg \cos \theta$

where θ is the angle between the horizontal axis and the surface on which the object rests.

Force of gravity affects the frictional force

How does the force of gravity affect the force of friction? An object on an incline has three forces acting on it: the normal force, the gravitational force, and the frictional force:

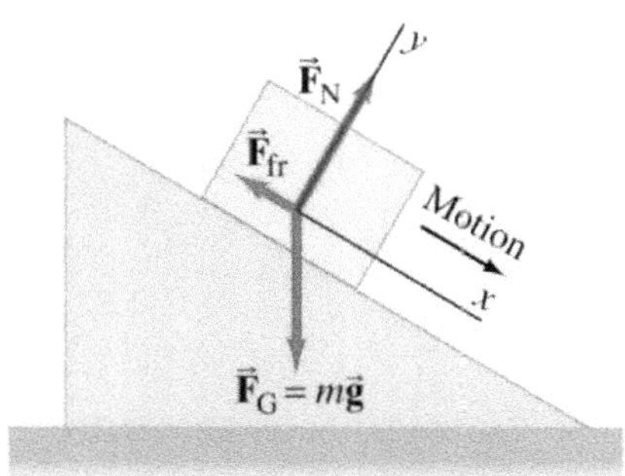

Inclined plane object has normal, gravitational, and frictional forces acting upon it

Parallel and perpendicular components of gravitational force

Gravitational force is split into *parallel* and *perpendicular components* for the incline.

For an object moving down the inclined plane at a constant velocity, the *parallel component* of gravity is equal and opposite to the *kinetic friction force*:

$F_{\parallel} = F_k$

$F_{\parallel} = \mu_k F_N$

$mg \sin \theta = \mu_k mg \cos \theta$

For stationary objects, the parallel component of gravity is equal to and opposite to the force of static friction:

$F_{\parallel} = Fs$

$F_{\parallel} = \mu_s F_N$

$mg \sin \theta = \mu_s mg \cos \theta$

Frictional forces are always *less than or equal* to the forces causing an object to move; frictional force is *never greater* than the forces causing movement.

If frictional force exceeds the other forces involved, it causes the object to move in the opposite direction. Friction does *not* produce movement; it *inhibits* it.

When an object is pushed or pulled up an inclined plane, the parallel component of gravity and the force of friction must be overcome.

x and *y* components of forces

Forces are isolated into *x* and *y* components for calculations. Only the force component *parallel* to the plane contributes to motion.

For example, if a block of mass *m* accelerates down an inclined plane of angle θ with friction present, what is its acceleration along the *x*-axis?

$F_{\parallel} = mg \sin \theta$

$F_N = mg \cos \theta$

$F_{fr} = \mu_k F_N$

$F_x = F_{\parallel} - F_{fr}$

$ma_x = mg \sin \theta - \mu_k mg \cos \theta$

$a_x = g \sin \theta - \mu_k g \cos \theta$

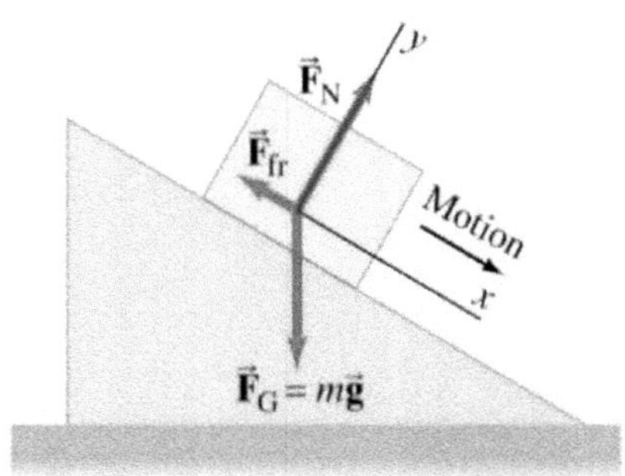

Calculating forces for objects on an inclined plane

Uniform Circular Motion and Centripetal Force

Uniform circular motion

An object moving in a circle at constant speed is in a *uniform circular motion*; this is an example of simple harmonic motion.

Uniform circular motion is when an object travels in a circle at a *constant speed*. For example, a point on a propeller spinning at a constant rate and the second hour hands of a watch.

Uniform circular motion has constant speed and travels in a circle with a fixed radius. If the speed of the particle is changing, then an additional acceleration in the direction tangential to the circle occurs.

Velocity *vs.* speed

Distinguishing between velocity and speed is essential.

Velocity (i.e., vector with magnitude and direction) is displacement over time.

Speed (i.e., scalar with magnitude but not direction) is the distance over time.

Displacement *vs.* distance

For uniform circular motion, displacement is the shortest straight-line distance between two points on the perimeter of a circle (i.e., a *chord*).

However, distance is the path that the object travels and can be calculated as the product of the radius and arc angle:

$$s = \theta r$$

For circular motion problems, instantaneous velocity is often given and equal to the speed.

Circular motion considerations

Typical issues for circular motion:

- Instantaneous velocity equals the speed for displacements and distances that approach zero.
- For a quarter circle ($\pi/2$ radians or 90°), displacement is the hypotenuse of a right-angled triangle with the radius as the other two sides.

  By Pythagorean Theorem, the displacement is $\sqrt{2r^2}$.

  Distance is the length of the arc or ¼ of the circumference.

- For a half-circle (π radians or 180°), displacement is the diameter.

  Distance is half the circumference.

- Pythagorean Theorem obtains the displacement for a three-quarters circle ($3\pi/2$ radians or 270°).

  Magnitude of the displacement is the same as a quarter of a circle ($\sqrt{2r^2}$) but the direction is different.

  Distance is ¾ of the circumference.

- Going around the circle entirely (2π radians or 360°) has zero displacement.

  Average velocity is zero.

  Distance equals the circumference.

- Velocity is always less than or equal to the speed.

- Displacement is always less than or equal to the distance.

  Displacement and velocity are vectors.

  Distance and speed are scalars.

Frequency and period

Frequency (f) is the number of times an object makes a revolution in one second.

Frequency is measured in Hertz (Hz = s^{-1}).

Period (T) of an object in a circular motion is the time it takes the object to make one revolution, measured in seconds (s).

Frequency and period of an object are *inversely* related:

$$f = 1 / T$$

Circular motion formulas

Centripetal acceleration: $a_c = \frac{v^2}{r}$

Circumference: $C = 2\pi r$

Arc: $arc = \frac{\theta}{2\pi} \times C = r\theta$

Area of a circle: $A = \pi r^2$

Theta (θ) is in radians.

2π radians = 360 degrees

Centripetal force, acceleration and instantaneous velocity

Number of forces (whether changing or constant) for circular motion depends on the situation.

For circular motion, centripetal force is from centripetal acceleration.

Centripetal force is calculated by ($F = \frac{mv^2}{r}$).

Centripetal force is found when the mass is multiplied by the centripetal acceleration.

Centripetal force may be provided by friction, gravity, tension, normal force, or other forces.

Centripetal force:

$$F = ma_c = \frac{mv^2}{r}$$

Centripetal acceleration is due to changes in velocity for an object revolving around a circle.

Centripetal acceleration:

$$a_c = \frac{v^2}{r}$$

Changes in velocity (vector) are from a constant change in direction.

Velocity changes due to a change in direction, and the resulting centripetal acceleration are shown diagrammatically.

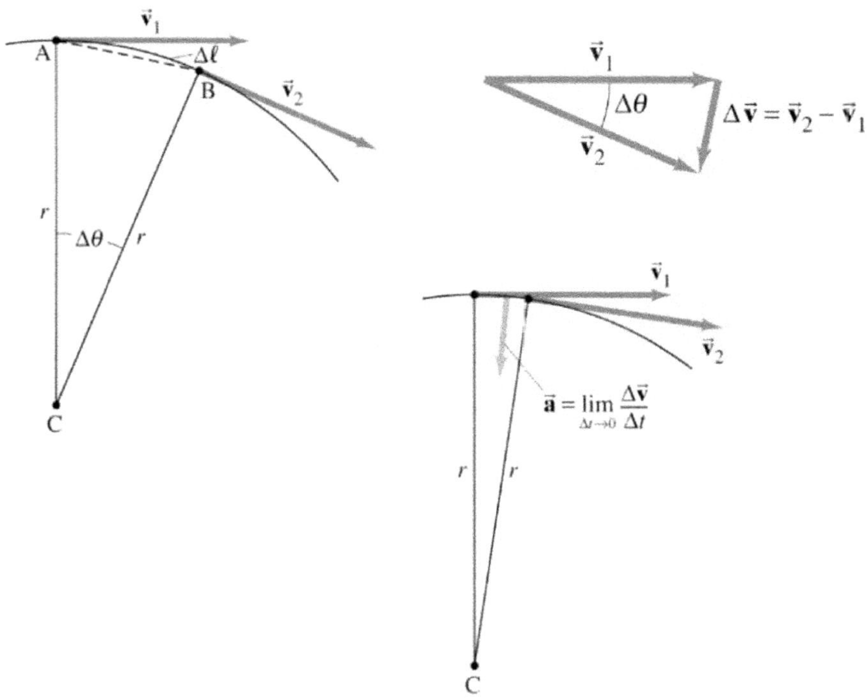

Velocity and centripetal acceleration depend on changes in direction

Negative sign is used for centripetal force to indicate that the direction of the force is toward the center of the circle.

In centripetal motion, the acceleration is always inward.

Force is always in the *same direction* as the acceleration.

Direction of acceleration and force is *toward the center* of the circle.

Instantaneous velocity is tangent to the circle.

Centripetal (or radial) *acceleration* points inward along the radius.

Centripetal (radial) acceleration for uniform circular motion is:

$$a_c = \frac{v^2}{r}$$

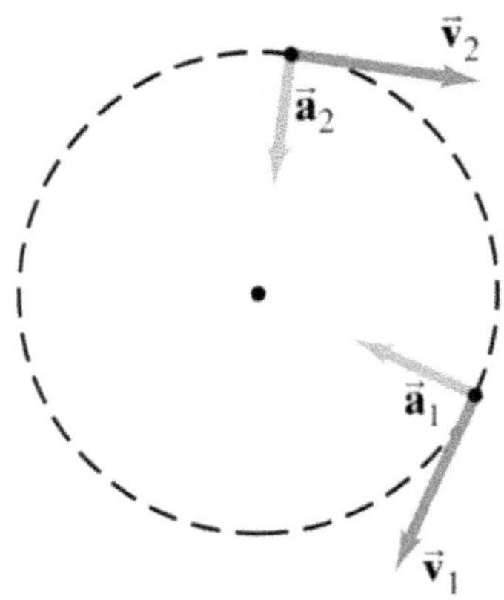

Instantaneous velocity and centripetal acceleration noted

Ball tethered to a string

For example, a ball attached to a string is spun around, and suddenly the string breaks. The ball flies off, following a line of motion tangent to its previous circular path. Why is this true?

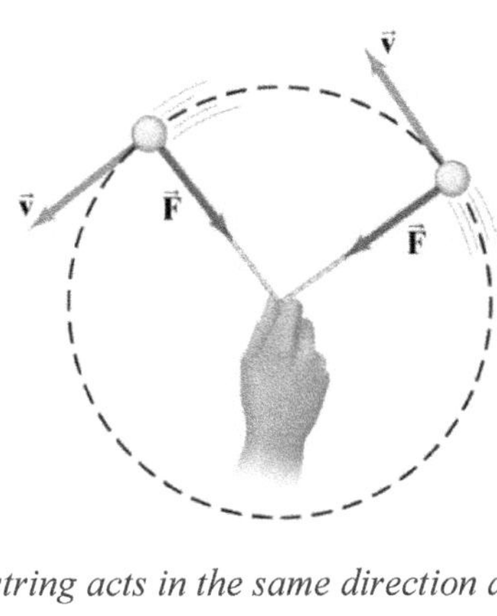

Net force of the string acts in the same direction as the acceleration

For an object to be in a uniform circular motion, a net force (string) must act on it in the same direction as the acceleration (inward).

Ignoring gravity, the centripetal force is the only force acting on the ball and keeping it the same distance from the center.

If the centripetal force vanishes (the string breaks), the object retains its instantaneous velocity at the point where the string broke (following Newton's laws) and flew off, tangent to the circle.

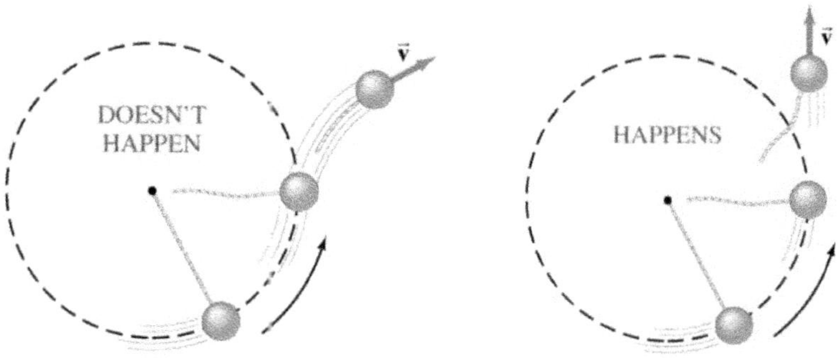

Object retains its instantaneous velocity at the point where the string broke

Banked roadway curves

Banked curves (sloped inwards) exemplify centripetal force.

When a car goes around a curve, a net force must point toward the center of the circle, of which the curve is an arc.

If the road is flat, that force is supplied by friction.

If frictional force is insufficient, the car moves more closely in a straight line.

If tires do not slip, friction is static.

If tires do start to slip, friction is kinetic.

Loss of static friction is detrimental in two ways:

1. Kinetic frictional force is smaller than the static frictional force, and the tires have less grip on the road.
2. Static frictional force points toward the center of the circle. Kinetic frictional force opposes the direction of motion, making it difficult to regain control of the car and continue around the curve.

   Banking the curve on the road helps keep cars from skidding.

For banked curves, there is one speed where the entire centripetal force is supplied by the horizontal component of the normal force, and no friction is required.

This occurs when:

$$F_N \sin\theta = \frac{mv^2}{r} = ma_c$$

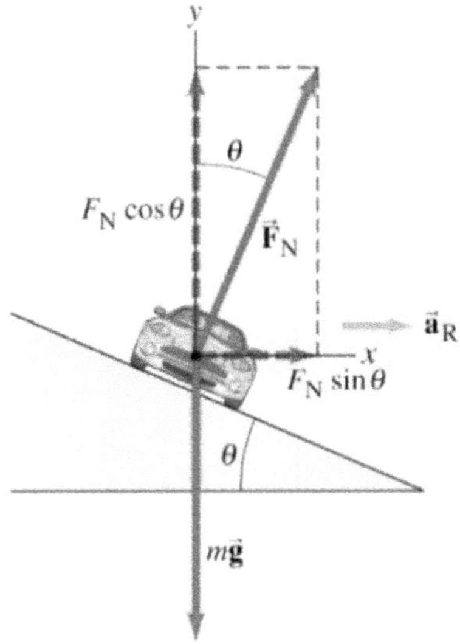

Centripetal force is supplied by the horizontal component of the normal force

Non-uniform circular motion

Uniform circular motion is idealized; most systems exhibit some non-uniform circular motion.

Objects with non-uniform circular motion have varying speeds.

The object has an instantaneous velocity tangent to its circular motion, but acceleration does not point inward along the radius.

Tangential component to acceleration (and force) and a radial one because of changing speed.

Direction of net acceleration, and thus net force, depends on combining these two components.

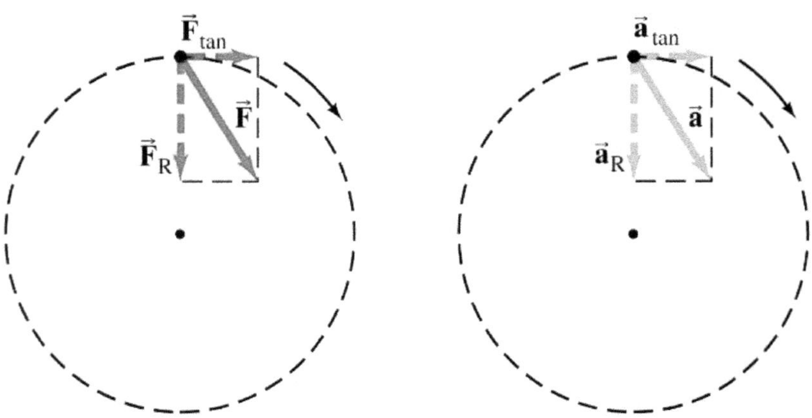

Tangential and radial components produce net acceleration and force

Law of Gravitation

Equal and opposite reaction forces

Newton's Third Law of Motion states that *every force has an equal and opposite reaction force*, indicating that objects pull on every other object.

If the force of gravity is being exerted on objects on Earth, what is the origin of that force? Newton realized that the force must come from the Earth, and this force keeps Moon in its orbit.

Newton's Law of Gravitation

Newton's *Law of Gravitation* explains this relationship.

Law of Gravitation states that any two bodies attract with force related to the square of the distance between them, the product of their masses, and the gravitational constant *G*.

Force, for the Law of Gravitation, is defined as:

$$F_G = G \times \frac{m_1 m_2}{r^2}$$

Force of gravity on an object acts through its *center of gravity*.

Newton derived this relationship

Gravitational force is one-half of an action-reaction pair. For example, Earth exerts a downward force on a person, and they exert an upward force on the Earth.

The reaction force they exert on the Earth is undetectable because a person's mass is negligible compared to the mass of the Earth.

However, the reaction force can be significant for bodies with less disparity in mass (such as the Earth and the Moon).

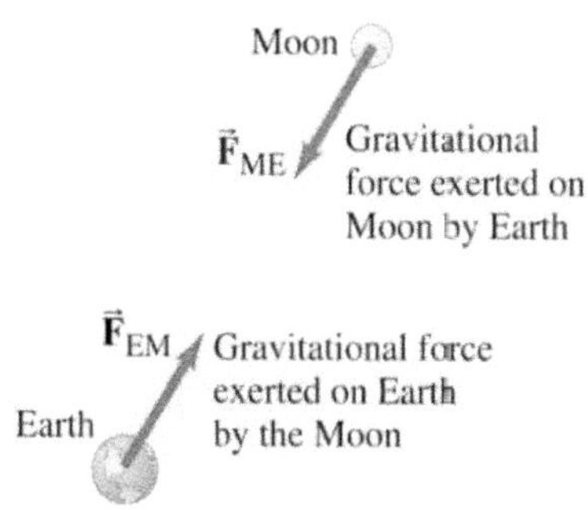

Gravitational force is proportional to both masses

Newton concluded that the gravitational force decreases as the inverse of the square of the distance between the masses by observing planetary orbits.

Gravitational constant G

Earlier in this chapter, the acceleration of an object toward Earth's surface due to the force of Earth's gravity equaled 9.8 m/s^2 and was represented by the variable g.

The Law of Gravitation uses variable G, different from lowercase g.

G is a constant of proportionality, the *universal gravitational constant.*

Note the term "universal" because all objects in the universe are pulling on all other objects.

Examine the value of G and how it affects the equation.

$$F_G = G \times \frac{m_1 m_2}{r^2}$$

Since Newton could measure force, he assigned a value for G by rearranging the equation and using known values.

For two 1 kg masses, 1 m apart of known force:

$$G = 0.000000000067 \text{ N}\cdot\text{m}^2/\text{kg}^2$$

$$G = 6.67 \times 10^{-11} \text{ N}\cdot\text{m}^2/\text{kg}^2$$

Magnitude of gravitational constant G is experimentally measured by the Cavendish experiment (1797-1798), pictured below:

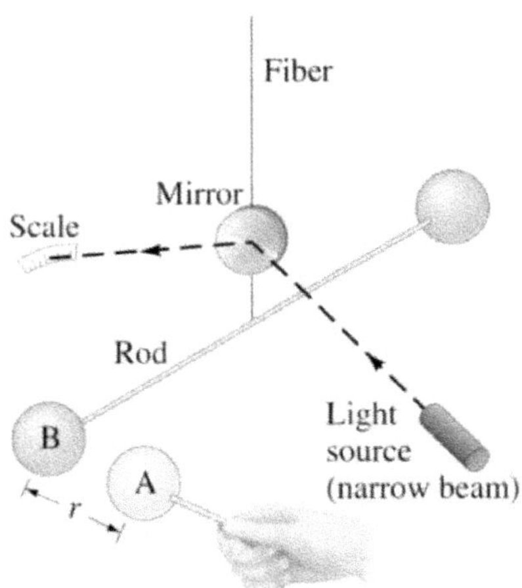

Cavendish experiment measured the gravitational constant

G is an exceedingly small value; it minimizes the value for:

$m_1 m_2 / r^2$

For example, a person pulls on the stars and vice versa.

Stars are a considerable distance apart, making $m_1 m_2 / r^2$ a small value (mainly since the distance is squared in this equation).

Multiplying by G makes it infinitely small, approaching zero.

Considering a person's pull on their neighbor's house, where the value of "r" is relatively small, G diminishes it.

A person weighs 70 kg, house weighs 70,000 kg, and the distance between them is 70 m, then:

m_1m_2 / r^2

$(70\text{ kg})\cdot(70{,}000\text{ kg}) / (70\text{ m})^2$

$1{,}000\text{ kg}^2/\text{m}^2 = 1 \times 10^3\text{ kg}^2/\text{m}^2$

Multiplied by $6.67 \times 10^{-11}\text{ N}\cdot\text{m}^2/\text{kg}^2$ (G), resulting force equals 6.67×10^{-8} N, an undetectable value.

Mass of Earth causes gravitational force

Earth causes the gravitational force felt on Earth.

The concept of G determined the mass of the Earth mathematically.

Force of gravity at Earth's surface was known (9.8 N), and radius of Earth known (approximates the distance between the center of Earth and the center of a person standing on Earth).

An object can be substituted for m and solved for Earth's mass as m_E:

On the surface of the Earth,

$$F = ma$$

so

$$F_G = mg = G\frac{mm_E}{r_E^2}$$

Determining the mass of Earth

Solving for g gives:

$$g = G\frac{m_E}{r_E^2}$$

From g and the radius of Earth ($r_E = 6.38 \times 10^6$ m), the mass of the Earth is:

$$m_E = \frac{gr_E^2}{G}$$

$$\frac{gr_E^2}{G} = \frac{(9.8\text{ m/s}^2)\cdot(6.38 \times 10^6\text{ m})^2}{6.67 \times 10^{-11}\text{ N}\cdot\text{m}^2/\text{kg}^2}$$

$$m_E = 5.98 \times 10^{24}\text{ kg}$$

Earth has a mass of about 6×10^{24} kg. Since this is a large number, multiplying it by G does not decrease it to a negligible amount.

Consideration for gravitational force

The resulting value is the gravitational force felt by objects on Earth.

Like G, the value of r significantly impacts the calculations.

Two aspects to note about the distance component, r, of this equation:

$$F_G = G\frac{m_1 m_1}{r^2}$$

First, it refers to the distance between the center of mass of the two objects (this becomes critical in discussing Earth's pull on objects).

The two objects are treated as *point masses* as if they were the size of one-dimensional points at the locations of their centers of mass but with the same amount of mass.

Second, the distance is squared; small changes in distance result in substantial changes in F, as demonstrated by the *inverse square law*.

Distance impacts force

Law of Gravitation equation, the force is inversely proportional to the square of the distance.

Therefore, by increasing the distance between two objects, the force between those two objects is decreased by the inverse square.

If distance increases by a factor of two (i.e., doubles), the force of gravity decreases by a factor of $2^2 = 4$.

The distance between Earth and another planet is considerable. This is why the gravitational forces of other planets are not felt on Earth, even though they have large masses equal to or greater than Earth's mass.

Distance is squared in the calculation of gravitational force.

Planetary effects on Earth

Force felt on Earth due to planets is negligible and ignored.

Mars is 2.25×10^{11} m away and has a mass of 6.4×10^{23} kg.

For a 1 kg mass on Earth, how much force would it exert on Mars?

$$F_G = G\frac{m_1 m_1}{r^2}$$

$$F = (6.67 \times 10^{-11}\ \text{Nm}^2/\text{kg}^2) \times \frac{(1\ \text{kg})\cdot(6.4 \times 10^{23}\ \text{kg})}{(2.25 \times 10^{11}\text{m})^2}$$

$$F = 6.83 \times 10^{-10}\ \text{N}$$

In most calculations thus far, a gravitational acceleration value of 9.8 m/s^2 has been used. For most calculations, this approximation is acceptable.

Acceleration due to gravity varies over Earth's surface because of altitude, local geology, and the nonspherical shape of Earth.

Acceleration due to gravity at specific locations

Location	Elevation (m)	g (m/s^2)
New York	0	9.803
San Francisco	0	9.800
Denver	1,650	9.796
Mount Everest	8,800	9.770
Sydney	0	9.798
Equator	0	9.780
North Pole (calculated)	0	9.832

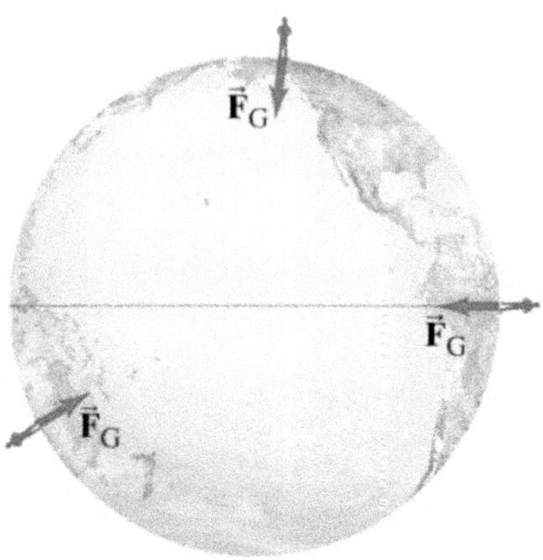

Force of gravitational acceleration points to Earth's core

Notes for active learning

Concept of a Field

Force fields

Forces often act on a body in a single direction (e.g., car accelerating in a straight-line path).

However, forces can act in many directions, from a singular point outward or all directions inward towards a point (e.g., gravitational force).

On Earth, people experience gravity in a single direction (downward) because people are so small relative to the size of Earth that the Earth is a flat surface below them.

Since Earth is an approximate sphere, its gravitational pull comes from all directions towards its center of mass.

Gravitational field lines

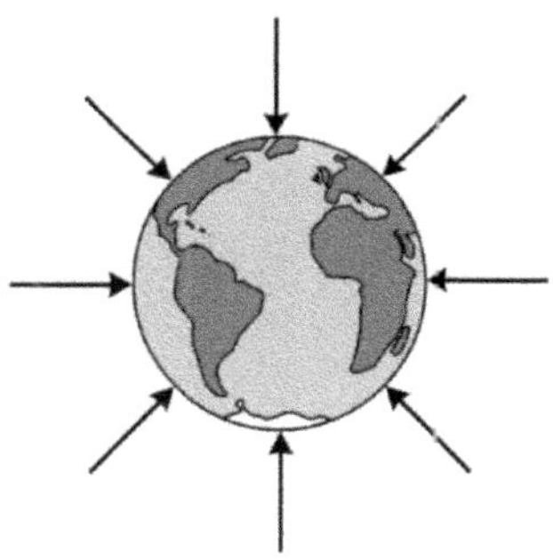

Gravitational field lines radiate towards the Earth's core

These are field lines, more specifically, gravitational field lines. Field lines are often used to depict situations where multiple vectors act in a plane or 3D space.

Notice above that only eight arrows are used to depict the gravitational field around the Earth, even though gravity acts continuously all around the Earth.

It is impossible to draw enough field lines to accurately represent the continuity of Earth's gravitational pull, so it suffices to draw enough lines to show that Earth's gravity acts from all directions.

Arrows closer to each other are closer to Earth, and farther apart are farther from the Earth.

Field lines drawn close together represent a stronger field, and field lines drawn farther apart represent a weaker field.

From the figure above, the field lines closer to Earth signify the more potent gravitational force closer to Earth's surface. This supports the Law of Gravitation (discussed earlier).

Gravitational field lines as vectors

In general, the gravitational field lines represent vectors; typically, these are force vectors.

Force vectors used in calculations, such as projectile motion, a mass-pulley system, or a mass on a pendulum, are usually single arrows that indicate the direction of a single force.

This is because the masses involved in these problems are treated as point masses or masses with an infinitesimally small size, reducing the complexity of the calculations.

A boulder-sized projectile or car-sized object on a pendulum would have a force vector field acting on its entire volume, with all the vectors pointing in the same direction.

Force vector fields

For example, when a car accelerates in a straight-line path, typically, this action is represented by a single force vector originating from the center of the car's mass in the direction of motion.

More accurately, a force vector would be drawn from every point in the car in the direction of motion, creating a force vector field.

Since this creates a far more complex calculation to determine a negligibly more accurate value of the force involved, the vector field is simplified to a single vector originating from the object's center of mass.

Chapter Summary

- Newton's First Law:

  An object at rest will remain at rest, and an object in motion (with constant velocity) will remain in motion (with constant velocity) unless acted upon by an external force.

- Newton's Second Law:

  Force is the vector product of mass and acceleration:

  $$F = ma$$

- Newton's Third Law:

  For every action, there is an equal and opposite reaction.

  When a person interacts with an object by exerting a force on it, that object exerts the same amount, or magnitude, of force back.

- Free-Body Diagrams using Newton's Laws

  1. Sketch the scenario described in the problem.
  2. For every object, draw a free-body diagram showing all the forces acting *on* the object.

     Make magnitudes and directions representative of the scenario.

     Label each force.

     If there are multiple objects, draw separate diagrams.
  3. Resolve vectors into directional components.
  4. Apply Newton's Second Law to each component.
  5. Solve for the net force in each component direction and combine each component to determine the total net force.

- Gravitational force:

  $$F_g = mg$$

  where g is the acceleration due to gravity at 9.8 m/s^2.

- Equations to find friction, the force that always opposes the direction of motion:

  Static friction: $F_s = \mu_s F_N$

  Kinetic friction: $F_k = \mu_k F_N$

  where μ is the coefficient of friction, and F_N is the normal force.

- Motion on an inclined plane:

  Parallel Force: $F_{\parallel} = mg \sin\theta$

  Perpendicular Force: $F_{\perp} = F_N$

  $F_N = mg \cos\theta$

  where θ is the angle between the horizontal axis and surface on which the object rests.

- Objects moving in a circle at constant speed have *uniform circular motion*, an example of simple harmonic motion.

  Instantaneous velocity is always tangent to the circle, and the *centripetal* acceleration (or radial acceleration) points inward along the radius.

- Centripetal acceleration for the uniform circular motion:

  $a_c = v^2 / r$

- Centripetal force:

  $$F = ma_c = \frac{mv^2}{r}$$

- Law of Gravitation:

  $$F_G = G \times \frac{m_1 m_2}{r^2}$$

Practice Problems: Force, Motion, Gravitation

1. Two bodies of different masses are subjected to identical forces. Compared to the body with a smaller mass, the body with greater mass experiences:

A. less acceleration because the ratio of force to mass is smaller
B. greater acceleration because the ratio of force to mass is greater
C. less acceleration because the product of mass and acceleration is smaller
D. greater acceleration because the product of mass and acceleration is greater
E. equal acceleration because the forces are identical

2. An object is propelled along a straight-line path by force. If the net force were doubled, the object's acceleration would:

A. halve
B. stay the same
C. double
D. quadruple
E. none of the above

3. A 500 kg rocket ship is firing two jets at once. The two jets are at right angles, with one firing with a force of 500 N and the other firing with a force of 1,200 N. What is the magnitude of the acceleration of the rocket ship?

A. 1.4 m/s^2
B. 2.6 m/s^2
C. 3.4 m/s^2
D. 5.6 m/s^2
E. 4.2 m/s^2

4. Assume the strings and pulleys in the diagram below have negligible masses, and the coefficient of kinetic friction between the 2 kg block and the table is 0.25. What is the acceleration of the 2 kg block? (Use acceleration due to gravity $g = 9.8$ m/s^2)

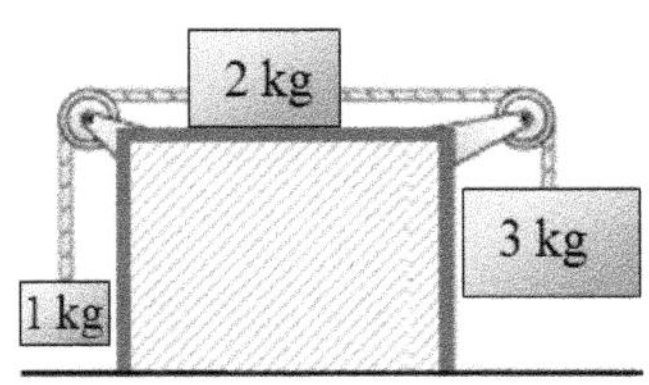

A. 3.2 m/s^2
B. 4 m/s^2
C. 0.3 m/s^2
D. 1.7 m/s^2
E. 2.5 m/s^2

5. Which statements must be valid for an object moving with constant velocity in a straight line?

A. The net force on the object is zero
B. No forces are acting on the object
C. A constant force is being applied in the direction opposite of motion
D. A constant force is being applied in the direction of motion
E. There is no frictional force acting on the object

6. A person gives a shopping cart an initial push along a horizontal floor to get it moving and then releases the cart. The cart travels forward along the floor, gradually slowing as it moves. Consider the horizontal force on the cart while it is moving forward and slowing. Which of the statements is correct?

A. Only a forward force is acting, which diminishes with time
B. Only a backward force is acting; no forward force is acting
C. Both a forward and a backward force are acting on the cart, but the forward force is larger
D. Both a forward and a backward force are acting on the cart, but the backward force is larger
E. No forces are acting because it has been released

7. A truck uses a hook to tow a car whose mass is one-quarter that of the truck. If the force exerted by the truck on the car is 6,000 N, then the force exerted by the car on the truck is:

A. 1,500 N
B. 24,000 N
C. 6,000 N
D. 12,000 N
E. Need to know if the car is accelerating

8. How large is the force of friction impeding the motion of a bureau when a 120 N bureau is pulled across the sidewalk at a constant speed by a force of 30 N?

A. 0 N
B. 30 N
C. 120 N
D. 3 N
E. 15 N

9. An object maintains its state of motion because it has:

A. mass
B. acceleration
C. speed
D. weight
E. all of the above

10. The force of gravity attracts the Earth and the Moon. The Earth's radius is 3.7 times that of the Moon, and the Earth's mass is 80 times greater than the Moon's. Acceleration due to gravity on the surface of the Moon is 1/6 the acceleration due to gravity on the Earth's surface. If the distance between the Earth and the Moon decreases by a factor of 4, how would the force of gravity between the Earth and the Moon change?

A. Remain the same
B. Increase by a factor of 16
C. Decrease by a factor of 16
D. Decrease by a factor of 4
E. Increase by a factor of $1/\sqrt{4}$

11. Two forces acting on an object have magnitudes $F_1 = -6.6$ N and $F_2 = 2.2$ N. Which third force causes the object to be in equilibrium?

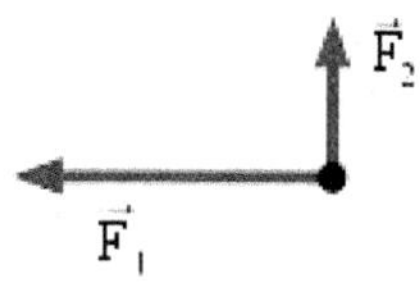

A. 4.4 N at 162° counterclockwise from F_1
B. 4.4 N at 108° counterclockwise from F_1
C. 7 N at 162° counterclockwise from F_1
D. 7 N at 108° counterclockwise from F_1
E. 3.3 N at 162° counterclockwise from F_1

12. What are the readings on the spring scales when a 17 kg fish is weighed with two spring scales if each scale has negligible weight?

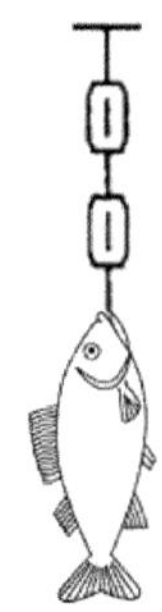

A. The top scale reads 17 kg, and the bottom scale reads 0 kg
B. Each scale reads greater than 0 kg and less than 17 kg, but the sum of the scales is 17 kg
C. The bottom scale reads 17 kg, and the top scale reads 0 kg
D. The sum of the two scales is 34 kg
E. Each scale reads 8.5 kg

13. Two forces of equal magnitude are acting on an object. If the magnitude of each force is 2.3 N and the angle between them is 40°, which third force causes the object to be in equilibrium?

A. 1.8 N pointing to the right
B. 2.2 N pointing to the right
C. 3.5 N pointing to the right
D. 6.6 N pointing to the right
E. 4.3 N pointing to the right

14. An object at rest on an inclined plane starts to slide when the incline is increased to 17°. What is the coefficient of static friction between the object and the plane? (Use acceleration due to gravity $g = 9.8$ m/s^2)

A. 0.37
B. 0.43
C. 0.24
D. 0.31
E. 0.17

15. What is the force exerted by the table on a 2 kg book resting on it? (Use acceleration due to gravity $g = 10$ m/s^2)

A. 100 N
B. 20 N
C. 10 N
D. 0 N
E. 40 N

Solutions: Force, Motion, Gravitation

1. A is correct.

Newton's Second Law ($F = ma$) is rearranged:

$$a = F / m$$

If F is constant, then a is inversely proportional to m.

A larger mass implies a smaller F / m ratio; the ratio is the acceleration.

2. C is correct.

$F = ma$, so doubling the force doubles the acceleration.

3. B is correct.

Pythagorean Theorem ($a^2 + b^2 = c^2$) to calculate the net force:

$$F_1^2 + F_2^2 = F_{net}^2$$

$$(500\text{ N})^2 + (1{,}200\text{ N})^2 = F_{net}^2$$

$$250{,}000\text{ N}^2 + 1{,}440{,}000\text{ N}^2 = F_{net}^2$$

$$F_{net}^2 = 1{,}690{,}000\text{ N}^2$$

$$F_{net} = 1{,}300\text{ N}$$

Newton's Second Law:

$$F = ma$$

$$a = F_{net} / m$$

$$a = 1{,}300\text{ N} / 500\text{ kg}$$

$$a = 2.6\text{ m/s}^2$$

4. E is correct.

Acceleration of the 2 kg block is the acceleration of the system because the blocks are linked together. Balance forces and solve for acceleration:

$$F_{net} = m_3g - m_2g\mu_k - m_1g$$

$$(m_3 + m_2 + m_1)a = m_3g - m_2g\mu_k - m_1g$$

$$a = (m_3 - m_2\mu_k - m_1)g / (m_3 + m_2 + m_1)$$

$$a = [3\text{ kg} - (2\text{ kg})\cdot(0.25) - 1\text{ kg}]\cdot(9.8\text{ m/s}^2) / (3\text{ kg} + 2\text{ kg} + 1\text{ kg})$$

$$a = 2.5\text{ m/s}^2$$

5. A is correct.

Objects moving at constant velocity experience no acceleration and, therefore, no net force.

6. B is correct.

The cart decelerates, which is an acceleration in the opposite direction caused by the force of friction in the opposite direction.

7. C is correct.

Newton's Third Law states that when two objects interact by a mutual force, the force of the first on the second is equal in magnitude to the force of the second on the first.

8. B is correct.

If the bureau moves in a straight line at a constant speed, its velocity is constant. Therefore, the bureau is experiencing zero acceleration and zero net force.

The force of kinetic friction equals the 30 N force that pulls the bureau.

9. A is correct.

Newton's First Law of Motion (Law of Inertia) states that an object at rest tends to stay at rest, and an object in motion tends to maintain that motion unless acted upon by an unbalanced force. This law depends on the property of an object's inertia, which is inherently linked to the object's mass. More massive objects are more challenging to move and manipulate than less massive objects.

10. B is correct.

$$F_g = Gm_{Earth}m_{moon} / d^2$$

d is the distance between the Earth and the Moon.

If d decreases by a factor of 4, F_g increases by a factor of $4^2 = 16$

11. C is correct.

Find equal and opposite forces:

$$F_{Rx} = -F_1$$

$$F_{Rx} = -(-6.6\text{ N})$$

$$F_{Rx} = 6.6\text{ N}$$

$$F_{Ry} = -F_2$$

$$F_{Ry} = -2.2\text{ N}$$

continued...

Pythagorean Theorem ($a^2 + b^2 = c^2$) to calculate the magnitude of the resultant force:

The magnitude of F_R:

$$F_R^2 = F_{Rx}^2 + F_{Ry}^2$$

$$F_R^2 = (6.6\text{ N})^2 + (-2.2\text{ N})^2$$

$$F_R^2 = 43.6\text{ N}^2 + 4.8\text{ N}^2$$

$$F_R^2 = 48.4\text{ N}^2$$

$$F_R = 7\text{ N}$$

The direction of F_R:

$$\theta = tan^{-1}\,(-2.2\text{ N} / 6.6\text{ N})$$

$$\theta = tan^{-1}\,(-1 / 3)$$

$$\theta = 342°$$

The direction of F_R with respect to F_1:

$$\theta = 342° - 180°$$

$$\theta = 162° \text{ counterclockwise of } F_1$$

12. D is correct.

Each scale weighs the fish at 17 kg, so the sum of the two scales is:

$$17\text{ kg} + 17\text{ kg} = 34\text{ kg}$$

13. E is correct.

If θ is the angle with respect to a horizontal line, then:

$$\theta = ½(40°)$$
$$\theta = 20°$$

Therefore, for the third force to cause equilibrium, the sum of all three forces' components must equal zero. Since F_1 and F_2 mirrors are equal in the y-direction:

$$F_{1y} + F_{2y} = 0$$

Therefore, for F_3 to balance the forces in the y direction, its y component must also equal zero:

$$F_{1y} + F_{2y} + F_{3y} = 0$$

$$0 + F_{3y} = 0$$

$$F_{3y} = 0$$

continued...

Since the y component of F_3 is zero, the angle that F_3 makes with the horizontal is zero:

$$\theta_3 = 0°$$

x component of F_3:

$$F_{1x} + F_{2x} + F_{3x} = 0$$

$$F_1 \cos \theta + F_2 \cos \theta + F_3 \cos \theta = 0$$

$$F_3 = -(F_2 \cos \theta_2 + F_3 \cos \theta_3)$$

$$F_3 = -[(2.3\ \text{N}) \cos 20° + (2.3\ \text{N}) \cos 20°]$$

$$F_3 = -4.3\ \text{N}$$

$$F_3 = 4.3\ \text{N to the right}$$

14. D is correct.

At $\theta = 17°$, the force of static friction is equal to the force due to gravity:

$$F_f = F_g$$

$$\mu_s mg \cos \theta = mg \sin \theta$$

$$\mu_s = \sin \theta / \cos \theta$$

$$\mu_s = \tan \theta$$

$$\mu_s = \tan 17°$$

$$\mu_s = 0.31$$

15. B is correct.

The force of the table on the book, the normal force (F_N), is a result of Newton's Third Law of Motion, which states that there is an *equal and opposite reaction for every action.*

A book sitting on the table experiences a force from the table equal to the book's weight:

$$W = mg$$

$$F_N = W$$

$$F_N = mg$$

$$F_N = (2\ \text{kg})\cdot(10\ \text{m/s}^2)$$

$$F_N = 20\ \text{N}$$

Notes for active learning

CHAPTER 4

Work & Energy

Concept of Work and Derived Units

Work Done by a Constant Force

Mechanical Advantage

Conservative *vs*. Non-Conservative Forces

Path-Independent Work in Gravitational Field

Concept of Energy

Kinetic Energy

Work-Kinetic Energy Theorem

Potential Energy

Conservation of Energy

Power, Units

Chapter Summary

Concept of Work and Derived Units

Work requires distance

Work is the relationship between a force and the *distance* traveled by an object acted upon by force in a *direction parallel* to the force.

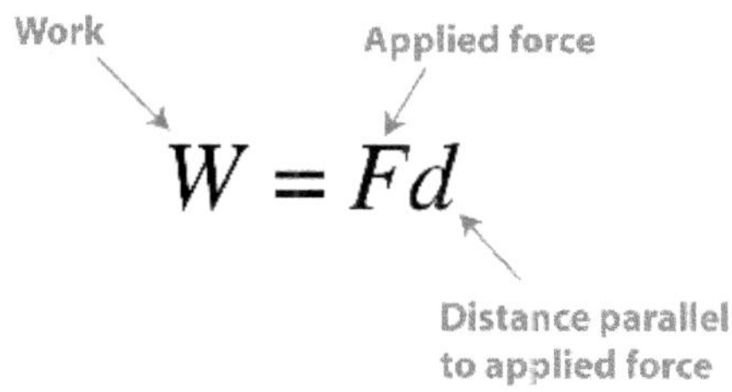

Work results from a force applied over a distance

Work equation is for an applied force and a *distance parallel to the force*.

If the distance traveled is not parallel to the force but at an angle, the angle must be accounted for by the formula:

$$W = \vec{F}d\cos\theta$$

where $\vec{F}$ is force, d is the distance over which the force is applied, and θ is the angle between the applied force and distance).

Force parallel to displacement

This formula divides the forces into components, and the only value used in the work calculation is the *force parallel* ($cos\ \theta$) *to displacement.*

Some instances when work done is zero:

- When distance traveled is zero

  $$\boldsymbol{W = \vec{F} \cdot 0 = 0}$$

- When distance is perpendicular to the applied force (the object moves at 90° to the applied force)

  $$W = \vec{F}d\cos\ (90°)$$

  $$W = \vec{F} \cdot 0$$

  $$W = 0$$

- When an object is moving at a constant velocity, with no forces acting in the direction of movement

  $$W = 0 \cdot d = 0$$

Units for work

Work is calculated using the formula:

$$W = \vec{F}\vec{d}$$

where W is work, F is force, and d is displacement.

Since $\vec{F} = m\vec{a}$,

$$W = m\vec{a}\vec{d}$$

$$m\vec{a}\vec{d} = (\text{kg}) \cdot \left(\frac{\text{m}}{\text{s}^2}\right) \cdot (\text{m})$$

$$W = \frac{\text{kg} \times \text{m}^2}{s^2}$$

$$\frac{\text{kg} \times \text{m}^2}{s^2} = \text{N} \cdot \text{m}$$

$$\text{N} \cdot \text{m} = \text{J}$$

$$W = \text{J}$$

Newton-meter represents the combined units of force and distance.

Magnitude of work (W) is expressed in standard SI units *Joules* (J) since the Newton-meter is the unit of measurement for torque.

Joule is named after English physicist James Prescott Joule (1818-1898) and is also the unit for energy measurement.

Positive and negative work

Work is *positive* if the applied force and the displacement are in the same direction.

For example, pushing a crate across rough terrain involves doing positive work (pushing forward as the crate moves forward).

Work is negative if the force opposes the direction of motion (e.g., friction).

Work Done by a Constant Force

Work in direction of displacement

Work done by a constant force is defined as *the distance moved multiplied by the component of the force in the direction of displacement*:

$$W = \vec{F}\vec{d} \cos\theta$$

For example, a crate on the ground has a rope attached to the side, which a woman pulls to move the crate a distance *d*. How much work did she perform if she pulled with a constant force and the rope formed an angle θ with the crate?

$$W = \vec{F_{\parallel}}\vec{d}$$

$$\vec{F_{\parallel}} = \vec{F} \cos\theta$$

$$W = \vec{F}\,\vec{d} \cos\theta$$

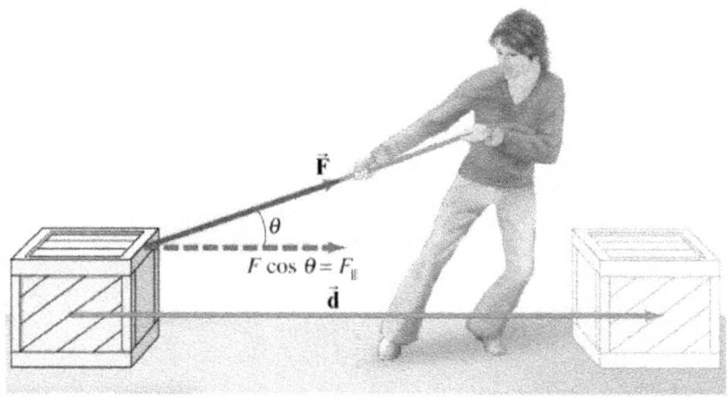

Angle of the applied force over a distance determines work

Five steps for solving work problems

1. Draw a free-body diagram.
2. Choose a coordinate system.
3. Apply Newton's Laws to determine any unknown forces.
4. Find the work done by a specific force.
5. To calculate the net work done, find:

   net force and then find the work it does, or

   work done by each force and add them together.

Perpendicular force and displacement

If the force and displacement are perpendicular, the work done is zero.

Centripetal force does no work.

In a uniform circular motion, the centripetal force is *constant* when an object moves in a *circle at a constant speed.*

Centripetal force is directed toward the center of a circular path and is perpendicular to the linear movement of the object.

For circular motion at constant speed, the velocity is tangent to the direction of motion.

How much work is performed if a ball attached to a string is spun in a circle?

$$W = \vec{F}_{\parallel}\vec{d}$$

$$\vec{F}_{\parallel} = \vec{F}\cos(90°) = 0\text{ N}$$

$$W = 0\text{ J}$$

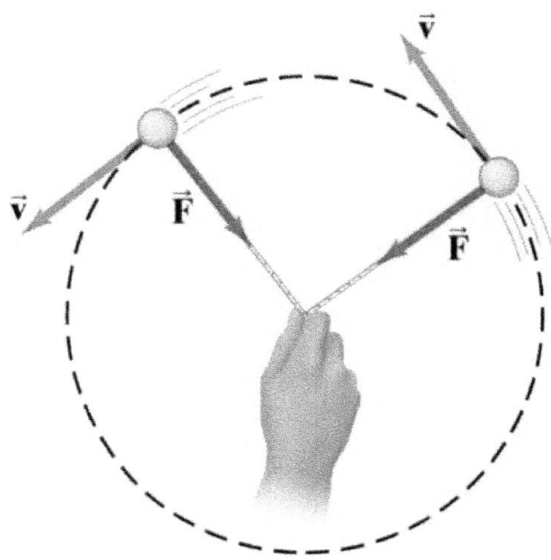

Centripetal force points toward the center, and velocity is tangent to the circle

Mechanical Advantage

Lever arms and pulleys

Mechanical advantage takes an input force and produces a much larger force.

Lever arms or *pulleys* achieve such a mechanical advantage.

Mechanical advantages are useful in construction and moving large objects, which are otherwise too heavy to manipulate.

Simple machines

Simple machines are any mechanical device that is used to *apply a force*.

The premise of a simple machine is that the work in equals the work out, which refers to the amount of force multiplied by a ratio of distances.

Examples of simple machines:

- Pulleys
- Inclined planes
- Wedges
- Screws
- Levers
- Wheels and axles

Pulley systems are conventional simple machines.

As simple machines, a pulley system does not reduce the amount of work done on an object but reduces the amount of force needed to manipulate it; the distance over which the force acts is increased.

The work equation shows that the force and distance are inversely related and that the pulling distance increases by the same factor as the force decreases.

$$\frac{W}{\vec{d}} = \vec{F}$$

Solving stationary pulley problems

A method for solving pulley problems is to realize that the ropes on either side of a moving pulley contribute to pulling the load.

A stationary pulley, however, does not contribute to the load.

The above rules apply to problems with simple pulleys.

Complicated pulleys have additional ropes that contribute to pulling the load.

For example, a pulley is attached to the ceiling to lift a box. The pulley is stationary, so no mechanical advantage is imparted to the system. Therefore, if the weight of the box is 100 N, the force needed to pull the box up is 100 N. For every 1 meter pulled, the box moves up 1 meter.

$$W = \vec{F}\vec{d}$$

$$W = (100\text{ N})\cdot(1\text{ m})$$

$$W = 100\text{ Joules}$$

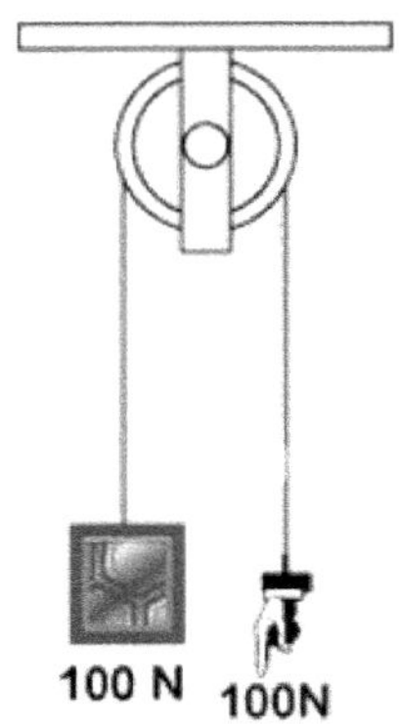

Stationary pully system imparts no mechanical advantage

Solving problems with moving pulleys

As stated, moving pulleys impart a mechanical advantage to the system.

When there is one moving pulley, the force needed to pull is halved because the rope on each side of the pulley contributes equally; 50 N force is transmitted to the right-hand rope, while the left-hand rope contributes to the other 50 N.

However, for a moving pulley system, the distance required to pull the rope is different than for a simple pulley system.

If the box is to be pulled up one meter (e.g., in the previous example), the distance of the rope required to pull the box is doubled because the force is halved.

Equivalent pulley systems

In the diagrams below, there is one moving pulley in each system; the pulley system on the right is equivalent to the pulley system on the left.

Although two pulleys are on the right, only the mobile pulley halves the force needed to pull the box.

For any system, the work required to lift the box a specific distance is the same regardless of the reduction in force.

$$W = \vec{F}\vec{d}$$

$$100\ \text{Joules} = (50\ \text{N}) \cdot (\vec{d})$$

$$\vec{d} = 2\ \text{meters}$$

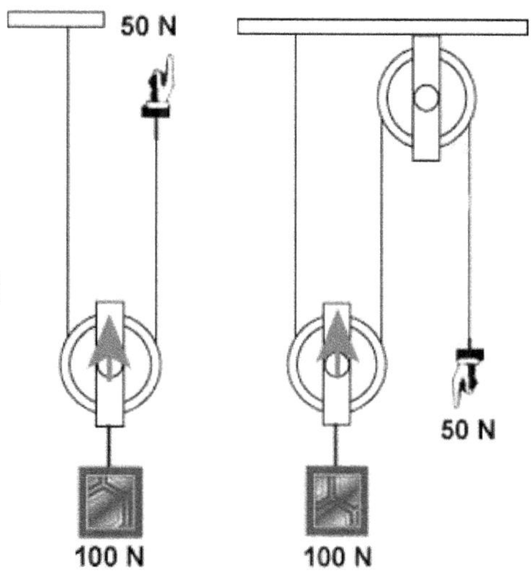

Pulley system on the right is equivalent system to the pulley system on the left

Systems with two mobile pulleys

In the diagram below, the pulley arrangement differs.

The system has two mobile pulleys, which reduce the force needed to lift the box.

If each mobile pulley reduces the force by half, the total force needed to lift the box will be reduced to a quarter of the original value.

100 N box can be lifted with only 25 N.

Work required to lift the box one meter must always be the same, so for every 1 m the box moves, 4 m of the rope must be pulled.

$W = \vec{F}\vec{d}$

$100 \text{ Joules} = (25 \text{ N}) \cdot (\vec{d})$

$\vec{d} = 4 \text{ meters}$

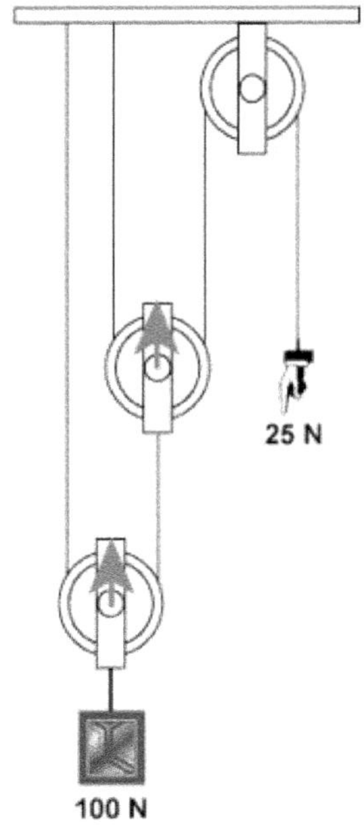

Mobile pulleys reduce the force needed to lift the box

Below is an example of a complex pulley system.

Like the simple pulleys, the ropes on both sides of the moving pulley contribute.

However, the left-most rope also contributes. This makes three contributing ropes, which reduces the effort required by a factor of three.

Distance needed to pull is three times the distance the box travels.

$W = \vec{F}\vec{d}$

$100 \text{ Joules} = (33 \text{ N}) \cdot (\vec{d})$

$\vec{d} \cong 3 \text{ meters}$

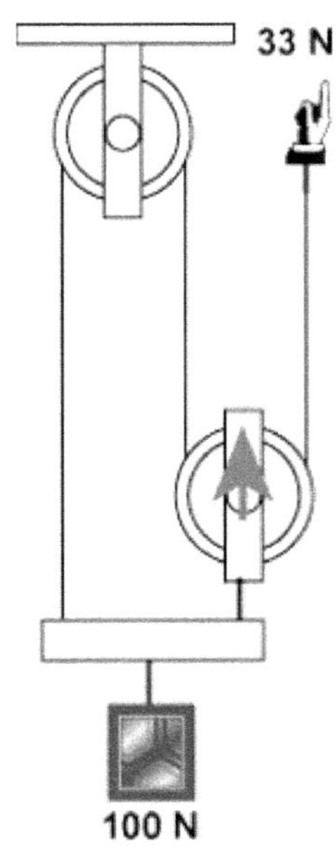

Conservative vs. Non-Conservative Forces

Conservative forces

Force can be *conservative* or *non-conservative*. Determining how to label a force can be confusing. However, a few qualifications can be checked to determine what type of conservation a force follows.

Conservative force does not dissipate heat, sound, or light. Furthermore, if the work done by a force is *path-independent*, then that force is conservative.

Conservative forces are associated with potential energy (PE), as potential energy can only be defined for conservative forces.

Force from a spring can be stored as spring potential energy, and gravitational force can be stored as gravitational potential energy.

Electromagnetic forces are also conservative.

If only conservative forces are acting, mechanical energy is conserved.

Particles moving along a path

Consider a particle moving along a path:

- Force is conservative if the work done to move a particle in any round-trip path is zero.
- Force is conservative if the work needed to move a particle between two points is the same regardless of the path taken.

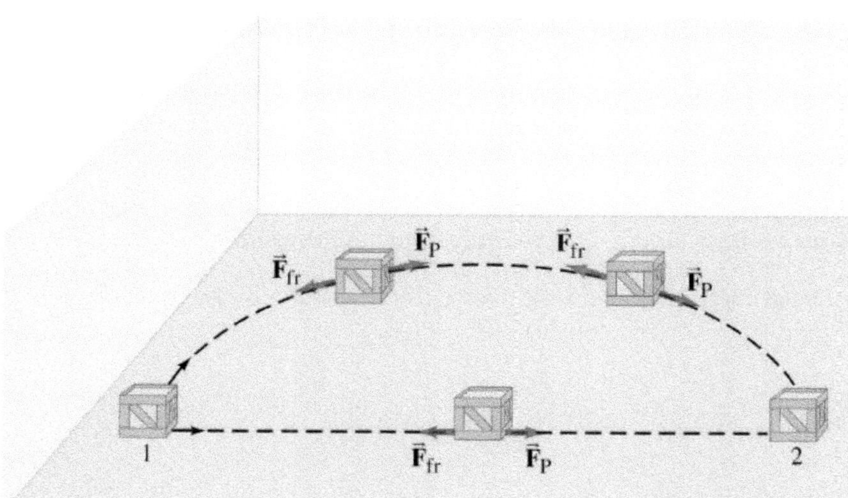

Conservative forces for specific particle movements

Non-conservative forces

Most forces are conservative; non-conservative forces include friction and human exertion.

If friction is present, the work done depends not only on the starting and ending points but also on the path taken, as frictional forces directly oppose the motion of an object.

Heat and sound energy are released when an object encounters friction, and the total energy is not conserved.

Heat energy is also lost in human exertion.

When muscles are flexed during exercise, the heat from the muscles rises off the skin and cannot be recovered or reabsorbed.

Conservative and non-conservative forces

Conservative forces	Non-conservative forces
Gravitational	Friction
	Air resistance
Elastic	Tension in a cord
	Motor or rocket propulsion
Electric	Push or pull by a person

Work by conservative and non-conservative forces

Work done by conservative forces must be distinguished from that done by non-conservative forces.

Work done by non-conservative forces equals total change in kinetic and potential energies:

$$W_{NC} = \Delta KE + \Delta PE$$

For all forms of energy, energy is conserved if the total energy neither increases nor decreases.

Conservation of energy with mechanical forces is discussed in a subsequent chapter.

Path-Independent Work in Gravitational Field

Path-independent work

Work done in a gravitational field is *path-independent* since gravitational forces always act downward.

Sideward motion, perpendicular to the gravitational force, involves no work.

For example, a man lifting a bag of groceries is positive work in a gravitational field.

Lifting a bag of groceries is positive work because a force is exerted (to overcome the force of gravity), and the distance the bag is lifted is in the direction of the force.

However, once the bag is held up, it is at rest (as in, no longer traveling), and no work is done on the bag. If the man does not lift or lower the bag, he does no work on it because there is no vertical displacement.

Force perpendicular to displacement

Walking with the bag of groceries does no work on the bag because horizontal displacement is perpendicular to force of gravity, and the resulting work must be zero.

$\vec{F}_P = m\vec{g}$

Work to lift the bag: $W = \vec{F}_P\vec{d}_y > 0\ \text{J}$

Work while walking: $W = \vec{F}_P\vec{d}_x = 0\ \text{J}$

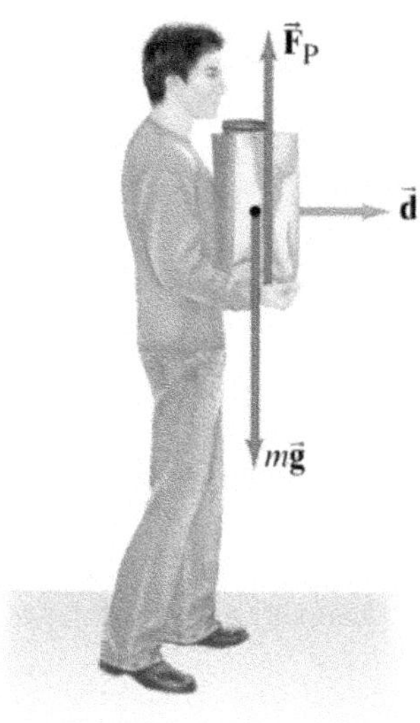

Lifting the bag exerts work relative to gravity but none while walking

Inclined plane examples

Another example of work in a gravitational field is pushing an object at constant speed up a frictionless, inclined plane.

This system uses the same amount of work as directly lifting the object to the same height at a constant speed because the displacement perpendicular to the force of gravity contributes no work.

Sliding an object down a frictionless, inclined plane involves the same gravitational work as the object undergoing a free fall from the same height.

Since the only forces involved in these motions are acting on a vertical plane, the work done is only dependent on the height of the ramp.

Concept of Energy

Work and energy

Work and energy are closely related concepts.

Energy is defined as the ability to do work; like work, the unit of energy is the Joule.

However, although work and energy have the same units, the two concepts are not always interchangeable.

Energy takes many forms, while mechanical work is limited to the definition above.

For example, energy may be exerted to move a heavy box, but no work has been done if the box does not budge.

Another important concept of energy is that the sum of the energy of the universe in its current state is the same amount of energy that was present at its inception. There is infinite space for it to go, but it never disappears; the *Law of Conservation of Energy*.

Energy forms

Energy created at the beginning of the universe is divided into types of energy.

Mechanical energy is associated with motion and position, equal to kinetic plus potential energy.

Mechanical energy = PE + KE

Electrical energy consists of the current and potential provided by a circuit. Any charged particle within an electrical field contains electrical energy.

Chemical energy participates in all chemical reactions, the potential of a substance to undergo a transformation or reaction.

Radiant energy originates from electromagnetic waves, the energy stored in a photon or the motion of an electromagnetic wave. Visible light is a small portion of the full spectrum of electromagnetic waves; the human eye cannot see many radiant energy sources.

Nuclear energy is released during reactions involving the fusion (joining) or fission (splitting) of the nucleus. Examples are nuclear bombs and nuclear power plants.

Although energy can take different forms, there are two main types of energy within mechanical physics problems:

- *Kinetic energy*: the energy of motion (energy consumed or released)
- *Potential energy*: stored energy (energy that could be released by a body)

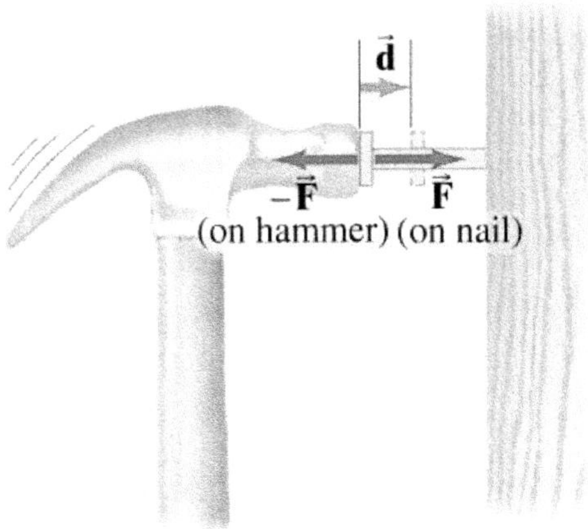

Hammer imparts kinetic energy to drive the nail into the wood

Kinetic Energy

Energy of motion

Kinetic energy (*KE*) is the energy associated with the motion of an object and results from work or a change in potential energy.

Kinetic energy is only dependent on the *mass* of an object as well as its *speed.*

As with energy, the unit of kinetic energy is a Joule (kg m^2/s^2).

The equation for kinetic energy is:

$$KE = \frac{1}{2}mv^2$$

KE is directly proportional to mass and the square of the velocity.

Kinetic energy is proportional to mass

An important feature of kinetic energy is its relation to the mass and speed of an object.

If cars travel at the same speed, the car with twice the mass has twice as much KE as the other car.

If the mass (m) of an object is doubled, *KE* is also doubled.

Mass and velocity influence KE

If the velocity (v) is doubled, the *KE* is quadrupled (because velocity is squared).

For example, two cars are on the road, but one is twice as heavy as the other.

For objects of equal mass, objects with a higher velocity always have greater *KE.*

For objects with the same velocity, objects with a greater mass have greater *KE.*

If two cars of equal mass are on the road, and one travels twice as fast as the other, the faster car has a *KE* four times that of the slower car.

This is important because a change in velocity is more significant regarding kinetic energy than a change in mass because velocity is squared.

Work–Kinetic Energy Theorem

Work transforms into kinetic energy

Work on an object can be transformed into kinetic energy.

If the net work on the object is positive, kinetic energy increases; if the net work is negative, kinetic energy decreases.

Thus, work (W) and kinetic energy (*KE*) are directly related by the equation:

$$|W| = |\Delta KE| \quad \text{(change in kinetic energy)}$$

Given the definition of kinetic energy:

$$KE = \frac{1}{2}mv^2$$

Work relates to kinetic energy

Work can be written regarding a change in the kinetic energy as:

$$W = \Delta KE$$

$$\Delta KE = \frac{1}{2}mv_2^2 - \frac{1}{2}mv_1^2$$

$$W = \frac{1}{2}mv_2^2 - \frac{1}{2}mv_1^2$$

Positive work increases kinetic energy

For example, the force *F* provided by a car's engine moves the car a distance *d*, thus performing positive work on the car. Work is transformed into kinetic energy and increases the total kinetic energy of the car.

The car has a higher final velocity than before work was performed.

Positive work increases the kinetic energy of objects

$$+W = +\Delta KE$$

$$v_2 > v_1$$

Negative work decreases kinetic energy

If the same car applies its brakes at a stop sign, the force provided by friction in the car's brakes does negative work on the car.

In this example, the negative work on the car would reduce its kinetic energy, and the change in kinetic energy would be negative.

As a result, the final velocity of the car would be less than its initial velocity.

$$-\mathrm{W} = -\Delta KE$$

$$v_2 < v_1$$

Kinetic energy performs work

Kinetic energy of an object can also do work when forces are involved.

For example, a moving object on a frictionless surface can convert all its kinetic energy to work if it slides up an inclined plane before stopping.

Kinetic energy of the object is converted to work to lift the object a distance against the force of gravity.

$$-\Delta KE = +\mathrm{W}$$

In summary, energy can perform work against:

- inertia
- gravity
- friction
- deformation of shape
- combinations of the above

Notes for active learning

Potential Energy

Stored potential energy

Potential energy (*PE*) is the energy stored in an object. Regardless of whether an object is in motion, an object always retains energy. This energy can be stored in many forms, such as the chemical potential energy of food.

Chemical potential energy

Living organisms eat because food has chemically bonded substances.

Organisms store ingested substances in cells until needed. Then, the cells degrade (i.e., catabolism) bonded substances into individual components.

This digestive process of breaking chemical bonds releases potential energy.

Mechanical potential energy

Unlike chemical potential energy, potential energy in mechanical physics problems depends solely upon the position or configuration of objects.

Objects have potential energy from their surroundings. For example:

- An object at some height above the ground
- A wound-up spring
- A stretched elastic band

Potential energy includes *gravitational* and *spring* (elastic) potential energy.

Local gravitational energy ($PE = mgh$)

Local gravitational potential energy is stored in an object due to its position (height) above the ground. It depends upon the mass (m) of the object, its height (h) above the ground, and the acceleration of gravity (g):

$$PE = mgh$$

From the equation, potential energy is directly proportional to mass and height. If two objects of equal mass are at different heights above the ground, the higher object has more potential energy. Likewise, if two objects of unequal mass are at the same height, the object with a larger mass will have higher potential energy.

On Earth, $g = 9.8$ m/s^2 and is assumed constant regardless of the location or height relative to the Earth (unless stated otherwise).

Relative height

Gravitational potential energy measures h relative to a reference height. Usually, the reference height is evident in a problem (i.e., the ground).

However, if not, a reference height can be established at any arbitrary location if the reference height $h = 0$; thus, the potential energy is zero.

For example, if a rock of mass m is held at a height h_1 above a table, its potential energy concerning the table (reference height) is:

$$PE = mgh_1$$

If the surface of the table is a height h_2 above the ground, then the rock's potential energy (PE) concerning the ground (reference height) is:

$$PE = mg(h_2 + h_1)$$

Work changes potential energy

Like kinetic energy, work can be performed to increase or decrease potential energy.

If all the work (W) is transformed into potential energy (or vice versa), then work and the equation can relate potential energy:

$$|\, \mathrm{W} \,| = |\, \Delta PE \,|$$

Work can be written regarding a change in the potential energy, as:

$$\mathrm{W} = \Delta PE = mgh_2 - mgh_1$$

For example, the diagram shows a block of mass m raised at height h from position y_1 to position y_2.

Work performed results in a positive change in potential energy because the new position of the block is at a greater height than its original position.

$$+\mathrm{W} = +\Delta PE$$

$$y_2 > y_1$$

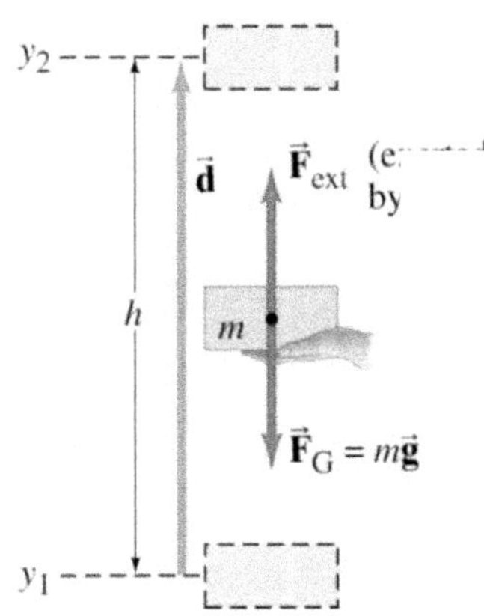

Work increases potential energy by a change in distance

Gravitational potential energy ($PE = -GmM/r$)

Gravitational potential energy uses Newton's *Law of Universal Gravitation,* which defines g as:

$$g = \frac{GM}{r^2}$$

where M is the mass of the attracting object, r is the distance between the two objects. and the gravitational constant G equals 6.67×10^{-11} N·m²/kg².

To find the general form of gravitational potential energy, g is substituted into the equation for local gravitational energy, and the equation results:

$$PE = m \times \frac{GM}{r^2} \times h$$

Height h measures the distance between two objects and equals the radius r. Canceling an r in the denominator, the equation for general gravitational potential energy is:

$$PE = \frac{GmM}{r}$$

Spring potential energy ($PE = \frac{1}{2}kx^2$)

Springs can also store potential energy, which is dependent upon the spring material and the stretch or compression of the spring:

$$PE = \frac{1}{2}kx^2$$

The spring constant k is a measure of the stiffness of the spring; stiffer springs have a larger k because they require more energy to stretch or compress.

Compression of the spring from its original equilibrium position is represented as x.

PE of a spring is the elastic potential energy (PE) and is sometimes notated as PE_{el}.

Force required to compress a spring is described by:

$$F_s = -kx$$

Notice the negative sign in front of the constant; this indicates that the spring force is a *restoring force* and acts *opposite* to the direction of the stretch.

Restorative force is observed in the diagram below:

Spring restoring forces

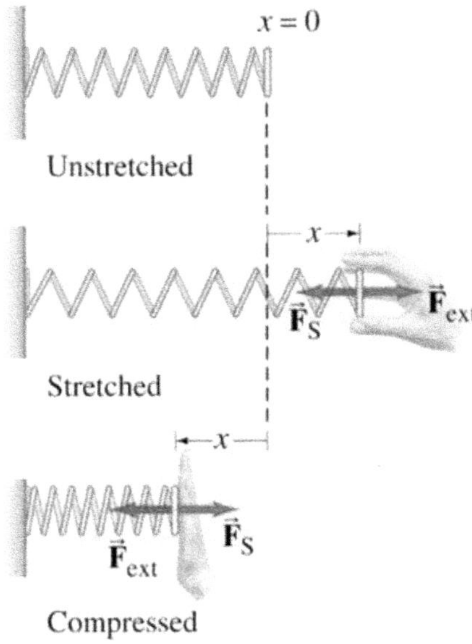

Spring at equilibrium, stretched and compressed positions

For the figure labeled *unstretched*, the spring is at *equilibrium.*

For the figure labeled *stretched*, if the spring is stretched a distance x from the equilibrium length, then the spring force *acts opposite to the direction of the stretch.*

For the figure labeled *compressed*, if the spring is compressed a distance x from the equilibrium length, the spring *force acts opposite to the compression.*

Conservation of Energy

Kinetic and potential energy changes

Conservative forces include *gravity*, *spring forces*, and *electrostatic forces*.

Non-conservative forces include *friction* and *human exertion*; mechanical energy, like all energy, must be conserved.

Sum of kinetic and potential energy changes is zero without non-conservative forces.

Kinetic and potential energy changes are equal but opposite in sign.

Total mechanical energy can be defined as:

$$\Sigma E = KE + PE$$

Moreover, its conservation is:

$$E_2 = E_1 = \text{constant}$$

Initial energy equals final energy

The total amount of initial energy equals the total amount of final energy.

Gravitational potential energy is converted to kinetic energy as an object falls, but the total amount of energy stays the same.

For instance, when a crate slides to a stop on a rough surface, its kinetic energy is converted (or dissipated) into heat (i.e., friction) and sound energy.

For example (diagram below), a rock is stationary above the ground at a distance of y, and its energy is potential. When the rock is released, it drops to the ground; half of its PE has been converted to KE at precisely the halfway point of its descent.

As the figure shows, all initial potential energy is converted to kinetic energy before impacting the ground.

Consider a man dropping a rock some distance y. By conservation of energy, the total mechanical energy of the rock at any given instant is:

$$\Sigma E = KE + PE$$

$$KE + PE = ½\, mv^2 + mgh$$

$$\Sigma E = ½\, mv^2 + mgh$$

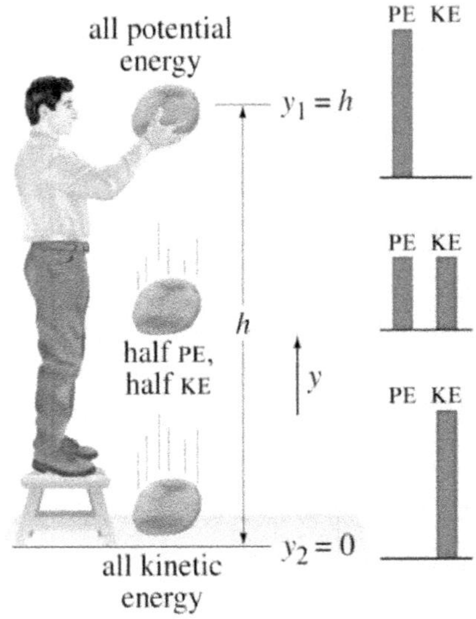

Falling object converts potential energy into kinetic. Energy bar graphs next to the figure show how the energy moves from all potential to all kinetic.

Potential energy becomes kinetic energy

Another example of energy conservation is a rollercoaster.

Speed of a roller coaster only depends on its *height compared* to its starting height.

At the top of the drop, all the rollercoaster's energy is potential energy.

At the bottom, the rollercoaster converted the potential energy into kinetic energy.

Without friction, the kinetic energy at the bottom is converted back to potential energy to bring the roller coaster up the second hill of equal height to the first.

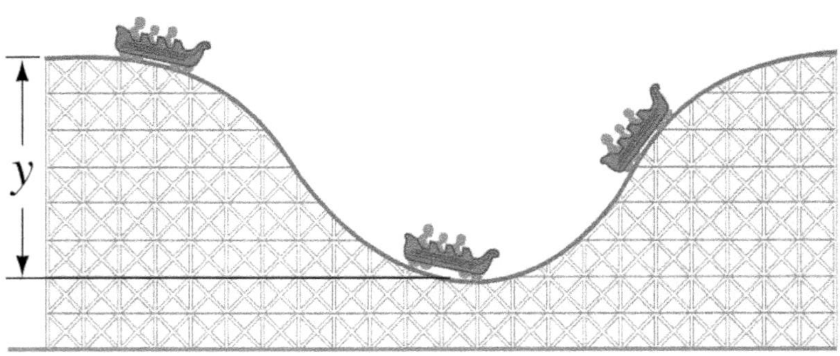

Rollercoasters exhibit conservation of energy

$$PE_{1st\ hill} = KE = PE_{2nd\ hill}$$

Springs conserve energy

Conservation of energy for a spring is illustrated by a ball being shot using a spring:

(a) (b) (c)

(a): $PE = 0$ Joules

(b): $W_{hand} = PE_{spring}$

(c): $PE_{spring} = KE_{ball}$

Figure (a), the spring is at its equilibrium length and has no potential energy.

Figure (b) shows that work has been performed against the spring's restoring force. Work performed is converted entirely into spring potential energy for an ideal system.

Figure (c), the ball is released, and all the spring's potential energy is converted into the ball's kinetic energy.

Friction reduces kinetic and potential energy

However, where does the kinetic and potential energy go if there is a non-conservative force, such as friction?

Energy cannot be destroyed; it is converted into different forms; in this case, friction performs work, transforming the initial energy into heat.

Remember, in the presence of friction:

$PE_{initial} > PE_{final}$

$KE_{initial} > KE_{final}$

Solving conservation of energy problems

To solve problems that involve conservation of energy follow these steps:

1. Draw a picture depicting the scenario under investigation.
2. Determine the system for which energy will be conserved
3. Determine the initial and final positions
4. Choose a logical reference frame
5. Apply the conservation of energy
6. Perform calculations

Power, Units

Power as work per unit time

Often, problems do not give the total energy or input work but instead give the power a system consumes.

A key concept of power is the *time at which work* is performed.

Power is the work performed over time and can be written as:

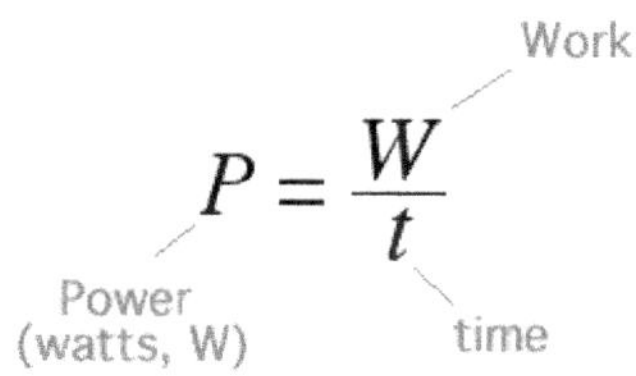

Power is work (W) per unit time

Electric companies use power because it describes energy use (work done) over time.

Lifting an object of mass *m* one meter off the ground always requires the same work. However, lifting the object in one minute requires more power than lifting the object in an hour.

The longer the time interval, the less power is required. The difference between walking and running upstairs is power—the change in gravitational *PE* is the same.

$$P = average\ power = \frac{work}{time} = \frac{energy\ transformed}{time}$$

Work measures power consumed

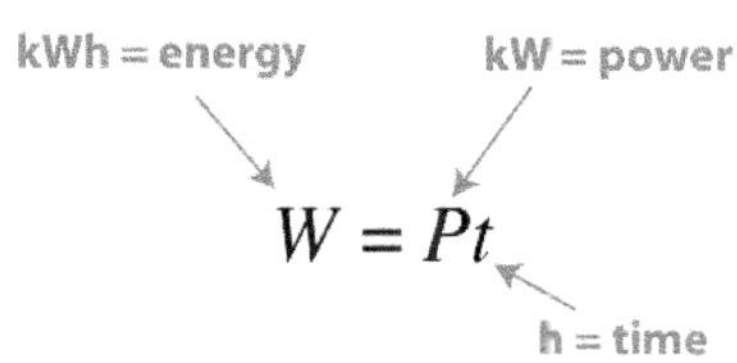

Work measures power consumed per unit time

Power can be derived as the product of the force on an object times its velocity:

$$P = \vec{F}\vec{v}$$

The SI unit for power is the *watt* (Joules/second).

Horsepower

Horsepower (hp) is another unit of power commonly used for high-powered objects such as engines.

1 hp equals ~745.7 Watts

Power predicts the amount of work that can be performed over a given time.

For example, if there are two engines, the more powerful engine can do the same amount of work in less time than the less powerful engine.

Sports cars are described by their engine's horsepower and are compared by how they accelerate from 0 mph to 60 mph. Sports cars have powerful engines that can perform work, which is, in turn, converted into kinetic energy.

Chapter Summary

- Work is force applied with distance and can cause a change in energy.

  Positive work puts energy into a system.

  Negative work takes energy out of a system.

  Equations for work include:

  $W = \vec{F}\vec{d}\cos\theta$

  $W = \Delta KE$

  W = area under an F vs. d graph

Conservative Forces	Non-conservative Forces
Gravitational	Friction
	Air resistance
Elastic	Tension in a cord
	Motor or rocket propulsion
Electric	Push or pull by a person

- Energy is the ability to do work and is a conserved quantity.

  Total initial energy equals total final energy

  Equations for energy:

  $KE = \frac{1}{2}mv^2$ $\quad PE_g = mgh$ $\quad PE_s = \frac{1}{2}kx^2$

- *Kinetic energy* is energy of motion (consumed or released at a given time)
- *Potential energy* is stored and can be released for a certain body)

- Types of energy:

  Mechanical energy – associated with motion and position. It is equal to kinetic plus potential energy.

  Electrical energy – comprises the current and potential provided by a circuit. Any charged particle within an electrical field contains electrical energy.

  Chemical energy – involved in all chemical reactions; the potential of a certain substance to undergo a transformation or reaction.

  Radiant energy – from electromagnetic waves; the energy is stored in a photon or the motion of an electromagnetic wave. Visible light is only a small part of the full spectrum of electromagnetic waves—many other radiant energy sources cannot be seen with the unaided eye.

  Nuclear energy – released during reactions involving the fusion or fission of the nucleus. Examples include nuclear bombs and nuclear power plants.

- Gravitational energy:

  $$PE = GmM / r$$

- Spring energy:

  $$PE = \frac{1}{2}kx^2$$

- Conservation of Energy Law states that the total initial energy equals the total final energy. Often, interactions are limited to mechanical energy with no heat lost or gained.

  $$KE_i + PE_i \pm \mathrm{W} = KE_f + PE_f$$

- Power is the rate at which work is performed and is given by:

  $$P = \frac{W}{t} \quad \text{or} \quad P = \vec{F}\vec{v}$$

Practice Problems: Work & Energy

1. When a pebble is dropped from height *h*, it reaches the ground with kinetic energy. Ignoring air resistance, from what height should the pebble be dropped to reach the ground with twice the *KE*?

A. $\sqrt{2}h$ **B.** $2h$ **C.** $4h$ **D.** $8h$ **E.** $16h$

2. Which of the following situations requires the greatest power?

A. 50 J of work in 20 minutes
B. 200 J of work in 30 minutes
C. 10 J of work in 5 minutes
D. 100 J of work in 20 minutes
E. 100 J of work in 10 minutes

3. A kilowatt-hour is a unit of:

I. work II. force III. power

A. I only
B. II only
C. III only
D. I and II only
E. I and III only

4. A hydraulic press (like a simple lever), properly arranged, is capable of:

I. multiplying the energy input
II. multiplying the output force
III. exerting force only vertically

A. I only
B. II only
C. III only
D. I and II only
E. I and III only

5. 4.5×10^5 J of work is done on a 1,150 kg car, accelerating from 10 m/s to some final velocity. What is this final velocity? (Use acceleration due to gravity $g = 10$ m/s^2)

A. 30 m/s
B. 37 m/s
C. 12 m/s
D. 19 m/s
E. 43 m/s

6. Which of the following is not a unit of work?

A. N·m
B. kw·h
C. J
D. kg·m/s
E. W·s

7. The law of conservation of energy states that:

I. the energy of an isolated system is constant
II. energy cannot be used faster than it is created
III. energy cannot change forms

A. I only
B. II only
C. III only
D. I and II only
E. I and III only

8. A crane lifts a 300 kg steel beam vertically upward a distance of 110 m. Ignoring frictional forces, how much work does the crane do on the beam if the beam accelerates upward at 1.4 m/s^2? (Use acceleration due to gravity $g = 9.8$ m/s^2)

A. 2.4×10^3 J
B. 4.6×10^4 J
C. 3.7×10^5 J
D. 6.2×10^5 J
E. 8.8×10^2 J

9. Steve pushes twice as hard against a stationary brick wall as Charles. Which of the following statements is correct?

A. Both do the same amount of positive work
B. Both do positive work, but Steve does one-half the work of Charles
C. Both do positive work, but Steve does four times the work of Charles
D. Both do positive work, but Steve does twice the work of Charles
E. Both do zero work

10. What is the change in gravitational potential energy of an object if the height of the object above the Earth is doubled? (Assume that the object remains near the surface)

A. Quadruple
B. Doubled
C. Unchanged
D. Halved
E. Tripled

11. If 1 N is exerted for a distance of 1 m in 1 s, the amount of power delivered is:

A. 3 W
B. 1/3 W
C. 2 W
D. 1 W
E. ½ W

12. A brick is dropped from a roof and falls a distance *h* to the ground. If *h* were doubled, how would the maximal *KE* of the brick change just before it hits the ground?

A. It doubles
B. It increases by $\sqrt{2}$
C. It remains the same
D. It increases by 200

E. Requires more information

13. A helicopter with single landing gear descends vertically to land with a speed of 4.5 m/s. The helicopter's shock absorbers have an initial length of 0.6 m. They compress to 77% of their original length, and the air in the tires absorbs 23% of the initial energy as heat. What is the ratio of the spring constant to the helicopter's mass?

A. 0.11 kN/kg·m
B. 1.1 N/kg·m
C. 0.8 kN/kg·m
D. 11 N/kg·m
E. 0.11 N/kg·m

14. A 21-metric-ton airplane is observed to be a vertical distance of 2.6 km from its takeoff point. What is the gravitational potential energy of the plane with respect to the ground? (Use acceleration due to gravity $g = 9.8 \text{ m/s}^2$, the metric ton = 1,000 kg)

A. 582 J
B. 384 J
C. 535 MJ
D. 414 MJ
E. 773 J

15. A tennis ball bounces on the floor. It loses 31% of its energy during each bounce due to heating. How high does the ball reach after the third bounce if it is initially released 4 m from the floor?

A. 55 cm
B. 171 mm
C. 106 cm
D. 131 cm
E. 87 cm

Solutions: Work & Energy

1. B is correct.

When the pebble falls, *KE* at impact = *PE* before impact.

$PE = KE$

$mgh = KE$

$mg(2h) = 2KE$

$2PE = 2KE$

If mass and gravity are constant, the height must be doubled.

2. E is correct.

Power = Work / time

A: Power = 50 J / 20 min = 2.5 J/min

B: power = 200 J / 30 min = 6.67 J/min

C: power = 10 J / 5 min = 2 J/min

D: power = 100 J / 20 min = 5 J/min

E: power = 100 J / 10 min = 10 J/min

Typically, power is measured in watts or J/s.

3. A is correct.

kilowatt = unit of power

hour = unit of time

kW·h = power × time

power = work / time

kW·h = (work / time) × time

kW·h = work

4. B is correct.

Mechanical advantage: δ_1 / δ_2

where d_1 and d_2 are the effort arm and load arm, respectively.

If d_1 is greater than d_2, force output is increased.

5. A is correct.

Find the final speed using the conservation of energy.

Let the energy added as work be represented by W. Then, conservation of energy requires:

$E_f = E_i + W$

$½mv^2_f = ½mv^2_i + W$

Solving for v_f:

$v_f = \sqrt{v^2_i + (2W) / m}$

$v_f = \sqrt{[(10\ \text{m/s})^2 + (2)\cdot(4.5 \times 10^5\ \text{J}) / (1150\ \text{kg})]}$

$v_f = 29.7\ \text{m/s} \approx 30\ \text{m/s}$

6. D is correct.

Work = Force × distance

W = Fd

The unit kg·m/s cannot be manipulated to achieve this.

7. A is correct.

The Law of Conservation of Energy states that for an isolated system (no heat or work transferred), the energy of the system is constant.

Energy is neither created nor destroyed; it can only transform from one form to another.

8. C is correct.

The work done by this force is:

$W = Fd$

where F = force applied by the crane and d = distance over which the force is active.

Solve for F using Newton's Second Law.

Two forces on the beam – the applied force due to tension in the crane's cable and gravity.

$F_{net} = ma$

$F - mg = ma$

Therefore:

$W = m(a + g)d$

$W = (300\ \text{kg})\cdot(1.4\ \text{m/s}^2 + 9.8\ \text{m/s}^2)\cdot(110\ \text{m})$

$W = 3.7 \times 10^5\ \text{J}$

9. E is correct.

Work = Force × displacement × $\cos \theta$

$W = Fd \cos \theta$

If $d = 0$, then work = 0

10. B is correct.

Relative to the ground, an object's gravitational $PE = mgh$, where h is the altitude.

PE is proportional to h, doubling h doubles PE.

11. D is correct.

$W = Fd$

$P = W / t$

$P = Fd / t$

$P = (1\ \text{N})\cdot(1\ \text{m}) / 1\ \text{s}$

$P = 1\ \text{W}$

12. A is correct.

$KE = PE$

$KE = mgh$

h is directly proportional to KE.

If h doubles, then the KE doubles.

13. C is correct.

Initial kinetic energy of the helicopter is entirely converted to the potential energy of the landing gear and heat. By conversation of energy:

$\Delta KE = \Delta PE + Q$

Let the equilibrium length of the landing gear's spring be L_0 and the compressed length be L. The change in potential energy of the landing gear is:

$\Delta PE = ½k(\Delta x)^2$

$\Delta PE = ½k(L_0 - L)^2$

$\Delta PE = ½k(L_0 - (0.23)L_0)^2$

$\Delta PE = ½k(0.77)L_0^2$

continued...

The energy lost to heat is 23% of the initial kinetic energy. The final kinetic energy is zero, therefore:

$$Q = (0.23)\ \Delta KE$$

$$Q = (0.23)\ ½mv^2$$

Conservation of energy becomes:

$$½mv^2 = (0.77)½kL_0^2 + (0.23)½mv^2$$

Solving for k / m:

$$k / m = [(0.77)\ v^2] / [(0.23)^2 L_0^2]$$

$$k / m = 819\ \text{s}^{-2}$$

$$k / m = 0.8\ \text{kN m}^{-1}\ \text{s}^{-1}$$

14. C is correct.

$$PE = mgh$$

$$PE = (21 \times 10^3\ \text{kg})\cdot(9.8\ \text{m/s}^2)\cdot(2.6 \times 10^3\ \text{m})$$

$$PE = 535\ \text{MJ}$$

15. D is correct.

$$h_0 = 3.5\ \text{m}$$

$$PE_0 = mgh$$

$PE_1 = mgh(0.69) \rightarrow$ after first bounce

$PE_2 = mgh(0.69)^2 \rightarrow$ after second bounce

$PE_3 = mgh(0.69)^3 \rightarrow$ after third bounce

Because mass and gravity are constant, the final height is:

$$\text{final height} = h(0.69)^3$$

$$\text{final height} = (4\ \text{m})\cdot(0.69)^3$$

$$\text{final height} = 1.31\ \text{m} = 131\ \text{cm}$$

Notes for active learning

CHAPTER 5

Waves & Periodic Motion

Page intentionally left blank

Amplitude, Period and Frequency

Periodic motion

Periodic motion (i.e., sinusoidal motion) is a motion that repeats. If an object moves back and forth over the same path at a constant speed, each cycle takes the same time; the resulting motion is periodic. In the periodic motion, this back-and-forth activity is *oscillation* or *vibration.*

Many systems display oscillatory or vibratory motion; there are several types of periodic motion as well. Periodic motion can be damped (e.g., friction is present), undamped, driven (an outside force is acting upon the system), or free.

Regardless of the motion, any oscillatory system has several inherent characteristics that define its periodic motion. These characteristics are *amplitude*, *frequency*, and *period.*

Most questions refer to several oscillatory systems but only one motion:

undamped and free periodic motion (simple harmonic motion)

Amplitude and frequency

Amplitude (A): In periodic motion, *amplitude* expresses the displacement of the system about its equilibrium (center or resting) position.

Amplitude may have different units depending on the system.

Frequency (f): *Frequency* is the rate of oscillation, or the number of cycles per second, where a cycle is one complete vibration (when the displacement is zero).

Frequency is usually measured in Hertz (Hz) and calculated by:

$$f = \frac{\#\ of\ cycles}{time}$$

Angular frequency

In some situations, the frequency is expressed as angular (ω).

Angular frequency is related to frequency; however, angular frequency is measured in radians per second (rad/s).

Angular frequency ω is:

$$\omega = 2\pi f$$

Period

Period (T) is the time required to complete one cycle or oscillation.

Period is often measured in seconds and calculated by:

$$T = \frac{time}{\#\ of\ cycles}$$

Period is sometimes measured in *degrees* or *radians.*

One rotation around a circle requires 360° or 2π radians.

Period and frequency relationship

The 360° (or 2π radians) value represents one complete oscillating system cycle.

From the above equation, the period (T) and frequency (f) are inversely related.

The relationship between the period and frequency is expressed as:

$$T = \frac{1}{f}$$

Displacement as a Sinusoidal Function of Time

Simple harmonic motion

An object or system exhibiting *simple harmonic motion* (SHM) is a particular case of a periodic motion that is undamped and free.

No force of friction is present, and no external, time-dependent force is acting upon it.

The system requires a *restoring force* (not a driving force), which must be proportional to the displacement (i.e., amplitude).

Proportionality allows the system to oscillate about its equilibrium position while retaining the same amplitude.

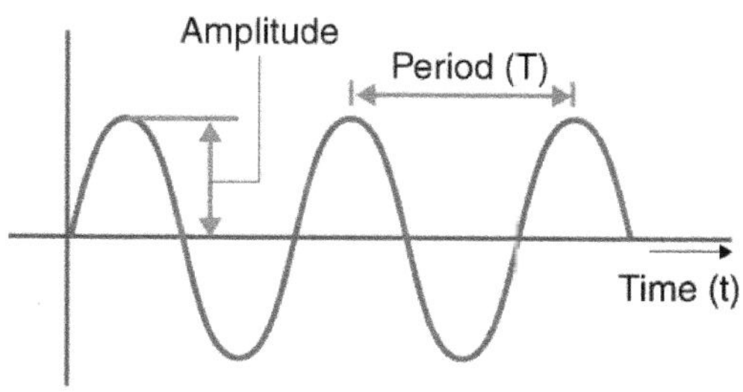

Height vs. time graph of simple harmonic motion, noting amplitude and period

Amplitude is the distance between the resting position and the maximum wave displacement.

Frequency (f) is the number of waves passing by a specific point per second.

Period (T) is the time for one complete wave cycle.

Graphing SHM

Simple harmonic motion (SHM) can be depicted as a sine or cosine graph:

If the motion starts at *equilibrium*, use *sine*

If the motion starts at the *maximum or minimum*, use *cosine*.

The system oscillates about its equilibrium position with the same amplitude and a fixed period (and thus constant frequency).

Regardless of the magnitude of the amplitude, the period and frequency *do not change*.

In simple harmonic motion, frequency and period are *independent* of amplitude.

Sine function

An object's simple harmonic motion may be determined using the equation:

$$x = A \sin(\omega t)$$

$$x = A \sin(2\pi f t)$$

where x is displacement, A is amplitude, t is time, ω is the angular frequency, and f is frequency.

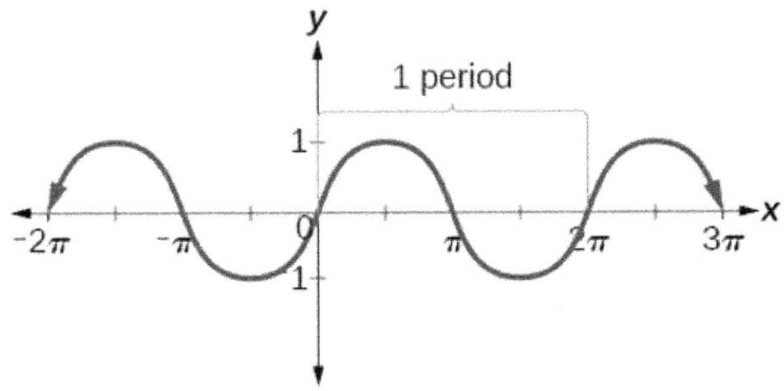

Motion starts at equilibrium for y = sin(x) graph

Cosine function

Sometimes, the equation is written as:

$$x = A \cos(\omega t)$$

$$x = A \cos(2\pi f t)$$

Sine or cosine may be used interchangeably if the translation of these graphs is accounted for.

A cosine graph has a y-intercept of A (maximum amplitude), whereas a sine graph has a y-intercept of 0 (equilibrium). These are translations, but the translations must be accounted for in the equation. Otherwise, the solution will be incorrect.

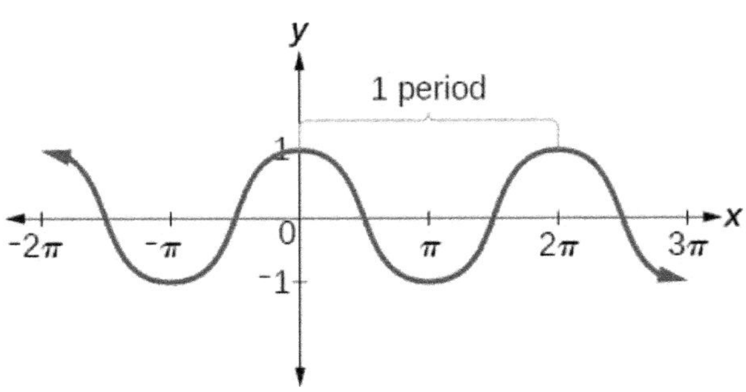

Motion starts at maximum or minimum for y = cos(x) graph

Hooke's Law

Oscillatory motion of a mass–spring system

Oscillatory motion of a mass–spring system exhibits *simple harmonic motion* (SHM).

When a mass is attached to a spring, the only forces acting upon it are the tension and weight of the mass. As such, there is no driving force, which is assumed to be completely undamped.

An essential characteristic of a mass–spring system is the *spring constant*. Robert Hooke (1635-1703) discovered that, given a specific spring, a constant can be found when different masses are attached, and the equilibrium position is measured.

This is an inherent constant of all springs and relates the spring's displacement (change in length) to the restoring force the spring exerts.

Hooke's Law

The relationship is Hooke's Law and is expressed as:

$$F = -k\Delta x$$

where k is spring constant measured in (N/m), Δx is displacement, and F is restoring force (N).

Restoring force

Restoring force is responsible for restoring the spring's original size and shape.

The minus sign in Hooke's Law is by convention, and the restoring force is in the opposite direction to the displacement (because force and displacement are vectors).

For example, an idealized spring exerts a force proportional to the spring's deformation from its equilibrium length, exerted in a direction that opposes the deformation.

Ideal spring

In the diagram below, a spring is attached to a wall, and the other side is attached to a mass (m).

For an ideal system (i.e., no friction present), when the mass is displaced and the spring is stretched, the direction of the restoring force is opposite to the displacement and acts to the left (middle image).

Conversely, if the spring were compressed the same distance, the restoring force would be equal in magnitude but in the opposite direction (bottom image).

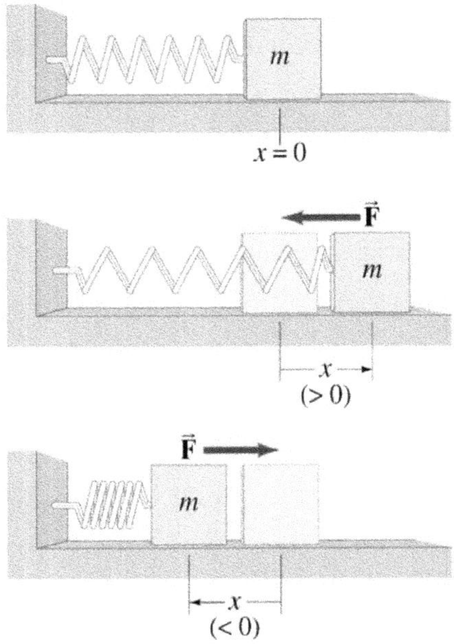

Restoring spring force (F) is opposite to the displacement, toward equilibrium

Vertical springs

If the spring is hung vertically (shown below), the only change is in the equilibrium position, which is now at a point where the restoring spring force is equal and opposite the weight of the mass.

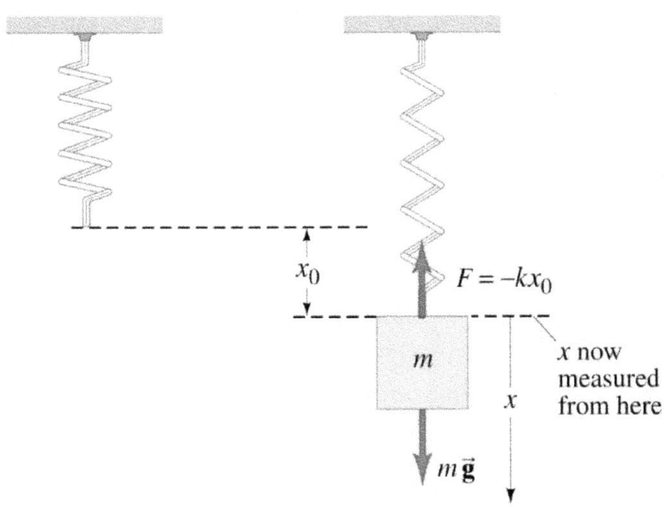

Restoring spring force is equal and opposite the weight of the mass

Acceleration of mass

Hooke's Law ($F = -k\Delta x$) can be used to find the acceleration of the mass attached to the spring.

Regardless of whether the mass–spring system is horizontally or vertically positioned, forces always follow Newton's Second Law ($F = ma$), therefore (assuming no friction):

$$| ma | = | k\Delta x |$$

$$| a | = \left| \frac{k\Delta x}{m} \right|$$

where k is the spring constant measured in (N/m), and Δx is displacement.

Hooke's Law demonstrates that mass–spring systems are examples of simple harmonic motion.

Restoring force returns the spring to *equilibrium* when disturbed from the equilibrium position.

Restoring force (F) is proportional to the stretched or compressed distance, which makes the system a simple harmonic.

Amplitude of oscillation

Amplitude (A) of this oscillation is the maximum displacement (Δx):

$$A = \Delta x$$

For example, if a mass–spring system is oriented vertically, and the mass is displaced a certain distance, the resulting motion follows a simple harmonic oscillator.

A simple harmonic oscillator is seen if a pen is attached to a mass and a roll of paper is moved across the surface of the pen during the motion of the mass:

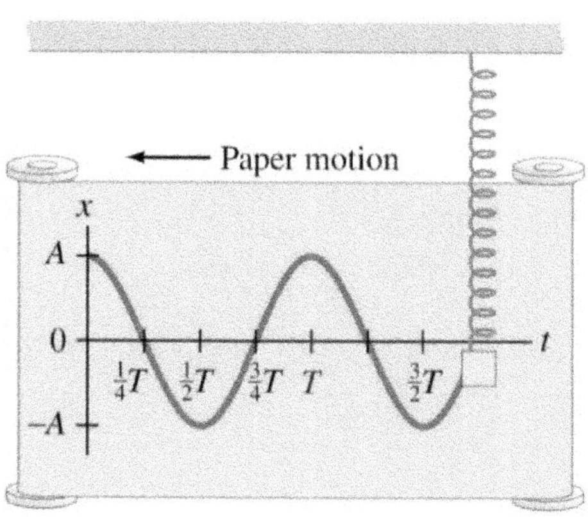

$x = A\ cos\ (\omega t)$

$F = -Ka$

Simple harmonic motion from the motion of a mass

Angular frequency

In simple harmonic motion of a mass–spring system, *angular frequency* (ω) and *frequency* (f) of the system can be calculated by:

$$\omega = \sqrt{\frac{k}{m}}$$

$$f = \frac{\sqrt{k/m}}{2\pi}$$

or

$$f = 2\pi\sqrt{\frac{m}{k}}$$

where k is the spring constant measured in (N/m).

Period

Period (T) can be calculated as:

$$T = \frac{2\pi}{\sqrt{k/m}}$$

or

$$T = 2\pi\sqrt{\frac{m}{k}}$$

where k is the spring constant measured in (N/m).

Mass and spring constants determine f and T

A mass–spring system's frequency (f) and period (T) are based on the mass and spring constant.

It is important to note that a vertically angled mass–spring system has the same *frequency* (f) and period (T), regardless of the value for (g) gravitational acceleration.

Frequency (f) and period (T) do not change, even if the system were on the moon or in space.

Energy of a Mass–Spring System

Kinetic and potential energy of oscillating mass

An oscillating mass accelerates, decelerates, changes direction, and repeats.

An oscillating mass contains motion (*KE*) and stationary (*PE*) energy (velocity is zero at the peaks and troughs of movement).

PE and *KE* of a mass on a spring:

Potential energy:

$$PE = \frac{1}{2}kx^2$$

Kinetic energy

$$KE = \frac{1}{2}mv^2$$

Equilibrium and displacement

When the mass is in motion and passes through the equilibrium position, the displacement is zero, as is (PE). However, kinetic energy will be at its maximum because the mass is at its maximum velocity. This occurs at the point where the net force is zero.

At maximum displacement, x equals the amplitude (A). Potential energy is at its maximum, and KE is zero (because the velocity is zero as the object changes its direction of motion). This occurs at the moment of maximum force, which is also the point of *maximum acceleration* of an oscillating particle.

System energy is constant

At any point along its motion, PE + KE of the particle will be the same value; energy (potential plus kinetic) is constant.

The sum is the same as the total energy in the system:

Total Energy = *E*

$$E = PE + KE$$

$$E = \frac{1}{2}kx^2 + \frac{1}{2}mv^2$$

Kinetic and potential energy relationship

Kinetic energy is zero when potential energy is at its maximum (at the amplitude). All energy in the system must equal only the potential energy.

The same relationship applies when kinetic energy is at its maximum:

$$PE_{max} = KE_{max} = E = Total\ Energy$$

Maximum potential and kinetic energy are summed by:

$$PE_{max} = \frac{1}{2}kA^2$$

$$KE_{max} = \frac{1}{2}mv^2 \qquad \text{at } x = 0$$

Maximum compression

In the following diagram, a spring has been compressed but not in motion.

Energy is stored PE:

$$PE_{max} = \frac{1}{2}kA^2 \text{ when } v = 0$$

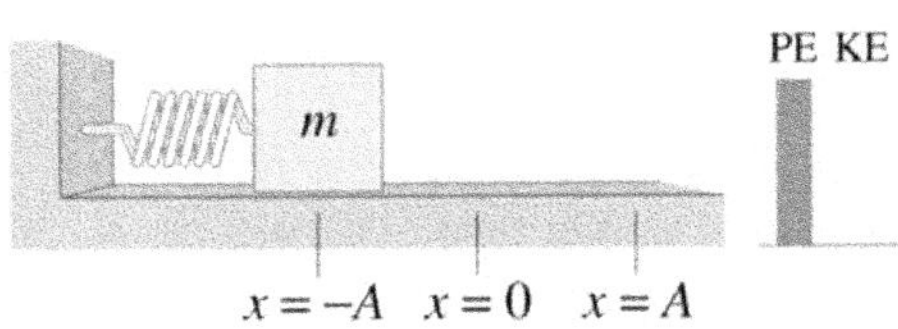

Compressed stationary spring has only potential energy; E = ½ kA²

Maximum velocity

As seen below, if the mass (m) can move, the spring expands toward equilibrium.

Once it has reached its equilibrium position, all PE is transformed into KE:

$$KE_{max} = \frac{1}{2}mv^2$$

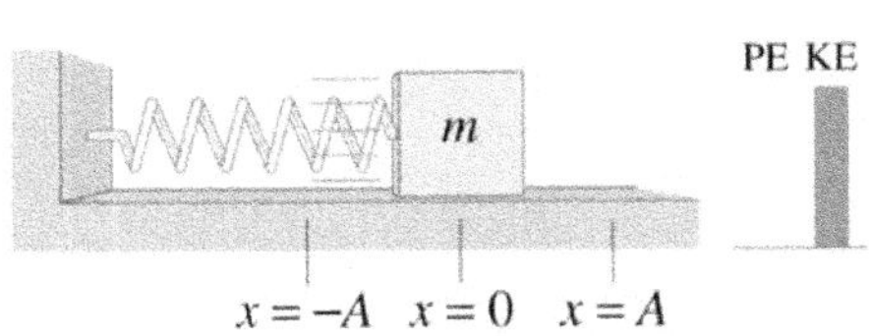

Moving spring has only kinetic energy at the equilibrium position; $E = \frac{1}{2} mv^2_{max}$

Maximum amplitude

In the following diagram, the mass (m) reaches the maximum stretched amplitude.

Kinetic energy (KE) is converted back to potential energy (PE):

$$PE_{max} = \frac{1}{2}kA^2 \text{ when } v = 0$$

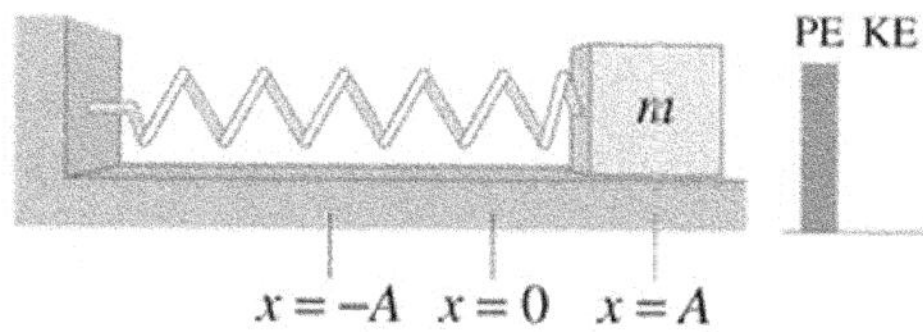

At maximum stretched amplitude, spring has only potential energy; $E = ½\ kA^2$

Exchanging potential for kinetic energy

The mass (m) then moves back toward equilibrium but does not yet reach equilibrium position.

Therefore, most energy is kinetic (KE), but some is stored as potential energy (PE).

In the following diagram, as the mass (m) gets closer to the equilibrium position, PE decreases, and KE increases:

$$KE_{max} = \frac{1}{2}mv^2 + PE_{max} = \frac{1}{2}kx^2$$

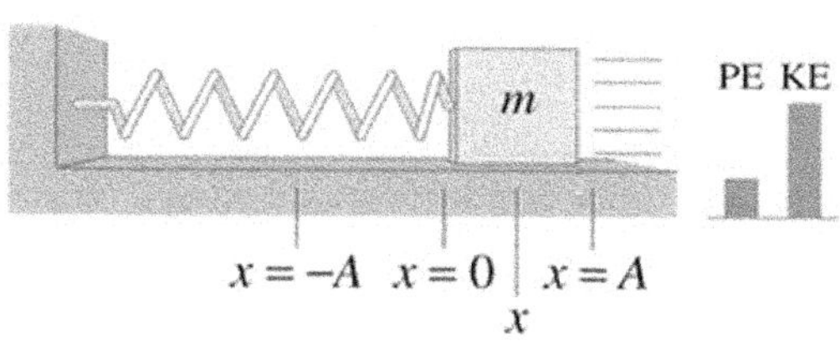

Mass approaching equilibrium position converts PE to KE; $E = ½\ mv^2 + ½\ kx^2$

Calculating velocity during motion

Velocity of the mass (m) on a spring can be calculated at various points during its motion.

Set the maximum PE and KE equal; the maximum velocity of mass (m) is:

$$v_{max} = \sqrt{\frac{k}{m}} \times A$$

where k is the spring constant measured in (N/m), Δx is displacement, and A is amplitude.

Velocity can be solved as a function of position:

$$v = \pm v_{max}\sqrt{1 - \frac{x^2}{A^2}}$$

Motion of a Pendulum

Simple harmonic motion of pendulums

Thus far, the examples have been oscillating masses attached to springs. This chapter discusses the simple harmonic motion of a pendulum.

A pendulum undergoing simple harmonic motion (i.e., simple pendulum) consists of a mass at the end of an assumed massless cord or rod.

To be in simple harmonic motion, the restoring force on the pendulum must be proportional to the negative of the displacement.

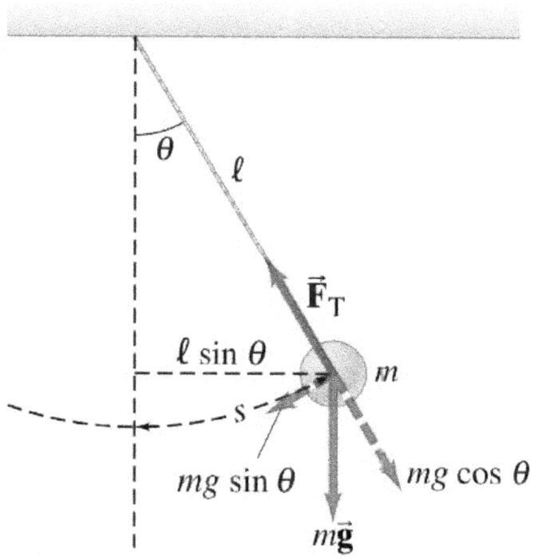

Angle θ measures the displacement of the pendulum

Restoring force

The restoring force is:

$$F = -mg\ sin\ \theta$$

$$|\ a\ | = |\ g\ sin\ \theta\ |$$

For small displacement angles, sin $\theta \approx \theta$.

Thus, the restoring force becomes:

$$F = -mg\theta$$

$$|\ a\ | = |\ g\theta\ |$$

Frequency

Unlike mass–spring systems, a simple pendulum's frequency (*f*) depends on the acceleration due to gravity (*g*).

Frequency (*f*) of a simple pendulum depends on the length of the cord attached to the mass:

$$\omega = \sqrt{\frac{g}{l}}$$

$$f = \frac{\sqrt{\frac{g}{l}}}{2\pi}$$

where g is the acceleration of gravity (9.8 m/s^2), and l is the length of the cord (or rod).

Period

Period (*T*) of a simple pendulum's oscillation is:

$$T = 2\pi\sqrt{\frac{l}{g}}$$

where g is the acceleration of gravity (9.8 m/s^2), and l is the length of the cord (or rod).

Pendulum total energy

Like mass–spring systems, simple pendulums conserve energy.

Total energy of the pendulum oscillation is the sum of potential energy and kinetic energy:

$$E = PE + KE$$

Unlike the mass–spring system, *PE* is dependent upon gravity:

$$PE = mgh$$

where h is the height above the equilibrium position.

Maximum potential and kinetic energy

Maximum *PE* is when the height (or x) equals the amplitude.

$$PE_{max} = mgl(1 - \cos\theta)$$

Maximum *KE* is the same for a simple pendulum as it is for mass–spring systems:

$$KE_{max} = \frac{1}{2}mv^2$$

Waves

Waves transport energy

Waves are periodic disturbances that transport energy (but not matter) through a medium.

For waves, energy can move from one location to another, yet the particles of matter in the medium return to their fixed position.

Examples of waves include light, sound, and ocean waves breaking on the beach. While different, they share characteristics intrinsic to all waves.

Wave characteristics

Like periodic motion, waves are defined by:

- Amplitude, A
- Frequency f and period T
- Wavelength, λ
- Wave velocity

Amplitude, frequency, and period were discussed earlier; their definition has not changed, only the context of what they describe.

Amplitude represents energy, and the magnitude is not dependent upon frequency or period.

Wavelength (λ) is the distance of one wave cycle and is measured in meters (m).

Wave velocity (v) is the speed and direction of the wave.

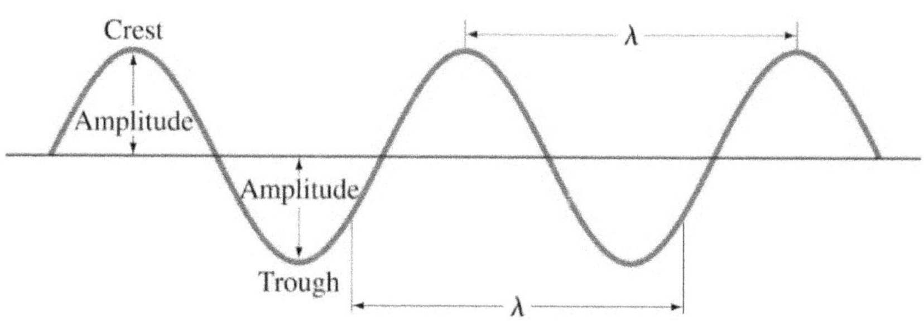

Simple wave with characteristic notions for crest, trough, amplitudes, and wavelength

Wavelength (λ) and *wave velocity* (v) are related to the frequency (f) and period (T) by:

$$f = \frac{v}{\lambda}$$

$$T = \frac{1}{f}$$

$$\frac{1}{f} = \frac{\lambda}{v}$$

$$T = \frac{\lambda}{v}$$

where v is the wave velocity (units of m/s), and λ is the wavelength (units of m).

Transverse, Longitudinal and Surface Waves

Wave motion

Three main types of wave motion are *transverse*, *longitudinal*, and *surface waves.*

These have common characteristics but differ in the motion of particles within the wave.

Waves do not transport particles. Instead, as the wave passes through a particle in a medium, the particle undergoes a specific periodic motion.

Transverse and longitudinal waves are shown for a hand moving a slinky (figures below).

Transverse waves

Transverse waves oscillate particles with motion *perpendicular* to the wave direction.

Transverse waves have *crests and troughs.*

Identify what is oscillating and in what direction. Imagine there is a mark on one of the crests.

Imaginary marks move *up and down* while energy travels from left to right through the medium.

Wave displacement is *perpendicular* to the direction of the wave motion for transverse waves.

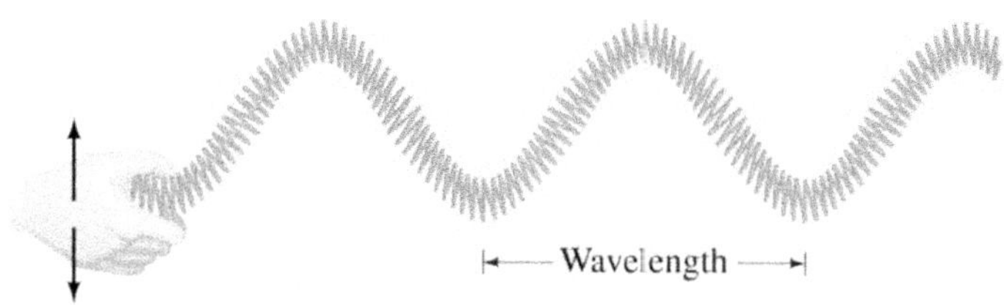

Particles are displaced perpendicular to wave direction

Longitudinal waves

Longitudinal waves oscillate particles in motion *parallel* to the wave direction.

Longitudinal waves have *compressions and expansions,* with particles moving closer and then farther apart.

Identify what is oscillating and in what direction by imagining a mark on one of the crests.

As a hand pushes the slinky through the medium from left to right and back, the imaginary mark travels *left to right* and vice versa along with the propagated wave.

There is *no vertical motion* along the wave; displacement is *parallel* to the direction of motion.

Compression and expansion only occur in longitudinal waves.

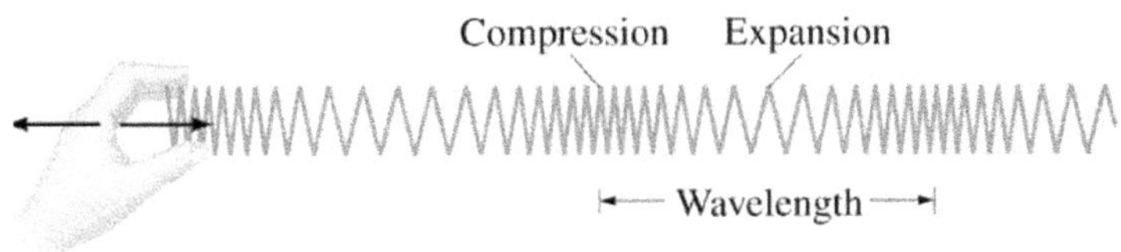

Particles are displaced parallel to the wave direction for longitudinal waves

Surface waves

Surface wave particles oscillate in a circular motion rather than vertically or horizontally.

Surface waves only occur at the interface of two mediums, where wave travels in only one medium.

For example, surface waves occur in the ocean. The diagram below shows a wave propagating through the ocean with a velocity *v*.

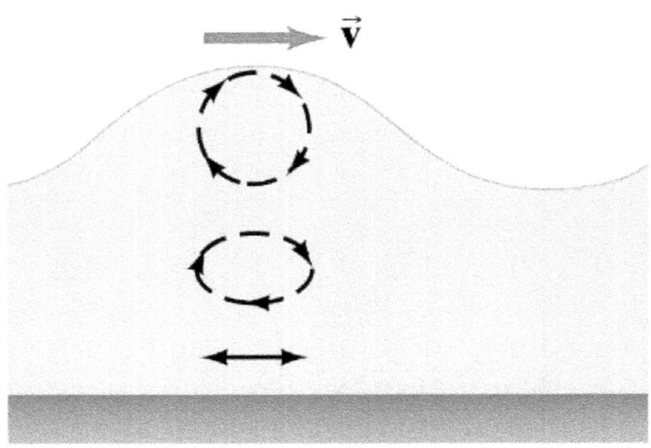

Surface waves oscillate in a circular motion

For surface waves, the interface of the two mediums (e.g., air and water) causes the particles to move in a circular pattern (illustrated above).

Deeper in the ocean, particles are further from the interface, and motion resembles an ellipse.

At greater depths, the particle motion is only horizontal, and the wave behaves as longitudinal.

For surface waves, the further (i.e., deeper) from the interface of the two mediums, the more longitudinal the wave becomes.

Phase, Interference and Wave Addition

Phases

Phases occur when two waves interact. A phase difference between two waves means that the crests and troughs do not occur at the same points.

The diagrams below have three waves: one in-phase and two out-of-phase:

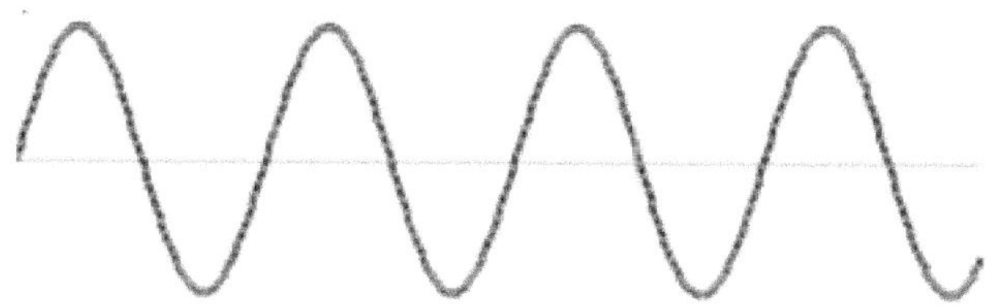

In-phase waves have crests and troughs at the same point

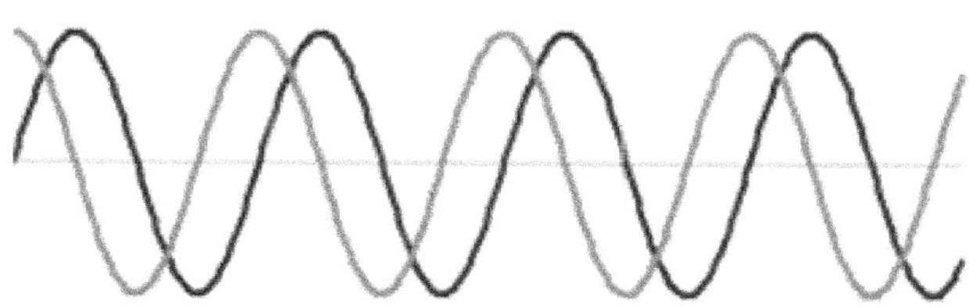

Out-of-phase waves have crests and troughs at different points

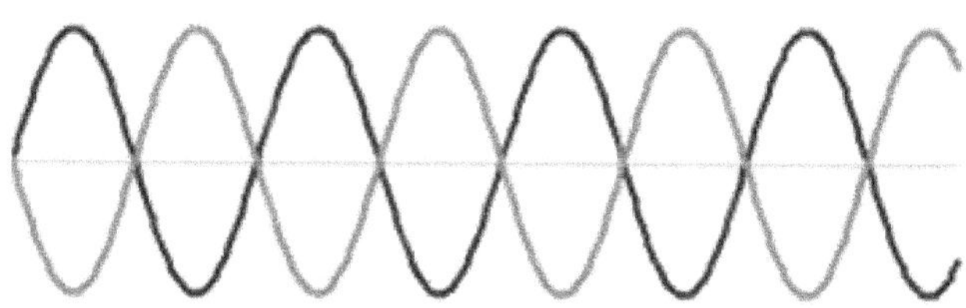

Completely out-of-phase waves have crests and troughs at different points

Interference

The phase of two waves is important because it gives rise to interference effects.

Interacting waves combine to create a combination wave, the *amplitude* which depends on the interaction between the component waves.

The resulting *amplitude* of the interference depends upon the phase of the waves:

constructive interference (in-phase) and

destructive interference (out-of-phase).

Constructive interference

In the diagram, two transverse waves of equal magnitude travel toward each other along a rope.

When waves meet and interfere, they combine.

In-phase occurs when waves are 0 or 2π radians (0 or 360°) apart in oscillations.

Peaks and troughs of one wave are aligned with the peaks and troughs of another.

Amplitude of the resulting wave is the sum of the amplitudes of the interfering waves.

When in-phase waves meet, they experience *constructive interference* and produce a maximum amplitude (middle panel below) as the sum of the magnitude of the incoming waves.

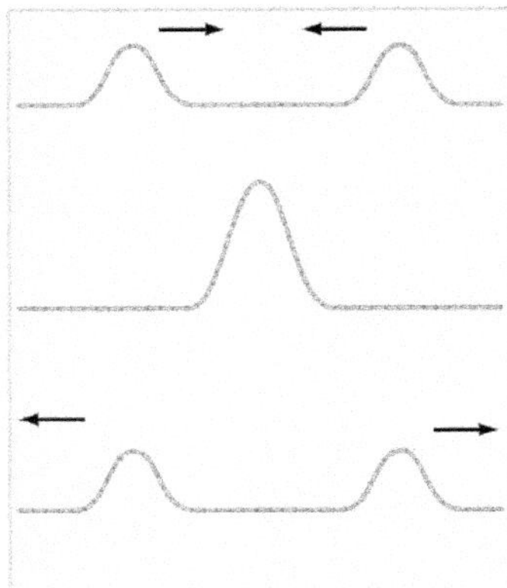

Three-panel diagram of in-phase transverse waves of equal magnitude meeting

Constructive interference:

$$A_1 + A_2 = A_{net}$$

Destructive interference

In the diagram, two transverse waves of equal magnitude travel toward each other along a rope.

The amplitude of the resulting wave is the sum of the amplitudes of interfering waves.

Out-of-phase occurs when waves are not 0 or 2π radians (0 or 360°) apart in oscillations.

When entirely out-of-phase equal waves meet, they experience *destructive interference* and produce no amplitude (middle panel below).

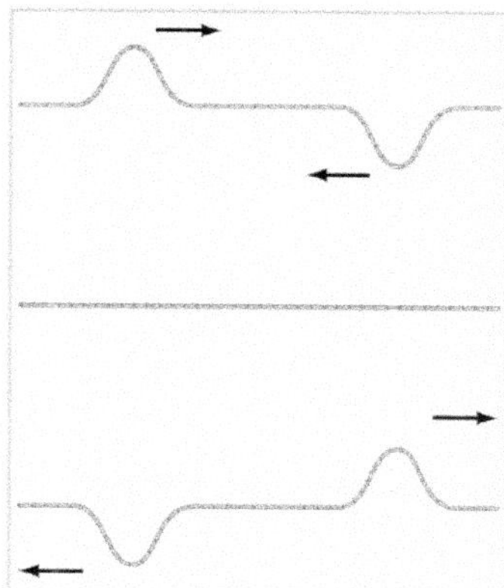

Three-panel diagram of out-of-phase transverse waves of equal magnitude meeting

Out-of-phase waves produce destructive interference, with the resulting amplitude of the combined waves being less than the sum of the interfering waves.

$$0 < A_{net} < A_1 + A_2$$

Completely out-of-phase waves

Completely out-of-phase occurs when the waves are π radians (180°) apart in their oscillations, and the crests of one wave are aligned with the troughs of the other wave.

When two waves are completely out of phase, the amplitude of the resulting wave is the absolute value of the difference between the two separate amplitudes.

If the waves have the same amplitude, the interfering waves completely destructively interfere, and no wave exists (amplitude = 0).

$$A_{net} = | A_1 - A_2 |$$

The diagram below illustrates two waves that are not entirely out-of-phase.

The resulting waves destructively interfere to produce a wave with an amplitude less than that of the wave with a greater amplitude.

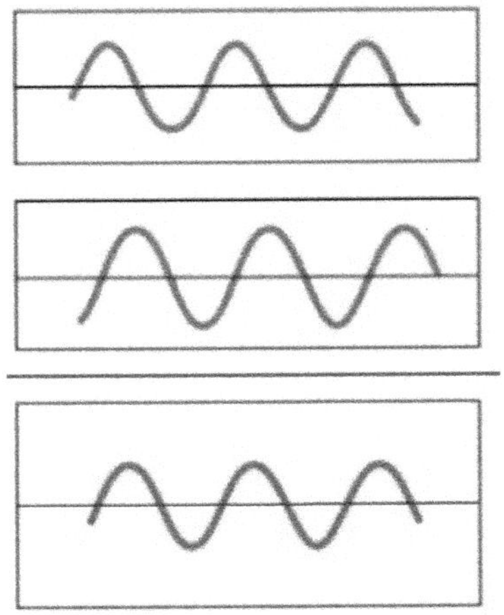

Waves not completely out-of-phase combine for resulting wave with intermediate amplitude

Reflection and Transmission

Reflection

Reflection occurs when waves bounce off the surface of an object or medium.

All waves exhibit reflection. Examples include reflection of light off a mirror to see an image, echo produced by the reflection of sound waves and glare from sunlight reflecting off water.

Reflections occur when a wave encounters an obstacle or an interface between two different mediums; the wave is forced to change direction.

Reflections of waves depend upon the object or medium the wave reflects off.

Specular and diffuse reflections

Reflection of light can be categorized into specular and diffuse reflection.

Specular reflection is light reflected from a smooth surface at a definite angle.

Diffuse reflection is produced by rough surfaces that reflect light in all directions.

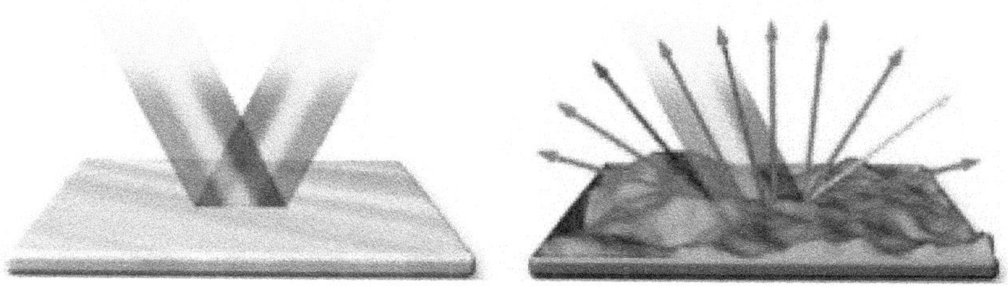

Specular reflection (on the left) and diffuse reflection (on the right)

Inverted or upright reflections

Imagine a wave traveling down a string, as in the figure below.

In the left diagram, the wave hitting an immobile object is reflected, but its reflection is inverted.

In the right diagram, the end of the string is connected to a point, so it is free to move. The wave on the string is reflected in the opposite direction, but its reflection is upright (not inverted).

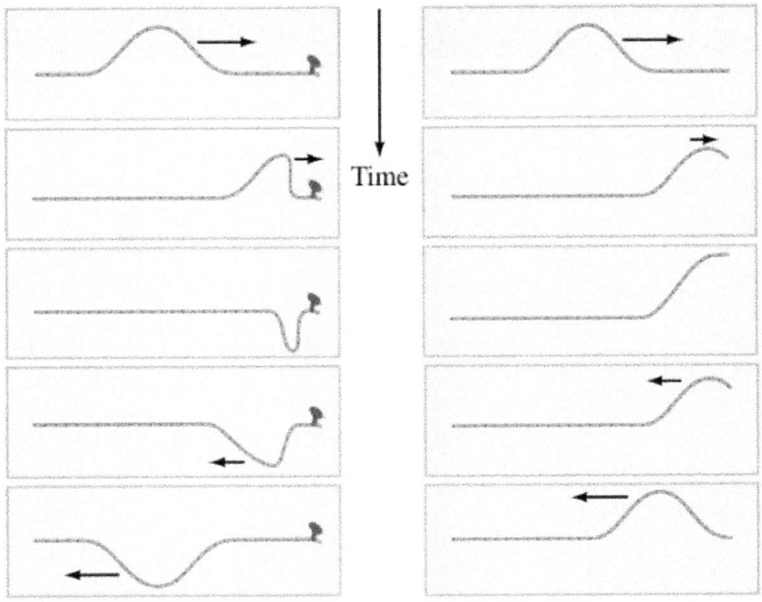

Inverted reflection from an immobile object (on the left) compared to unrestrained reflection

Law of Reflection

Law of Reflection governs light reflection off surfaces with interfaces of two mediums. It states that *the angle of incidence of the incoming light ray is equal to the angle of reflection of the outgoing light ray concerning the normal line of the reflecting surface.*

$$\theta_{incident} = \theta_{reflected}$$

Ideal reflectors obey the Law of Reflection; thus, the angle of reflection equals the angle of incidence, allowing an observer to see the source of the light (e.g., a mirror).

Billiards is an example, in which the balls are waves and the edges of the table surface from which they reflect.

Reflection angle of the ball off an edge equals the incoming angle of the ball relative to the normal (perpendicular) of the surface.

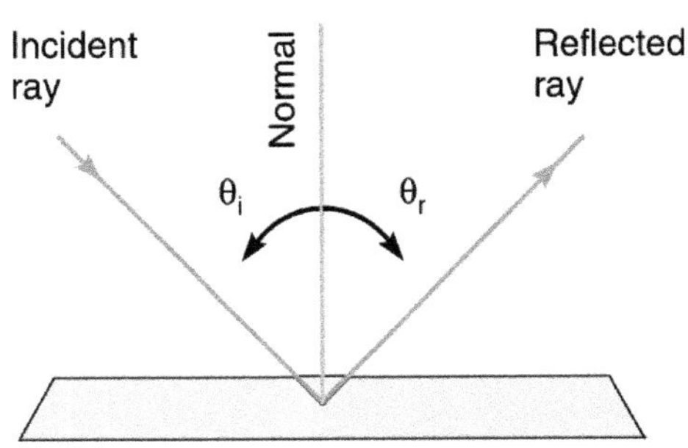

Incident angle of light equals the reflected angle

Law of Reflection only applies to smooth surfaces, which are assumed to be ideal reflectors.

Plane mirrors are modeled as ideal reflectors.

If the surface is rough and not smooth, the angle of reflection of one wave is not equal to the angle of reflection of the other wave, even if they left the same point at the same angle. Rough surfaces scatter incoming waves.

The rough surface scatters the incoming light rays, and the angle of reflection is not equal to the angle of incidence. The observer would not see the light source but only the surface of the rough-surfaced reflector.

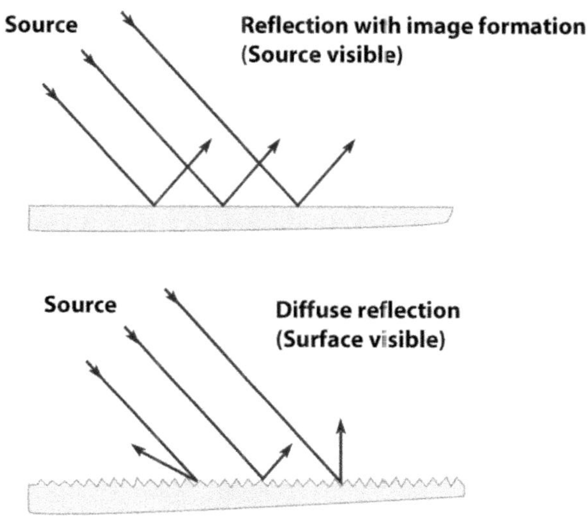

Smooth surfaces reflect light at same angle of incident while rough surfaces do not

Transmission

Transmission occurs when a wave encounters a different medium, and some or all wave energy passes (transmitted) into a different medium.

Transmission is usually not total.

Incoming waves are usually partially reflected from an interface of two different mediums, while the remaining portion is transmitted through the new medium.

Transmission and reflection of water are examples.

When light travels through the atmosphere and encounters the surface of a lake, some of the light is reflected, but some is transmitted through the water.

Glare on lakes indicates reflection.

However, when a person enters the water, it becomes see-through, indicating light transmission.

Consider two strings of different densities attached (shown below).

For a wave traveling along the lower-density section, it both reflects off and transmits some of its energy to the higher-density section.

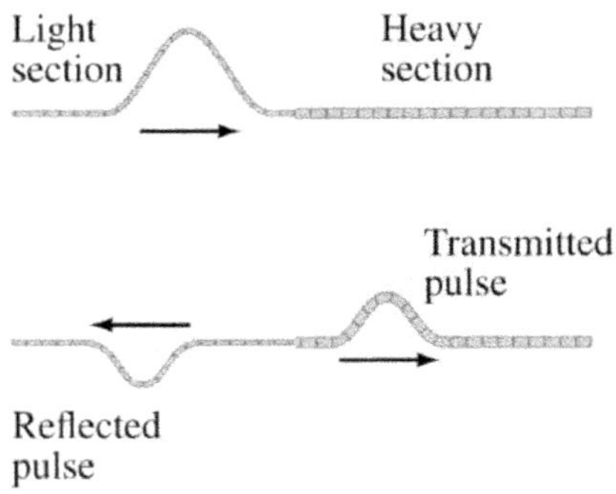

Two strings with different densities transfer some energy from lighter to heavier density string

Amplitude and intensity

Like the oscillation that starts the wave, the energy transported by a wave is proportional to the square of the amplitude.

Recall that the amplitude is the maximum height of a crest or the maximum displacement from the equilibrium position.

Amplitude correlates with the energy of the wave.

A greater amplitude means the wave has higher energy.

A lesser amplitude means the wave has lower energy.

Intensity is the energy per area per time, which translates to power (Watts) per area (m^2).

$$I = \frac{P}{A}$$

Thus, amplitude and intensity are correlated similarly to amplitude and energy.

A higher amplitude (and therefore greater energy) leads to a higher intensity.

Refraction and Diffraction

Refraction

When a wave encounters a boundary between two mediums or enters a medium where the wave speed differs, the wave is refracted.

Specifically, when waves are transmitted from one medium to another, the different physical properties cause the light to bend and change direction within the new medium. This is *refraction* and is commonly observed in light waves.

Refraction is possible with any wave that encounters such a boundary.

For light, the speed at which the wave travels is the *light propagation speed* and depends upon the medium through which light propagates.

In a vacuum, light travels at a constant 3×10^8 m/s.

In different mediums (e.g., glass or water), light propagating speed is less than in a vacuum.

Direction in which the light bends during refraction depends on the light speed of the mediums.

Snell's law

Snell's Law expresses the direction in which the light bends when traveling through mediums with different wave propagation speeds.

Snell's Law relates the *speed and angle* of the incoming wave to that of the refracted wave:

$$n_1 \sin(\theta_1) = n_2 \sin(\theta_2)$$

where n is index of refraction, and θ is angle of the wave concerning the normal of the medium interface.

Index of refraction is the relation of the light propagation speed in a medium to that of the speed of light propagation in a vacuum.

n is always greater than 1 because light cannot travel faster than in a vacuum.

$$n = \frac{c}{v}$$

where n is index of refraction, c is the speed of light in a vacuum, v is wave propagation speed for that medium.

Indices of refraction

Substance	Index of refraction	Light speed
Air	Approx. 1	~c
Water	1.333	0.75c
Glass	1.5	0.67c
Diamond	2.4	0.42c

Indices of refraction (n) for some common materials. c is the speed of light in a vacuum

Waves moving to a more dense medium (higher refractive index) bend *toward the normal.*

Waves moving to a less dense medium (smaller refractive index) bend *away from the normal.*

Refraction between mediums

Light refracts from water to air and bends *away* from the normal.

Light refracts from air to water and bends *toward* the normal.

Given the angle of incidence, the angle of refraction is computed by Snell's Law.

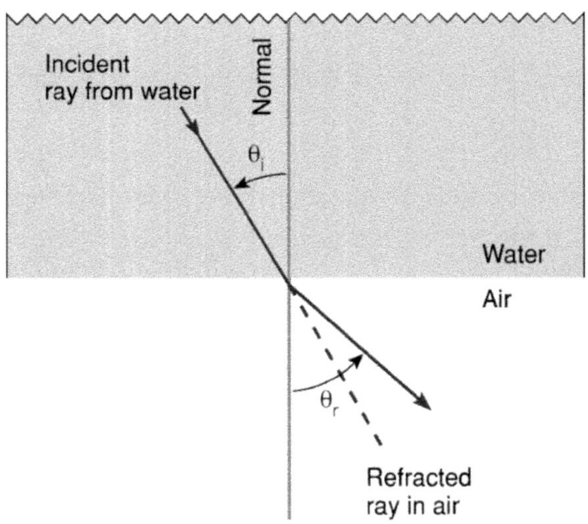

Refraction of light from water to air

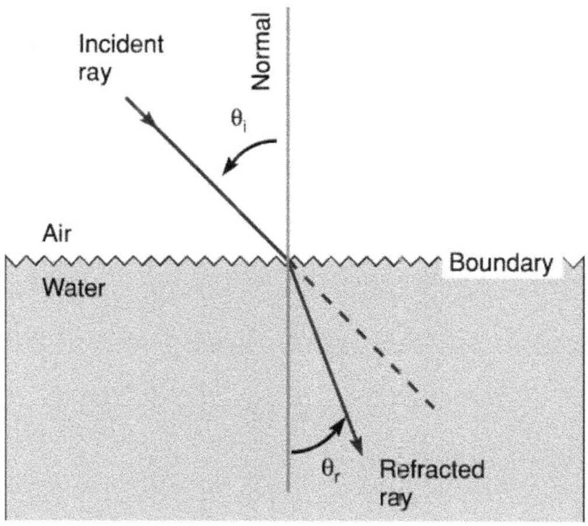

Refraction of light from air to water

Prisms

Consider the refraction of white light through a prism. The prism is an exciting example of refraction because it separates the mixture of colors in white light into a rainbow.

Observing a rainbow through a prism is possible because the index of refraction (n) varies with speed (and wavelength).

Different wavelengths refract at different angles. Differences in refraction produce assorted colored light (e.g., a rainbow), where refraction occurs in suspended water droplets instead of through a prism.

Violet light has the shortest wavelength for visible light and is most refracted, skies as blue.

Red light has the longest wavelength for visible light and is refracted the least. When the angle of refraction is greater, red sunsets are observed in the sky.

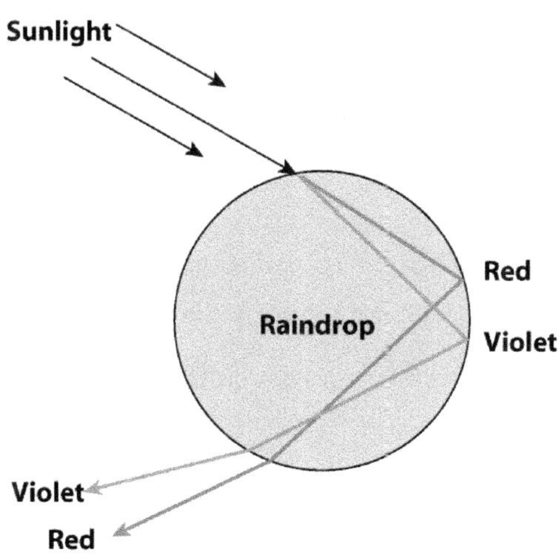

Visible light wavelengths refract at different angles

Critical angle

An essential aspect of refraction is the *critical angle.*

Critical angle is the greatest angle a wave can make with the boundary between two mediums and not be refracted into the second medium.

At the critical angle, a wave is refracted *parallel* to the surface of the interface boundary.

Angles > than the critical angle causes the incident wave to reflect off the boundary completely.

Thus, at angles greater or equal to the critical angle, none of the waves passes through the boundary; there is *total internal reflection.*

Total internal reflection includes fiber optics and gemstone brilliance.

$$n_1 \sin(\theta_{critical}) = n_2 \sin(90°)$$

Diffraction

Diffraction occurs when a wave encounters an obstacle in the medium it travels.

A wave spreads out (or dissipates) to move around the obstacle or through an opening (*aperture*) in the obstacle, producing a "*shadow region.*"

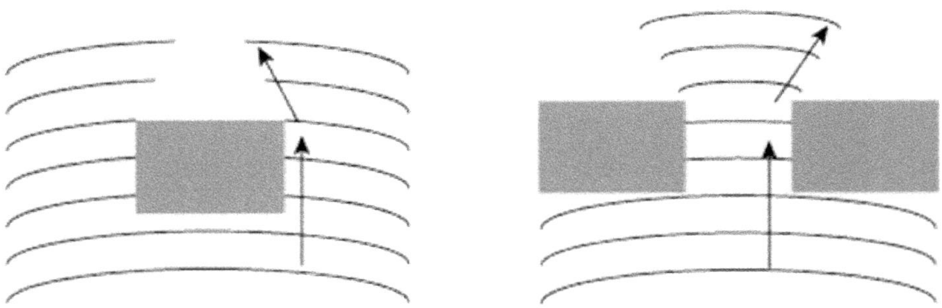

Diffraction around obstacle (left) vs. diffraction through aperture (right)

Some types of waves diffract more efficiently than others. For example, consider a sound wave compared to a light wave.

Music (i.e., sound waves) can be heard from around the corner of a building, but the speaker (light waves as the visual image) emitting cannot be seen around the corner.

Conversely, shining light through a hole does not produce a dot but a diffused circle of light.

Wavelength determines diffraction patterns

Diffraction depends on how the wavelength and obstacle size (or aperture) compare.

If the opening is *larger* than the wavelength, diffraction is *minimal*, and mostly, the wave continues as before.

If the opening is much *smaller* than the wavelength, the diffraction is noticeable, and the wave bends into a *circular orientation.*

If the wave encounters an obstacle, its reaction is the opposite. For example, the figures below depict different waves diffracting off different objects.

If the obstacle is much smaller than the wavelength, the wave barely diffracts, as in the diagram for water waves passing blades of grass (top left).

If the object is comparable to or larger than the wavelength, diffraction is more significant. This is seen in the diagrams for a stick in the water (top right), short-wavelength waves passing a large log (bottom left), or long-wavelength waves passing the same log (bottom right).

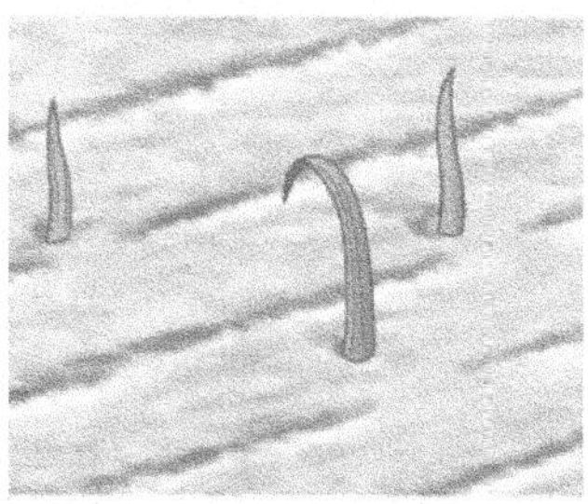

Water diffraction passing a blade of grass with minimal diffraction

Large objects produce significant diffraction.

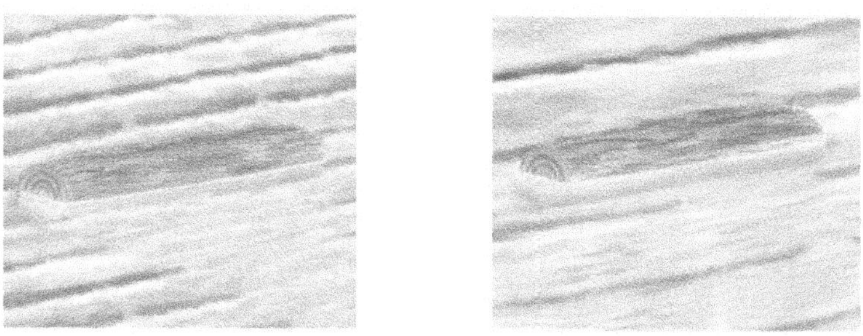

Short-wavelength waves passing log (on the left) compared to long-wavelength waves

Notes for active learning

Chapter Summary

Periodic motion summary

- *Periodic motion* is a motion that repeats
- *Amplitude* relates the displacement and potential energy
- *Frequency* gives cycles per second
- *Period* is the time to complete one cycle
- At equilibrium position:

  $PE = 0$, $KE =$ maximum

- At maximum displacement (amplitude)

  $x = \text{A}$, $PE =$ maximum, $KE = 0$

- At any point, $PE + KE =$ maximum $PE =$ maximum $KE =$ constant
- From the velocity at the equilibrium position, amplitude is calculated by setting the maximum $KE =$ maximum PE
- From amplitude, velocity at the equilibrium position is calculated by setting the maximum $PE =$ maximum KE

Wave characteristics summary

- Wave speed depends on the medium through which it travels.

  Wave speed is determined by $v = f\lambda$

  where f is the frequency ($f =$ # cycles/time), and λ is wavelength.

  $f = 1/T$, the speed $v = \lambda/T$

- Changing a wave's period, frequency, or wavelength affects the other two quantities. This change *does not* affect the speed of the wave if it remains in the same medium.
- Superimposition is when waves interact constructively or destructively (e.g., interference creates larger or smaller amplitudes).
- Waves *reflect and transmit* when traveling through *different mediums*.

- Waves transmitted through mediums with different indices of refraction (n) bend.

  Snell's Law: $n_1 \sin(\theta_1) = n_2 \sin(\theta_2)$

- Angles equal to or greater than the critical angle exhibit total internal reflection:

  $n_1 \sin(\theta_{critical}) = n_2 \sin(90°)$

- Diffraction occurs for all waves around objects or through apertures.

- Standing waves on a string *fixed at both ends* form with *wavelengths* of

  $\lambda_n = 2L/n$

  where L is the length of the string and n is a whole positive number.

- Standing waves on a string fixed at both ends form with *frequencies* of

  $f_n = nv/2L$

  where L is the length of the string, v is the speed of the wave, and n is a whole positive number.

Practice Problems: Waves & Periodic Motion

1. Consider a vibrating mass on a spring; what effect on the system's mechanical energy is caused by doubling the amplitude only?

A. Increases by a factor of two
B. Increases by a factor of four
C. Increases by a factor of three
D. Produces no change
E. Increases by a factor of $\sqrt{2}$

2. Which of the following is an accurate statement?

A. Tensile stress is measured in N·m
B. Stress is a measure of external forces on a body
C. Stress is inversely proportional to strain
D. Tensile strain is measured in meters
E. The ratio stress/strain is the elastic modulus

3. The efficient transfer of energy taking place at a natural frequency occurs in a phenomenon is:

A. reverberation
B. the Doppler effect
C. beats
D. resonance
E. the standing wave phenomenon

4. A simple pendulum and a mass oscillating on an ideal spring both have period T in an elevator at rest. If the elevator now accelerates downward uniformly at 2 m/s^2, what is true about the periods of these two systems?

A. The period of the pendulum increases, but the period of the spring remains the same
B. The period of the pendulum increases, and the period of the spring decreases
C. The period of the pendulum decreases, but the period of the spring remains the same
D. The periods of the pendulum and the spring both increase
E. The periods of the pendulum and the spring both decrease

5. All of the following are true of a pendulum that has swung to the top of its arc and has not yet reversed its direction, EXCEPT:

A. The *PE* of the pendulum is at a maximum
B. The displacement of the pendulum from its equilibrium position is at a maximum
C. The *KE* of the pendulum equals zero
D. The velocity of the pendulum equals zero
E. The acceleration of the pendulum equals zero

6. The explanation for refraction must involve a change in:

I. frequency   II. speed   III. wavelength

A. I only
B. II only
C. III only
D. I and II only
E. I and III only

7. Consider the wave shown in the figure. The amplitude is:

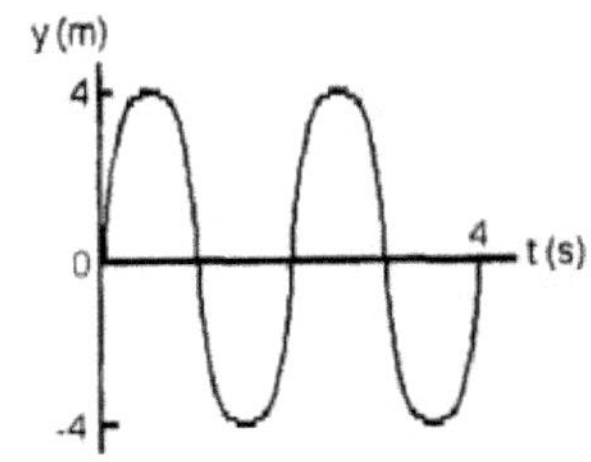

A. 1 m
B. 2 m
C. 4 m
D. 8 m
E. √2 m

8. Increasing the mass *m* of a mass-and-spring system causes what kind of change on the resonant frequency *f* of the system?

A. The *f* decreases
B. There is no change in the *f*
C. The *f* decreases only if the ratio k / m is < 1
D. The *f* increases
E. The *f* decreases only if the ratio k / m is > 1

9. A simple pendulum with a bob of mass *M* has a period T. What is the effect on the period if *M* is doubled while all other factors remain unchanged?

A. T/2
B. T/√2
C. 2T
D. √2T
E. T

10. A skipper on a boat notices wave crests passing the anchor chain every 5 s. The skipper estimates that the distance between crests is 15 m. What is the speed of the water waves?

A. 3 m/s
B. 5 m/s
C. 12 m/s
D. 9 m/s
E. Requires more information

11. For an object undergoing simple harmonic motion, the:

A. maximum potential energy is larger than the maximum kinetic energy
B. acceleration is greatest when the displacement is greatest
C. displacement is greatest when the speed is greatest
D. acceleration is greatest when the speed is greatest
E. total object energy oscillates at frequency $f = \frac{1}{2\pi}(\sqrt{k / m})$

12. As the frequency of a wave increases, which of the following must decrease?

A. The speed of the wave
B. The velocity of the wave
C. The amplitude of the wave
D. The cycles per second
E. The period of the wave

13. What is the period for a weight on the end of a spring that bobs up and down one complete cycle every 2 s?

A. 0.5 s
B. 1 s
C. 2 s
D. 3 s
E. 2.5 s

14. After rain, one sometimes sees brightly colored oil slicks on the road. These are due to:

A. selective absorption of different λ by oil
B. diffraction effects
C. polarization effects
D. interference effects
E. birefringence

15. If two traveling waves with amplitudes of 3 cm and 8 cm interfere, which of the following best describes the possible amplitudes of the resultant wave?

A. Between 5 and 11 cm
B. Between 3 and 8 cm
C. Between 3 and 5 cm
D. Between 8 and 11 cm
E. Less than 8 cm

Solutions: Waves & Periodic Motion

1. B is correct.

$PE = ½kx^2$

Doubling the amplitude x increases the PE by a factor of 4.

2. E is correct.

The elastic modulus is given by:

E = tensional strength / extensional strain

$E = σ / ε$

3. D is correct.

Resonance is when one system transfers its energy to another at that system's resonant frequency (natural frequency). A forced vibration produces the highest amplitude response for a given force amplitude.

4. A is correct.

Period of a pendulum:

$T_P = 2\pi\sqrt{(L / g)}$

Period of a spring:

$T_S = 2\pi\sqrt{(m / k)}$

The period of a spring does not depend on gravity and is unaffected.

5. E is correct.

The pendulum rests momentarily at the top of its arc; the KE and the velocity equal zero.

Since its height above the bottom of its arc is at a maximum at this point, its (angular) displacement from the vertical equilibrium position is also at a maximum.

The pendulum continually experiences the forces of gravity and tension and is, therefore, continuously accelerating.

6. B is correct.

Refraction is the bending of a wave when it enters a medium where its speed is different.

Refraction occurs in sound waves and light waves.

7. C is correct.

Amplitude of a wave is the magnitude of its oscillation from its equilibrium point.

8. A is correct.

$$f = (1/2\pi)\sqrt{(k/m)}$$

An increase in m causes a decrease in f.

9. E is correct.

$$T = 2\pi[\sqrt{(L / g)}]$$

No effect on the period because T is independent of mass.

10. A is correct.

speed = wavelength × frequency

period = 1 / frequency

$v = \lambda f$

$v = \lambda / T$

$v = 15\ \text{m} / 5\ \text{s} = 3\ \text{m/s}$

11. B is correct.

When the displacement is greatest, the force on the object is greatest.

When force is maximized, then acceleration is maximum.

12. E is correct.

$\lambda = vf$

$f = v / \lambda$

$f = 1 / T$

The period is the reciprocal of the frequency.

If f increases, then T decreases.

13. C is correct.

The definition of the period is the time required to complete one cycle.

Period = time / # cycles

$T = 2\ \text{s} / 1\ \text{cycle}$

$T = 2\ \text{s}$

14. D is correct.

When it rains, the brightly colored oil slicks on the road are due to thin-film interference effects. This is when light reflects from the upper and lower boundaries of the oil layer and forms a new wave due to interference effects. These new waves are perceived as different colors.

15. A is correct.

Constructive interference: the maximal magnitude of the amplitude of the resultant wave is the sum of the individual amplitudes:

$$3\text{ cm} + 8\text{ cm} = 11\text{ cm}$$

Destructive interference: the minimal magnitude of the amplitude of the resultant wave is the difference between the individual amplitudes:

$$8\text{ cm} - 3\text{ cm} = 5\text{ cm}$$

Magnitude of the amplitude of the resultant wave is between 5 and 11 cm.

Notes of active learning

Notes of active learning

CHAPTER 6

Sound

Page intentionally left blank

Production of Sound

Sound

Sound is a wave that is experienced in everyday life. Everything heard results from sound waves, including traffic noise, the wind through trees, and a jet flying overhead.

However, many sounds are not audible.

Sound, like other waves, is produced by vibrations in a medium.

Longitudinal waves occur when the oscillations are parallel to the direction of propagation.

Mechanical waves can be transverse and longitudinal; electromagnetic waves are transverse.

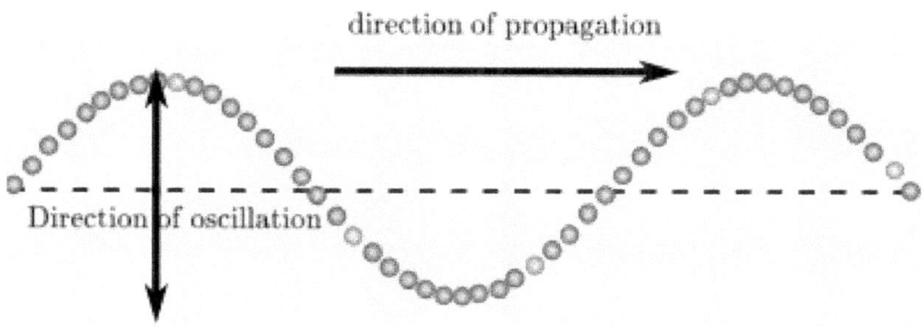

Sound produces longitudinal waves

Sound is a *mechanical longitudinal wave* that travels through a medium as *vibrations*.

Vibrations produce *pressure waves*, which oscillate *parallel to the direction of propagation*.

Sound perception

Vibration frequencies too low to hear are *infrasound* (below 20 Hz) and *ultrasound* if the frequency is above 20,000 Hz.

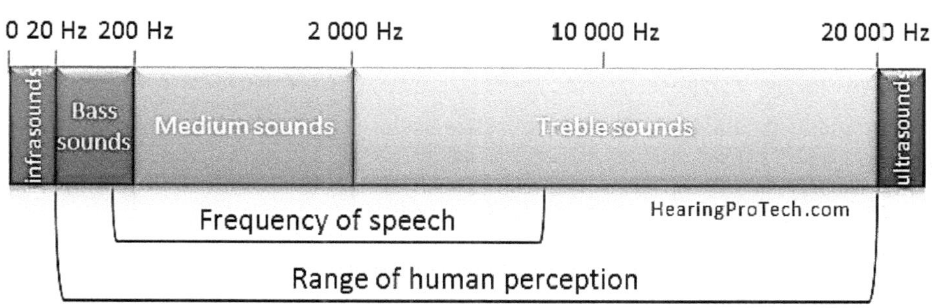

Human perception ranges from 20 Hz to 20,000 Hz with bass, medium, and treble sounds

Musical instruments

Musical instruments produce sounds through vibrations in several ways—strings, membranes, metal or wood shapes, or air columns.

Vibrations may be started by plucking, striking, bowing, or blowing which are transmitted to the air and then to the perceivers' ears.

For example, when struck, a drum produces sound by vibrating a thin membrane.

The vibrating membrane creates a longitudinal wave through the air to a person's ears, which results in the perception of sound.

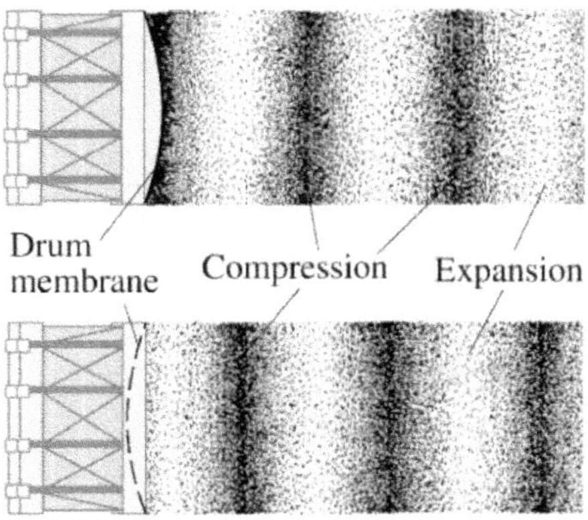

Longitudinal waves produced by the beating of a drum

Strings on a guitar can be shortened by fingering, which raises the fundamental pitch.

Pitch and harmonics

Pitch of a string at a given length can be altered by using a different *density* string.

Strings on stringed instruments produce a *fundamental tone* whose wavelength is twice the length of the string. There are also various harmonics present.

Pianos use both methods (string length and string density) to cover more than its seven-octave range—the lower notes are produced by strings that are much longer and much thicker than the strings that produce the higher notes.

Wind instruments create sound through *standing waves in a tube*, with a vibrating column of air when played.

During a storm, lightning is seen before thunder is heard, even though they occur at the same time because the speed of light (300,000,000 m/s or 3.0×10^8 m/s) is much faster than the speed of sound (300 m/s or 3.0×10^2 m/s).

Speed of Sound in Solids, Liquids and Gases

Medium vibrates

Like all waves, sound waves do not transport matter.

Even though a sound wave travels, the medium through which the sound wave propagates does not travel.

Particles within the medium vibrate around their initial position (horizontally because the wave is longitudinal) and return to their initial position after the wave has propagated.

It is important to remember that waves carry energy, *not matter*.

Elasticity and density

Two values affect the speed of sound: the *elasticity* and *density* of the medium.

Sound travels through a medium by transferring energy from one particle to another.

A rigid medium, or one with little elasticity, has a strong attraction between its molecules. Therefore, molecules vibrate faster and transfer energy faster than a more elastic material.

Speed of sound depends upon the *density* and *elasticity* of the medium.

Speed of sound

Speed of sound is faster through solids than through liquids and faster through liquids than through gases.

Sound cannot travel in a vacuum; sound requires a medium, and there are no particles to vibrate and transmit the wave in a vacuum.

Speed of sound equation:

$$v = \sqrt{\frac{\beta}{\rho}}$$

where β is the bulk modulus, which indicates how the medium responds to compression (how elastic or inelastic it is), and ρ is the density of the medium.

Density of medium

Generally, a substance with a large density is made of larger molecules.

Larger molecules mean more mass, thus requiring more kinetic energy to vibrate them.

Waves are kinetic energy; they travel slower through a medium with larger molecules and a higher mass and faster through a medium with smaller molecules and lower mass.

If the medium is a gas, temperature can also change the speed of sound. Say two equally sized jars contain the same gas but with different densities. Since there is so much space between particles in a gas, the ideal gas laws play a much larger role than individual particle characteristics, such as mass.

Higher density at the same volume means an increase in temperature.

Increasing temperature means adding heat or kinetic energy to the system.

Temperature

More kinetic energy means the gas molecules bounce around faster than at cooler temperatures and transfer energy quickly.

Sound waves travel faster through the warmer medium.

$$v = \sqrt{\frac{\gamma R T}{M}}$$

where γ is the adiabatic constant of the gas, R is the ideal gas constant (8.314 J/mol K), T is the temperature in Kelvin (K), and M is the molecular mass of the gas (kg/mol).

Speed comparison

Speed comparison through different media:

- Sound travels faster through wood than through air. Wood is inelastic, whereas air is exceptionally compressible.
- Sound travels faster through American Redwood, which has a density of 28 lb/ft^3, than through African Teak, which has a density of 45 lb/ft^3.

  Elasticities are similar, which is why density matters.
- Even though gases are less dense than solids, sound travels slower in a gas (e.g., helium) because it is too compressible.
- Sound travels much faster over a hot desert than a cold tundra; the temperature is higher. The speed of sound is higher in the hot desert.

At room temperature and atmospheric pressure, the speed of sound is approximately 343 m/s.

Speed of sound in sample materials

Table of sound speed through mediums at constant temperature and pressure unless noted:

Material	Speed (m/s)
Air	343
Air (0 °C)	331
Helium	1,005
Hydrogen	1,300
Water	1,440
Seawater	1,560
Concrete	≈3,000
Hardwood	≈4,000
Glass	≈4,500
Iron and steel	≈5,000
Aluminum	≈5,100

Sound propagation speed in common materials (20 °C and 1 atm)

Notes for active learning

Sound Intensity

Human perception

Intensity of a wave is equal to the energy transported per unit of time across a unit area.

Human ear detects sounds with intensity between 10^{-12} W/m^2 and 1 W/m^2.

Perceived loudness is not proportional to intensity; the two are related, as in the table below.

Intensity of Sample Sounds

Sound source	Sound level (dB)	Intensity (W/m^2)
Jet plane at 30m	140	100
Threshold of pain	120	1
Siren at 30m	100	1×10^{-2}
Heavy street traffic	80	1×10^{-4}
Noisy restaurant	70	1×10^{-5}
Talk, at 50cm	65	3×10^{-6}
Quiet radio	40	1×10^{-8}
Whisper	30	1×10^{-9}
Rustle of leaves	10	1×10^{-11}
Threshold of hearing	0	1×10^{-12}

Loudness

Loudness of a sound is much more closely related to the logarithm of the intensity.

Sound level is measured in decibels (dB):

$$\beta = 10log\left(\frac{I}{I_0}\right)$$

where β is the sound level in decibels (dB), and I is the intensity in power per area, or energy per unit time per unit area, given in Watts per meter squared (W/m^2).

I_0 is the threshold of hearing at 10^{-12} W/m$^{2.}$

Intensity follows an inverse square law of $I \alpha 1 / r^2$, where r is the distance between the source and detector (observer).

Decibels

From the above equation, as the distance increases, the decibels decrease.

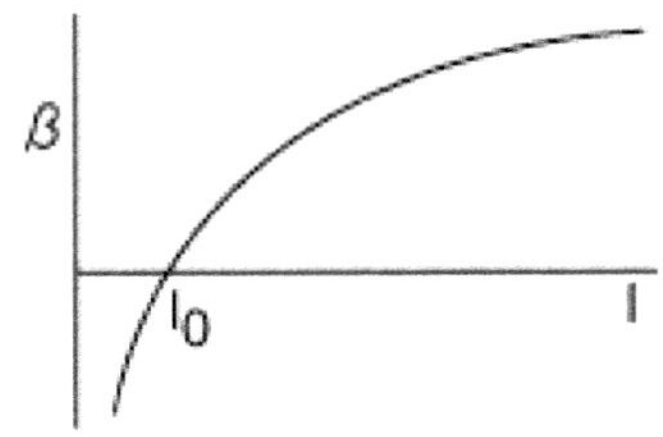

Sound vs. intensity graph where horizontal line signifies audible sounds

Intensity	**Decibels**
I_0	0
$10\ I_0$	10
$100\ I_0$	20
$1{,}000\ I_0$	30

Intensity and decibel relationship

An increased sound level of 3 dB, which doubles in intensity, is a slight change in loudness. In a large area (substantial value of *r*), this change in intensity is insignificant.

In a smaller space (smaller value of *r*), any change in intensity will be highly noticeable.

Decibel systems are based on human perception.

Decibel value for sound with an intensity of I_0 is zero; below this, the sound is not audible.

As intensity increases, the perception of loudness increases, but to a much lesser degree.

Attenuation: Damping

Sound attenuation

Sound attenuation is the gradual loss of intensity as sound travels through a medium and is the greatest for soft, elastic, viscous, or less dense material.

Vibration causes the sound, and the characteristics of this vibration determine how the sound wave acts as it travels through the medium.

Waves can oscillate in free harmonic motion, where the amplitude remains unchanged throughout the vibration.

Wave damping

A wave can exhibit damped simple harmonic motion (SHM), in which a frictional or dragging force inhibits the vibrations' oscillation.

If the damping is small, it can be treated as an "envelope" that modifies undamped oscillations.

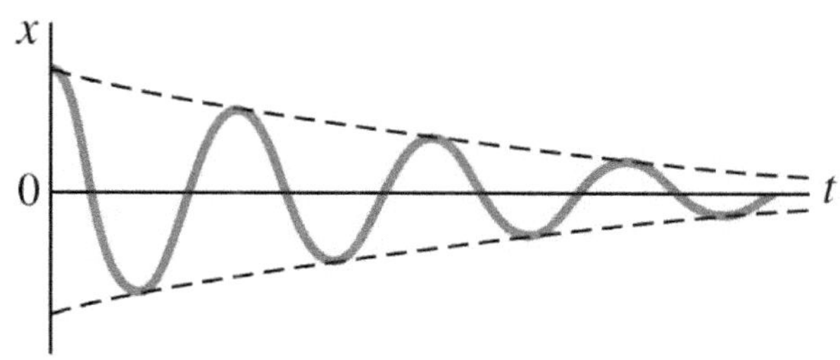

Displacement vs. time graph shows damping

Damping variations

If damping is significant, it no longer resembles a general SHM oscillation wave:

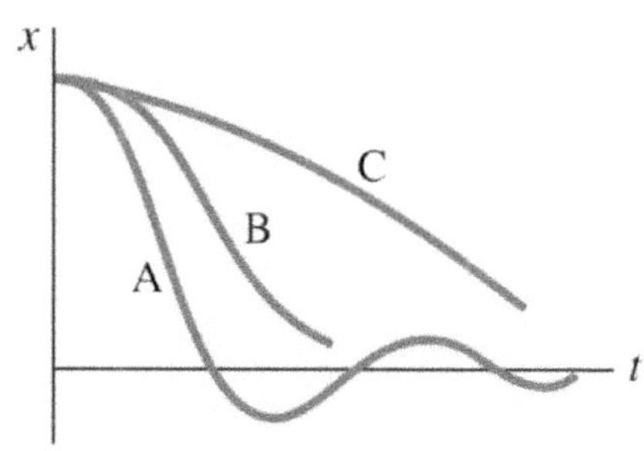

Displacement vs. time with underdamping (A), critical damping (B), and overdamping (C)

There are a few representative shapes on a graph, such as distance *vs.* time graph, when a large damping force is acting on the system (critical damping as *B* in the image above).

Underdamping (labeled as line A in graph above) is when there are a few small oscillations before the oscillator rests.

Critical damping (labeled B in the graph above) is the fastest way to equilibrium.

Over-damping (labeled C in the graph above) occurs when the system is slowed so much that it takes a long time to reach equilibrium.

There are systems where damping is unwanted, such as clocks and watches.

Critical damping

Some systems require dampening, which must achieve *critical damping.*

Critical damping is the threshold between overdamping and underdamping; the oscillator returns to equilibrium position quickly, without oscillating, and passes once at most.

For example, critical damping is desirable for earthquake protection for buildings and automobile shock absorbers (shown below).

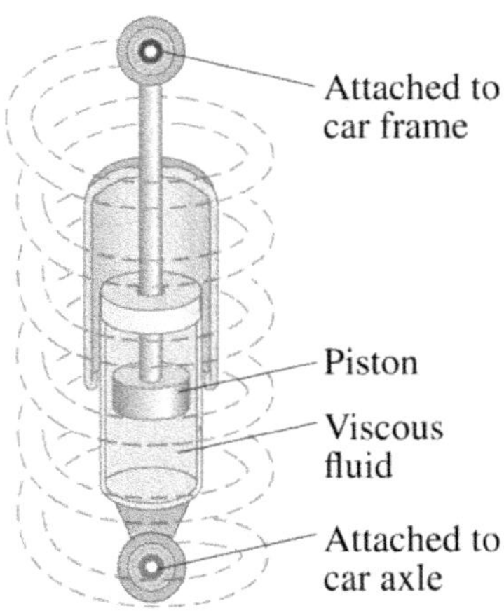

Automobile shocks dampen the vibrations from the bumps on the road surface

Doppler Effect

Relative motion

Doppler shift describes the relationship between the frequency detected and frequency emitted by a source when a source and detector are in *relative motion.*

$$f_d = f_e[(v_s \pm v_d) / (v_s \mp v_e)]$$

where f_d is the frequency detected by the observer, f_e is the frequency emitted by the source v_d is the velocity of the detector, v_e is the velocity of the source, and v_s is the speed of the wave.

When the source and a detector move with respect to each other, the f of the detected wave is shifted from the f of the emitted wave (Doppler shift).

f_d *increases* when the source and detector *approach.*

f_d *decreases* when the source and detector *recede.*

When calculating the Doppler shift between a source and detector, the individual velocities are with respect to each other.

If the source and detector move away at equal speeds, this is equivalent to the source moving at double its speed away from a stationary detector. This allows one of the velocities to be set to zero in the calculation.

Frequency shifts

Doppler Effect states that *the frequency of the sounds varies if there is relative motion between the sound source and the observer.*

Frequency is *higher* if the relative motion between the sound source and observer approach.

Frequency is *lower* if the relative motion between the sound source and observer recedes.

Therefore, oncoming sirens seem to sound louder than when they have already passed.

The equation expresses this:

$$f' = \frac{v \pm v_o}{v \mp v_s} \times f$$

where f' is the perceived frequency, f is the wave frequency, v_o is the velocity of the observer or detector, v_s is the velocity of the source, and v is a reference value equal to the velocity of sound through the given medium.

Approaching or receding source

Decide whether to add or subtract the velocities from the speed of sound through the medium.

If the observer walks away from the source, the wavelength becomes longer, and the perceived pitch decreases. The velocity of the observer must be subtracted in the numerator.

If the source gets closer to an observer, wavelengths shorten, and perceived frequency increases. The denominator decreases, and velocity of the source must be subtracted in the denominator.

Use care when analyzing the movement of the source and remember how fractions react when their denominator is either increased or decreased.

Higher observed frequency

Observed frequency is *higher* than the actual when:

- Source moving *toward* stationary observer:

$$f' = \frac{v}{v - v_s} \times f$$

- Observer moving *toward* stationary source:

$$f' = \frac{v + v_o}{v} \times f$$

- Source and observer both moving *toward* each other:

$$f' = \frac{v + v_o}{v - v_s} \times f$$

Lower observed frequency

Observed frequency is *lower* than the actual when:

- Source moving *away* from the stationary observer:

$$f' = \frac{v}{v + v_s} \times f$$

- Observer moving *away* from the stationary source:

$$f' = \frac{v - v_o}{v} \times f$$

- Source and observer both moving *away* from each other:

$$f' = \frac{v - v_o}{v + v_s} \times f$$

Moving source and observer

Observed frequency could be higher or lower than the actual when:

- Source moves toward the observer while the observer moves away:

$$f' = \frac{v - v_o}{v - v_s} \times f$$

- Source moves away from the observer while the observer moves toward:

$$f' = \frac{v + v_o}{v + v_s} \times f$$

Reference frames

There is no frequency shift when the source and observer move at the same speed and direction. Their relative motion is zero. This instance shows that reference frames are important.

When the source of waves and a detector are moving with respect to each other, the frequency of the detected wave shifts from the frequency of the emitted wave (*Doppler shift*).

Detected frequency *increases* when the source and detector are *approaching*.

Detected frequency *decreases* when the source and detector are *receding*.

Detected frequency

f_{det} is the detected frequency, v_s is the speed of the wave in the medium, v_{det} is the speed of the detector, v_{em} is the speed of the emitter, and f_{em} is the emitted frequency.

Numerator sign reflects the direction the *detector* is moving (positive if approaching, negative if receding).

Denominator sign reflects the direction the *source* is moving (negative if approaching, positive if receding).

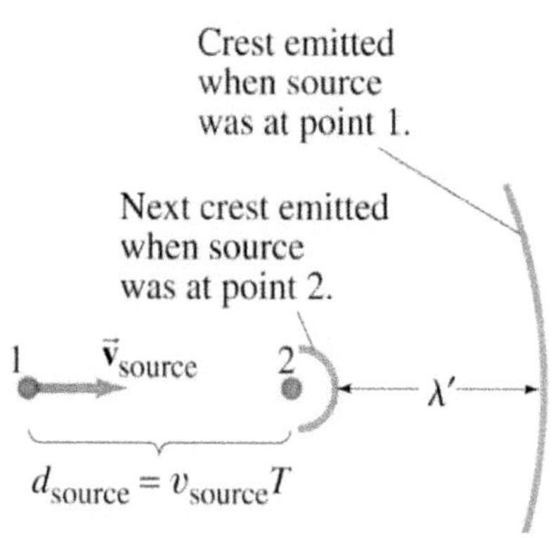

Sound source approaches the observer

Wavelength change

Change in wavelength is:

$$\lambda' = d - d_{source}$$

$$\lambda' = \lambda - v_{source}$$

$$\lambda' = \lambda - v_{source} \times \frac{\lambda}{v_{snd}}$$

$$\lambda' = \lambda(1 - \frac{v_{source}}{v_{snd}})$$

If the observer moves relative to the source, the wavelength and frequency are different for the observer:

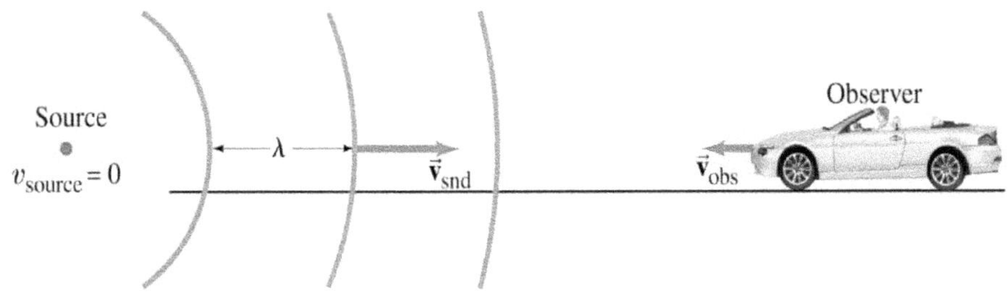

Observer moving toward the source notices a higher frequency and longer wavelength

Pitch

Pitch is the human perception of the frequency of sound.

High pitch has high vibrations and shorter wavelengths, resulting in a higher frequency.

Low pitch has low vibrations and longer wavelengths, resulting in a lower frequency.

Humans hear sound waves with frequencies between 20-20,000 Hz.

Upper limit of this range decreases with age.

Waves with frequencies above 20,000 Hz are *ultrasonic.*

Waves with frequencies below 20 Hz are *infrasonic.*

Ultrasound

Sound has three fundamental properties: *reflection*, *refraction*, and *diffraction*.

Ultrasound imaging uses the reflective property of sound.

A source emits a sound wave with an ultrasonic frequency, which reflects off a surface (e.g., an internal organ) and travels back into the detector to form an image.

Ultrasound has 20,000 Hz or higher frequencies and is used for medical imaging.

Repeated traces are made as the transducer is moved, and a complete picture is constructed.

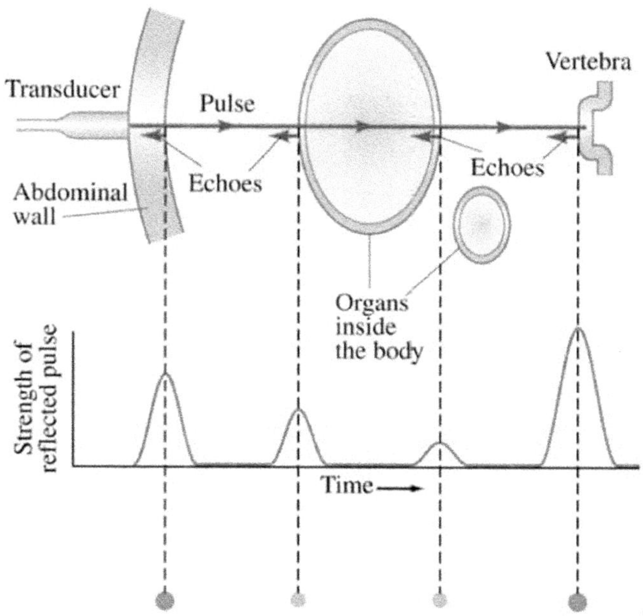

Ultrasound has many applications for medical imaging

Sonar

Sonar was originally an acronym for SOund Navigation And Ranging.

Sonar is used to find and identify objects in the water or underground; sonar techniques are used to learn about the Earth's internal structure.

Sonar calculates the distance to objects by measuring the time between when a sound was generated and when the reflected sound is detected (i.e., echo received).

Researchers use sonar to develop nautical charts, locate underwater hazards for navigation, search for and identify objects in the water column and on the seafloor (e.g., shipwrecks), and map the seafloor.

Sonar usually uses ultrasound waves, as the shorter wavelengths are less likely to be diffracted by obstacles.

Active sonar is the most effective means for locating objects underwater.

Active sonar sends a pulse of energy that travels through water, reflects off an object, and returns to the receiver. Active sonar can locate objects that are too quiet to be detected using passive sonar technology.

Passive sonar uses specialized transducers called hydrophones (or underwater microphones) to listen to sounds in the ocean.

Hydrophones convert received sounds into electrical signals that are sent to a computer for a sonar operator to analyze. Passive sonar does not transmit sound into the water and allows sonar operators to detect objects without giving away their position.

Some marine mammals, such as dolphins, use echolocation. By emitting clicks (or short pulses of sound), marine mammals listen for echoes and detect objects underwater. Echolocation can be used to locate prey, navigate environments, and avoid obstacles in low-visibility conditions.

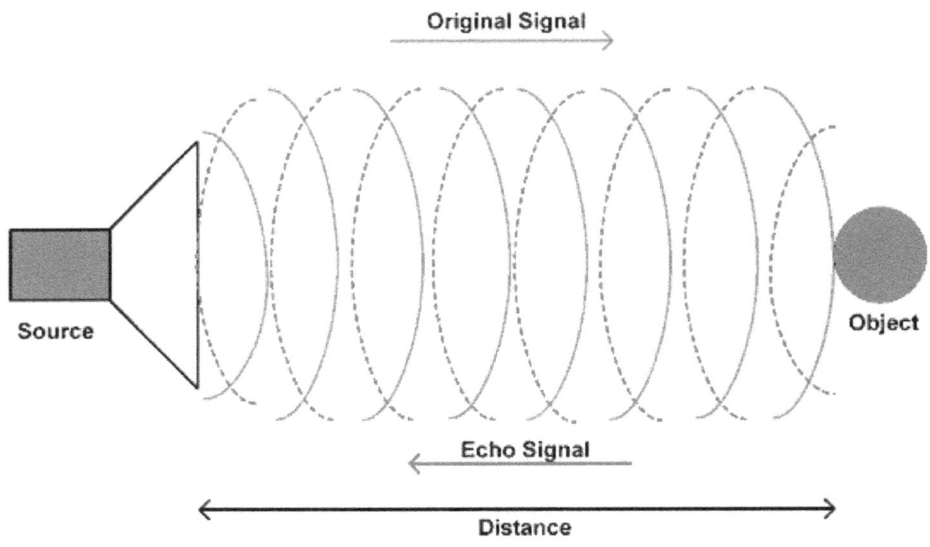

Active sonar sends an energy pulse through water, which reflects and returns to a receiver

Resonance in Pipes and Strings

Standing wave patterns

In standing wave patterns, points along the medium appear to be still.

These still points are *nodes* and are described as points of no displacement.

Some points along the medium undergo maximum displacement during each vibrational cycle.

Maximum displacement points are *antinodes,* and they are opposite of nodes.

Standing wave patterns consist of an alternating pattern of *nodes* and *antinodes*.

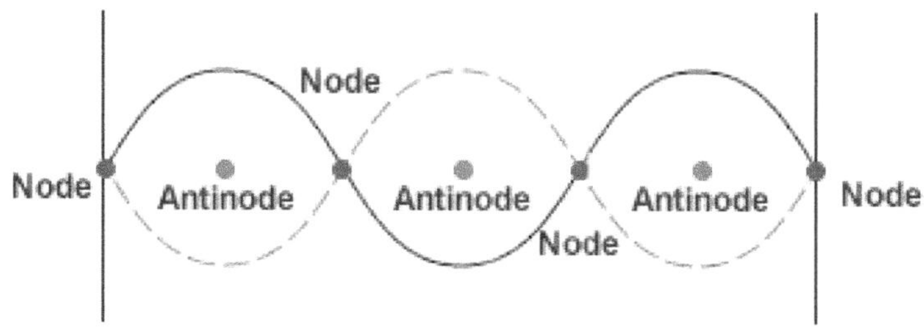

Standing waves consist of alternating nodes and antinodes

Tubes

If a tube is open at both ends (e.g., most wind instruments), and a wave of sound passes through it, the characteristics of the wave can be analyzed if the number of nodes or antinodes the wave has can be determined.

In a tube:

open end has a displacement *antinode.*

closed end has a displacement *node.*

Pressure in the air varies in the same way air displacement does but shifts.

Everywhere there is a *displacement antinode*, there is a *pressure node.*

In standing waves, displacement nodes and pressure antinodes are the opposite of each other.

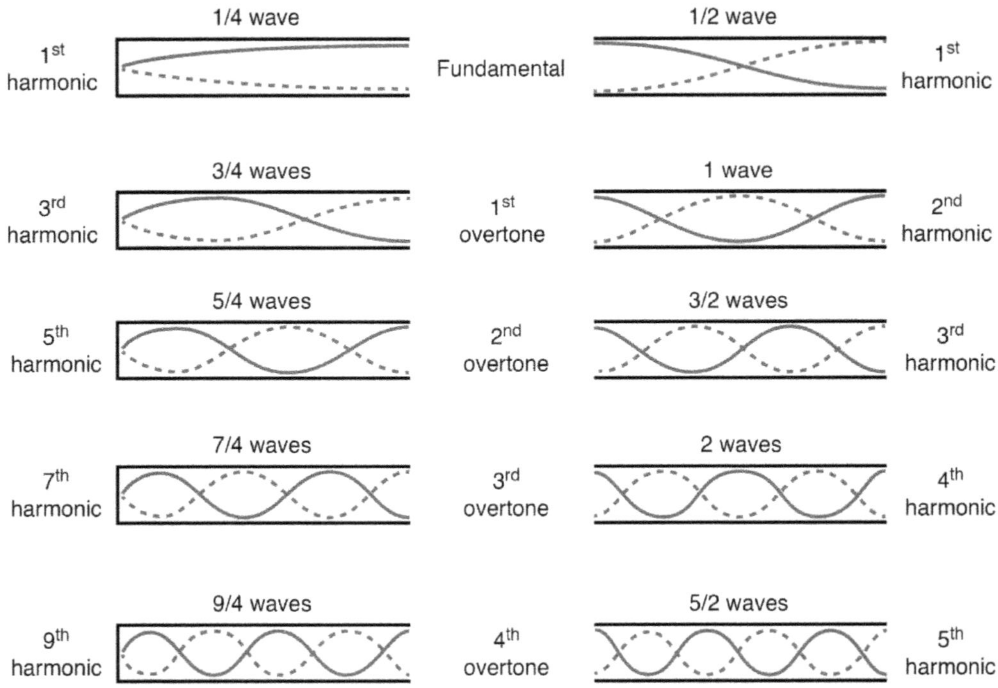

Standing sound waves in open-ended tubes

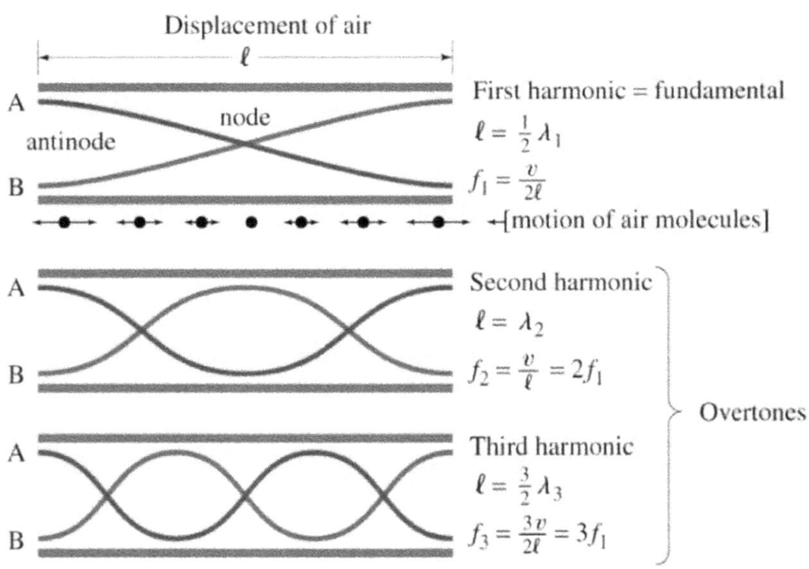

Tube open at both ends, illustrating the displacement of air

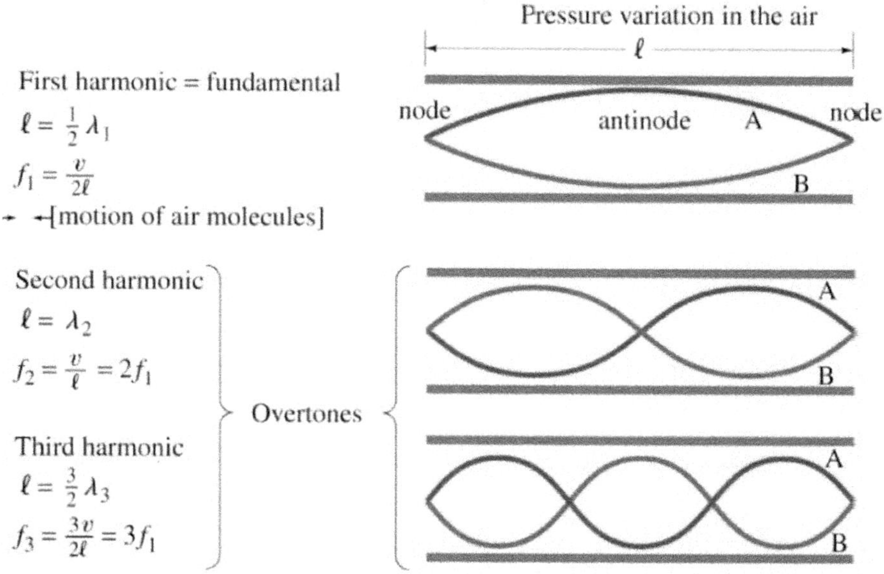

Tube open at both ends illustrating the pressure variation of air

Harmonics

Tube closed at one end (e.g., organ pipe) has a displacement node and a pressure antinode at the closed end.

The number of each node and antinode contained within the pipe depends on the harmonic level (e.g., first harmonic, second harmonic, etc.).

Fundamental frequency (or fundamental) is the first harmonic ($n = 1$).

Second harmonic ($n = 2$) is the next frequency; it continues to the nth harmonic.

Higher harmonics in pipes (open or closed) have shorter wavelengths and higher frequencies but the same wave speed.

For the nth harmonic on a string, an n number of half wavelengths fit precisely along the length of the string.

Frequency and wavelength

Frequency of the wave can be obtained by:

$$f = \frac{v}{\lambda}$$

If a pipe is open at both ends or if a string is unfixed at both ends, its length is:

$$L = \frac{n}{2} \times \lambda$$

If a pipe has one closed-end, or if a string has one fixed end, its length is:

$$L = \frac{2n - 1}{4\lambda}$$

Resonance

For an open-ended tube, resonance occurs when:

$$f_n = n \times \frac{v}{2L}$$

where n can be any integer

Resonance occurs when an outside force acts at a frequency close to the standard frequency of some system component.

Resonance depends on whether the tube is closed at one or both ends.

For a tube with one closed end, resonance occurs when:

$$f_n = n \frac{v}{4L}$$

and

$$\lambda = \frac{4L}{n}$$

where n must be an *odd* integer.

An instrument does not play just one sound and is known for its overtones, which combine with other instruments' overtones to create music instead of merely noise.

Trumpets sound different from flutes, and the reason is overtones—which ones are present and how strong they are.

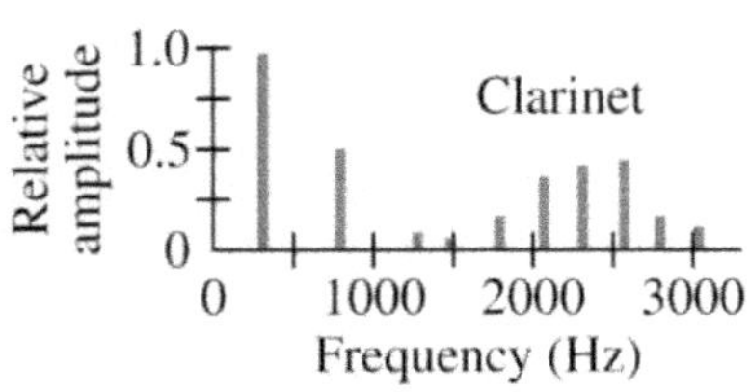

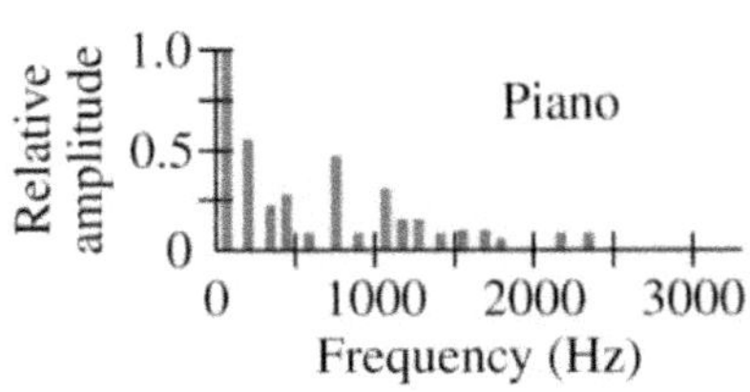

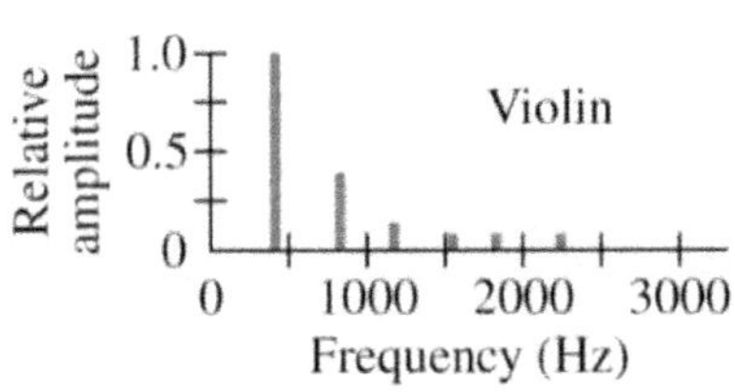

Frequency spectra for a clarinet, piano, and violin, with differences in overtone strength

Beats and Shock Waves

Sound wave interference

Sound waves interfere like other waves do, causing *beats*.

Beats are the envelope around two waves close in frequency, as the two speakers exhibit below.

Beats are produced from the region of overlapping sound waves.

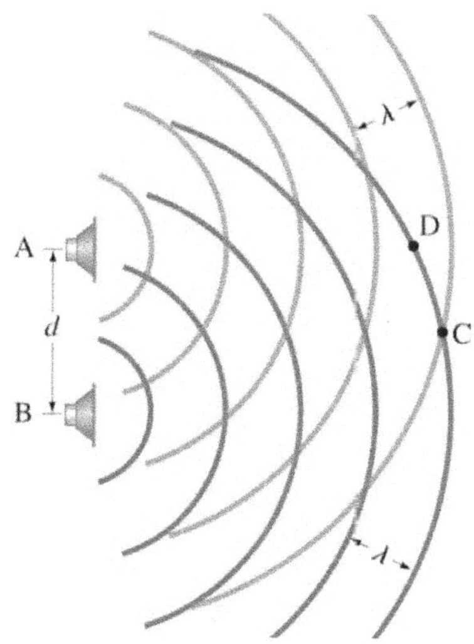

Regions of overlapping sound waves produce beats

Frequency variation

Two frequencies are graphed on a linear axis. Although they start in the same place, one frequency is higher.

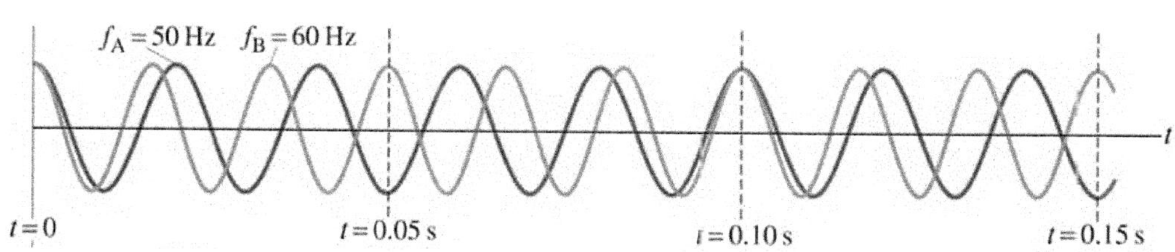

Two waves with different frequencies

Beat period

The combination (or interference) of these two results in the wave shown.

This is an out-of-phase interference, meaning the amplitude of this combined wave is equal to the addition of the two individual amplitudes of the two waves shown above.

Beat period is the time it takes for the beat to start again.

Waves in the graph do not oscillate with a constant amplitude as in the graph above; it repeats if the component waves stay constant.

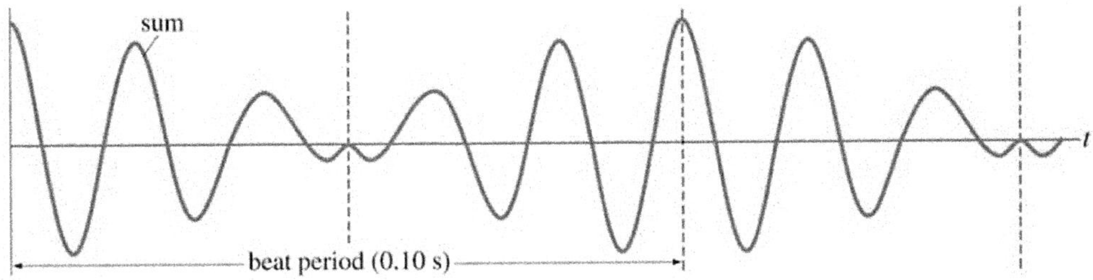

Interference of two waves causes beat period shown

Beat frequency

Beat frequency equation:

$$f_{beat} = |f_1 - f_2|$$

where f_1 and f_2 are the frequencies of the interfering waves.

If two notes are played at the same time close in pitch but not the same, a pulse or "wobble" is heard. This is the beat, and its frequency is f_{beat}.

Musicians tune their instruments by listening to this beat. They try to eliminate it, thereby matching the frequency of their instrument with the frequency of the reference pitch.

Shock waves

Shock waves form if a source moves faster than the speed of the waves it produces in a given medium.

Diagram shows an object with a velocity of zero (top left)

Velocity is less than the speed of sound (top right).

Velocity is equal to the speed of sound (bottom left).

Velocity higher than the speed of sound (bottom right) required for shock waves.

Object in the bottom left create a shock wave front, but the waves do not pile up as they do in the bottom right.

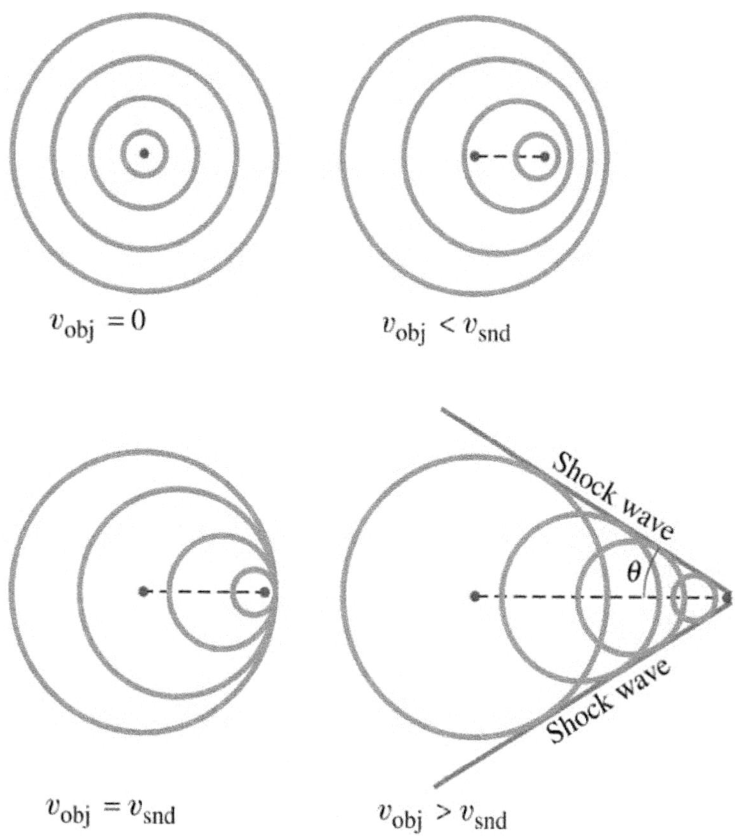

Shock waves occur when the source moves faster than the speed of the waves produced

Wavefront

Shock waves are analogous to the bow waves produced by a boat going faster than the wave speed of the water.

An airplane exceeding the speed of sound in the air produces two sonic booms, one from the nose and one from the plane's tail.

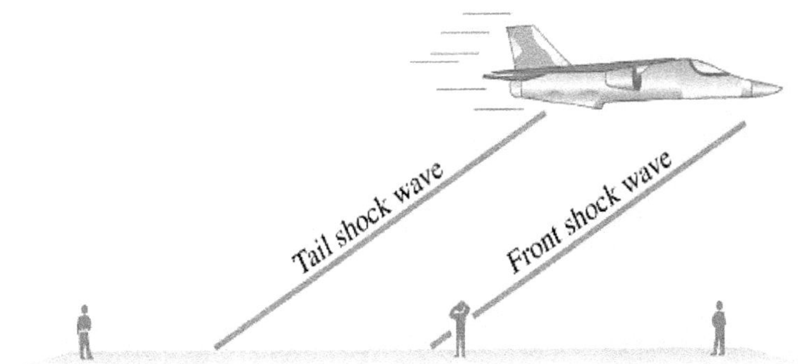

Shock waves produce sonic boom

Chapter Summary

- Sound is a mechanical longitudinal wave traveling through a medium as vibrations.

  Vibrations produce pressure waves, oscillating parallel to the direction of propagation.

  Elasticity and density of the medium affect the speed of sound.

  $$v = \sqrt{\frac{\beta}{\rho}}$$

  where β is the bulk modulus, which indicates how the medium responds to compression (how elastic or inelastic it is), and ρ is the density of the medium.

- *Intensity* of a wave equals energy transported per unit time per unit area.
- Loudness of a sound is much more closely related to the logarithm of the intensity.

  Sound level is measured in decibels (dB) and is defined as:

  $$\beta = 10log\left(\frac{I}{I_0}\right)$$

  where β is the sound level in decibels (dB), and I is the intensity in power per area, or energy per unit time per unit area, given in Watts per meter squared (W/m^2).

- Doppler shift describes the relationship between the frequency detected and the frequency emitted by the source when the source and detector are in relative motion.

  $$f_d = f_e[(v_s \pm v_d) / (v_s \mp v_e)]$$

  where f_d is frequency detected by the observer, f_e is frequency emitted by the source v_d is velocity of the detector, v_e is velocity of the source, and v_s is speed of the wave.

- Standing wave patterns have still points along the medium.

  No-displacement points are *nodes*.

  Other points along medium have maximum displacement during vibrational cycle.

  Maximum displacement points are *antinodes*.

- Frequency of a wave can be obtained by:

  $$f = \frac{v}{\lambda}$$

- For a pipe open at both ends or a string unfixed at both ends, length can be found by:

  $$L = \frac{n}{2} \times \lambda$$

- If a pipe has one closed-end, or if a string has one fixed end, length can be found by:

$$L = \frac{2n - 1}{4\lambda}$$

- For an open-ended tube, resonance occurs when:

  $f_n = n \times \frac{v}{2L}$ where n can be any integer

- Sound waves interfere as other waves do, causing a phenomenon of *beats*.

  Beat frequency is given by the equation:

  $$f_{beat} = |f_1 - f_2|$$

  where f_1 and f_2 are the frequencies of the interfering waves.

If a source is moving faster than the speed of the waves that it is producing in a given medium, then a *shock wave* is formed.

Practice Problems: Sound

1. Two tuning forks have frequencies of 460 Hz and 524 Hz. What is the beat frequency if both are sounding simultaneous and resonating?

A. 52 Hz
B. 64 Hz
C. 396 Hz
D. 524 Hz
E. 588 Hz

2. Which of the following is true of the properties of a light wave as it moves from a medium of the lower refractive index to a medium of the higher refractive index?

A. Speed decreases
B. Speed increases
C. Frequency decreases
D. Frequency increases
E. None of the above

3. Resonance can be explained as forced vibration with the:

A. matching of constructive and destructive interference
B. matching of wave amplitudes
C. maximum amount of energy input
D. least amount of energy input
E. minimum beat frequency

4. The Doppler effect occurs when a source of sound moves:

I. toward the observer II. away from the observer III. with the observer

A. I only
B. II only
C. III only
D. I and II only
E. I and III only

5. Which of the following increases when a sound becomes louder?

A. Amplitude
B. Period
C. Frequency
D. Wavelength
E. Velocity

6. Sound intensity is defined as the:

A. sound power per unit volume
B. sound power per unit time
C. sound energy passing through a unit of area
D. loudness of a sound
E. sound energy passing an area per unit time

7. A violin with a string length of 36 cm and a string density of 3.8 g/cm resonates with the first overtone of an organ pipe with one end closed. The pipe is 3 m. What is the tension in the string so that the sound wave resonates at its fundamental frequency? (Use speed of sound v = 340 m/s)

A. 1,390 N
B. 1,946 N
C. 1,414 N
D. 987 N
E. 1,216 N

8. A speaker produces 10 W of sound, and Rahul hears the music at 20 dB. His roommate turns up the power to 100 W. What level of sound does Rahul now hear?

A. 15 dB
B. 30 dB
C. 40 dB
D. 100 dB
E. 120 dB

9. Thunder shakes a house seven seconds after a flash of lightning. Approximately how far was the lightning strike from the house? (Use speed of sound v = 340 m/s)

A. Requires more information
B. About five kilometers away
C. About one kilometer away
D. About ten kilometers away
E. About two kilometers away

10. The natural frequencies for a stretched string of length L and wave speed v are nv / ($2L$), where n equals:

A. 0, 1, 3, 5, ...
B. 1, 2, 3, 4, ...
C. 2, 4, 6, 8, ...
D. 0, 1, 2, 3, ...
E. 1, 3, 4, 5, ...

11. What is the source of all electromagnetic waves?

A. electric fields
B. vibrating charges
D. magnetic fields
E. none of the above
C. heat

12. When a radio is tuned to a radio station, the frequency of the internal electrical circuit is matched to the frequency of that radio station. What is being affected when tuning the radio?

A. Beats
B. Reverberation
C. Forced vibrations
D. Resonance
E. Wave interference

13. A standing wave is oscillating at 670 Hz on a string, as shown. What is the wave speed?

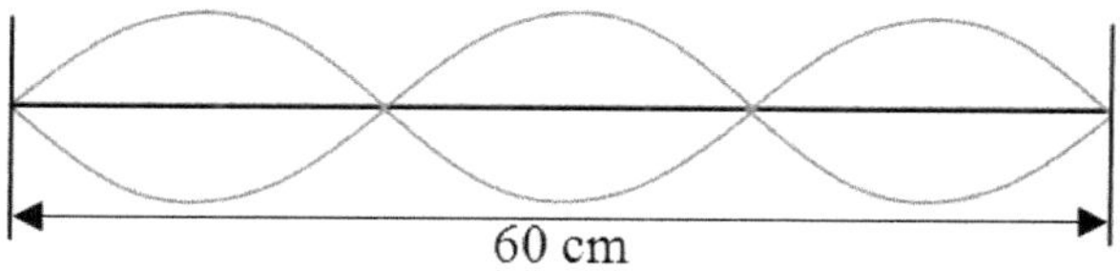

A. 212 m/s
B. 178 m/s
C. 268 m/s
D. 404 m/s
E. 360 m/s

14. A 50 m/s train moves directly toward Carlos, standing near the tracks. The train emits a whistling sound at 415 Hz. What frequency does Carlos hear? (Use speed of sound $v = 340$ m/s)

A. 290 Hz
B. 476 Hz
C. 481 Hz
D. 487 Hz
E. 473 Hz

15. Two tuning forks are struck simultaneously, and beats are heard every 500 ms. What is the frequency of the sound wave produced by one tuning fork if the other produces a sound wave of 490 Hz frequency?

A. 490 Hz
B. 498 Hz
C. 492 Hz
D. 506 Hz
E. 510 Hz

Solutions: Sound

1. B is correct.

Beat frequency equation:

$f_{beat} = | f_2 - f_1 |$

$f_{beat} = | 524\ Hz - 460\ Hz |$

$f_{beat} = 64\ Hz$

2. A is correct.

Frequency of a wave does not change when it enters a new medium.

$v = (1 / n)c$

where c = speed of light in a vacuum and n = refractive index.

Speed v decreases when light enters a medium with a higher refractive index.

3. D is correct.

Resonance is when one system transfers its energy to another at that system's resonant frequency (natural frequency).

Resonance is the forced vibrations with the least energy input.

4. D is correct.

Doppler effect is the observed change in frequency when a sound source is in motion relative to an observer (away or towards). If the sound source moves with the observer, there is no relative motion between the two, and the Doppler effect does not occur.

5. A is correct.

When a sound becomes louder, the sound wave's energy becomes higher.

Amplitude is directly related to sound wave energy:

more energy = higher amplitude

less energy = lower amplitude

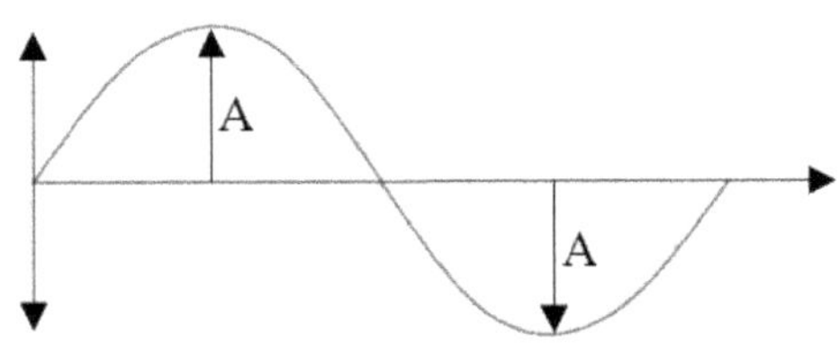

6. E is correct.

Intensity is the power per unit area:

$I = W/m^2$

Loudness is a subjective measurement of the strength of the ear's perception to sound.

7. C is correct.

An overtone is any frequency higher than the fundamental.

In a stopped pipe (i.e., open at one end and closed at the other):

Harmonic #	Tone
1	fundamental tone
3	1st overtone
5	2nd overtone
7	3rd overtone

The first overtone is the 3rd harmonic.

Find wavelength:

$\lambda_n = (4L / n)$, for stopped pipe

where n = 1, 3, 5, 7…

$\lambda_3 = [(4)\cdot(3\text{ m}) / 3)]$

$\lambda_3 = 4\text{ m}$

Find frequency:

$f = v / \lambda$

$f = (340\text{ m/s}) / (4\text{ m})$

$f = 85\text{ Hz}$

Find velocity of a standing wave on the violin:

$f = v / 2L$

$v = (2L)\cdot(f)$

$v = (2)\cdot(0.36\text{ m})\cdot(85\text{ Hz})$

$v = 61\text{ m/s}$

continued…

Convert linear density to kg/m:

$\mu = (3.8 \text{ g/cm})\cdot(1 \text{ kg}/10^3 \text{ g})\cdot(100 \text{ cm}/1 \text{ m})$

$\mu = 0.38 \text{ kg/m}$

Find the tension:

$v = \sqrt{(T / \mu)}$

$T = v^2\mu$

$T = (61 \text{ m/s})^2 \times (0.38 \text{ kg/m})$

$T = 1{,}414 \text{ N}$

8. B is correct.

$I = \text{Power} / \text{area}$

The intensity I is proportional to the power, so an increase by a factor of 10 in power leads to an increase by a factor of 10 in intensity.

$I \text{ (dB)} = 10\log_{10} (I / I_0)$

dB is related to the logarithm of intensity.

If the original intensity was 20 dB, then:

$20 \text{ dB} = 10\log_{10} (I_1 / I_0)$

$2 = \log_{10} (I_1 / I_0)$

$100 = I_1 / I_0$

New intensity is a factor of 10 higher than before:

$I_2 = 10\, I_1$

$1{,}000 = I_2 / I_0$

$I \text{ (dB)} = 10\log_{10} (1{,}000)$

$I \text{ (dB)} = 30 \text{ dB}$

9. E is correct.

velocity = distance / time

$v = d / t$

$d = vt$

$d = (340 \text{ m/s})\cdot(7 \text{ s})$

$d = 2{,}380 \text{ m} \approx 2 \text{ km}$

10. B is correct.

A stretched string has all harmonics of the fundamental.

11. B is correct.

All electromagnetic waves arise from accelerating charges.

When a charge vibrates, it accelerates (a change in the direction of motion is acceleration).

12. D is correct.

When tuning a radio, the tuner picks up specific frequencies by resonating at those frequencies. This filters out the other radio signals, so only that specific frequency is amplified.

By changing the tuner on the radio, a frequency is chosen so that the tuner resonates at that frequency, and the signal is amplified.

13. C is correct.

From the diagram, the wave is a 3rd harmonic standing wave.

Find the wavelength:

$\lambda_1 = 2L$

$\lambda_n = 2L / n$

where n = 1, 2, 3, 4…

$\lambda_n = (2 \times 0.6 \text{ m}) / 3$

$\lambda_n = 0.4 \text{ m}$

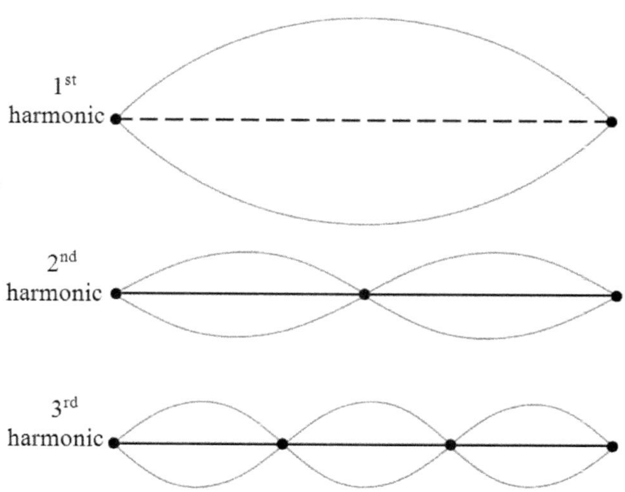

Find the speed:

$\lambda = v / f$

$v = \lambda f$

$v = (0.4 \text{ m})\cdot(670 \text{ Hz})$

$v = 268 \text{ m/s}$

14. D is correct.

Doppler equation for an approaching sound source:

$$f_{observed} = [v_{sound} / (v_{sound} - v_{source})]f_{source}$$

$$f_{observed} = [(340 \text{ m/s}) / (340 \text{ m/s} - 50 \text{ m/s})]\cdot(415 \text{ Hz})$$

$$f_{observed} = 487 \text{ Hz}$$

The observed frequency is always higher when the source is approaching.

15. C is correct.

Beats are heard every 500 ms (i.e., ½ s), or 2 beats per second.

Since $f_{beat} = 2$ Hz, the frequencies of the two tuning forks differ by 2 Hz.

One tuning fork has an f of 490 Hz:

$$f \text{ is } (490 \text{ Hz} - 2 \text{ Hz}) = 488 \text{ Hz}$$

or

$$f \text{ is } (490 \text{ Hz} + 2 \text{ Hz}) = 492 \text{ Hz}$$

Notes of active learning

Notes of active learning

CHAPTER 7

Light & Optics

Page intentionally left blank

Electromagnetic Radiation Spectrum

Visible light spectrum

Electromagnetic radiation is a form of energy that oscillates electric and magnetic fields (without requiring a medium to propagate).

Electromagnetic radiation spectrum spans from long radio waves to high-energy gamma rays with wavelengths from 10^4 m to 10^{-14} m, respectively.

Visible light spectrum is electromagnetic radiation with *wavelengths between* 400 and 750 nm, consisting of all colors humans can perceive.

Visible light spectrum is an essential topic within physics because it contributes extensively to how humans perceive and interact with the world.

Perceiving colors

Human eyes perceive electromagnetic radiation wavelengths between 400 and 750 nm within the electromagnetic radiation spectrum.

Frequencies of light have energy expressed as $E = hf$

Visible light is a narrow spectrum of electromagnetic radiation that humans perceive as colors.

For example, red stop signs, masterpiece paintings, and reading glasses characterize the diverse range of visible light perceived by humans.

Electromagnetic radiation of *specific frequencies* is perceived as distinct colors.

When an object is perceived as a particular color, it usually does not emit light but reflects a specific frequency to the observer.

Mnemonic ROY G BIV A is the visible light spectrum with increasing frequency (and energy).

Red, **O**range, **Y**ellow, **G**reen, **B**lue, **I**ndigo and **V**iolet

Lower-frequency light consists of red, orange, and yellow hues.

Higher-frequency light consists of greens, blues, and violets.

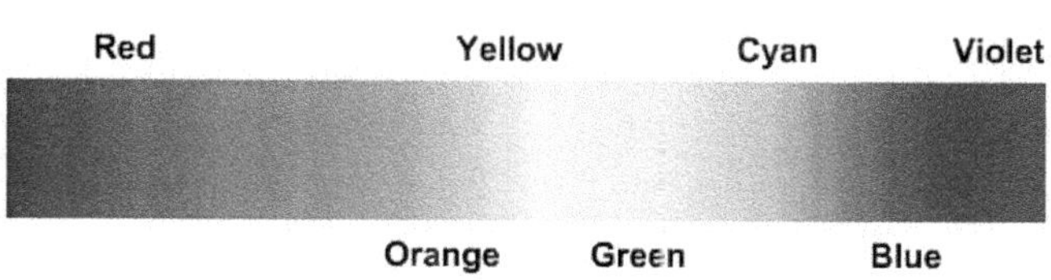

Longest wavelengths, shortest frequency, and low energy are on the left, and shortest wavelengths, longest frequencies, and high energy on the right

Absorbing *vs.* emitting energy

The sun, candles, and light bulbs emit.

However, most objects absorb and reflect some light but do not emit light.

Absorption versus *reflection* of visible light determines the *perceived colors* of objects.

If an object absorbs specific frequencies of light, the observer will *not* perceive those colors.

Sunlight is considered white light because it contains all the visible light spectrum frequencies and is perceived as white.

Red tablecloths absorb most visible light frequencies but reflect predominately red light.

White tablecloths reflect most visible light frequencies and appear white.

Black tablecloths absorb most visible light frequencies and appear black.

Color and energy relationship

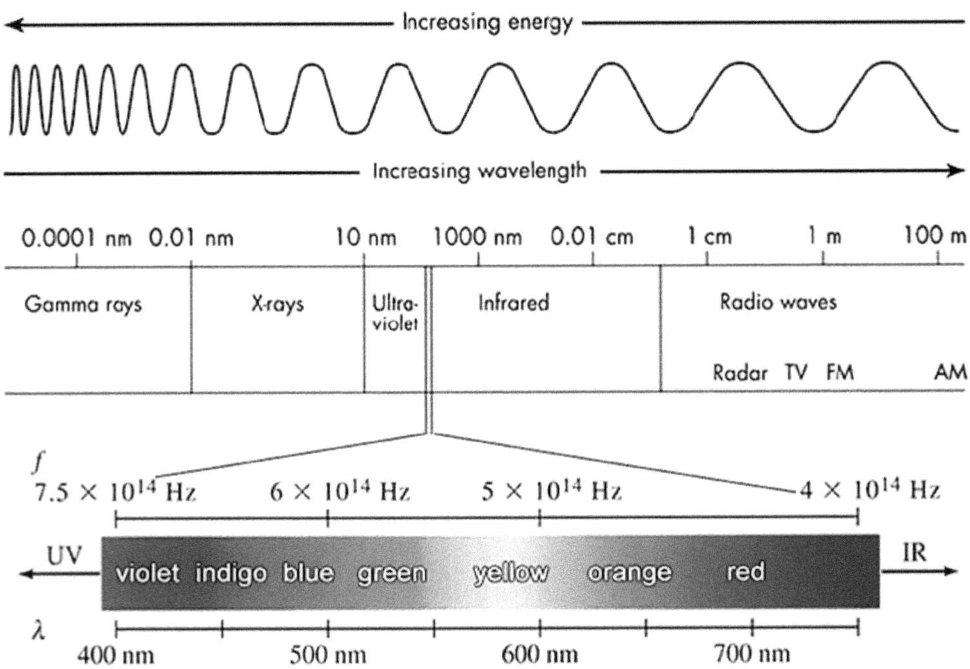

Electromagnetic spectrum with visible light frequency, wavelengths, and perceived colors

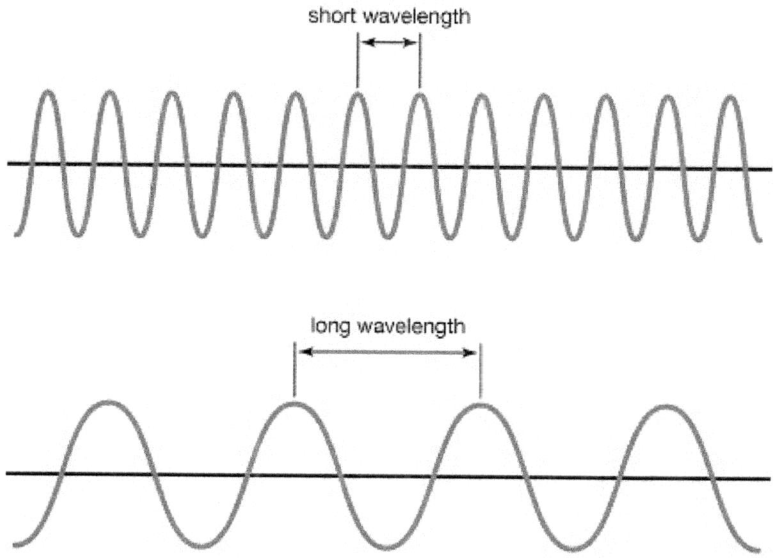

Higher-frequency waves have shorter wavelengths (top image), while low-frequency waves have longer wavelengths (lower image)

Atmospheric scattering

Rayleigh scattering causes the blue sky because atmospheric blue light is scattered.

Sunlight strikes nitrogen and oxygen molecules, distorts, and is reemitted at a high frequency.

Violet wavelengths scatter the most and red the least.

Human eyes are not violet-sensitive; the next color is perceived along the spectrum (blue).

Scattering of light is affected by the composition of the atmosphere. For example, a deep blue is seen in the Mediterranean, where the air is dry.

When dust, water vapor, or other large particles are present, lower-frequency light is scattered, and the sky appears paler or hazier.

In cities, large smog particles absorb light instead of scattering it, resulting in a brownish haze.

Sunsets appear red because red, orange, and yellow are transmitted better than colors at the other end of the visible light spectrum.

When the sun is lower on the horizon, its light must travel through the largest distance of Earth's atmosphere. The more atmosphere it travels through, the more that violet and blue are scattered, causing red to be perceived more intensely.

Primary and secondary colors

Red, green, and blue are the *primary* colors (in physics) because they are detected by the eye independent of intensity but based on frequencies and ocular photoreceptor pigment receptors in cone cells.

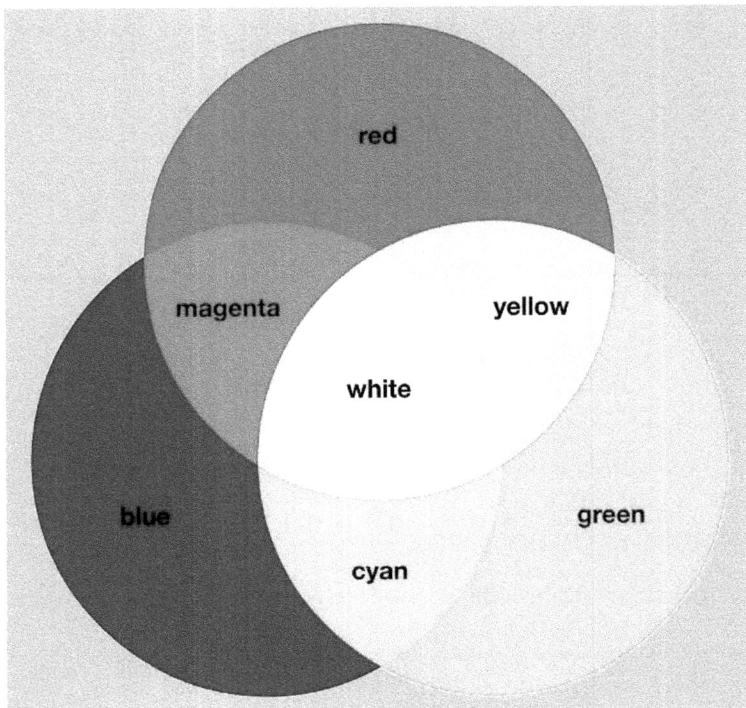

Primary and secondary colors result from addition

Magenta, yellow, and cyan are the *secondary* colors produced by adding primary colors.

Other colors can be made if these primary colors are added.

Combining primary colors yields:

Red + blue = magenta

Red + green = yellow

Blue + green = cyan

Double-Slit Diffraction

Particle–wave duality

Like all waves, visible light can constructively and destructively interfere, as Thomas Young's (1773-1829) *double-slit diffraction* experiment observed.

Young's *double-slit diffraction* experiment proved that light was indeed a wave, as it displayed the *wave property* of interference.

Double-slit diffraction experiment established the wave nature of light in *particle-wave duality*.

Double-slit diffraction experiment

Double-slit diffraction experiments have two slits (S_1 and S_2) positioned on an opaque wall at a known distance. A viewing screen is placed at a known distance from the wall with the slits.

When a light source is shone at the wall with the slits, the expected behavior of light, according to particle theory, mimics the slit screen (top image below).

Light passing through the slits constructively and destructively interfere, producing the pattern seen on the viewing screen (top image below).

Expected observation on viewing screen for particle theory predictions

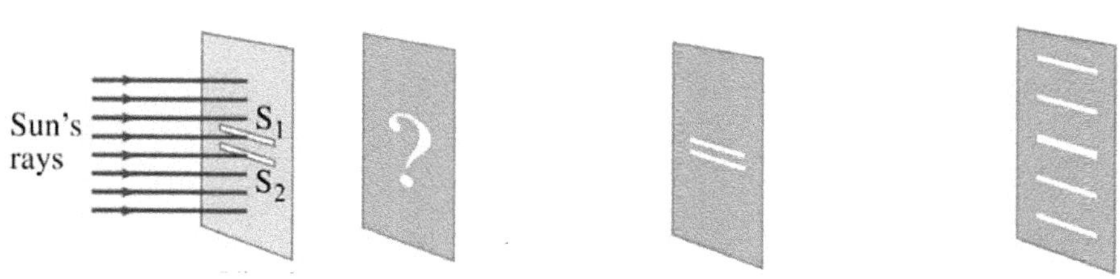

Experimental observation on a viewing screen supports the wave nature of light

Constructive and destructive interference

Double-slit diffraction experimental results support the theory that light is indeed a *wave* because the resulting pattern is consistent with the *interference patterns* of waves.

Interference occurs because each point on the screen is not the same distance from both slits.

Waves interfere *constructively* to produce a *bright spot* on the wall or *destructively* to produce a *dark spot*, depending on the path length difference.

Constructive interference spots on the viewing screen result from light waves coming through the screen, which are *in phase*, constructively interfere, and are visible.

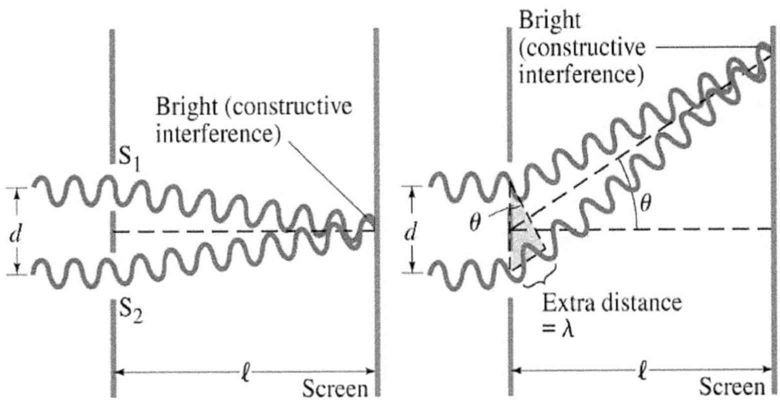

Constructive interference of parallel light normal to the slit surface (on the left) compared to light offset by wavelength distance (on the right)

Notice that the lower slit wave traveled an extra wavelength compared to the upper wave.

Path length difference

Path length difference determines if a spot will be bright or dark (constructive or destructive).

Geometry determines constructive and destructive interference:

Equation for constructive interference (*bright spots*).

$d \sin \theta = m\lambda$ $\quad$ m = 0, 1, 2, 3, …

Equation for *destructive interference* (*dark spots*).

$d \sin \theta = (m + ½)\, \lambda$ $\quad$ m = 0, 1, 2, 3, …

where d is slit separation distance (m), θ is the angle of the bright or dark spot with respect to the center between the two slits (degrees), λ is wavelength of the light passing through the slits (m), and m is the order value and determines whether the spot appears bright (constructive interference) or dark (destructive interference).

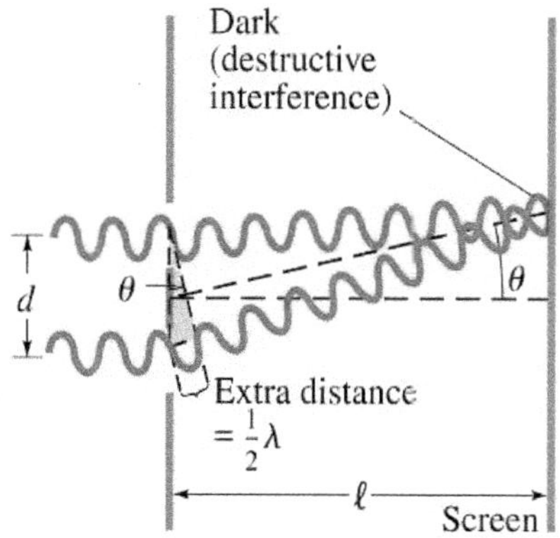

Destructive interference of light offset by ½ wavelength produces a dark spot

Order value for constructive and destructive interference

Order values m for *constructive interference.*

If $m = 0, 1, 2, 3, \ldots$ then the interference is constructive, and the spot is *bright.*

Order values m for *destructive interference.*

If $m = \frac{1}{2}, \frac{3}{2}, \frac{5}{2} \ldots$ then the interference is destructive, and the spot is *dark.*

Maxima represents *constructive* interference.

Minima represents *destructive* interference.

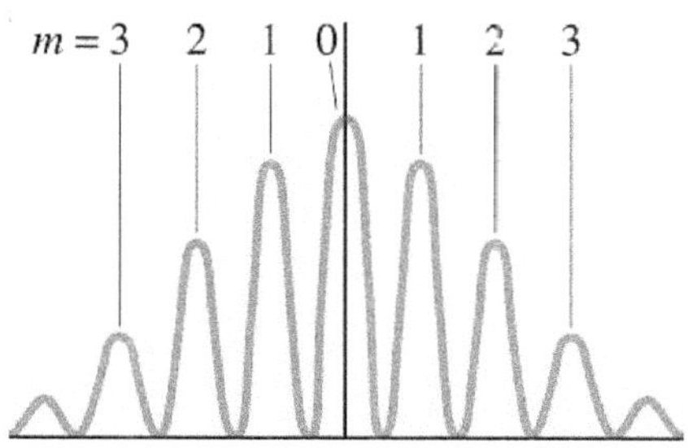

Maxima order number m represents constructive interference

Notes for active learning

Diffraction Grating

Interference

Diffraction grating is like the double-slit experiment, except that instead of two slits, the diffraction grating has many slits positioned close together.

Diffraction gratings operate using the same principle as double-slit experiments.

Diffraction grating interference spot equation is the same as double-slit experiments, including *order values* (m) for destructive and constructive interference.

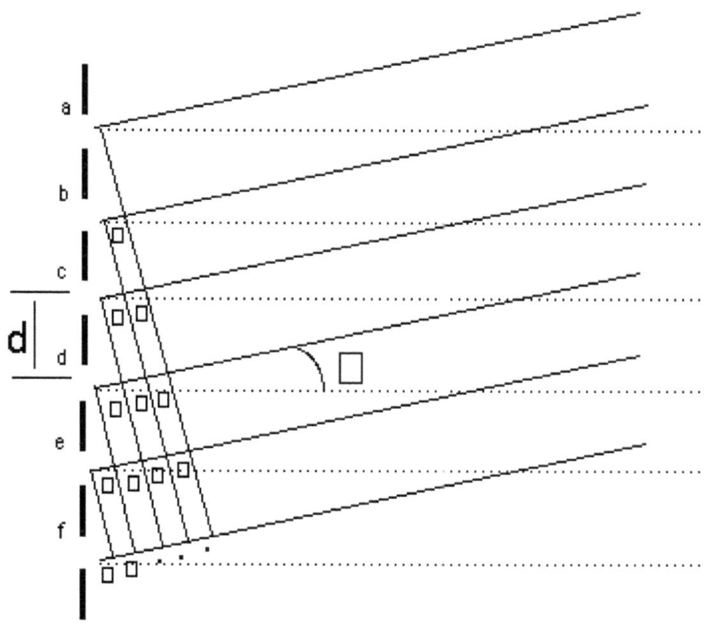

Diffraction grating has many evenly spaced parallel slits that produces an interference pattern similar to but sharper than that of a double slit.

Order values *m*

Order values m for *constructive* interference.

If $m = 0, 1, 2, 3, \ldots$ then the interference is *constructive*, and the spot is *bright*.

Order values m for *destructive* interference.

If $m = \frac{1}{2}, \frac{3}{2}, \frac{5}{2} \ldots$ then the interference is *destructive*, and the spot is *dark*.

Maxima represents *constructive* interference.

Minima represents *destructive* interference.

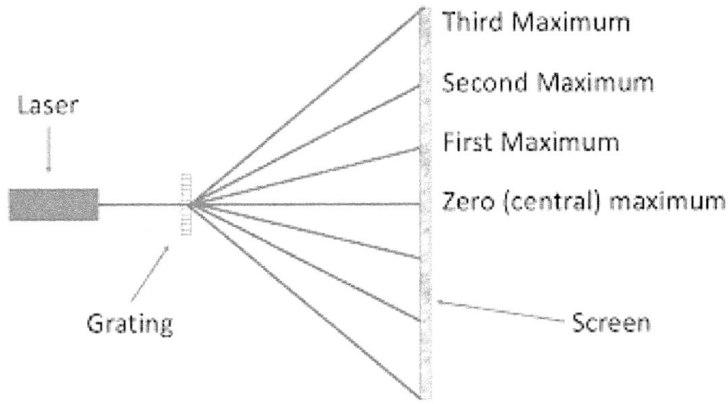

Diffraction pattern with maxima

Diffraction gratings *vs.* double slit

Diffraction gratings separate wavelengths of incoming light better than with double slit.

Double slit produces broader peaks of the maxima.

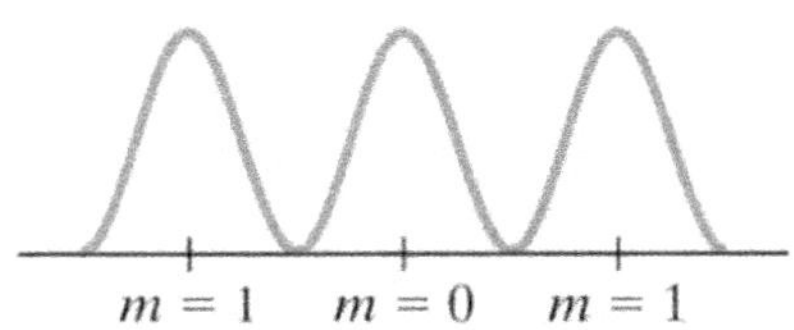

Maxima pattern produced by double slit diffraction

Diffraction grating produces sharper peaks of the maxima.

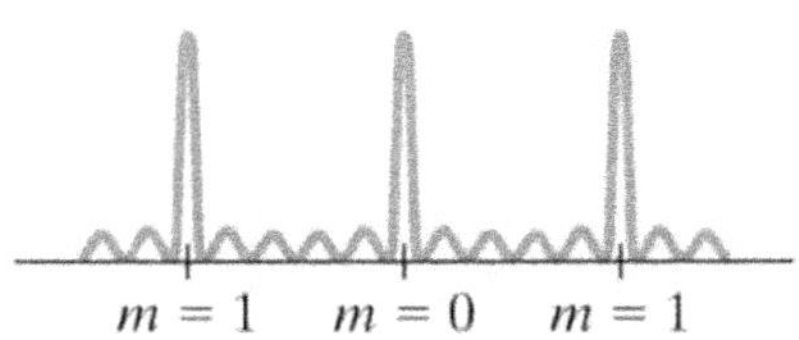

Maxima pattern produced by diffraction grating

Single-Slit Diffraction

Interference of incoming wave

Single-slit diffraction is a diffraction method where light is passed through a single slit, and the resulting diffraction pattern can be observed.

Like other diffraction techniques, single-slit diffraction is the result of interference of the incoming wave as it bends around the slit through which it passes.

Light waves in single-slit diffraction are *in-phase* (constructive) and *out-of-phase* (destructive) at different points than diffraction grates and double-slit diffraction.

Unlike double-slit and diffraction grates, the maxima pattern produced by single-slit diffraction does not follow the same order values or equation.

Maxima and interference

Maxima for a single slit diffraction pattern is:

$$a \sin \theta = m\lambda$$

where a is the width of the slit (m).

If $m = 0, \frac{3}{2}, \frac{5}{2} \ldots$, then the interference is constructive, and the spot is bright.

If $m = 1, 2, 3, \ldots$, then the interference is destructive, and the spot is dark.

Single-slit diffraction patterns

Single-slit diffraction patterns are different than those for diffraction grating and double-slit.

Single slit diffraction pattern produces the widest maxima; subsequent maxima after $m = 0$ are substantially reduced.

Pattern produced by diffraction grating and double slit is sharper than single slit diffraction pattern and does not fade nearly as much after a value of $m = 0$ (see the diagram below).

Diffraction pattern profiles

Figure below compares diffraction patterns produced by the three diffraction methods.

Double-slit diffraction (left), *diffraction grating* (middle), and *single-split diffraction* (on right)

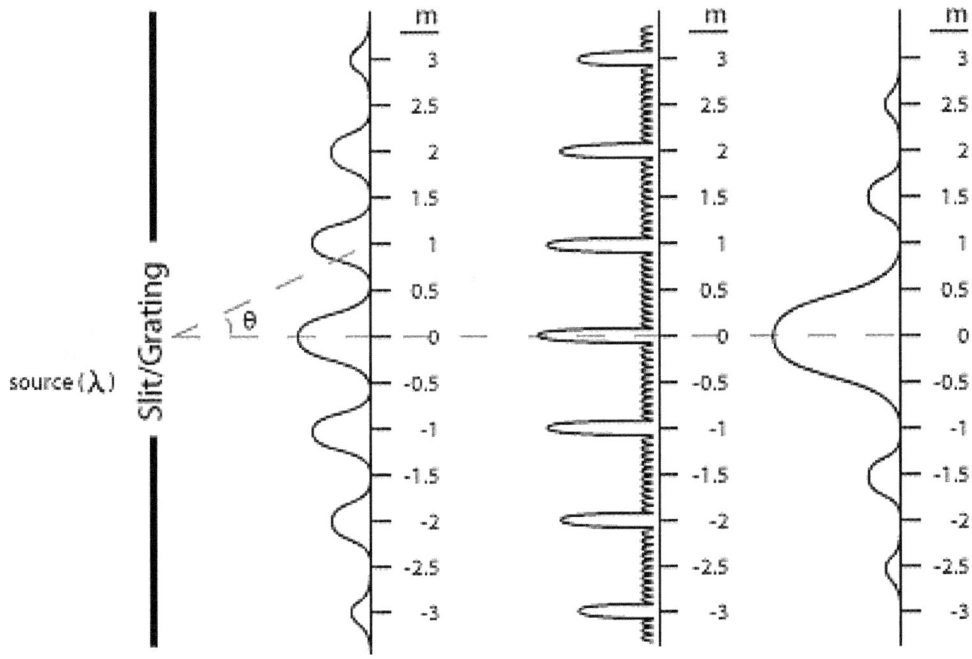

Double slit: $d \sin\theta = m\lambda$ (shown on the left); diffraction grating: $d \sin\theta = m\lambda$ (shown in the middle); single slit: $a \sin\theta = m\lambda$ (shown on the right)

Thin Films

Thin-film interference

When observing an oil slick on water or the ground, colors swirl on top of the slick.

Thin-film interference is observed in colors, not of the oil, but an effect of light interference reflecting off the slick.

Interference of the light (destructive and constructive) occurs because the incoming light partially reflects off the oil surface and partially transmits through it.

Transmitted light is reflected off the substance below the oil (water, in this case), and the observer sees both the oil-reflected and water-reflected light.

The two reflected light rays traveled different distances, had a path difference, and interfered.

Most visible swirling colors on top of the oil slick result from *constructive interference*, and the dim colors (or those not visible) result from *destructive interference*.

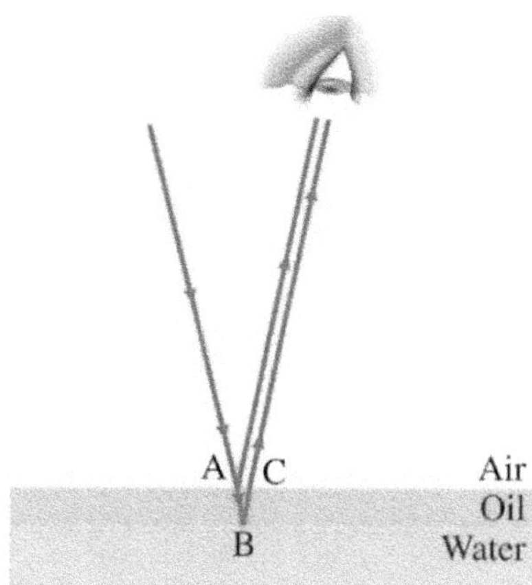

Thin-film interference with oil-reflected and water-reflected light

Light displayed

Wavelengths of light displayed:

$$(2)\cdot(n)\cdot(d) = m\lambda$$

where n is the index of refraction for the thin film,

d is the thickness of the thin film (m),

λ is the wavelength of the light source (m), and

m is the order value dependent upon the material underneath the thin film.

Interference patterns

Let $n_{thin\ film}$ be the *index of refraction* of the thin-film material.

Let n_{under} be the *index of refraction* of the material underneath the thin film.

If $n_{thin\ film} > n_{under}$ then:

Destructive interference: $m = 1,\ 2, 3, \ldots$

Constructive interference: $m = \frac{1}{2}, \frac{3}{2}, \frac{5}{2} \ldots$

If $n_{thin\ film} < n_{under}$ then:

Destructive interference: $m = \frac{1}{2}, \frac{3}{2}, \frac{5}{2} \ldots$

Constructive interference: $m = 1, 2, 3, \ldots$

X-Ray Diffraction

X-ray interference patterns

X-ray waves produce interference patterns caused by *phase differences*.

X-ray diffraction determines the structure of crystal lattices and molecular arrangements.

Due to the short wavelength of X-rays, high-resolution structural determinations are possible.

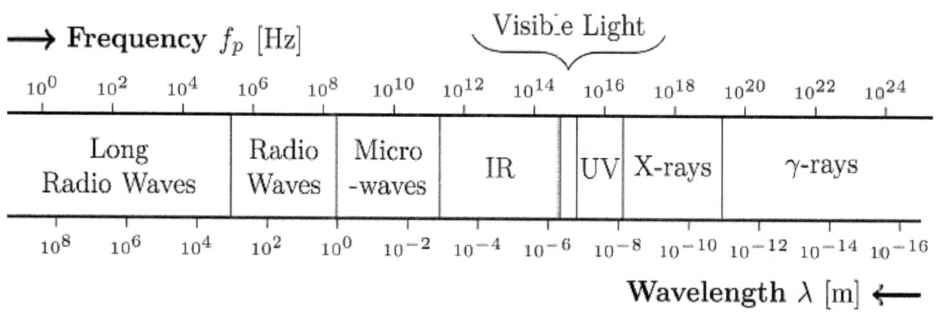

X-rays have wavelengths of 0.01 to 10 nanometers or 1.0×10^{-11} to 1.0×10^{-8} meters.

X-ray crystallography

X-rays diffract off objects, revealing molecular-level patterns of the investigated structure.

X-ray crystallography is the favored technique for structure determination of proteins and biological macromolecules.

X-ray crystallography provides a three-dimensional molecular structure of a crystal. A purified sample at a high concentration is crystallized, and the crystals are exposed to X-ray beams. The resulting diffraction patterns provide information about the crystal packing symmetry and the size of the repeating unit that forms the crystal.

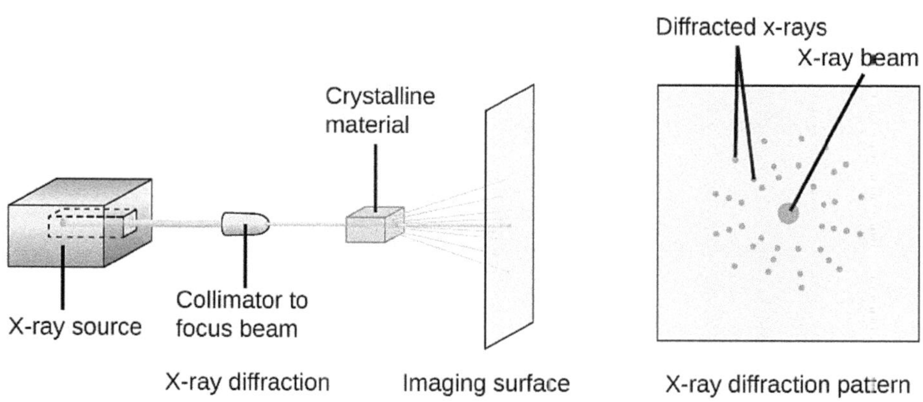

Experimental design for lattice structure in crystalline solids

Crystal structure

When X-rays are shone upon a crystal structure, the wavelengths are small enough such that the waves reflect off individual atoms within the crystal structure.

If waves reflect off the crystal surface, they interfere due to their path length difference. The resulting diffraction pattern gives information about the crystal structure of the specimen.

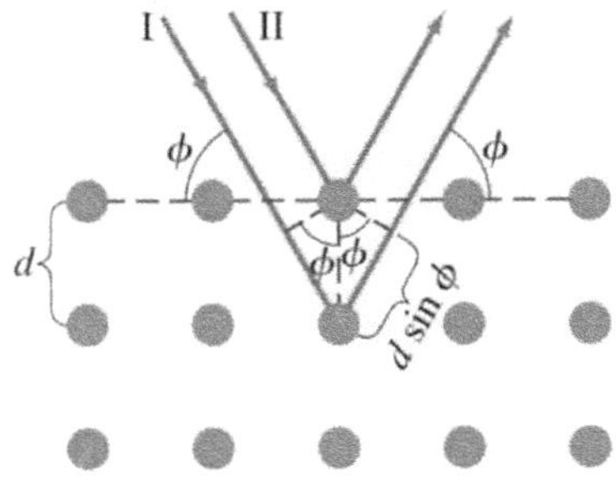

X-ray crystal diffraction experiments provide information about crystal structures

Bragg's Law

Maxima produced by X-ray diffraction is calculated via Bragg's Law (1913):

$$n\lambda = 2d \sin \phi$$

where n is the order number (1, 2, 3, …), λ is the wavelength of the X-ray (m), d is the lattice spacing of the crystal (m), and ϕ is the reflection angle (degrees).

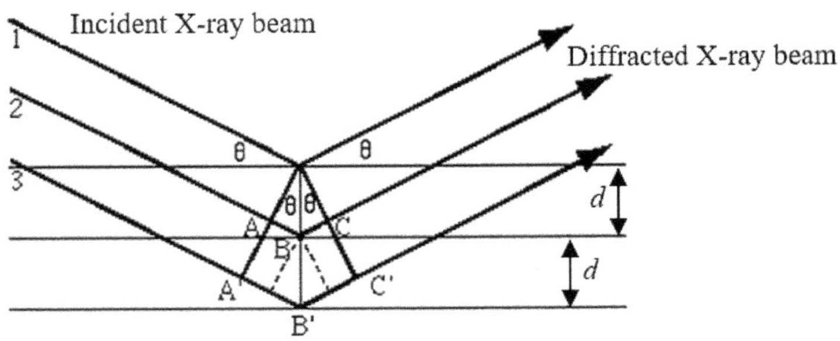

Pattern of maxima produced by X-ray diffraction according to $n(\lambda) = 2d \sin(\theta)$

Polarization of Light

Polarization

Polarization refers to the planes in which the electric and magnetic fields oscillate.

Light from the sun and most sources is un-polarized, with waves oscillating randomly.

Electric and magnetic field oscillation would be uniform if the light becomes polarized.

When the light strikes the water's surface, it reflects and becomes horizontally polarized. The horizontally polarized light has its electric field oscillating only in one direction rather than all.

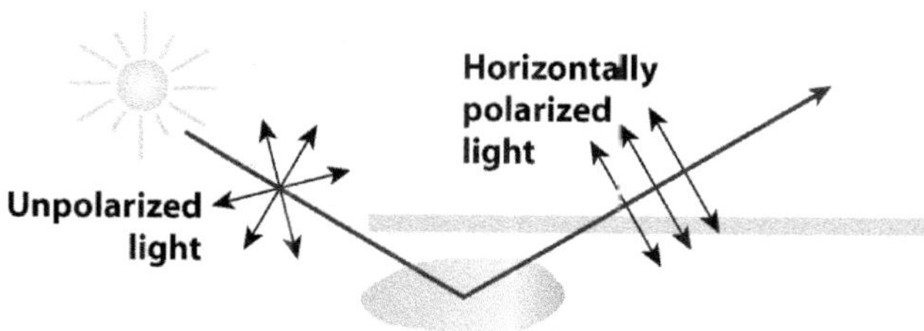

Light emitted from the sun is unpolarized, with an electric field oscillating in many directions, as the arrows indicate

Horizontal *vs.* vertical polarization

Polarized light and materials that polarize light have many applications.

Waves are horizontally polarized on the right below and cannot pass through the vertical polarizer. This is how polarized glasses remove glare from surfaces such as water.

Glare reflected off water is usually horizontally polarized and eliminated by a vertical polarizer.

Polarized sunglasses allow the wearer to see through water because they have a vertical polarizer, reducing horizontally polarized light reflecting off the water.

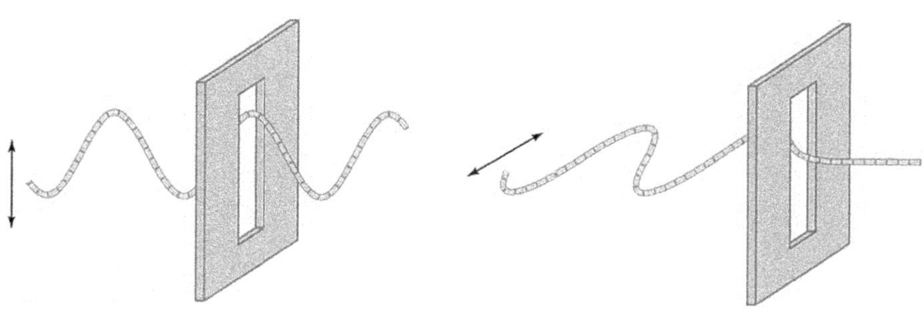

Rope represents polarized light passing through the slit when aligned with the direction of oscillation (on the left) but not when out of phase (on the right)

Polarized light intensity

Intensity of light after passing through a polarizer filter is calculated by:

$$I = I_0 \cos^2(\theta)$$

where I_0 is the initial intensity of the light (W/m^2), I is the intensity of the light after passing through the polarizer, and θ is the angle of the original light wave electric field concerning the axis of the polarizer, the transmission axis (degrees).

Unpolarized light intensity

Unpolarized light incident upon a polarizing filter has light intensity reduced by half:

$$I = \frac{1}{2} I_0$$

No light passes if the light is incident upon crossed polarizers (90° between transmission axis).

In the diagram below, the unpolarized light is initially polarized such that its electric field only oscillates vertically after it passes through a filter with a vertical transmission axis.

Light passes through a polarizer with a horizontal transmission axis; no light is transmitted because none of the light's electric field oscillates in the horizontal axis.

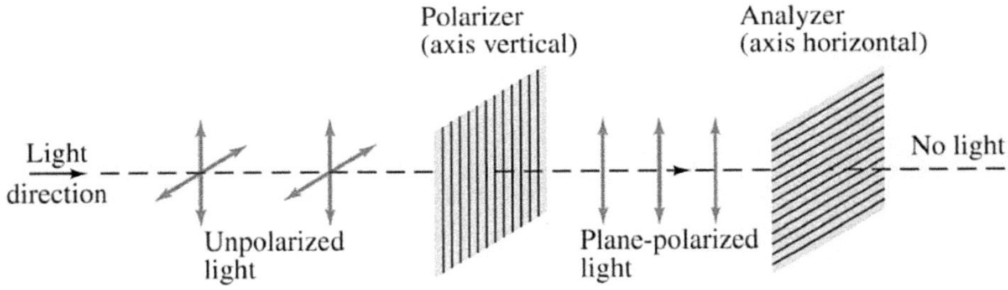

No light passes if the light is 90° between the transmission axis

Doppler Effect

Moving bodies shift light

Electromagnetic radiation (e.g., light, sound) can undergo the *Doppler effect.*

Doppler effect for light is prevalent in astronomy. When observing stars, galaxies, and other interstellar bodies, astronomers noted that sometimes the observed colors of these bodies shifted at different periods of the year.

Color change is because of the Doppler effect, which accounts for changes in relative motion.

Earth moves towards and away from these stellar bodies during the year.

The color shifts due to the Earth's speed relative to the observed body's.

Waves compress and expand

Wave compression and expansion of moving objects affect observed colors, known as redshift and blueshift.

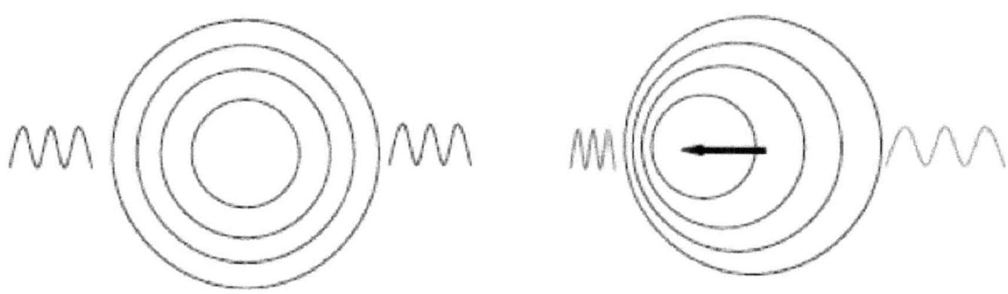

Stationary object (image on the left) and object moving to the left (image on the right). Observers in the same reference as objects on the right experience blueshift when observing motion toward the left and redshifts relative to objects on the right.

Redshift

Redshift occurs when the light source and the observer move *away* from each other.

Frequency of the observed wave *decreases* for the Doppler effect.

Observer sees a *lower frequency* than the source (e.g., light, sound).

Wavelength of the observed wave *increases* as waves expand from positive velocity.

Lower-frequency visible light is red; these sources appear redder than they are because the Doppler effect shifts the source frequency to a *lower frequency* for the observer.

Blueshift

Blueshift occurs when the light source and the observer move *towards* each other.

Frequency of the observed wave *increases* by the Doppler effect.

Observer sees a *higher frequency* than the source (e.g., light, sound).

Wavelength of the observed wave *decreases* as waves compact from negative velocity.

Blues and violets are higher-frequency colors in the visible color spectrum. Thus, the sources of light appear bluer than they are.

Reflection from Plane Surfaces

Reflection

Light striking a surface can be *absorbed, transmitted,* or *reflected.*

Mirrors are surfaces that *reflect* incoming light (all light for an ideal mirror). When a person looks in a mirror, they see their image because it reflects light.

Surfaces such as shiny metal or calm water reflect light. However, these images are not as clear as those reflecting from a mirror and tend to be blurry.

Blurry images form when reflecting on irregular surfaces according to the law of reflection.

Law of reflection

Law of reflection states that the *angle of incidence* of a light ray, with respect to the normal of the surface, equals the *angle of reflection.*

Angle of incidence = angle of reflection

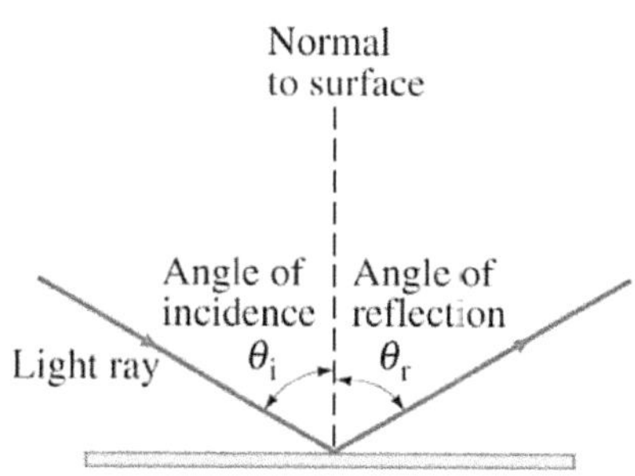

Angle of incidence θ_i equals the angle of reflection θ_r

Diffuse and specular reflection

When the reflecting surface is irregular (not smooth), light rays reflect off the irregularities; the reflected rays will not be parallel from irregular surfaces but radiate in many directions.

Therefore, the image is not clear and appears blurry to the viewer.

Diffuse reflection produces blurry images.

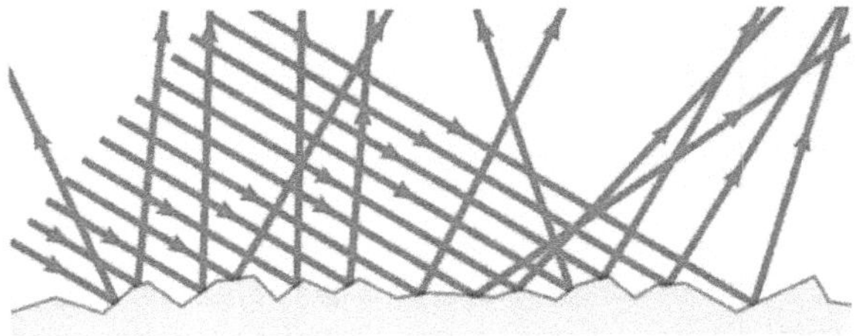

With diffuse reflection from irregular surfaces, the light is reflected at all angles

Visualize a person positioning themselves in the mirror to see something behind them—where they place themselves determines what they can see.

Diffuse reflection is when the observer sees reflected light from the source in multiple locations.

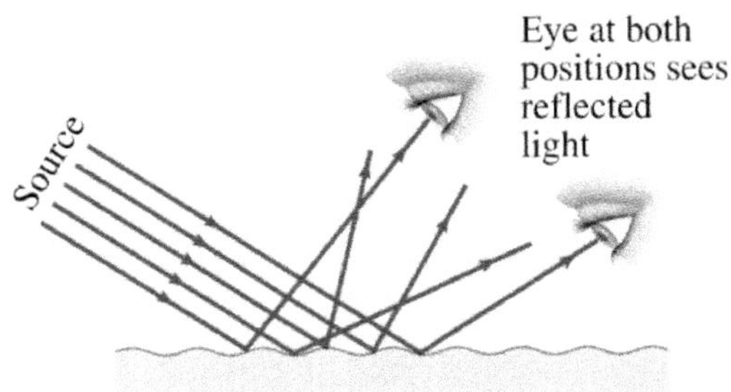

Diffuse reflection produces a clear image seen at multiple locations

With specular reflection (from a mirror), the eye must be in the correct position.

Specular reflection is observed in flat mirrors.

Specular reflection is when the reflected light can only be seen in a specific location.

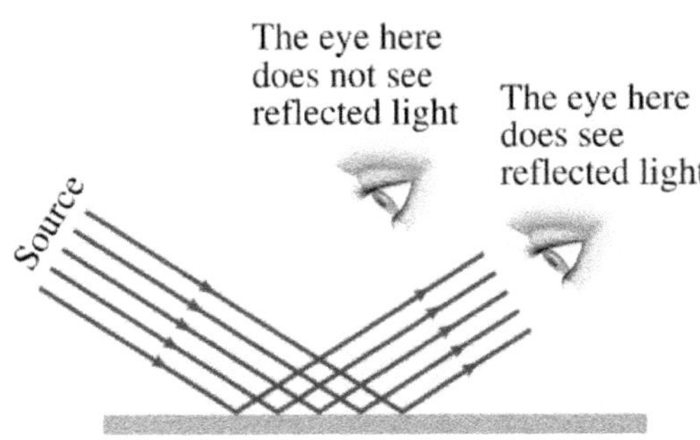

Specular reflection produces a blurry image seen at a single location

Refractive Index and Snell's Law

Refraction

Refraction is when light transmits from one medium to another and changes direction slightly based on the properties of the two mediums.

The angle at which the light ray bends, with respect to the normal, depends upon the *indices of refraction* (n) for each medium.

Snell's Law

Snell's Law gives the *angle of refraction* given the *angle of incidence:*

$$n_1 \sin(\theta_1) = n_2 \sin(\theta_2)$$

where n is the index of their respective refraction for each medium, and θ is the angle of the wave with respect to the normal of the medium interface.

Index of refraction

Index of refraction (n) is a relation of the light propagation speed v in a medium to that of the speed of light c in a vacuum.

N is always greater than 1 because light cannot travel faster than in a vacuum.

$$n = \frac{c}{v}$$

Waves moving to a *denser medium* (with a higher refractive index) bend *toward* the normal.

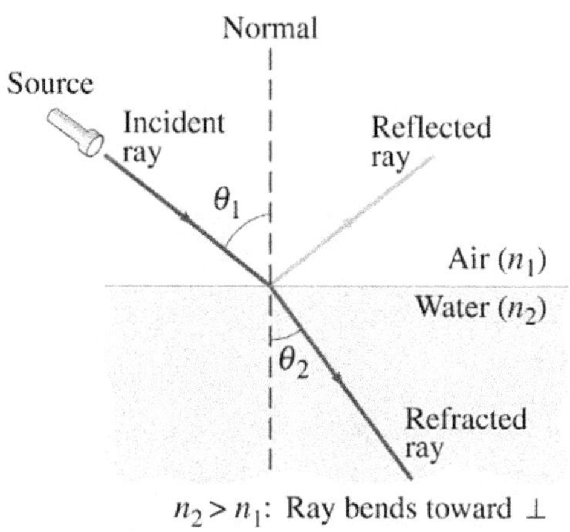

Light refracted from air to water bends toward the normal ($n_2 > n_1$)

Waves moving to a *less dense* medium (smaller refractive index) bend *away from* the normal.

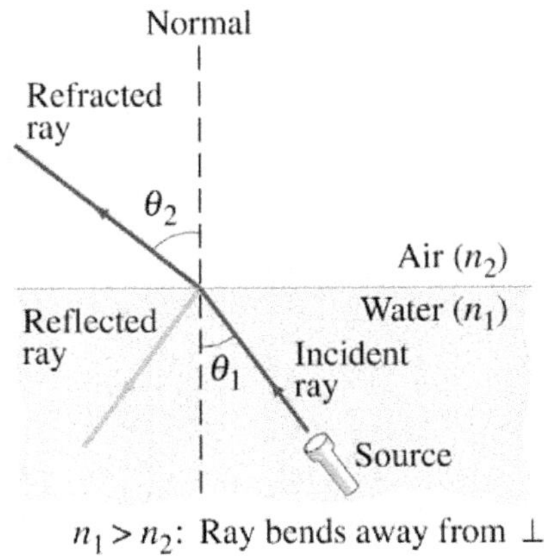

Light refracted from water into the air bends away from the normal ($n_1 > n_2$)

Mirages

Mirage (in optics) is the deceptive appearance of a distant object or object caused by bending light rays in layers of air of varying density.

A mirage is an optical phenomenon that creates the illusion of water and results from the refraction of light through a non-uniform medium.

There are two types of mirages: *superior mirages* and *inferior mirages*. Superior mirages refract, and inferior mirages reflect.

Superior mirages from cold surfaces

When an object apparently lifts into the air, superior mirages occur in cold places because of the density difference between the air near the ground and the air above.

Cold air is denser and has a higher refractive index than hot air. When light passes from the warmer air to the colder air near the ground, it bends upwards more than in the colder air above, producing mirage distortion.

Superior mirages are common in colder regions (hence the term "arctic mirage") during spring and summer and on the ocean.

Cooler air near the surface and warmer air above creates a refracting lens that elevates distant objects. Sometimes, objects appear out of focus; other times, they may appear upside-down.

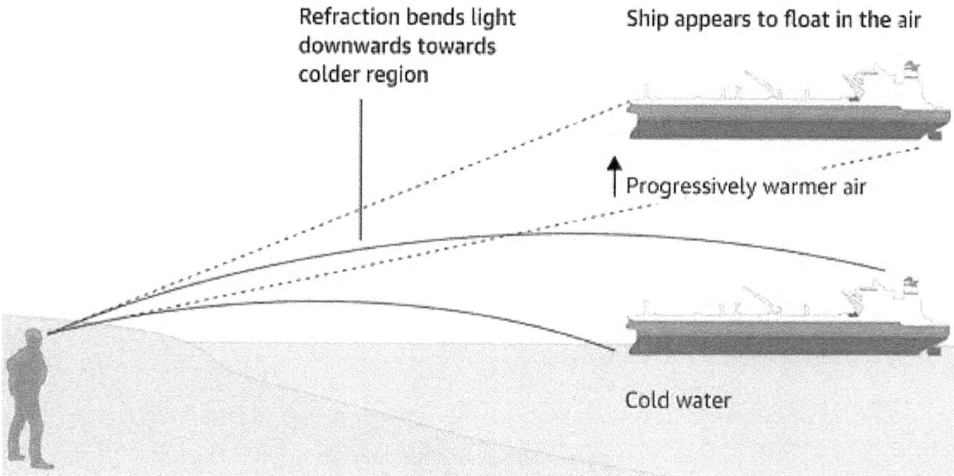

Warm air over cooler water produces an elevated image, and the ship appears higher

Inferior mirage from hot surfaces

Hot surfaces warm the air above them, producing air with a slightly lower index of refraction. When a light ray passes from cooler to warmer air, light curves so the observer sees an image reflecting off the ground.

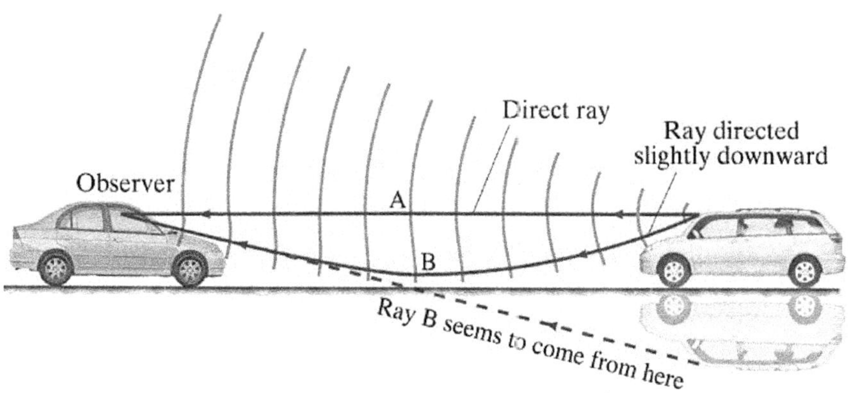

Warm air above the surface lowers the index of refraction, and observers see objects reflect

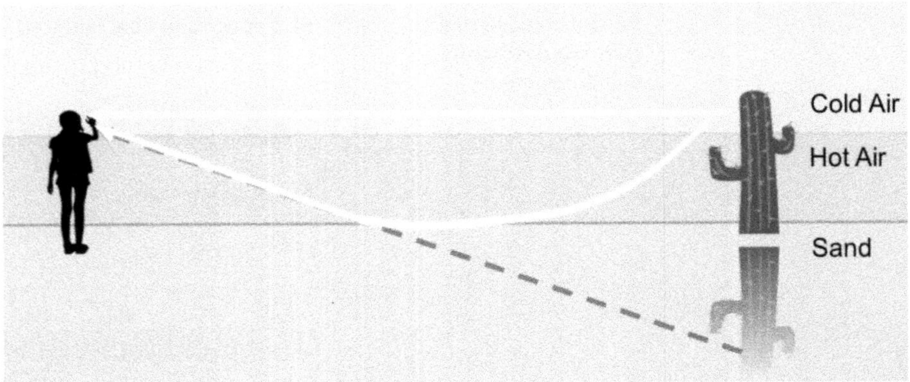

Cold air over hot air lowers the index of refraction and causes the image to appear lower on the horizon

Dispersion

Separating light into colors

Dispersion is the phenomenon that separates light into its colors.

Materials are distinguished from each other by their optical densities. Optical density is a measure of the tendency of a material to slow down light as it travels through it.

A light wave traveling through a transparent material interacts with the atoms of that material.

When a light wave impinges upon an atom of the material, it is absorbed by that atom. The absorbed energy causes the electrons in the atom to *vibrate*.

If the frequency of the light wave does not match the resonance frequency of the vibrating electrons, the atom *re-emits* the light at the same frequency at which it impinged upon it.

Light waves travel through the interatomic vacuum toward the next atom of the material. Once it impinges upon the next atom, the process of absorption and re-emission repeats.

Wavelengths

All wavelengths of light refract upon crossing a boundary between two mediums. However, different wavelengths of light refract at slightly different angles.

Lower wavelength light refracts the most. Thus, violet light refracts at a greater angle than blue light, which will refract at a greater angle than green light.

An example of dispersion is when white light passes through a prism. The diagram below represents a prism producing the visible spectrum of colors.

White light comprises the entire visible spectrum, in which each wavelength (color) refracts at slightly different angles.

Visible spectrum can be seen if a screen is at a specific distance from the prism because the white light is dispersed into its component wavelengths.

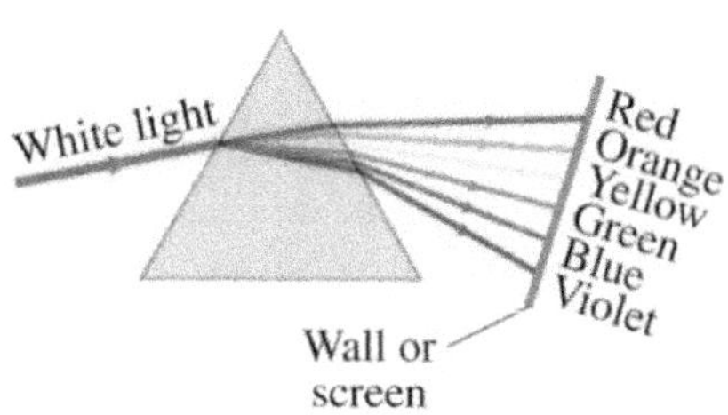

Prisms disperse white light into component colors based on wavelength

Optical density

Optical density of a material is the tendency of the atoms of a material to maintain the absorbed energy of the light wave as vibrating electrons before re-emitting it.

As the density of the medium increases, the number of particles per unit volume increases. Thus, more light gets obstructed, and the refractive index of the medium increases.

Increasing optical density increases the index of refraction.

Light waves travel through a vacuum at a speed of c (3.00×10^8 m/s) but travel through a transparent material at speeds less than c (i.e., speed of light in vacuum)

Index of refraction (n) is a quantitative expression of the optical density of a given medium.

Materials with a higher index of refraction *hold onto* absorbed light longer before re-emitting it to the interatomic void.

The more closely the frequency of the light wave matches the resonant frequency of the electrons of the atoms, the greater the *optical density* and the greater the *index of refraction.*

Prisms

Light waves slow more when passing through optically dense material.

Units for the index of refraction (n) depend upon the frequency of light.

For visible light, n does not show a slight variation in frequency. For example, some glass has (n) for frequencies of violet light at 1.53 and at 1.51 for red light.

Absorption and re-emission cause the higher frequency (lower wavelength) violet light to travel slower through glass than the lower frequency (higher wavelength) red light.

This difference in n for frequencies (and wavelengths) causes the dispersion of light by prisms.

Violet light, slowed more by absorption and re-emission, refracts more than red light.

Upon entry of white light at the first boundary of a triangular prism, there is a slight separation of the white light into the spectrum component colors.

The separation of wavelengths (or colors) becomes even more pronounced upon exiting the triangular prism at the second boundary, and ROYGBIV is observed.

Angle of deviation

Angle of deviation (delta, δ) is the angle between the direction of the refracted ray and the direction of the incident ray when a ray of light passes from one medium to another.

It is the angle between the incident ray of light entering the *first face* of the prism and the refracted ray emerging from the *second face* of the prism.

Because of the different indices of refraction for the different wavelengths of visible light, the angle of deviation varies with wavelength.

Colors of the visible light spectrum with shorter wavelengths (BIV; blue, indigo, and violet) deviate more from their original path than those with longer wavelengths (ROY).

Atmospheric rainbows

Atmospheric rainbows are created by dispersion, and water droplets are mini prisms.

Water droplets are roughly spherical and contain water with a refractive index that enables light to refract. Like for a prism, light passing through tiny raindrops or mist refracts and disperses.

When sunlight (white light) strikes water droplets suspended in air, it refracts and spreads into its constituent colors through dispersion.

Sunlight shines on a water droplet (at some angle), and light is refracted and dispersed.

The refracted light undergoes total internal reflection, which causes the light rays to fall on the front side of the droplet and emerge from the back.

Rainbow patterns are seven colors in a specific order.

Wavelength of red light is higher, it deviates least, while the wavelength of violet is lower and deviates the most. This is why the red light is at the bottom, and the violet light is at the top.

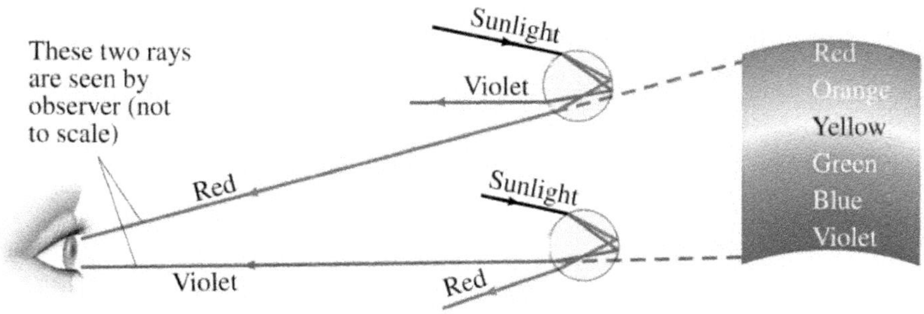

Atmospheric rainbows are created by the dispersion of light passing through tiny raindrops or mist refracts and disperses into the visible spectrum.

Notes for active learning

Total Internal Reflection

Critical angle

Critical angle is when the *angle of incidence* for which an incident light ray makes an *angle of refraction* equal to 90°.

Critical angle expression:

$$n_1 \sin(\theta_c) = n_2 \sin(90°)$$

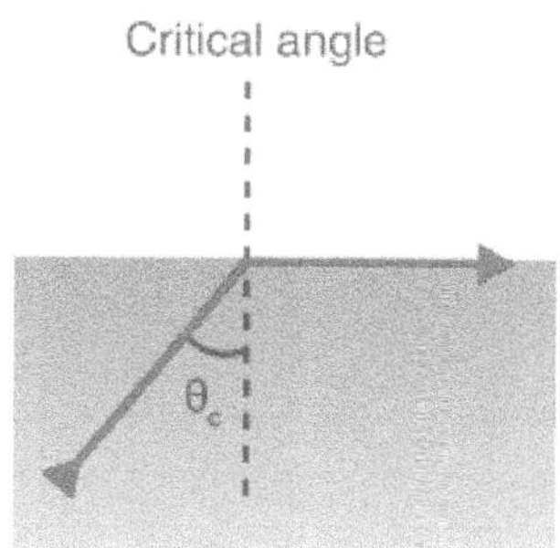

Critical angle when the angle of refraction equals 90°

Total internal reflection

Total internal reflection occurs when the *angle of incidence* is larger than the *critical angle* (point L below).

Angle of incidence > critical angle

No light is transmitted between the two mediums; light is only reflected in the original medium.

Total internal reflection occurs when light rays travel from a more optically denser medium to a less dense one.

Diagram below illustrates various incidence angles and their corresponding diffraction angles.

At points I and J, the angle of incidence < critical angle, and the light is refracted through the surface (a portion is also partially reflected).

Point K is the critical angle; the angle of refraction equals 90°.

At point L, the angle of incidence > critical angle. Therefore, all light is reflected, resulting in *total internal reflection.*

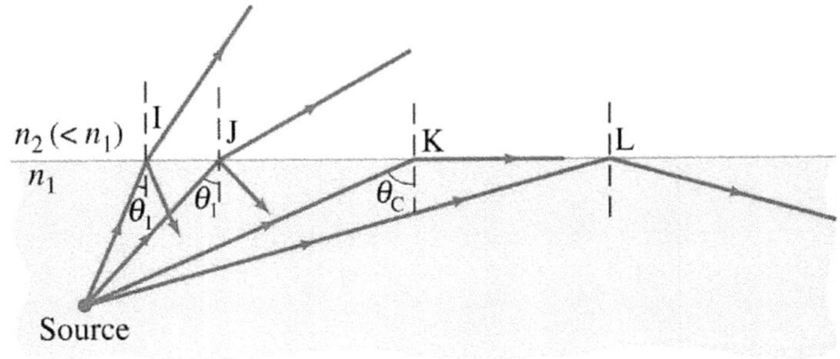

Total internal reflection is at L when the angle of incidence is greater than (>) critical angle

Critical angle *vs.* total internal reflection

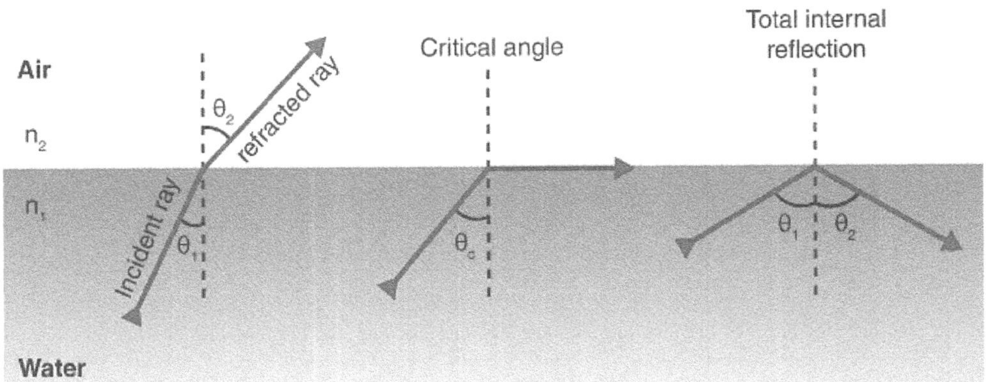

Incident vs. refracted ray with critical angle and total internal reflection shown

Fiber optics

Total internal reflection is essential because it is the principle behind fiber optics.

In a fiber optic cable, light is transmitted along the fiber regardless of loops or bends in the cable because the indices of refraction of the cable and outer sheath are such that the light always experiences total internal reflection.

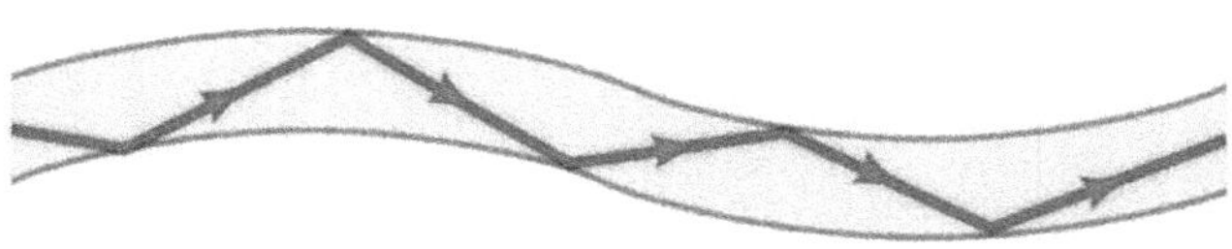

Fiber optic cable has total internal reflection when angle of incidence > critical angle

Mirrors

Spherical mirrors

Mirrors include *plane* and *spherical mirrors.*

Spherical mirrors allow users to see different images than possible with only flat mirrors.

Spherical mirrors have distinctive properties depending on the curvature of a reflective surface. For example, parking garages have convex mirrors mounted near blind corners; they allow drivers to see a much wider image of what is around a corner than a flat mirror.

Similarly, beauty mirrors are slightly concave and present a magnified image to the user.

Mirror is a reflective surface reflecting light, producing a *real or virtual image.*

Mirror curvature

Mirror curvature can be *concave* or *convex.*

Convex mirror is curved where the reflective surface *bulges towards* the light source and disperses light rays so they diverge.

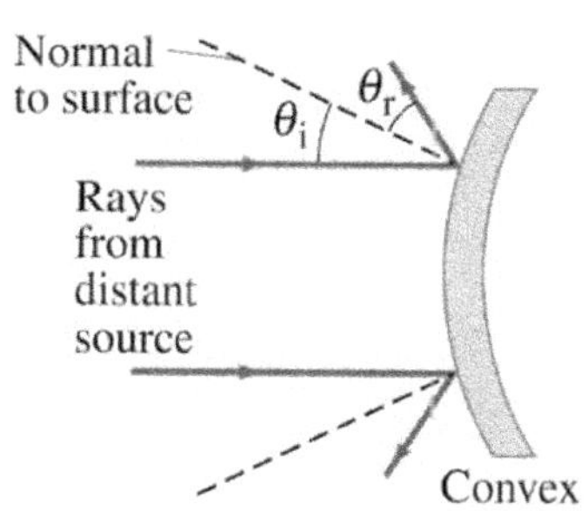

Convex mirrors disperse light rays such that they diverge

Concave mirror is curved where the reflective surface *bulges away* from the light source and focuses light rays so they converge.

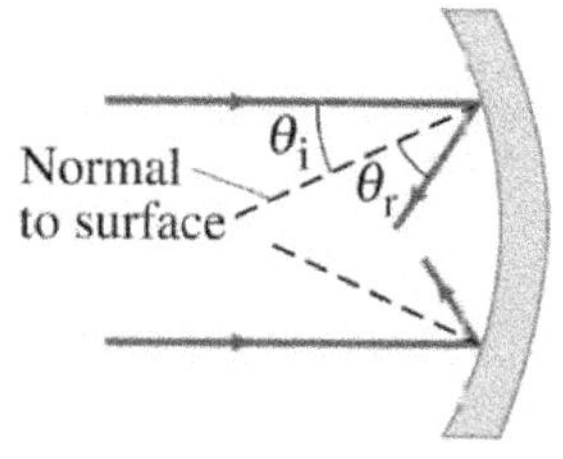

Concave mirrors focus light rays such that they converge

Radius of curvature

Convex and concave spherical mirrors have defined *radii of curvature* and *foci*.

Radius of curvature I is the radius of the mirror's curve, in spherical mirrors. Its curvature is constant throughout the mirror's surface.

Radius of curvature is the mirror's radius that forms a complete sphere.

Light ray AB is incident on a spherical mirror at point B and parallel to the principal axis.

Ray CB is normal to the surface at point B.

Ray CP equals ray CB, which equals the radius of curvature I.

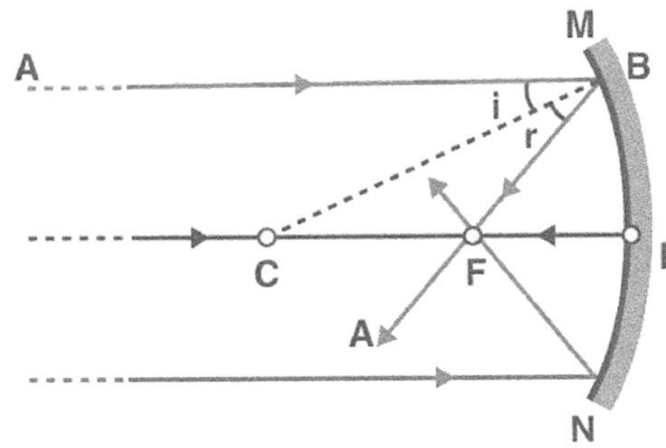

Focal point (f) and center of curvature I for concave mirror

Radius of curvature equals twice the focal length for spherical mirrors.

$$R = 2f$$

where f is the focal length (m), and R is the radius of curvature (m).

Focal length

Focal length is the distance from the mirror surface to where light rays converge (theoretically converge, in the case of convex mirrors).

Concave mirrors have a *positive focal length.*

Convex mirrors have a *negative focal length.*

Concave and convex mirrors have focal lengths equal to half the radius:

$$f = \frac{1}{2}R$$

where f is the focal length (m), and R is the radius of curvature (m).

Concave center of curvature

Concave mirrors have point C, the *center of curvature,* and where the radius of curvature originates (center of a circle).

Point F is the focus and demonstrates where all light rays focus if reflected off the mirror.

F and *r* are the focal length and radius of curvature, which are distances to points F and C.

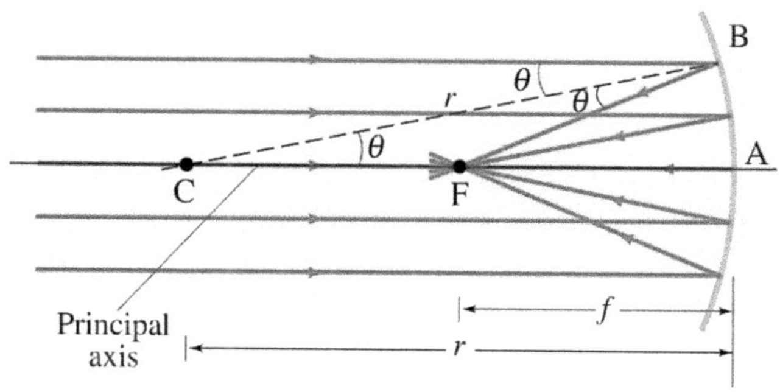

Concave mirror with focal length (f) and radius of curvature I noted

Convex center of curvature

Convex mirrors have centers of curvature, focal point, radius of curvature, and focal length; however, light is incident upon the outer edge of the mirror and *disperses* rather than converges.

Center of curvature I is the center of the circle (sphere) of which the mirror is an arc.

For a convex mirror, the focal point is the point from which *light appears to have originated after reflection* from the mirror.

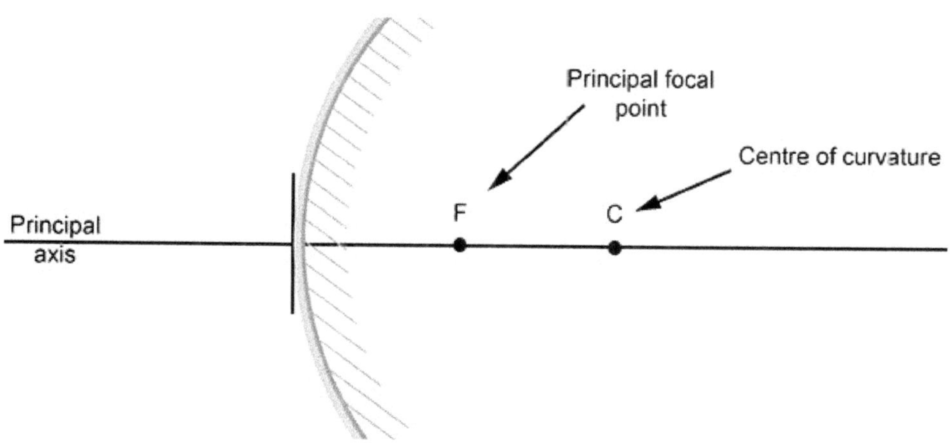

Principal focal length equals ½ radius, center of curvature equals ½ radius, and radius of curvature equals the distance from mirror to point C (center of curvature)

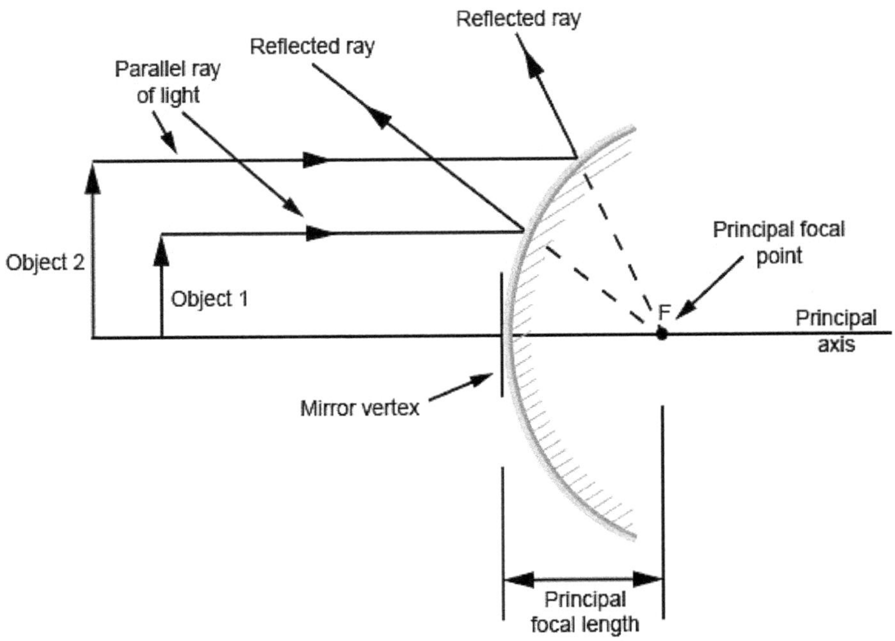

Multiple objects for convex mirrors produce specific reflected rays with a common focal point

Mirror ray tracing

Know the relationship between the object held up to the mirror and the image formed by it.

For example, a face reflected on the inside of a shiny spoon appears inverted and smaller.

Depending upon the object's distance from the mirror and characteristics of the mirror, the formed image can be:

smaller (or larger),

virtual (or real), and

erect (or inverted).

Real images form when light rays converge upon a point in a real location.

Virtual images form from diverging light rays. Virtual image is where light rays appear to diverge from (the point is not there, hence *virtual*).

Ray diagrams

Ray diagrams identify the image formed by a mirror and allow the user to visualize the image formed, depending upon the object's location with respect to the mirror.

Three rays are drawn for a concave mirror (shown below):

Ray 1 from the top of the object parallel to the axis; after reflection off the mirror, it passes through the focal point (F).

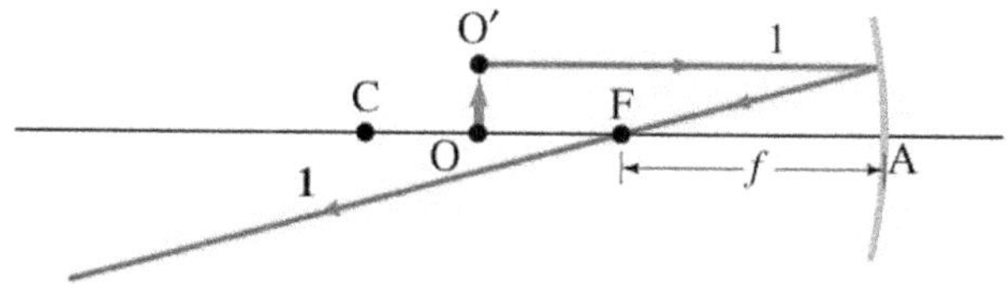

Ray 1 is from the top of the object (O') parallel to the axis, which, after reflecting, passes through the focal point (F)

Ray 2 from the top of the object through the focal point; after reflection, it is parallel to the axis.

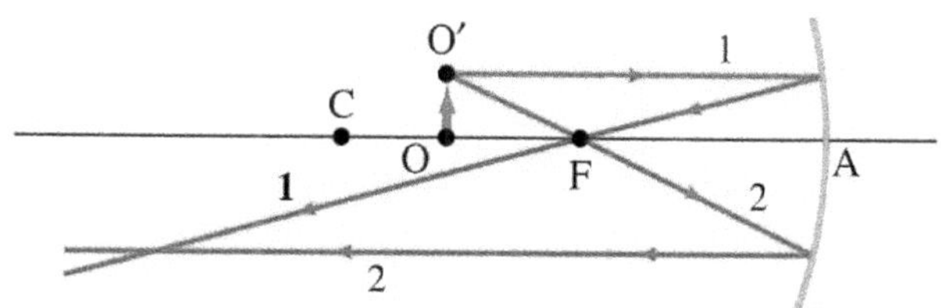

Ray 2 goes through the focal point (F) and reflects parallel to the axis

Ray 3 is perpendicular to mirror and passes through the object's height and center of curvature.

Convergence point (I') is the location of the image.

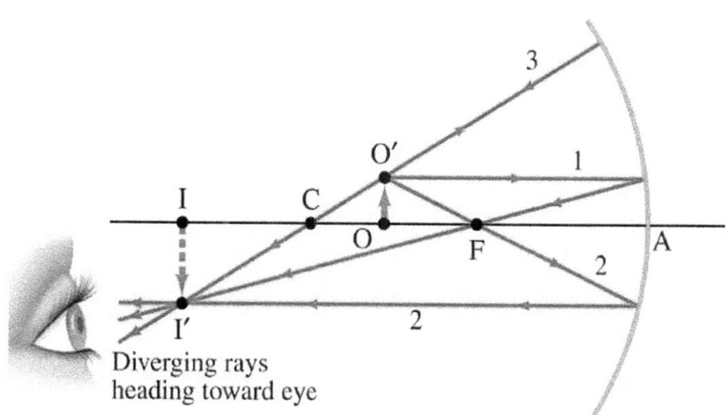

Ray 3 is perpendicular to the mirror, reflects on itself, and goes through the center of curvature I, with the convergence point (I') being the location of the image

Image height and orientation

A line from the center axis to the convergence point determines *image height* and *orientation.*

Image formed is *inverted, larger*, and *real* (image by convergence of light rays on a real point).

Object location determines the image.

Below is an image formed by an object *in front of* the focal point of a concave mirror.

Image is *erect, larger*, and *virtual* (formed by convergence of light rays on a virtual point)

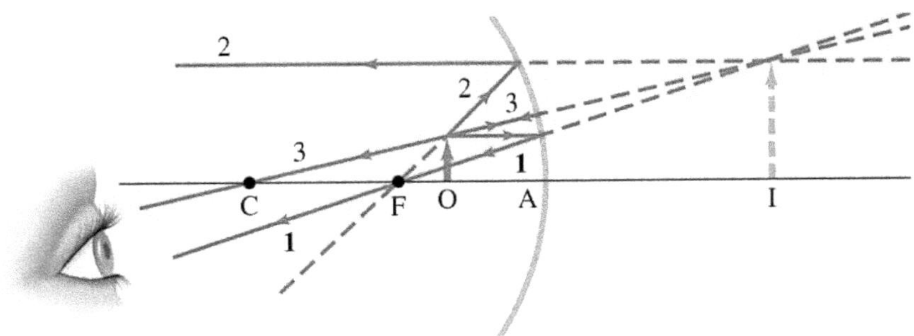

Concave mirror produces a larger, erect, virtual image when the object is placed within the focal point

Concave real and virtual images

Concave mirrors produce *real* and *virtual images* depending on where the object is located.

If the object is *outside the radius* of the sphere of the mirror, the image is *real.*

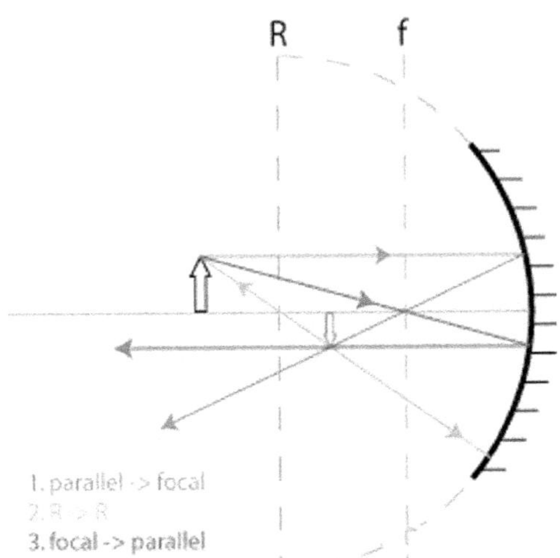

Concave mirror produces a real image if the object is beyond the radius of curvature

If the object is *inside the radius* of the sphere of the mirror, the image is *virtual.*

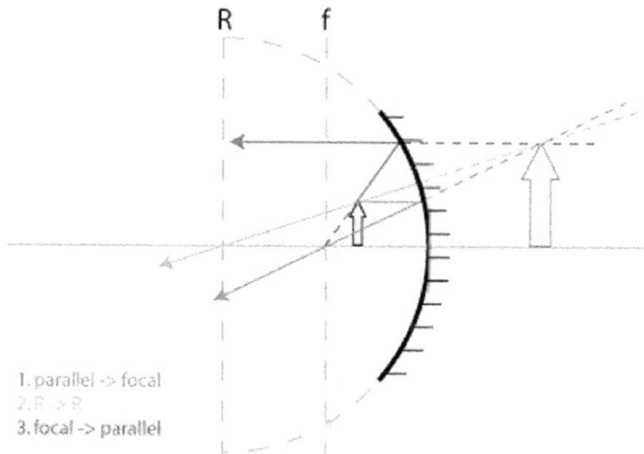

Concave mirror produces a virtual image if the object is inside the radius of curvature

Convex mirrors produce virtual images

Convex mirrors produce *virtual images* and are located behind the mirror because the curves of the mirrors are different.

When the object is far from the mirror, the image is upright and located at the focal point.

As the object approaches the mirror, the image approaches it and grows until its height equals that of the object.

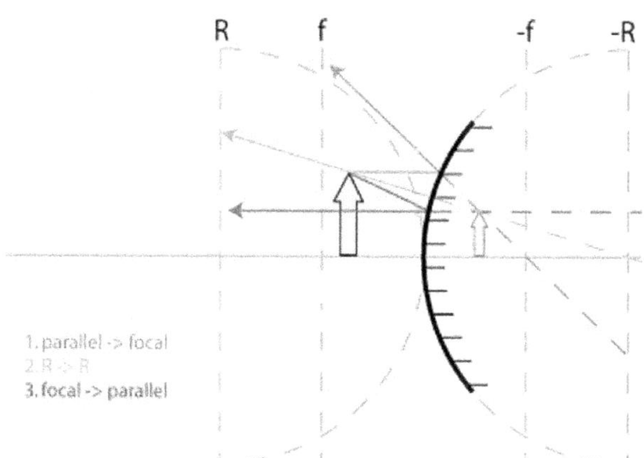

Convex mirrors produce virtual images

Object–image relationship for mirrors

Object distance	Concave mirror	Convex mirror
Before C	inverted, smaller, real	erect, smaller, virtual
At C	inverted, equal size, real	erect, smaller, virtual
Between C and F	inverted, larger, real	erect, smaller, virtual
At F	no image	erect, smaller, virtual
Before F	erect, larger, virtual	erect, smaller, virtual

* C is the radius of curvature, and F is the focal point.

Three-step problem-solving for mirrors

1) Draw a parallel line from the object as it reflects from the mirror, intersecting the focal point.

Now, which focal point should be intersected, left or right?

Concave mirrors focus the ray to the *left focal* point.

Convex mirrors diverge the ray, which cannot focus; extrapolate to the *right focal* point.

2) Draw a line intersecting point R on the principal axis (left or right?); the ray drawn should bounce back its original path and not reflect elsewhere. Analyze the mirror.

With two rays drawn, an intersection can be made to guide the drawing of the last ray.

3) Final ray intersects the focal point and bounces off the mirror parallel to the principal axis.

Which focal point intersects (and is extrapolation necessary)? Only one combination for the ray to fit the intersection made by the previous two rays.

Draw the parallel line to cross the intersection already made by the previous two rays.

Thin Lenses

Optical lens

Lenses bend and focus light using the *refractive properties* of the lens material. Lenses are essential in many devices and scientific instruments.

For example, lenses in a microscope bend and focus light, so objects become magnified.

Eyeglasses correct vision problems by focusing the light in front of the eyes so that those who wear glasses can see without blurry vision.

Similarities exist between lenses and mirrors for image formation, but differences exist.

Convex lenses are like a concave mirror (both converging) except:

- *Real images* are on the *opposite side of the lens* as the object because light travels through the lens and focuses on a screen behind the lens.
- *Virtual images* formed are on the *same side* of convex lenses as the object because light cannot focus in front of a lens and be cast on a screen.

Concave lenses are like a convex mirror (both diverging) except:

- *Virtual images* formed are on the *same side* of the concave lens as the object; light cannot focus in front of a lens and be cast on a screen.

Converging lenses

Most lens problems approximate the lens as a *thin lens.*

Thin lenses are those whose thickness is negligible compared to their *radius of curvature.*

Thin lenses may be converging (i.e., convex) or diverging (i.e., concave).

Like with mirrors, the light that passes through the lens converges on the focal point of the lens (two focal points, one on each side).

Radius of curvature is the same as for a spherical mirror, except the lens is double-sided and has two radii of curvature (most times, these are equal).

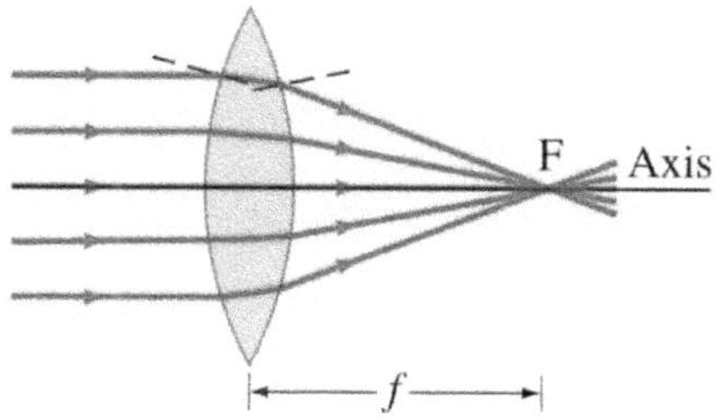

Converging lens

Diverging lenses

Diverging lens (concave) make parallel light diverge; the focal point is where the diverging rays would converge if projected back.

Like a converging lens, a diverging lens has two radii of curvature (most times, these are equal).

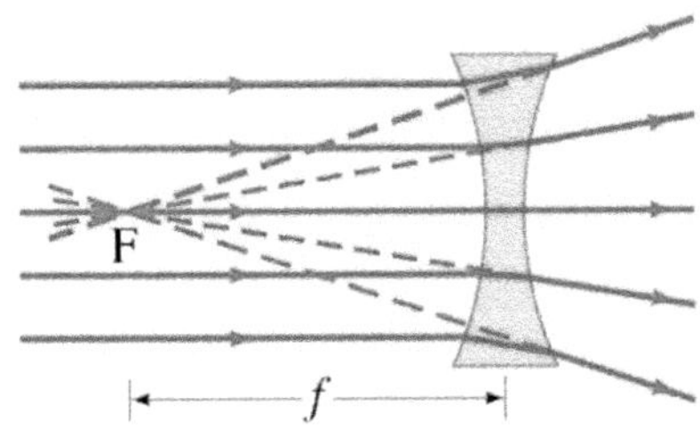

Diverging lens

Converging lens ray tracing

Ray tracing for thin lenses is like that for mirrors.

For example, the figure below demonstrates ray tracing through a converging lens.

After the rays have been drawn, the *convergence point* is the *location of the image*.

Draw a line from the center axis to the *image height* and *orientation* convergence point.

For example, the image formed is *inverted* and *real*.

Ray 1 is from the top of the object height parallel to the axis; after transmission through the lens, it *passes through the focal point*.

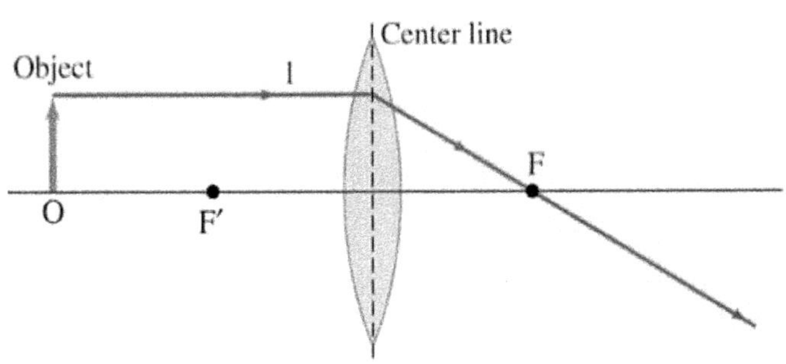

Ray 1 leaves object parallel to the axis and refracts through the focal point behind the lens

Ray 2 is from top of the object through focal point, after transmission, *parallel* to the *axis*.

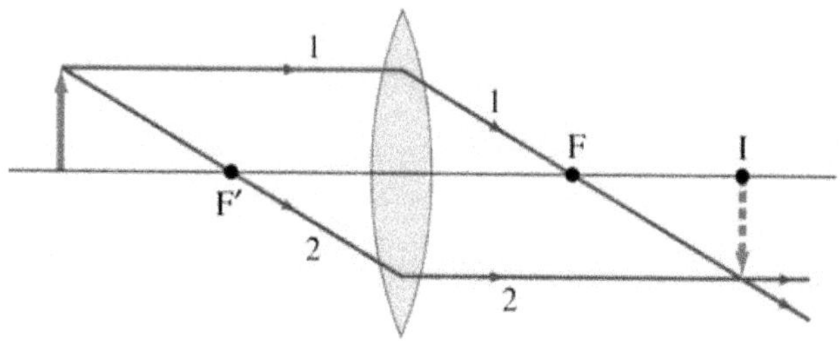

Ray 2 passes through the focal point (F) in front and parallel to the axis behind the lens

Ray 3 is from the top of the object height through the *center of the lens*.

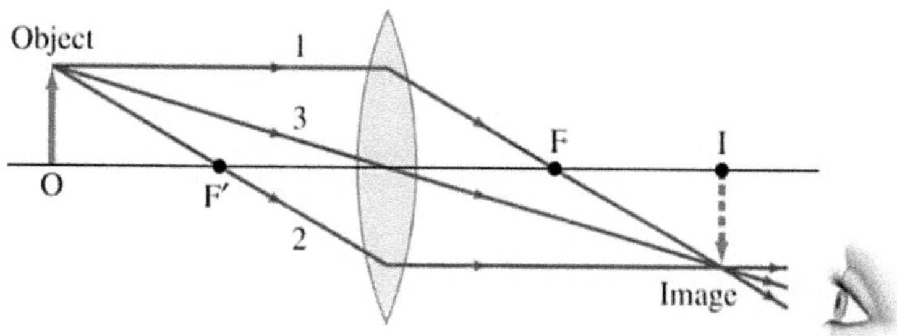

Ray 3 passes through the center of the lens

Diverging lens ray tracing

For a diverging lens, the same three rays can be used; the image is *upright* and *virtual*.

Convex lenses (concave mirrors) create a *real or virtual image*, depending on object's position.

Image beyond the focal length:

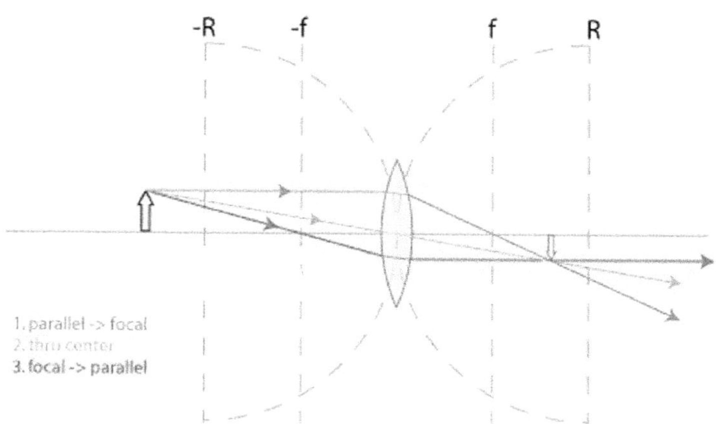

Convex lens produces real images if the object is beyond the focal length

Image within the focal length:

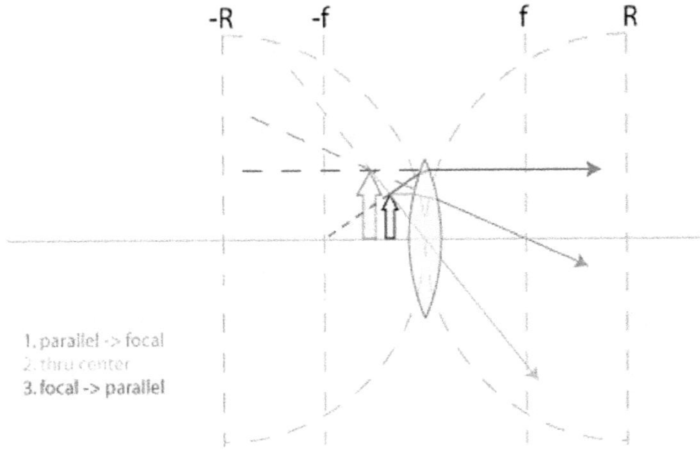

Convex lens produces a virtual image if the object is within the focal length

Concave lenses (like convex mirrors) produce only virtual images.

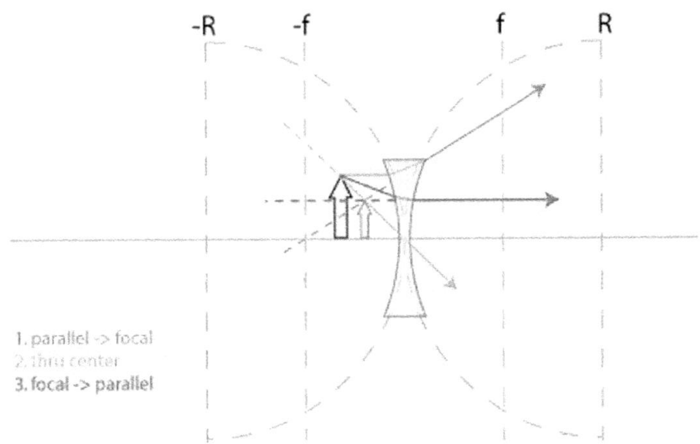

Concave lens produces a virtual image if the object is within the focal length

For lenses, use the method like drawing rays for mirrors:

1. Draw the parallel → focal point ray. Given the converging or diverging nature of the lens, it should make sense which focal point the ray should hit or extrapolate.
2. Draw a ray intersecting the center of the lens.
3. Using the intersection already made by the previous two rays as a guide, draw the focal point → parallel ray.

   Draw the parallel line first and force it to pass through the intersection already made by the previous two rays.

Sign conventions for mirrors and lenses

Determine the distance from the mirror or lens that an image forms by the thin lens equation:

$$\frac{1}{d_o} + \frac{1}{d_i} = \frac{1}{f}$$

where d_o is the distance of the object away from the lens or mirror (m), d_i is the distance of the image away from the lens or mirror (m), and f is the focal length of the lens or mirror (m).

Thin lens equation is relatively simple, but the proper sign conventions must be followed for each type of mirror and lens.

For lenses, differences depend on whether the lens is *converging* or *diverging*.

Focal length is *positive* for *converging* lenses.

Focal length is *negative* for *diverging* lenses.

Object distance is *positive* for objects on the *same side* as the light entering the lens.

Object distance is *negative* for objects on the *opposite side* as the light enters the lens.

Image distance is *positive* if the image is on the opposite side of the light entering the lens.

Image distance is *negative* if the image is on the same side of the light entering the lens.

For mirrors, sign conventions are different.

Focal length is *positive* for concave mirrors.

Focal length is *negative* for convex mirrors.

Object distance is *positive* when the object is *in front of* the mirror.

Object distance is *negative* if the object is *behind* the mirror (i.e., virtual object; this is rare).

Image distance is *positive* if the image is *in front of* the mirror (real image).

Image distance is *negative* if the image is *behind* the mirror (virtual image).

Perform as many steps as possible when solving problems using the thin lens equation.

Draw a ray diagram and place the image where the principal rays intersect.

Solve for unknowns by following the sign conventions and check that the answers are consistent with the ray diagram.

Magnification

Magnification of an object, in relation to image height, can be obtained for lenses and mirrors.

Magnification equation is expressed as:

$$M = \frac{h_i}{h_0} = -\frac{d_i}{d_0}$$

where M is magnification, h_i is height of the image (m), h_0 is height of the object (m), d_o is distance of the object away from the lens or mirror (m), and d_i is distance of the image away from the lens or mirror (m).

Like the lens equation, the magnification equation has specific sign conventions for the type of mirrors and lenses.

Magnification is *positive* for all lenses when the image is *upright* with respect to the object.

Magnification is *negative* when the image is *inverted.*

Magnification is expressed as positive for all types of mirrors when the image is upright with respect to the object.

Magnification is expressed as negative when the image is inverted.

Lens power

Power of a lens is not related to magnification.

Power of a lens is expressed as the inverse of focal length, where lens power has units of diopters (D) equal to m^{-1}.

Power of a lens is positive if it is converging.

Power of a lens is negative if it is diverging.

$$P = \frac{1}{f}$$

where P is the lens power (D).

Lensmaker's equation

Lensmaker's equation relates the radii of curvature of a lens and the produced focal length:

$$\frac{1}{f} = (n-1)\cdot\left(\frac{1}{R_1} + \frac{1}{R_2}\right)$$

where f is the focal length of the lens (m), n is the index of refraction of the lens material, R_1 is the radius of curvature of the first lens face (m), and R_2 is the radius of curvature of the second lens face (m).

Combination of Lenses

Multiple lenses

Combinations of lenses multiply the magnification of objects.

The simplest combination of lenses is when two lenses are in contact.

For example, light microscopes and telescopes magnify objects. These devices contain multiple lenses, allowing users to magnify objects beyond what a single lens can produce.

Real images formed by a lens can be used as the object for another lens.

Magnification

Magnification by multiple lenses is the *product of individual magnifications*:

$$M_{total} = M_1 M_2$$

where M_{total} is the total magnification, M_1 is the magnification of the first lens, and M_2 is the magnification of the second lens.

Total power

Total power of the combination of lenses is the sum of the two lens's powers:

$$P_{total} = P_1 + P_2$$

or

$$\frac{1}{f_{total}} = \frac{1}{f_1} + \frac{1}{f_2}$$

If some distance separates two lenses, then the total power of the combination is expressed as:

$$\frac{1}{f_{total}} = \frac{1}{f_1} + \frac{1}{f_2} - \frac{d}{f_1 f_2}$$

where d is the distance between the two lenses (m)

Notes for active learning

Lens Aberration

Spherical aberration

No lens is ideal because light rays converge in slightly different spots. This phenomenon is an *aberration*, and there are several types.

Spherical lenses and mirrors experience *spherical aberration* because the curvature of the lens/mirror is often large, and the rays striking the outer edges refract/reflect more and bend more than the rays striking the center portion of the lens/mirror.

Spherical aberration is reduced by utilizing only the *center portion* of the lens or mirror.

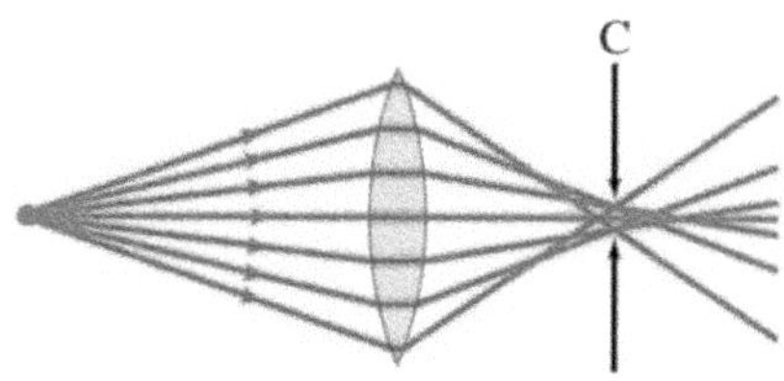

Spherical aberrations of converging lens may be reduced by using a mirror or lens

Chromatic aberration

Chromatic aberration is when light of different wavelengths focuses at different points.

Since different wavelengths of light refract at slightly different angles, this is *dispersion.*

When light shines through a lens, this results in different focus points, depending upon the wavelength of light.

Blue light is refracted more than red light, so blue light focuses at an earlier point than red light.

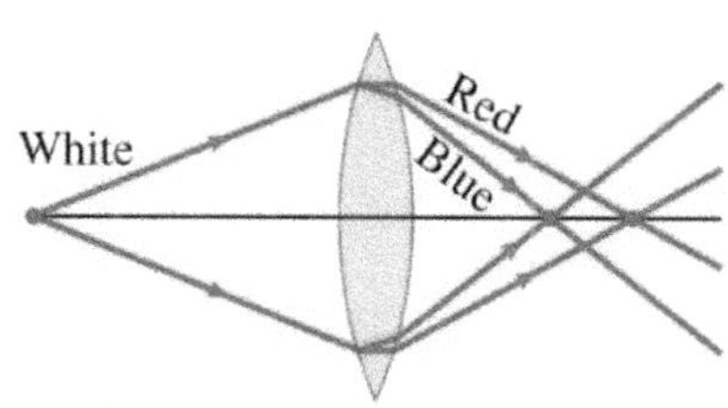

Chromatic aberrations focus light of different wavelengths to produce colors

Achromatic doublet

Chromatic aberration is reduced by an *achromatic doublet* where a second lens is attached to the first lens made of dissimilar materials.

When light passes through the chromatic aberration, it is corrected in the second lens, which focuses the wavelengths to one point.

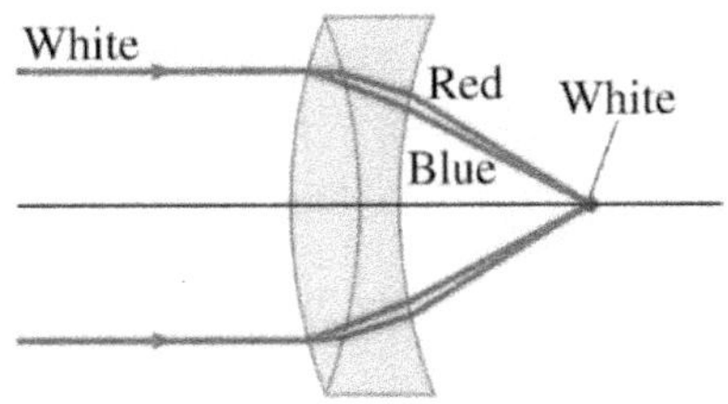

Chromatic aberrations are corrected for with a second lens

Distortion

Distortion is an aberration caused when lens parts have slightly different magnification factors.

Diagrams below illustrate two types of distortion.

Barrel distortion is where the magnification decreases with distance from the axis.

Pincushion distortion is where magnification increases with distance from the optical axis.

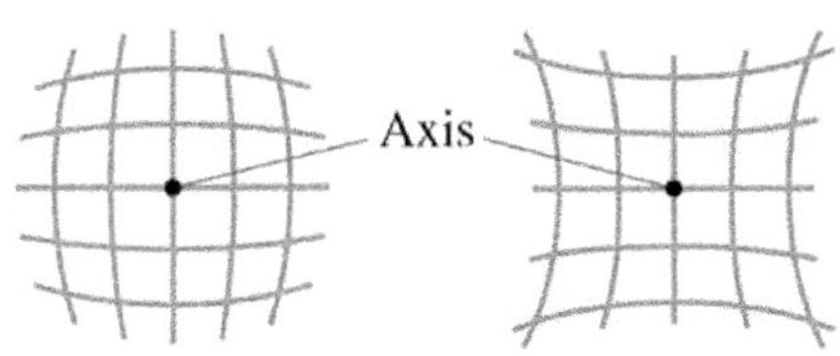

Barrell distortion *Pincushion distortion*

Optical Instruments

Human eye

Human eye manipulates electromagnetic waves with an *adjustable lens* and an *iris.*

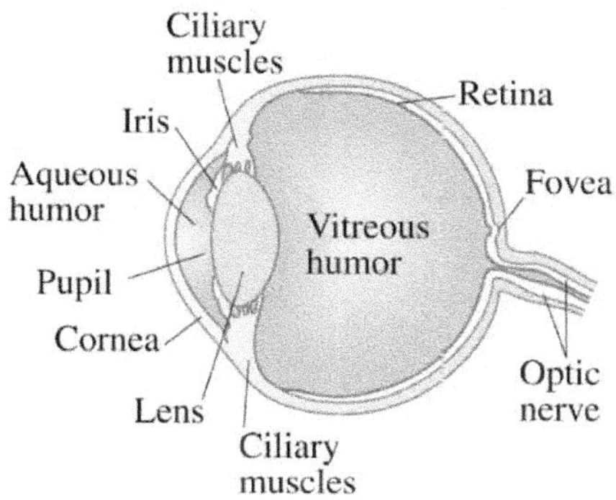

Anatomy of the human eye for vision perception

Human eye makes minor adjustments to the lens shape to allow objects at different distances to be focused on the retina.

Light refraction occurs at the *cornea's surface* while the lens makes fine-tuned adjustments.

Lens system

The eye lens system, adjusting to compensate for *object distance*, is shown below.

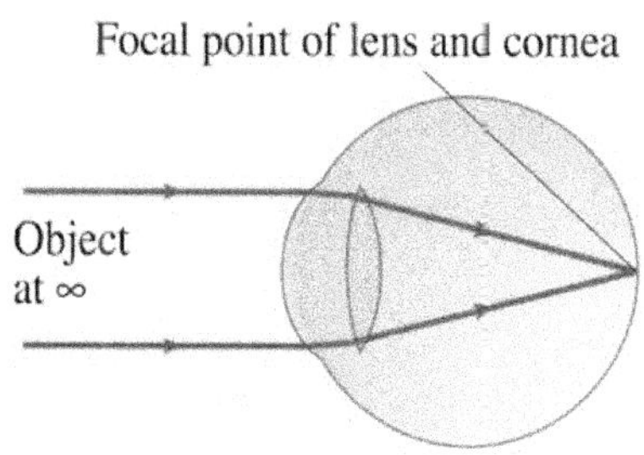

Parallel light rays are focused on the retina of the eye

When the object is far away (far away objects have rays assumed to be parallel), the lens of the eye adjusts so that its radius of curvature creates a focal point at the retina.

Near point is the closest distance the eye can focus clearly; the typical distance is about 25 cm.

Far point is the farthest distance an object can be seen clearly; the typical distance is infinity.

Corrective lenses for myopia and hyperopia

Nearsightedness (*myopia*) occurs when a person's far point is too close.

Farsightedness (*hyperopia*) occurs when a persons near point is too far away.

Nearsightedness can be corrected with a diverging lens.

Myopia (nearsightedness) is when the lens refracts light, so the light diverges before the cornea.

Consequently, the higher refraction of the eye is compensated for, and the focal point occurs at the retina, producing a clear image.

In the diagram on the left below, an eye with myopia refracts light too much, causing the focal point to be located before the retina.

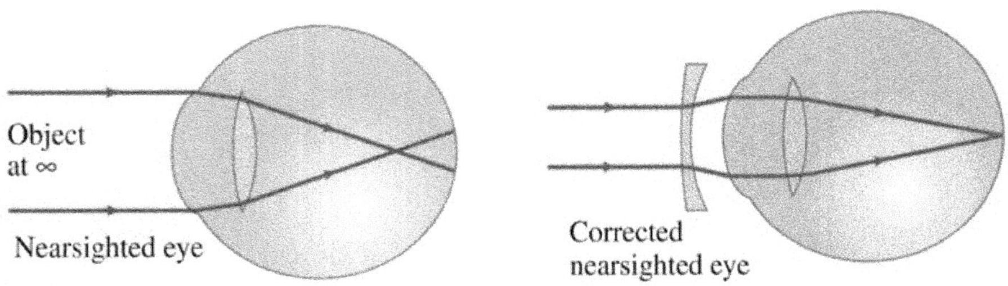

Myopia focuses light before the retina and is corrected with a diverging lens

Hyperopia (farsightedness) is corrected with a converging lens.

Hyperopia is when the eye refracts incoming light too little; the focal point is behind the retina.

Farsightedness can be corrected with a converging lens, which brings the light rays to a more parallel orientation before entering the eye.

The lower refractive index of the eye is compensated, and the focal point is at the retina.

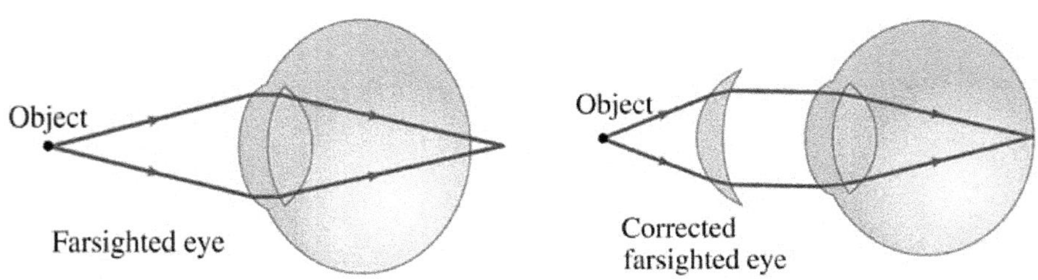

Hyperopia focuses light past the retina and is corrected with a converging lens

Magnifying glass

Magnifying glass is a converging lens that focuses objects closer than the near point so that they create a larger image on the retina.

Figure below demonstrates the use of a magnifying glass.

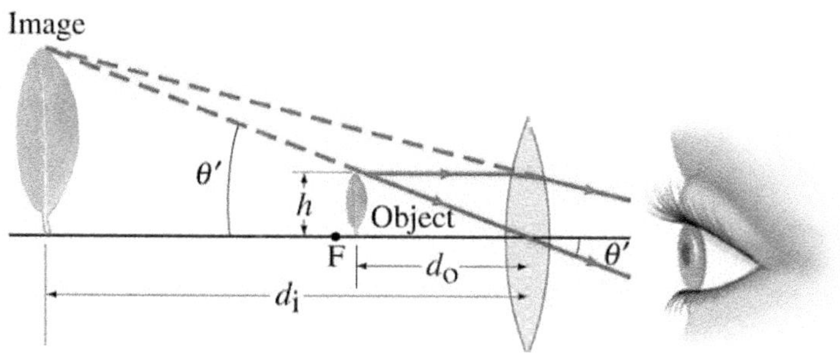

Magnifying glass enlarges objects

Angular magnification

Angular magnification describes the power of a magnifying glass:

$$M = \frac{\theta'}{\theta}$$

$$\theta \approx \frac{h}{f}$$

$$\theta' \approx \frac{25cm}{h}$$

If the eye is relaxed (*N* is the near point distance and *f* the focal length), magnification is:

$$M = \frac{\theta'}{\theta} = \frac{h/f}{h/N} = \frac{N}{f}$$

$$M = \frac{25cm}{f}$$

If the eye is focused at the near point, magnification is given as:

$$M = \frac{N}{f} + 1$$

$$M = \frac{25cm}{f} + 1$$

Refracting telescope

Refracting telescope consists of two lenses at opposite ends of a long tube.

Objective lens is closest to the object.

Eyepiece is closest to the eye.

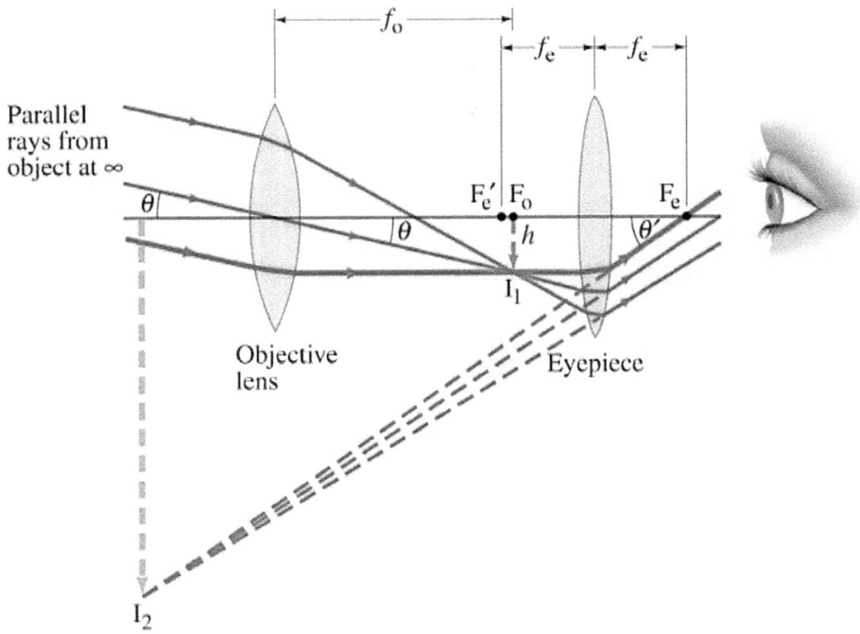

Parallel rays from an object at infinity enter the objective lens

Refracting magnification

Magnification is given by:

$$M = \frac{\theta'}{\theta} = \frac{(h/f_e)}{(h/f_o)}$$

$$M = -\frac{f_o}{f_e}$$

Chapter Summary

Light, refraction, and reflection

- Light paths are rays.
- Visible spectrum of light ranges approximately from 400 nm to 750 nm.
- Two light sources are coherent if they have the same frequency and maintain the same phase relationship.
- Index of refraction varies with wavelength, leading to dispersion.
- Boundaries between dissimilar materials and regions of the same material under different conditions (e.g., temperature or pressure) cause a change in the index of refraction.
- Index of refraction:

$$n = \frac{c}{v}$$

- Law of refraction (Snell's Law):

$$n_1 \sin(\theta_1) = n_2 \sin(\theta_2)$$

- Angle of reflection equals the angle of incidence.
- Total internal reflection occurs when the angle of incidence > critical angle:

$$\sin(\theta_C) = \frac{n_2}{n_1} \sin(90°) = \frac{n_2}{n_1}$$

Diffraction—single-slit and double-slit

- Wavelength of light in a medium with an index of refraction n:

$$\lambda_n = \frac{\lambda}{n}$$

- Wavelength can be measured precisely with a spectroscope.
- Young's double-slit experiment demonstrated interference:

  constructive interference occurs when:
  $d \sin(\theta) = m\lambda$, $m = 0, 1, 2, 3, \ldots$

  destructive interference occurs when:
  $d \sin(\theta) = (m + \frac{1}{2})\lambda$, $m = 0, 1, 2, 3, \ldots$

- *Diffraction grating* has many small slits or lines and the same condition for constructive interference.

- Light bends around obstacles and openings, producing diffraction patterns.
- Light passing through a narrow slit produces a bright central maximum of width:

$$\sin(\theta) = \frac{\lambda}{D}$$

- Interference occurs by reflections from the front and back surfaces of a thin film.

Polarization

- Light whose electric fields are all in the same plane is plane-polarized.
- Intensity of plane-polarized light is reduced after it passes another polarizer:

$$I = I_o \cos^2(\theta)$$

- Light can also be polarized by reflection; it is entirely polarized when the reflection angle is the polarization angle:

$$\tan(\theta_p) = n$$

Mirrors

- Plane mirror: image is *virtual*, *upright,* and *same size as object.*
- Real image means light passes through it.
- Virtual image means light does not pass through it.
- Spherical mirrors can be concave or convex.
- Focal length of the mirror:

$$f = \frac{r}{2}$$

- Mirror equation:

$$\frac{1}{d_o} + \frac{1}{d_i} = \frac{1}{f}$$

- Magnification:

$$m = \frac{h_i}{h_o} = -\frac{d_i}{d_o}$$

Lenses

- Converging lenses focus incoming parallel rays to a point.
- Diverging lenses spread incoming rays so they appear to come from a point.
- Power of a lens:

$$P = \frac{1}{f}$$

- Thin lens equation:

$$\frac{1}{f} - \frac{1}{d_i} = \frac{1}{d_o}$$

or

$$\frac{1}{d_o} + \frac{1}{d_i} = \frac{1}{f}$$

Lens aberrations

- *Spherical aberration* occurs when rays far from the axis do not go through the focal point.
- *Chromatic aberration* occurs when different wavelengths have different focal points.

Optical devices

- Resolution of optical devices is limited by diffraction.
- Nearsighted vision is corrected by a diverging lens.
- Farsighted vision is corrected by a converging lens.
- Simple magnifier: an object at the focal point

$$M = \frac{\theta'}{\theta} = \frac{h/f}{h/N} = \frac{N}{f}$$

- Magnification:

$$M = \frac{\theta'}{\theta} = \frac{(h/f_e)}{(h/f_o)} = -\frac{f_o}{f_e}$$

Comparing lenses and mirrors

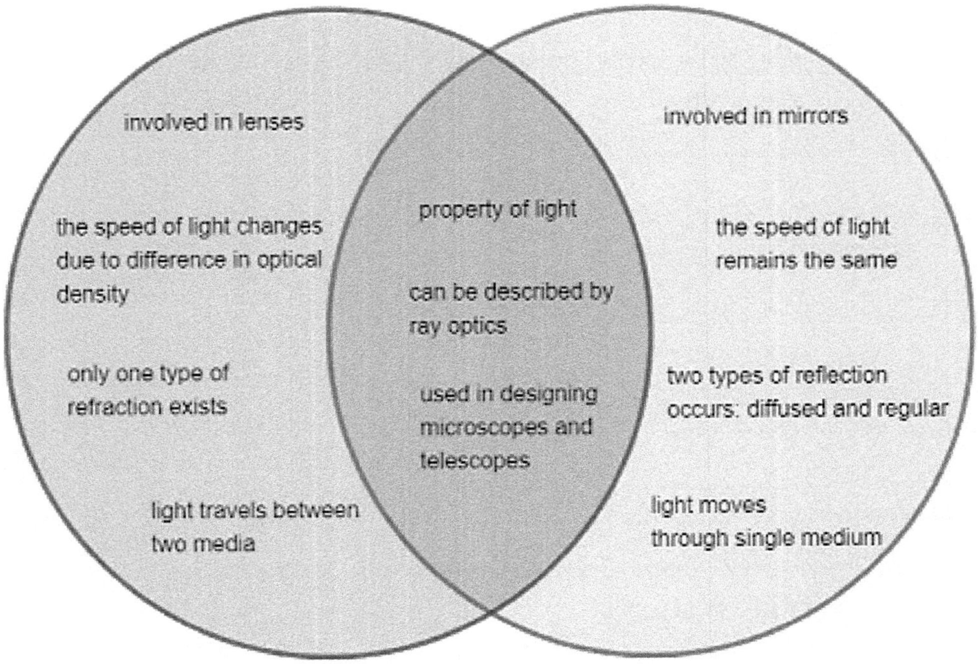

Practice Problems: Light & Optics

1. Which visible spectrum color has photons with the most energy?

A. Violet
B. Green
C. Orange
D. Red
E. Yellow

2. An index of refraction of less than one for a medium implies that:

A. the speed of light in the medium is lower than the speed of light in a vacuum
B. refraction is not possible
C. the speed of light in the medium is the same as the speed of light in a vacuum
D. the speed of light in the medium is higher than the speed of light in a vacuum
E. reflection is not possible

3. The colors on an oil slick are caused by reflection and:

A. refraction
B. polarization
C. diffraction
D. interference
E. diffusion

4. Which of the following describes the best scenario if a person wants to start a fire using sunlight and a mirror?

A. Use a concave mirror with the object to be ignited positioned at the center of the curvature of the mirror
B. Use a concave mirror with the object to be ignited positioned halfway between the mirror and its center of curvature
C. Use a plane mirror
D. Use a convex mirror
E. A fire cannot be started using a mirror since mirrors form only virtual images

5. A lens has a focal length of 2 m. What lens could be combined with it to get a combination with a focal length of 3 m?

A. A lens of power 1/6 diopters
B. A lens of power 6 diopters
C. A lens of power of –6 diopters
D. A lens of power of –1/6 diopters
E. A lens of power 3 diopters

6. An incident ray traveling in the air makes an angle of 30° with the surface of a medium with an index $n = 1.73$. What is the angle that the refracted ray makes with the surface? (Use the index of refraction for air $n = 1$)

A. 30°
B. 90 sin^{-1} (0.5)
C. 60°
D. sin^{-1} (0.5)
E. 90°

7. A mirage is produced because:

A. light travels faster through the air than through water
B. images of water are reflected from the sky
C. warm air has a higher index of refraction than cool air
D. warm air has a lower index of refraction than cool air
E. water from oceans and lakes is highly reflective

8. An object is viewed at various distances using a concave mirror with a focal length of 14 m. What happens when a candle is placed at the focus?

A. Light rays end up parallel, going to infinity
B. Light rays meet at the focus
C. An image is formed 7 m in front of the mirror
D. An image is formed 7 m behind the mirror
E. Light rays end up parallel, going out 7 m on either side

9. A material which can rotate the direction of polarization of linearly polarized light is said to be:

A. diffraction limited
B. dichroic
C. circularly polarized
D. birefringent
E. optically active

10. When light enters a material of a higher index of refraction, its speed:

A. increases
B. stays the same
C. first increases then decrease
D. decreases
E. first decreases, then increase

11. What is the correct electromagnetic spectrum order from the shortest to longest wavelength?

A. Radio waves → X-rays → Ultraviolet radiation → Visible light → Infrared radiation → Microwaves → Gamma rays
B. Gamma rays → X-rays → Visible light → Ultraviolet radiation → Infrared radiation → Microwaves → Radio waves
C. Gamma rays → X-rays → Ultraviolet radiation → Visible light → Infrared radiation → Microwaves → Radio waves
D. Visible light → Infrared radiation → Microwaves → Radio waves → Gamma rays → X-rays → Ultraviolet radiation
E. Gamma rays → X-rays → Infrared radiation → Visible light → Ultraviolet radiation → Microwaves → Radio waves

12. The angle of incidence can vary between zero and:

A. 2π radians
B. π radians
C. $\pi/2$ radians
D. 1 radian
E. $3\pi/2$ radians

13. An amateur astronomer grinds a double-convex lens whose surfaces have radii of curvature of 40 cm and 60 cm. What is the focal length of this lens in the air? (Use the index of refraction for glass $n = 1.54$)

A. 44 cm
B. 88 cm
C. 132 cm
D. 22 cm
E. 1.32 cm

14. When light reflects from a stationary surface, there is a change in its:

I. frequency II. speed III. wavelength

A. I and II only
B. I and III only
C. III only
D. I, II and III
E. none of the above

15. If a distant galaxy is moving away from the Earth at 4,300 km/s, how do the detected frequency (f_{det}) and λ of the visible light detected on Earth compare to the f and λ of the light emitted by the galaxy?

A. The f_{det} is lower, and the λ shifts towards the red end of the visible spectrum
B. The f_{det} is lower, and the λ shifts towards the blue end of the visible spectrum
C. The f_{det} is the same, but the λ shifts towards the red end of the visible spectrum
D. The f_{det} is the same, but the λ shifts towards the blue end of the visible spectrum
E. The f_{det} and the λ are the same as the sources

Solutions: Light & Optics

1. A is correct.

Violet light has the highest frequency of visible light colors, ranging from 668 to 789 THz. Because the frequency is related to photon energy by:

$$E = hf$$

Violet light is also the most energetic of the visible light spectrum.

2. D is correct.

The index of refraction for a given material is expressed as the speed of light in a vacuum divided by the speed of light in that material:

$$n = c / v$$

If the index of refraction is less than one, it implies that the speed of light in the material is greater than in a vacuum. However, this is never the case because the speed of light in a vacuum cannot be exceeded.

3. D is correct.

The colors observed on an oil slick pool are caused by reflection and thin-film interference. This is the process by which the incoming light is reflected off the oil's top and bottom layers, producing reflected waves in and out of phase. This phase change determines the interference (constructive or destructive) and colors form.

4. B is correct.

$$f = \tfrac{1}{2}R$$

where R is the radius of curvature

5. D is correct.

The first lens has a power:

$$P_1 = 1 / f$$

$$P_1 = \tfrac{1}{2}\ \text{D}$$

For a combination of total power:

$$P_{tot} = 1 / f_{tot}$$

$$P_{tot} = 1/3\ \text{D}$$

continued...

Thus,

$P_2 = P_{tot} - P$

$P_2 = 1/3\text{ D} - 1/2\text{ D}$

$P_2 = -1/6\text{ D}$

6. A is correct.

$n_1 \sin \theta_1 = n_1 \sin \theta_2$

$(1) \sin (90° - 30°) = (1.73) \sin \theta_2$

$\sin (60°) = 1.73 \sin \theta_2$

$\sin \theta_2 = \sin (60°) / 1.73$

$\sin \theta_2 = 0.5$

$\theta_2 = \sin^{-1} (0.5)$

$\theta_2 = 30°$

7. D is correct.

A mirage is produced due to the higher index of refraction of cold air compared to warm air.

As light from the sky goes through the denser cold air and approaches warm air (less dense and therefore with a lower refraction index) near the ground (generated by a hot surface of asphalt or sand), the light bends away from the warm air. Reflection of the sky can be observed on the ground.

8. A is correct.

Mirror equation:

$1 / f = 1 / d_o + 1 / d_i$

If the object is placed at the focus:

$d_o = f$

$1 / f = 1 / f + 1 / d_i$

$1 / f - 1 / f = 1 / d_i$

$0 = 1 / d_i$

Only if $d_i = \infty$ is this true.

This implies that no image is formed.

Thus, the light rays neither converge nor diverge and travel parallel to infinity.

9. E is correct.

Optically active material rotates the direction of polarization of linearly polarized light.

10. D is correct.

When light enters a material with a higher index of refraction, its speed *decreases*.

11. C is correct.

Correct order of the electromagnetic spectrum from shortest to longest wavelength is

Gamma rays → X-rays → Ultraviolet radiation → Visible light →

→ Infrared radiation → Microwaves → Radio waves

12. C is correct.

The angle of incidence ranges from 0° to 90°.

Convert degrees to radians:

$0° = 0$ radians

$(90° / 1) \cdot (\pi / 180°) = \pi/2$ radians

13. A is correct.

Double convex lens:

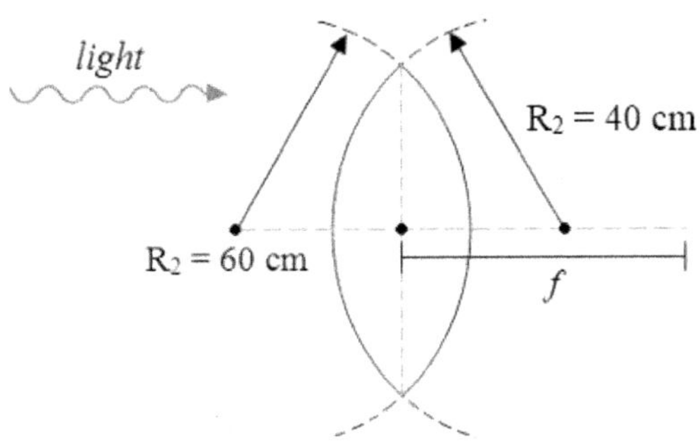

Lensmaker's formula:

$1 / f = (n - 1) \cdot (1 / R_1 - 1 / R_2)$

$1 / f = (1.54 - 1) \cdot (1 / 40\text{ cm} - 1 / {-60}\text{ cm})$

$1 / f = -0.0225\text{ cm}^{-1}$

$f = 44\text{ cm}$

If light passes through the center of the radii of curvature (as it does to R_2) before the curve itself in the lens, then that R is negative by convention.

14. E is correct.

Light does not experience a change in frequency, wavelength, or speed when it reflects from a stationary surface.

15. A is correct.

Doppler effect is qualitatively similar for both light and sound waves.

If the source and observer move *towards* each other, the f_{det} is higher than the f_{source}.

If the source and observer move *away*, the f_{det} is lower than the f_{source}.

Since the galaxy is moving away from the Earth, the f_{det} is lower.

The speed of light through space is constant ($c = \lambda f$).

A lower f_{det} means a longer λ_{det}, so the λ_{det} is longer than the λ_{source}.

λ has been shifted towards the red end of the visible spectrum because red light is the visible light with the longest λ.

Notes for active learning

CHAPTER 8

Fluids & Solids

Density and Specific Gravity

Archimedes' Principle of Buoyancy

Pressure

Continuity Equation ($Av = constant$)

Bernoulli's Equation

Viscosity and Poiseuille's Law

Surface Tension

Strain and Types of Stress

Elasticity: Elementary Properties

Thermal Expansion Coefficient

Chapter Summary

Page intentionally left blank

Density and Specific Gravity

States of matter

Three standard states of matter are solid, liquid, and gas.

Solids have a definite shape and volume.

Liquid has a fixed volume but can be any shape.

Gas can be any shape and can be compressed (variable volume).

Since liquids and gases can "flow," they are fluids.

Plasma

Plasma is the fourth state of matter, distinct from the other states.

Like liquids boiling, changing into a gas when energy is added, heating a gas will form plasma.

Plasma is a mixture of positively charged (ions) and negatively charged particles (electrons).

Moving charged particles generate electric currents, and any movement of a charged plasma particle affects and is affected by the fields created by the other charges.

Describing a low-density plasma as merely an *ionized gas* is incorrect despite being similar to the gas phase in that neither assumes a definite shape or volume.

State transitions

Changing states of matter occur when it loses or absorbs energy (i.e., heat or thermal energy).

When a substance absorbs energy, the atoms and molecules move more rapidly, and this increased kinetic energy pushes particles far enough that they change form. This energy is usually heat or thermal energy.

Usually, changes in the state of matter go in a specific order.

Adding thermal energy to a solid transforms it into a liquid and then a gas; when removing enough thermal energy from a gas, it changes back into a liquid and, finally, a solid.

Phase change also depends on factors like pressure and temperature.

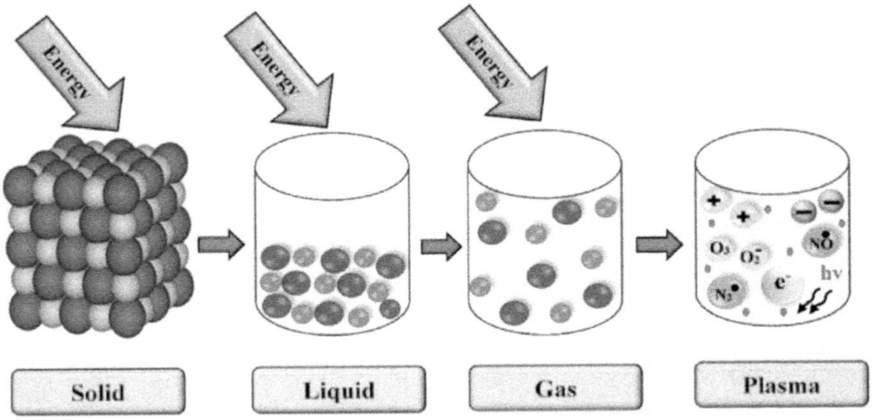

States of matter transition when energy is added

Determining density

Density (ρ) of an object is its mass per unit volume, expressed as:

$$\rho = \frac{m}{V}$$

where ρ is density, m is mass, and V is volume.

Density is expressed in SI units of kg/m^3; it may also be given in g/cm^3.

Density is the weight per unit volume of material at 23°C and is expressed as D 23°C in g/cm^3.

To convert g/cm^3 to kg/m^3, use dimensional analysis.

However, it is easier to multiply by 1,000 (cm^3kg/m^3g):

$$\left(\frac{\text{g}}{\text{cm}^3}\right) \cdot \left(\frac{100\text{ cm}}{1\text{ m}}\right)^3 \cdot \left(\frac{1\text{ kg}}{1{,}000\text{ g}}\right) = \left(\frac{\text{g}}{\text{cm}^3}\right) \cdot \left(\frac{1{,}000\text{ cm}^3\text{kg}}{\text{m}^3\text{g}}\right)$$

$$\left(\frac{\text{g}}{\text{cm}^3}\right) \cdot \left(\frac{1{,}000\text{ cm}^3\text{kg}}{\text{m}^3\text{g}}\right) = \frac{\text{kg}}{\text{m}^3}$$

Specific gravity

Specific gravity of a substance is the ratio of its density to the density of the reference substance.

Specific gravity is referenced against water. Thus, the specific gravity of water equals 1.

Specific gravity is the ratio of the weight of a given volume of material at 73 °F (23 °C) to that of an equal volume of water at the same temperature, expressed as specific gravity, 23/23°C.

Equation for specific gravity:

$$\text{specific gravity} = \frac{\rho_{\text{object}}}{\rho_{\text{water}}}$$

Archimedes' Principle of Buoyancy

Principle of buoyancy

Archimedes' Principle of buoyancy (250 B.C.) states that *an object wholly or partially submerged in a fluid will be buoyed up by force equal to the weight of the fluid it displaces.*

Buoyancy is:

$$F_{\text{buoyant}} = \rho_{\text{fluid}} V_{\text{object}} g$$

where F_{buoyant} is the buoyant force (N), V is the volume of the fluid displaced (m^3). ρ is the density of the fluid it displaces (kg/m^3), and g is the value of gravity (9.8 m/s^2).

For floating objects, the fraction of the submerged object is the ratio of the object's density to that of the fluid.

$$\frac{V_{\text{sub}}}{V_{\text{object}}} = \frac{\rho_{\text{object}}}{\rho_{\text{fluid}}}$$

Buoyancy force

Buoyancy force counteracts the weight of the object in the fluid, making objects in water feel lighter than when out of the water.

Objects float when their densities are less than the fluid in which they are submerged.

If an object's density is less than that of water, there is an upward net force on it, and the object rises until it is partially out of the water.

Calculating buoyancy force

The figure below depicts a log submerged in water $\rho = 1{,}000$ kg/m^2. When fully submerged, the net force on the log is greater than zero. Thus, the log floats until the buoyant force is equal to the weight of the log (shown below):

$$F_{net} = F_B - m_o g$$

$$F_{\text{net}} = \rho_F V_{\text{displaced}} g - m_o g$$

$$F_{\text{net}} = \left(1{,}000\,\frac{\text{kg}}{\text{m}^3}\right) \cdot (2\ \text{m}^3)g - (1{,}200\ \text{kg})g$$

$$F_{\text{net}} = (2{,}000\ \text{kg})g - (1{,}200\ \text{kg})g$$

$$F_{\text{net}} = (800\ \text{kg})g$$

$$F_{\text{net}} > 0$$

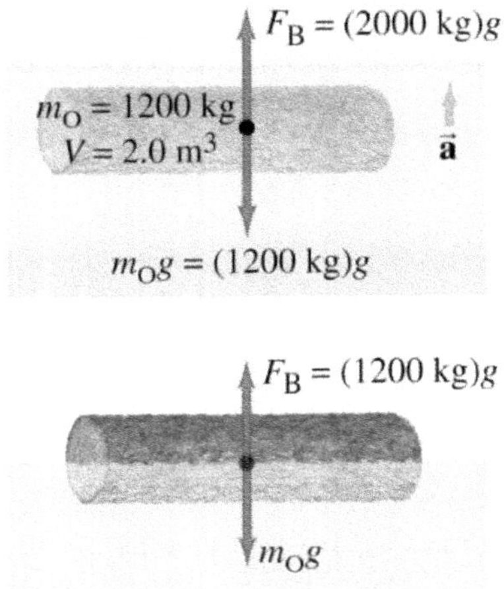

Log floats until the buoyant force is equal to the weight of the log

An object floats perfectly (i.e., the top is level with the surface of the fluid) when the buoyant force is equal to the weight of the object.

An object will rise upward out of the fluid when the buoyant force is greater than the object's weight, and an object will sink when the buoyant force is less than the object's weight.

Buoyant force in air

Principle of buoyancy works for air; hot-air and helium balloons rise. Both hot air and helium have densities less than air at room temperature. The difference in densities causes the hot air or helium to exert a buoyant force on the balloon.

The balloon rises because the buoyant force is greater than the weight of the object it is carrying. The graphic below shows that the net force on the object is the difference between the buoyant and the gravitational forces.

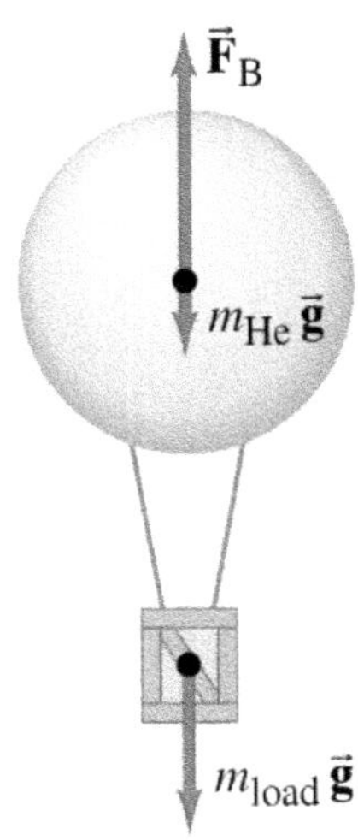

Buoyant force is greater than the weight of the object

Pressure

Calculating pressure

Pressure is the force exerted on the area over which the force acts.

Equation for pressure is:

$$P = \frac{F}{A}$$

where P is pressure (N/m^2), F is a force (N), and A is area (m^2).

Unit of pressure is N/m^2, which is the Pascal (Pa).

$$1\ \text{Pa} = 1\ \text{N/m}^2$$

Anything that exerts a force can exert pressure; therefore, the concept of pressure applies to solid objects, liquids, and gases.

When discussing fluids, the pressure exerted by the atmosphere and measurements of pressure relative to it are of concern.

At sea level, atmospheric pressure is about 1.013×10^5 N/m^2; this is one atmosphere (1 atm).

Another unit of pressure is the bar (1 bar = 1.00×10^5 N/m^2). Standard atmospheric pressure is slightly more than 1 bar.

Atmospheric pressure

Atmospheric pressure does not crush organisms because their cells maintain an internal pressure that balances it.

The pressure is measured using two common types.

Absolute pressure is pressure zero-referenced against a perfect vacuum.

Gauge pressure

Gauge pressure is pressure zero-referenced against atmospheric pressure.

Most gauges measure pressure above the atmospheric pressure. Hence, most systems measure gauge pressure.

For example, what would the reading be if there were two different pressure dials (with one displaying absolute pressure and the other displaying gauge pressure) at sea level?

Absolute pressure dial reads 1.013×10^5 Pa, and the gauge pressure dial reads 0 Pa. This difference is because the atmosphere exerts 1.013×10^5 Pa of pressure over a perfect vacuum; thus, the absolute pressure gives this value. Gauge pressure reads 0 Pa because it references atmospheric pressure and requires pressures above this value to give a reading.

Gauge pressure is the difference between absolute and atmospheric pressure.

$$P_{gauge} = P_{abs} - P_{atm}$$

Absolute pressure equation:

$$P_{abs} = P_{atm} + P_{gauge}$$

Pressure versus depth (*P* = *pgh*)

Below the surface of a liquid, pressure increases linearly with depth. Liquid pressure on an object is found only using the liquid's density and the object's distance below the surface.

The pressure on an object is not determined by using the object's density in the liquid or the shape of the container.

Liquid pressure is expressed as:

$$P = \frac{F}{A} = \frac{\rho A h g}{A}$$

Simplifies to:

$$P_{liquid} = \rho g h$$

where P_{liquid} is the pressure exerted by the liquid (N/m^2 or Pa), g is the acceleration of gravity, ρ is the density of the liquid, and h is the distance below the surface of the liquid (m).

Liquid pressure is useful in the operation of some types of pressure gauges.

For example, the gauge pictured below is an open-tube manometer. The pressure in the open end is open to atmospheric pressure, and the pressure that is being measured causes the fluid to rise until the pressures on both sides of the tube are equal.

From the liquid pressure equation, the pressure being measured is calculated as:

$$P_0 = P_{atm}$$

$$P = P_{atm} + P_{liquid}$$

$$P = P_{atm} + \rho g \Delta h$$

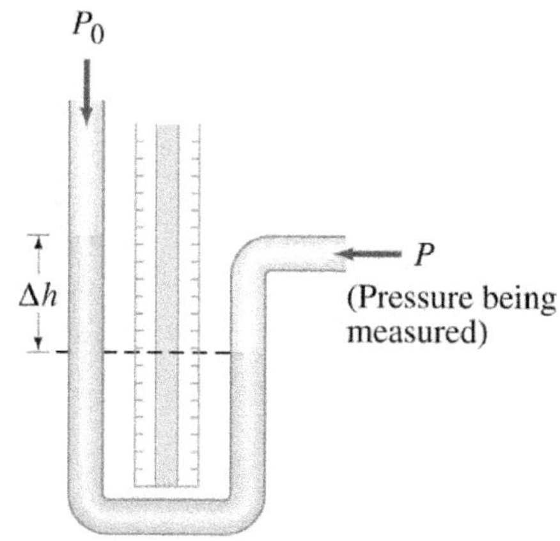

Open tube manometer includes atmospheric pressure

The measured pressure is absolute in this example because the equation includes atmospheric pressure. This is an important distinction because if the pressure on an object is evaluated at depth, the gauge pressure only gives the liquid pressure on the object.

In contrast, the absolute pressure gives the gauge pressure plus atmospheric pressure so that it would be more accurate.

Calculating gauge and atmospheric pressure

The figure below shows two containers with different shapes holding water at equal levels. If points 1 and 2 are equal height, what are the gauge and absolute pressure at points 1 and 2?

The containers are different shapes; the two points are at equal depths within the water. Therefore, they have the same gauge and absolute pressures.

The shape of the container does not influence the liquid pressure; only the depth below the liquid contributes to liquid pressure. Gauge and absolute pressure can be calculated as:

$$P_1 = P_2$$

$$P_{\text{gauge}} = \rho g \Delta h$$

$$P_{\text{absolute}} = P_{\text{atm}} + \rho g \Delta h$$

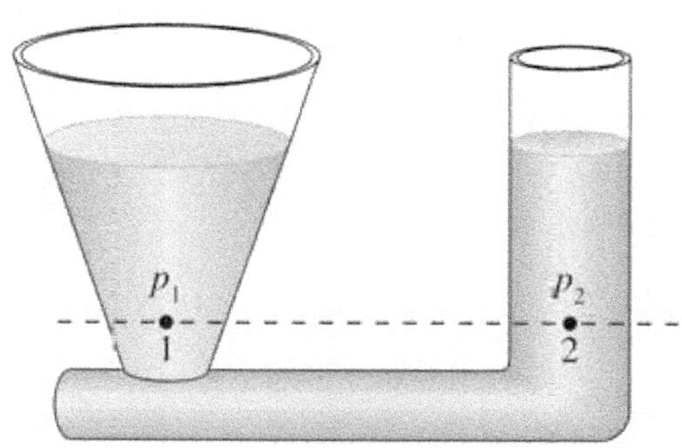

Two points are at equal depths, so they have the same gauge and absolute pressures

Pascal's Law

Pascal's Law (1647-1648) states that *a pressure exerted on an incompressible liquid transmits equally to all liquid parts.*

This Principle has important implications because it explains why pressure differentials are not created within a closed volume of liquid.

A small cube is within a beaker of water. For this example, consider the sides small enough that liquid pressure due to depth is equal on the top and bottom of the cube.

The pressure on all sides of the object is equal because the pressure is the same in every direction in a fluid at a given depth.

If this were not true, the fluid would flow because of the natural tendency of particle movement from high pressure to low pressure.

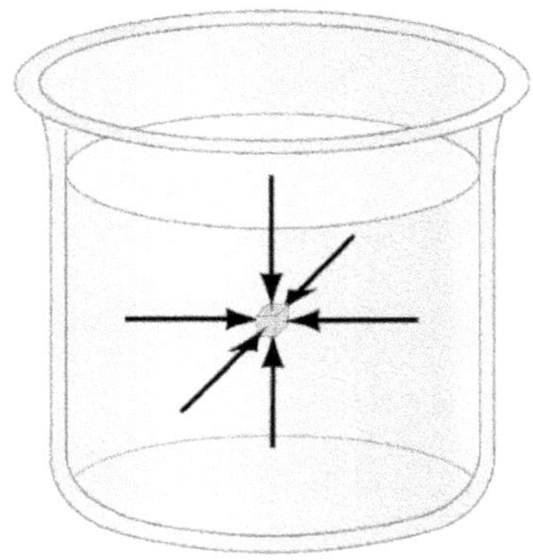

Pressure on all sides of the object is equal because the pressure is the same in every direction in a fluid at a given depth

Although the pressures across an incompressible liquid are the same, the forces vary depending on the area they are applied over (because of $P = F / A$). This observation is useful because it allows the development of systems that impart mechanical advantage.

If the external pressure is applied to a confined fluid, the pressure at every point within the fluid increases by that amount. Pascal's Principle is used in hydraulic lifts and hydraulic brakes. The fluid is confined, so when pressure is applied, the fluid is compressed, which raises the pressure throughout the system.

In the system below, a small force is applied over a small area (F_1 over A_1). The output force (F_2) at the location with the larger area (A_2) will be higher.

Energy must be conserved; thus, the work done on one end is the same as on the other.

The volume of liquid displaced must be conserved to equal the volume displaced at both the input and output.

Work Equivalence

$$W_1 = W_2$$

$$F_1 d_1 = F_2 d_2$$

Volume Equivalence

$$V_1 = V_2$$

$$A_1 d_1 = A_2 d_2$$

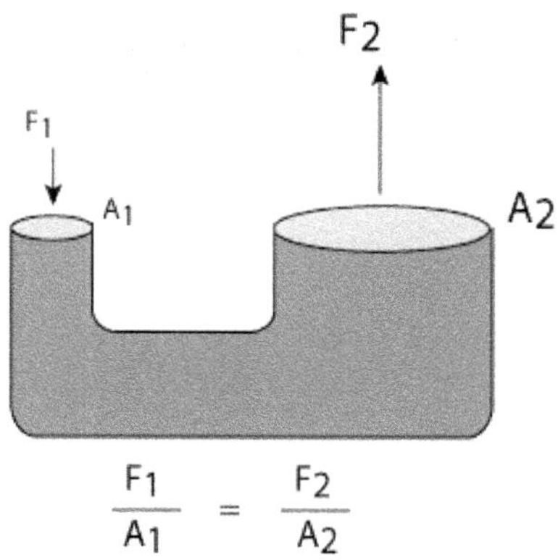

Pressure applied compresses the fluid, which raises the pressure throughout the system

Continuity Equation (*Av = constant*)

Conservation of volume

When liquids flow through pipes, the flow is considered incompressible, and the volume of the liquid remains constant regardless of pressure. As such, the volume of liquid flowing through a pipe in a given time must *always be equal.*

This Principle is the conservation of volume and can be expressed by the *continuity equation*, which describes the volume flow rate of an incompressible liquid.

Continuity equation for calculating flow rate

Often, the continuity equation is used to calculate the speed of the fluid flowing through pipes and is given as:

$$A_1v_1 = A_2v_2$$

where A is the area of the pipe's cross-section (m^2), and v is the velocity of the fluid through that point (m/s).

Imagine an incompressible liquid flowing through a pipe that has a diameter reduction. The volume of fluid passing through the constriction over time cannot change.

Continuity equation can solve for the resulting velocity of the liquid.

$$A_1v_1 = A_2v_2$$

$$v_2 = \frac{A_1v_1}{A_2}$$

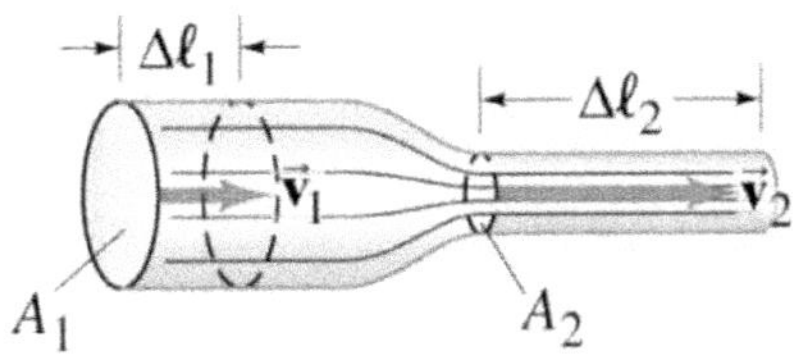

Continuity equation solves the resulting velocity of the liquid

Bernoulli's Equation

Energy conservation within flowing fluids

Bernoulli equation (1738) relates energy conservation within flowing fluids. It states that *a flowing fluid's pressure, kinetic, and potential energy is conserved.*

Bernoulli equation is:

$$P_1 + \rho g h_1 + \frac{1}{2}\rho v_1^2 = P_2 + \rho g h_2 + \frac{1}{2}\rho v_2^2$$

where P is pressure (Pa), ρ is the density of the liquid, g is the acceleration due to gravity, h is the height of the fluid (m), and v is the velocity of the fluid through that point (m/s).

Fluid flow across pressure or height differentials

Bernoulli equation analyzes fluid flow across pressure or height differentials and the energy needed to transport fluid volumes.

In the diagram below, water flows into the left side at a pressure of P_1, a velocity of v_1, through an area of A_1, and at the height of y_1.

As fluid moves to the right, the pipe rises in height and decreases in area (i.e., circumference).

Given a few conditions, Bernoulli's Principle can be used to find the other values.

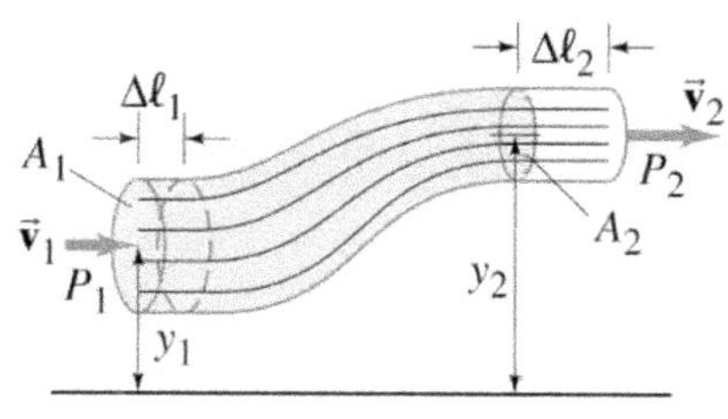

Bernoulli's Principle analyzes fluid flow across pressure or height differentials

Continuity equation

Continuity equation (or *transport equation*) describes the transport of some quantity.

It is often applied to a conserved quantity but can be generalized for any extensive quantity (e.g., mass, energy, momentum, electric charge, and other conserved quantities).

$$A_1 v_1 = A_2 v_2$$

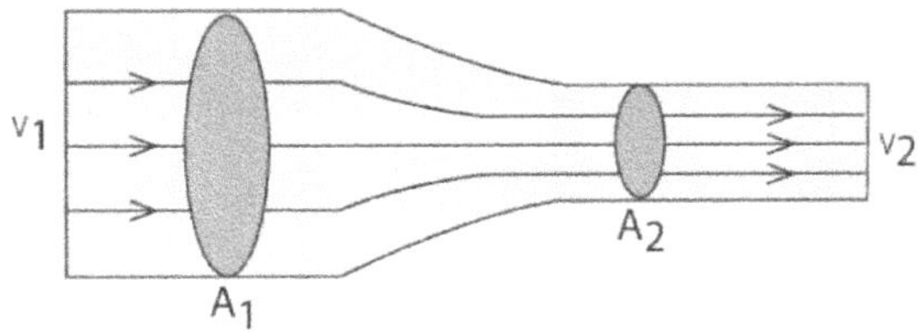

Continuity equation means flow rate through A_1 equals A_2

Solved problem example: Bernoulli's Principle

Bernoulli's Principle correlates the speed of a fluid through an opening to the height of the fluid above the opening. As shown below, suppose a spigot is at the bottom of an open water tank.

Speed of the fluid can be found using Bernoulli's Principle and making a few assumptions.

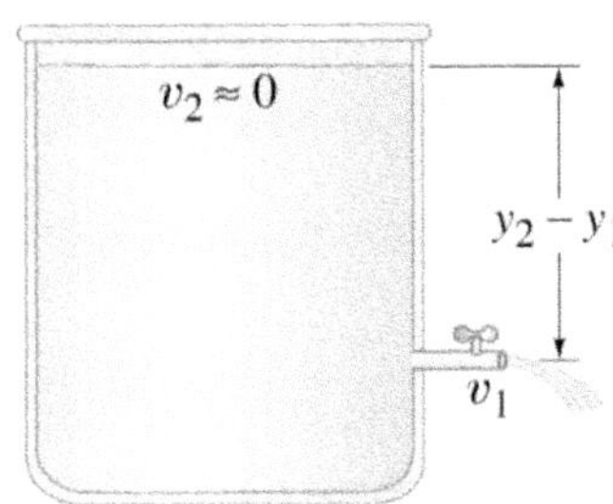

Bernoulli's equation solves for the speed of fluid through bottom spigot

$$P_1 + \rho g y_1 + \frac{1}{2}\rho v_1^2 = P_2 + \rho g y_2 + \frac{1}{2}\rho v_2^2$$

Assume that the top of the container is much larger than the opening of the spigot:

$$A_{top} \gg A_{spigot}$$

According to the *continuity equation*, fluid flow velocity through the top is much smaller than that through the spigot. Flow through the spigot can be set to zero:

$$v_{top} \ll v_{spigot} \qquad \text{so} \qquad v_{top} \approx 0$$

Atmospheric pressure acts on both the top of the container and at the end of the spigot. This pressure is equal on both sides of the equation. It cancels, leaving:

$$g y_1 + \frac{1}{2} v_1^2 = g y_2$$

Assume the height of the spigot as the reference height and set it equal to zero. Then, solve for the velocity of fluid flow out of the spigot:

$$\frac{1}{2} v_1^2 = g y_2 \quad \text{or} \quad v = \sqrt{2 g y_2}_1$$

Same expression as for the conservation of mechanical energy into kinetic energy.

Airplane lift

Bernoulli equation can be used to explain the lift process for airplane wings. When air strikes an airplane wing, the air must separate to flow over both the top and bottom of the wing.

Due to the wing curvature, air flowing over the top surface has a slightly higher speed than air flowing over the bottom. The speed difference causes a pressure imbalance.

The top of the wing experiences less pressure than the bottom.

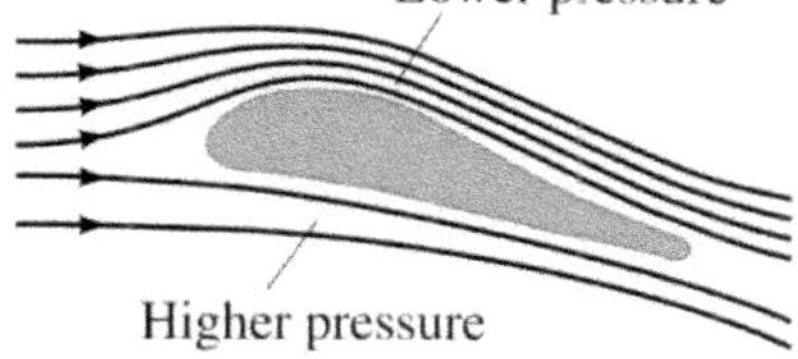

Air flowing over an airplane wing with a resulting pressure differential

If the height of the wing is negligible, then the Bernoulli equation reduces to:

$$P_T + \frac{1}{2}\rho v_T^2 = P_B + \frac{1}{2}\rho v_B^2$$

Airflow speed over the top of the wing is higher than the flow speed under the bottom.

Thus, by conservation of energy:

$v_T > v_B$ and $P_T < P_B$

Venturi effect and pivot tube

From the Bernoulli principle, higher velocity gas experiences a pressure drop. This is the Venturi effect, which measures fluid flow by measuring pressure differences.

Venturi effect (1797) describes the drop in fluid pressure that occurs when a fluid flows through a constriction in a pipe.

From the continuity principle, flow through a constriction in a tube increases the velocity of the flow to conserve volume.

Pitot tube is a flow measurement device that uses the Venturi effect to determine fluid flow velocity. It is commonly used to calculate an aircraft's airspeed.

Measuring fluid flow through constricted pipe

Measuring fluid flow velocity through a pipe uses a constriction placed in the pipe. Two pressure meters are attached, one before the constriction and the other at the constriction.

If the areas of the pipe and the constriction are known, the velocity of the fluid flow through the pipe is solved by both the continuity and Bernoulli equation.

$A_1v_1 = A_2v_2$ *continuity equation*

$P_1 + \frac{1}{2}\rho v_1^2 = P_2 + \frac{1}{2}\rho v_2^2$ *Bernoulli equation*

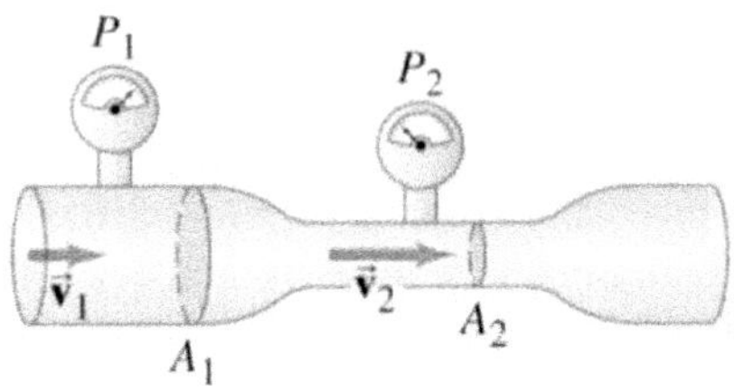

Flow through a constriction increases the velocity of the flow to conserve volume

Viscosity and Poiseuille's Law

Viscosity

When water poured out of a cup is compared to honey poured out of a jar, the flow pattern is substantially different. Although water and honey are both liquids, they behave differently when flowing due to their *viscosity*.

Viscosity of a fluid is a measure of the internal frictional force. All fluids have some measure of viscosity. Viscosity may be high (e.g., honey) or low (e.g., water).

The characteristic viscosity of a fluid is a known value and is represented by the symbol η with units of (Pa·s).

Flow patterns

When highly viscous fluids flow through a pipe, the behavior is markedly different than when a fluid with lower viscosity flows through the pipe.

The flow from highly viscous fluids forms a front shaped like a *parabola bulging outward.*

A parabola shape is observed in the fluid because the frictional forces on the fluid from the sides of the pipe are much higher than the frictional forces in the center of the fluid.

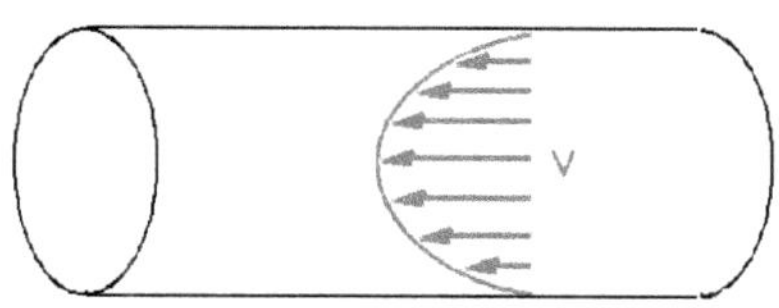

Parabola shape observed for high viscous fluids

Along with the parabola flow shape, the *flow rate* of a fluid through a pipe must account for the viscosity of the liquid.

Poiseuille's Law calculates flow rate

Poiseuille's Law is expressed as:

$$\dot{V} = \frac{\Delta P \pi r^4}{8 \eta L}$$

where $\dot{V}$ is the volumetric flow rate (m^3/s), ΔP is pressure differential across pipe (Pa), r is the radius of the pipe (m), η is the viscosity of the pipe (Pa·s), and L is the length of the pipe (m).

Intravenous flow rate

Poiseuille's Law (1838) has significant consequences for blood flow.

Hagen–Poiseuille equation determines vascular resistance and flow rate of intravenous (IV) fluids achieved by using various sizes of peripheral and central *cannulas* (i.e., tubes inserted into the body).

Flow rate is proportional to the radius to the *fourth power*, meaning that a slight increase in the internal diameter of the cannula yields a significant increase in the flow rate of IV fluids.

For example, if the radius of an artery is halved by cardiovascular occlusion (i.e., blockage), pressure must increase by a factor of 16 to maintain the same blood flow. Usually, the human heart cannot sustain 16 times the output.

Blood pressure increases (i.e., *hypertension*) as the heart pumps (cardiac stress) to satisfy this demand for typical blood flow with blockage.

Surface Tension

Hydrogen bonding produces surface tension

Small objects like paper clips can be placed lightly on the water's surface and float. However, if pushed down, they readily sink. A small object placed gently on top of the water may not weigh enough to break the surface tension. Therefore, it floats despite its higher density than water. Hydrogen bonding attractions between liquid water molecules produce *surface tension.*

Surface tension is due to the *cohesive forces* within water molecules. Cohesion occurs because water molecules are electrostatically attracted and stick to neighboring water molecules.

Some insects, for example, can walk on water, not because they are less dense than the water but because of the resistive force of surface tension.

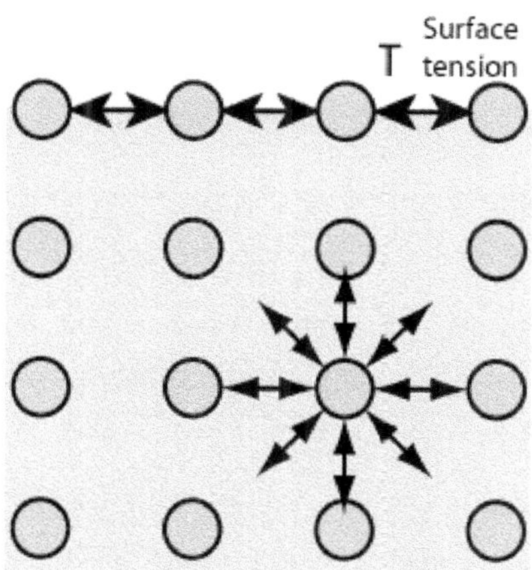

Surface tension as cohesive forces join adjacent water molecules by hydrogen bonding

Adhesive and cohesive forces

Adhesion is the property of a substance to be attracted to a different substance.

Adhesion, unlike cohesion, is caused by several mechanisms and may or may not be due to electrostatic forces, as discussed for water molecules.

Cohesive and *adhesive forces* of the liquid determine the behavior of a liquid in a container.

Often, the surface of a liquid in a container is not flat when in a small container (e.g., graduated cylinder), where the liquid curves up or down as it contacts the container walls. This curved surface for the liquid is the *meniscus* and results from surface tension and adhesion.

If the adhesive force between the liquid and the container is greater than the cohesive force between the liquid molecules, the liquid curves up to meet the edges of the container, creating a *concave meniscus*.

If the cohesive force exceeds the adhesive force, the liquid curves downward, creating a *convex meniscus*. For example, water molecules are more attracted to glass than to each other (adhesion > cohesion), so the water's surface curves upward toward the walls of its container.

Mercury has adhesion < cohesion, so its surface curves downward against the container walls.

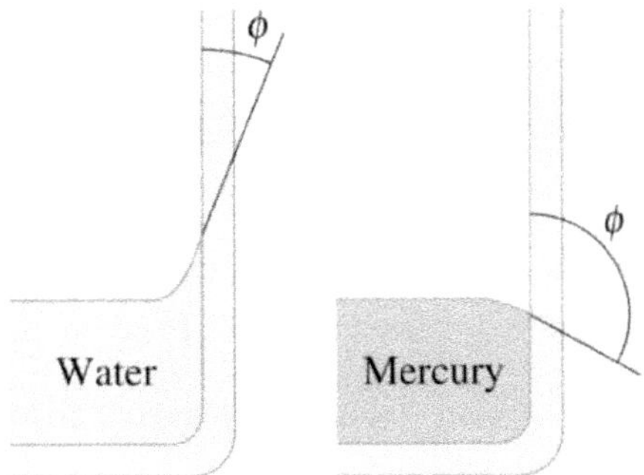

Cohesive and adhesive forces affect the liquid's position relative to the container's walls

Capillary action of liquids

Fluids exhibit capillary action if a narrow tube is placed in the fluid. This phenomenon is due to cohesive and adhesive forces.

If a narrow tube is placed in water (adhesion > cohesion), the water travels up the tube some distance (figure below on the left). This movement is because the adhesive forces pull the water along the edge, and the cohesive forces pull water up the tube.

Suppose the same tube is placed in mercury (adhesion < cohesion); then the cohesive forces of the mercury keep the mercury together rather than climb the tube wall. Mercury becomes lower in the capillary tube than the height of the mercury level around it (figure below on the right).

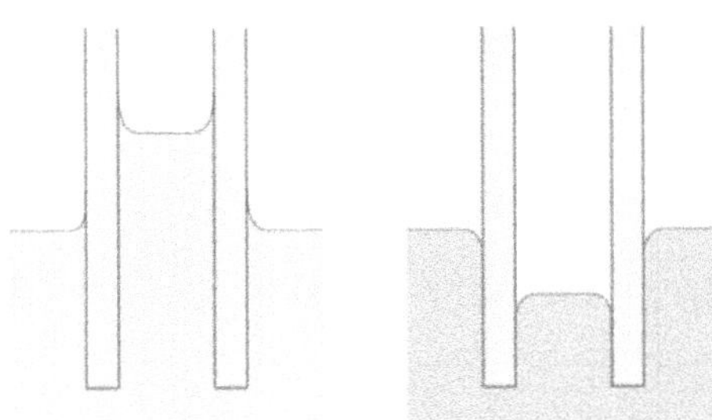

Central capillary tube in water (on the left) compared to in mercury (on the right)

Strain and Types of Stress

Deformation of solids

Solids (e.g., ceramics, metals, and composites) undergo the process of *stress* and *strain.*

When a solid is subjected to an external force, it experiences *stress* (or deformation), measured in terms of *strain.* When a solid material has a force applied to it, it resists deformation (up to a maximum applied force) and retains its shape.

Deformation resistance is from material's stress, which applies a restoring force per unit area.

Stress has the form:

stress = force / area

Normal, shearing and volumetric strain

Strain (or *deformation*) of a solid material generates an internal elastic stress analogous to the reaction force of a spring, which tends to restore the material to its original non-deformed state.

Normal strain (or longitudinal strain) is the ratio of the change in length with respect to its original length. It is a longitudinal strain since it is concerned with the change in the length of the body in relation to its original length.

$S_1 = \Delta V / V$

Shearing strain is the angular distortion made by the body, so it is equal to the displacement of the body surface in relation to the body height, but keep in mind that the body surface is the one that bears the strain.

$S_2 = l / H$

Volumetric strain (or bulk strain) is the ratio is the volume change of a body compared to its original volume. It is a bulk strain because it considers the body's entire volume.

$S_1 = \Delta V / V$

Solids experience stress

Solid materials undergo three basic types of stress: *tension*, *compression*, and *shear stress.*

Stress and strain are *linearly proportional* in many materials when the stress is small.

The material obeys Hooke's Law and the ratio of stress to strain is the *elastic modulus.*

Hooke's Law only holds for a range of stresses, referred to as the elastic region.

These types of stress are important because they dictate how structures are built, the loads they can tolerate, and how the failure of materials may occur.

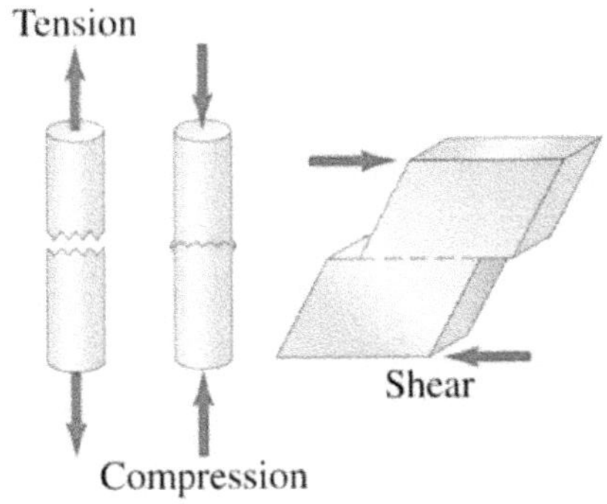

Applied stress includes tension, compression, and shear force

Tension

Tension force (or *tensile stress*) is transmitted through an object (e.g., string, rope, or wire) when pulled by forces acting from opposite ends. The tension force is directed along the length and pulls equally on *opposite ends* of the object.

Tensile stress is expressed as:

$$\sigma = F / A$$

where σ is the stress an object experiences (N/m^2), F is the force exerted on an object (N), and A is the area over which it is applied (m^2).

Compressive stress

Compressive stress results when a force acts to compress the material. The same formula expresses compressive stress as for tensile stress:

$$\sigma = F / A$$

where σ is the stress an object experiences (N/m^2), F is the force exerted on an object (N), and A is the area over which it is applied (m^2).

Shear stress

Shear stress occurs when two forces act opposite such that the material slides on a plane between the applied forces.

Shear stress is:

$$\tau = F/A$$

where τ is the stress an object experiences (N/m^2), F is the force exerted on an object (N), and A is the area over which it is applied (m^2).

Elasticity: Elementary Properties

Elastic properties influence shape distortions

Elasticity is an essential property in tensile and compressive stress.

When a material undergoes tensile or compressive deformation, elasticity measures its ability to return to its *original shape after distortion.*

For example, rubber bands have high elasticity because, after moderate stretching, a rubber band returns to its original shape.

Conversely, if a low-elasticity material is deformed, such as a piece of ceramic pottery, it experiences slight deformation before snapping.

Elastic properties determine the behaviors of these two materials.

Stretching and compression

When materials undergo *tension*, they stretch according to the force acting upon them. As mentioned earlier, all materials have some degree of elasticity.

After exceeding this inherent value, they stretch according to the tensile stress applied.

When a solid material undergoes tension, its elasticity is described as *ductile* or *brittle*.

Material is *ductile* if it tends to deform (stretch) under tensile stress.

However, if the material undergoes slight deformation before breaking, the material is *brittle*.

Compression is the inverse of tension and occurs when force points inward on a material, compacting it. Similarly, if the applied force is high enough, the material deforms.

Malleability is the propensity for a material to deform under compression.

Young's Modulus measures elasticity

Material can be compressed and deformed easily if it is highly *malleable.*

Non-malleable materials break rather than deform.

The two properties that indicate the elasticity of material in tension and compression are strain and Young's modulus.

Strain, given by the lowercase Greek letter epsilon (ε), is the change in length of an object in the direction of the applied force, divided by the original length.

$$\varepsilon = \frac{\Delta l}{l_0}$$

where ε is the strain, Δl is the change in length (m), and l_0 is the original length (m).

Young's modulus (elastic modulus) is a measure of the force per unit needed to stretch material.

Young's modulus is expressed as the ratio of stress divided by strain:

$$E = \frac{\sigma}{\varepsilon} = \frac{F l_0}{A \Delta l}$$

Most structural materials are considered somewhat elastic, and the behavior of these materials under *tensile* or *compressive stress* is important.

Strength and breaking point

A general force *vs.* elongation graph of an elastic material undergoes *tension* or *compression*:

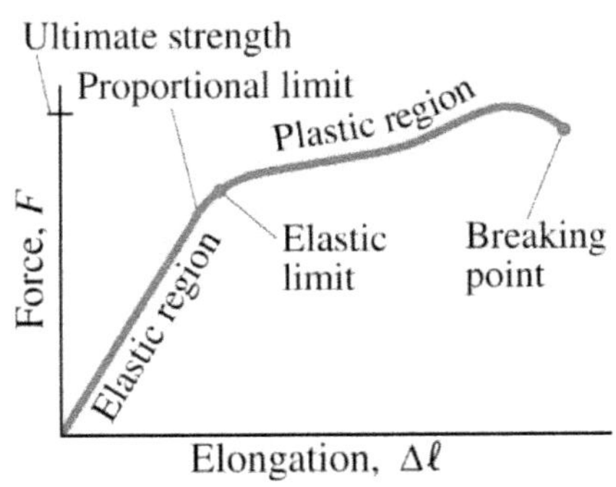

Force vs. elongation (Δl) graph shows an elastic and plastic region

Elastic materials under stress display a pattern of force *vs.* elongation (or compression), as seen above. When the maximum amount of stress is applied to the material, it no longer deforms *elastically* but deforms *plastically* (i.e., permanent change in shape without fracture).

Elastic limit and shear modulus

Elastic limit (yield strength) is the maximum stress a material can sustain before all subsequent deformation becomes permanent (i.e., plastic deformation).

After passing the elastic limit, the material continues to deform plastically until the ultimate strength is reached, after which the material breaks apart into two pieces.

Like tension and compression, all materials resist shearing to a certain degree. This resistance is governed by the shear strain and the *shear modulus*, which, like Young's modulus, is used to measure the stiffness of materials.

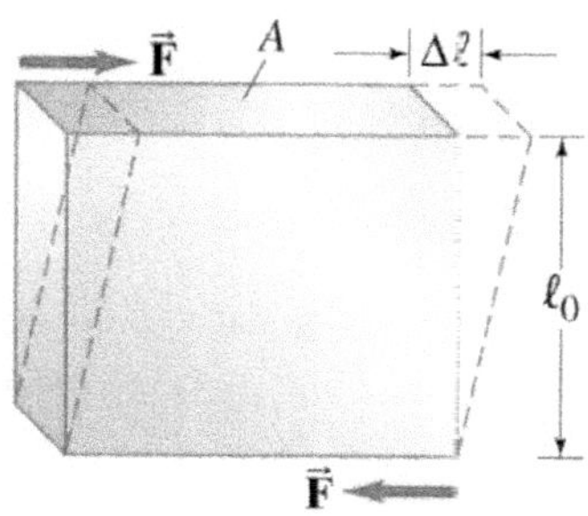

Block undergoing shearing forces and deforming according to shear forces

The shear strain is given by:

$$\varepsilon_{shear} = \frac{\Delta l}{l_0}$$

Shear Modulus can be written as:

$$S = \frac{\tau}{\varepsilon_{shear}}$$

$$\frac{\tau}{\varepsilon_{shear}} = \frac{Fl_0}{A\Delta l}$$

$$S = \frac{Fl_0}{A\Delta l}$$

When an object undergoes shear forces, the *shear strain* and *shear modulus* are calculated with Δl and l_0, which are perpendicular.

Thermal Expansion Coefficient

Thermal expansion of materials

Most materials expand when the temperature rises and contract when the temperature falls; this phenomenon is *thermal expansion.*

In solid materials, the magnitude of thermal expansion can be calculated to account for:

linear expansion,

area expansion

volumetric expansion

All three calculations require the thermal coefficient of the material.

Thermal expansion coefficient is an empirically derived constant that describes the expansion of a given material during a temperature rise and depends on the geometry of the material.

Linear, area and volume expansion

Thermal expansion equation for linear expansion is:

$$\Delta L = \alpha L_0 \Delta T$$

where ΔL is the change in length (m), L_0 is the initial length (m), ΔT is the change in temperature (K), and α is the coefficient of linear expansion(K^{-1})

Area expansion:

$$\Delta A = \gamma A_0 \Delta T$$

where ΔA is the change in the area (m^2), and γ is the coefficient of area expansion (K^{-1}).

Volume expansion:

$$\Delta V = \beta V_0 \Delta T$$

where ΔV is the volume change (m^3), and β is the coefficient of volume expansion (K^{-1}).

Chapter Summary

- States of matter: solid, liquid, and gas phases.

  Liquids and gases are fluids.

- Density is mass per unit volume and is given by:

  $$\rho = \frac{m}{v}$$

- Pressure is force per unit area and is given by:

  $$P = \frac{F}{A}$$

- External pressure applied to a confined fluid is transmitted throughout the fluid.

- Gauge pressure is the total pressure referenced at the atmospheric pressure:

  $$P_{\text{gauge}} = \rho g \Delta h$$

- An object submerged partly or wholly in a fluid is buoyed by force equal to the weight of the fluid that the object displaces.

- The buoyant force is an upward force experienced by an object submerged in a fluid due to displacement.

  Buoyant force is given by:

  $F_{\text{buoyant}} = \rho V g$, where V is the volume of the fluid displaced.

- Continuity Equation states that the flow rate through a pipe (area times velocity) is constant so that $A_1 \upsilon_1 = A_2 \upsilon_2$. This expresses the idea that a larger cross-sectional area of pipe experiences fluids traveling at a lower velocity.

- Bernoulli's Equation is a statement of conservation of energy:

  $$P + \rho g y + \frac{1}{2}\rho \upsilon^2 = \text{constant}$$

- Viscosity is an internal frictional force within fluids.

- Resistance to the deformation of solid material is due to stress.

  Three basic types of stress are *tension*, *compression*, and *shear*.

  Stress is given by:

  $$\text{stress} = \frac{\text{force}}{\text{area}}$$

- *Tensile stress* results when a force acts on the material to pull it apart.

  Compressive stress results when a force acts to compress the material.

  Shear stress occurs when two forces act opposite such that the material slides on a plane between the applied forces.

- *Elasticity* measures a material's ability to return to its original shape after distortion after undergoing tensile or compressive deformation.

- *Strain,* given by the lowercase Greek letter epsilon (ε), is the change in length of an object in the direction of the applied force, divided by the original length.

$$\varepsilon = \frac{\Delta l}{l_0}$$

  where ε is the strain, Δl is the change in length (m), and l_0 is the original length (m)

- *Young's modulus* (elastic modulus) is a measure of the force per unit needed to stretch material. It is expressed as the ratio of stress divided by strain, given as:

$$E = \frac{\sigma}{\varepsilon} = \frac{F l_0}{A \Delta l}$$

- Most materials expand when the temperature rises and contract when the temperature falls; this phenomenon is *thermal expansion.*

  Thermal expansion equation for linear expansion is given as:

$$\Delta L = \alpha L_0 \Delta T$$

  where ΔL is the change in length (m), L_0 is the initial length (m), ΔT is the change in temperature (K), and α is the coefficient of linear expansion (K^{-1})

Practice Problems: Fluids & Solids

1. The Bernoulli effect describes the lift force on an airplane wing. Wings must be designed to ensure those air molecules:

A. move more rapidly past the lower surface of the wing than past the upper surface
B. flow around wings that are smooth enough for an easy flow of the air
C. are deflected upward when they hit the wing
D. are deflected downward when they hit the wing
E. move more rapidly past the upper surface of the wing than past the lower surface

2. Two horizontal pipes (A and B) are the same length, but pipe B has twice the diameter of pipe A. Water undergoes viscous flow in both pipes, subject to the same pressure difference across the lengths of the pipes. If the flow rate in pipe A is Q, what is the flow rate in pipe B?

A. 2Q
B. 4Q
C. 8Q
D. 16Q
E. $\sqrt{2}$Q

Questions **3-5** are based on the following:

A pressurized cylindrical tank is 5 m in diameter. Water exits from the pipe at point C with a velocity of 13 m/s. Point A is 10 m above point B, and point C is 3 m above point B. The pipe's cross-sectional area at point B is 0.08 m^2, and the pipe narrows to a cross-sectional area of 0.04 m^2 at point C. Assume an ideal fluid in laminar flow. The density of water is 1,000 kg/m^3. (Use the acceleration due to gravity $g = 9.8$ m/s^2)

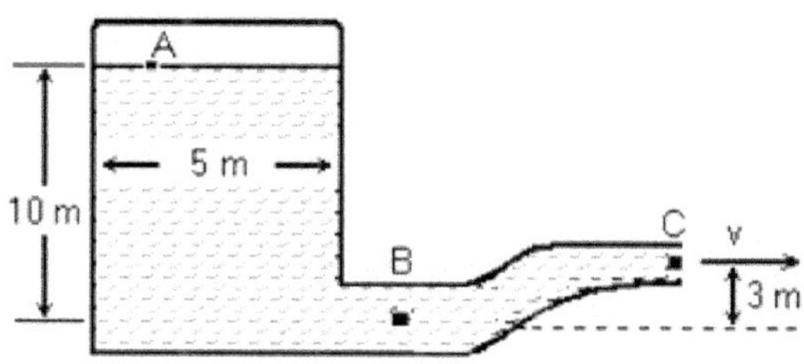

3. What is the mass flow rate in the pipe at point C?

A. 320 kg/s
B. 440 kg/s
C. 570 kg/s
D. 610 kg/s
E. 520 kg/s

4. What is the rate at which the water is falling into the tank?

A. 12 mm/s
B. 15 mm/s
C. 86 mm/s
D. 44 mm/s
E. 26 mm/s

5. What is the gauge pressure in the pipe at point B?

A. 82 kPa
B. 167 kPa
C. 71 kPa
D. 98 kPa
E. 45 kPa

6. A closed cubical chamber resting on the floor contains oil and a piston. If the piston is pushed down hard enough to increase the pressure just below the piston by an amount ΔP, which of the statements is correct?

A. The pressure at the top of the oil increases by less than ΔP
B. The pressure on the sides of the chamber remains the same.
C. The pressure in the oil increases by less than ΔP
D. The pressure at the bottom of the oil increases by more than ΔP
E. The increase in the force on the top of the chamber equals the increase in the force on the bottom of the chamber

7. What is the gauge pressure in the water at the deepest point of the Pacific Ocean, 11,030 m? (Use the density of seawater ρ = 1,025 kg/m^3 and the acceleration due to gravity g = 9.8 m/s^2)

A. 1.1×10^8 Pa
B. 3.1×10^8 Pa
C. 4.2×10^7 Pa
D. 7.6×10^7 Pa
E. 5.4×10^8 Pa

8. A cubical box with 25 cm sides is immersed in a fluid. The pressure at the top surface of the box is 108 kPa, and the pressure on the bottom surface is 114 kPa. What is the density of the fluid? (Use acceleration due to gravity g = 9.8 m/s^2)

A. 980 kg/m^3
B. 1,736 kg/m^3
C. 2,452 kg/m^3
D. 2,794 kg/m^3
E. 1,360 kg/m^3

9. A polar bear of mass 240 kg stands on floating ice 100 cm thick. What is the minimum area of the ice that will support the bear? (Use specific gravity of ice = 0.98 and saltwater = 1.03)

A. 2.6 m^2
B. 4.9 m^2
C. 4.8 m^2
D. 11.2 m^2
E. 9.1 m^2

10. A 14,000 N car is raised using a hydraulic lift. The lift consists of a U-tube with arms of unequal areas, initially at the same level. The lift is filled with oil with a 750 kg/m3 density with tight-fitting pistons at each end. The narrower arm has a radius of 6 cm, while the broader arm of the U-tube has a radius of 16 cm. The car rests on the piston on the broader arm of the U-tube. What force must be applied to the smaller piston to lift the car after it has been raised 1.5 m? (Ignore the weight of the pistons and use the acceleration due to gravity $g = 9.8$ m/s^2)

A. 4,568 N
B. 3,832 N
C. 2,094 N
D. 1,379 N
E. 1,748 N

11. An object has a volume of 4.2 m^3 and weighs 41,800 N. What is its apparent weight in water? (Use acceleration due to gravity $g = 9.8$ m/s^2 and density of water $\rho = 1{,}000$ kg/m^3)

A. 1,140 N
B. 230 N
C. 800 N
D. 640 N
E. 2,360 N

12. A solid sphere of mass 9.2 kg, made of metal whose density is 3,650 kg/m^3, hangs by a cord. When the sphere is immersed in a liquid of unknown density, the tension in the cord is 42 N. What is the density of the liquid? (Use acceleration due to gravity $g = 9.8$ m/s^2)

A. 1,612 kg/m^3
B. 1,468 kg/m^3
C. 1,950 kg/m^3
D. 1,742 kg/m^3
E. 1,144 kg/m^3

13. The tensile strength of a steel wire is 2,860 × 10^6 N/m^2. What is the maximum load that can be applied to a wire made of this steel if its diameter is 3.5 mm?

A. 46.4 kN
B. 19.7 kN
C. 33.2 kN
D. 58.5 kN
E. 27.5 kN

14. A grade of motor oil with a viscosity of 0.3 N·s/m2 flows through a 1 m long tube with a radius of 3.2 mm. What is the average speed of the oil if the drop in pressure over the length of the tube is 225 kPa?

A. 0.82 m/s
B. 0.96 m/s
C. 1.2 m/s
D. 1.4 m/s
E. 1.7 m/s

15. Ice has a lower density than water because ice:

A. molecules vibrate at lower rates than water molecules

B. is made of open-structured, hexagonal crystals

C. is denser and, therefore, sinks when in liquid water

D. molecules are more compact than when in the solid state

E. density decreases with decreasing temperature

Solutions: Fluids & Solids

1. E is correct.

According to the Bernoulli effect, air moving with a higher velocity exerts less pressure against a surface than air with a lower velocity.

Thus, to produce lift, the pressure on the underside of a wing should be higher, and the air should be slower on the bottom surface of the wing compared to air on the top surface.

2. D is correct.

Poiseuille's Law:

$Q = \pi \Delta P r^4 / 8\eta L$

$D_B = 2D_A$

$r_B = 2r_A$

$Q_A = \pi \Delta P r_A{}^4 / 8\eta L$

$Q_B = \pi \Delta P(2r_A)^4 / 8\eta L$

$Q_B = 16(\pi \Delta P r_A{}^4 / 8\eta L)$

$Q_B = 16Q_A$

3. E is correct.

Mass flow rate $\dot{m}$ is:

$\dot{m} = \rho v A_C$

where ρ = density of the fluid, v = velocity of flow, A_C = cross-sectional area

Dimensions of the tank are irrelevant to this answer, so $\dot{m}$ is calculated as:

$\dot{m} = (1{,}000 \text{ kg/m}^3)\cdot(13 \text{ m/s})\cdot(0.04 \text{ m}^2)$

$\dot{m} = 520 \text{ kg/s}$

4. E is correct.

Relationship between area and velocity of both exists as:

$A_1 v_1 = A_2 v_2$

$v_1 = A_2 v_2 / A_1$

$v_1 = [(0.04 \text{ m}^2)\cdot(13 \text{ m/s})] / [(\pi / 4)\cdot(5 \text{ m/s})]$

$v_1 = 0.26 \text{ m/s} = 26 \text{ mm/s}$

5. D is correct.

Gauge pressure is the pressure reference against the surrounding air pressure. Because of this, it is valid to ignore both the pressure in the tank and atmospheric pressure.

$$P = \rho gh$$

$$P = (10^3\ \text{kg/m})\cdot(9.8\ \text{m/s}^2)\cdot(10\ \text{m})$$

$$P = 98{,}000\ \text{Pa} = 98\ \text{kPa}$$

6. E is correct.

According to Pascal's Law, the pressure is transmitted undiminished in an enclosed static fluid. Thus, the pressure increases by ΔP everywhere in the oil. If the chamber is cubic, the top and bottom sides have the same area and experience the same increase in force:

$$A_{\text{top}} = A_{\text{bottom}}$$

$$\Delta \text{P}_{\text{top}} = \Delta \text{P}_{\text{bottom}}$$

$$P = F / A$$

Thus, force is directly proportional to pressure, and if the area of the top equals the bottom:

$$\Delta F_{\text{top}} = \Delta F_{\text{bottom}}$$

7. A is correct.

Gauge pressure is referenced relative to atmospheric pressure and is calculated by:

$$P_{\text{gauge}} = \rho gh$$

$$P_{\text{gauge}} = (1{,}025\ \text{kg/m}^3)\cdot(9.8\ \text{m/s}^2)\cdot(11{,}030\ \text{m})$$

$$P_{\text{gauge}} = 1.1 \times 10^8\ \text{Pa}$$

8. C is correct.

$$P_{\text{top}} = \rho gh$$

$$P_{\text{top}} = 108 \times 10^3\ \text{Pa}$$

$$h = (1 / \rho)\cdot(108 \times 10^3\ \text{Pa} / 9.8\ \text{m/s}^2)$$

$$h = (1 / \rho)\cdot(11{,}020\ \text{kg/m}^2)$$

$$P_{\text{bottom}} = \rho g(h + 25\ \text{cm})$$

$$\rho g(h + 25\ \text{cm}) = 114 \times 10^3\ \text{Pa}$$

$$h = (1 / \rho)\cdot(114 \times 10^3\ \text{Pa} / 9.8\ \text{m/s}^2) - 0.25\ \text{m}$$

$$h = (1 / \rho)\cdot(11{,}633) - 0.25\ \text{m}$$

continued...

Set equal and solve for ρ:

$$(1 / \rho)\cdot(11{,}020\ \text{kg/m}^2) = (1 / \rho)\cdot(11{,}633\ \text{kg/m}^2) - 0.25\ \text{m}$$

$$(11{,}020\ \text{kg/m}^2) = (11{,}633\ \text{kg/m}^2) - 0.25\ \text{m}(\rho)$$

$$0.25\ \text{m}(\rho) = 613\ \text{kg/m}^2$$

$$\rho = (613\ \text{kg/m}^2) / (0.25\ \text{m})$$

$$\rho = 2{,}452\ \text{kg/m}^3$$

9. C is correct.

Solve for density of ice and saltwater:

$$\text{SG} = \rho_{\text{substance}} / \rho_{\text{water}}$$

$$\text{SG}_{\text{ice}} = 0.98 = \rho_{\text{ice}} / 10^3\ \text{kg/m}^3$$

$$\rho_{\text{ice}} = 980\ \text{kg/m}^3$$

$$\text{SG}_{\text{saltwater}} = 1.03 = \rho_{\text{saltwater}} / 10^3\ \text{kg/m}^3$$

$$\rho_{\text{saltwater}} = 1{,}030\ \text{kg/m}^3$$

Solve for volume of ice:

$$F_{\text{B}} = F_{\text{bear}} + F_{\text{ice}}$$

$\rho_{\text{saltwater}}\text{V}_{\text{ice}}g = m_{\text{bear}}g + \rho_{\text{ice}}\text{V}_{\text{ice}}g$, cancel g from all terms

$$V_{\text{ice}} = m_{\text{bear}} / (\rho_{\text{saltwater}} + \rho_{\text{ice}})$$

$$V_{\text{ice}} = (240\ \text{kg}) / (1{,}030\ \text{kg/m}^3 - 980\ \text{kg/m}^3)$$

$$V_{\text{ice}} = 4.8\ \text{m}^3$$

Solve for area of ice:

$$A = V / h$$

$$A = (4.8\ \text{m}^3) / (1\ \text{m})$$

$$A = 4.8\ \text{m}^2$$

10. C is correct.

$P_1 = P_2 + \rho gh$

$F_1 / A_1 = F_2 / A_2 + \rho gh$

$F_1 = A_1(F_2 / A_2 + \rho gh)$

$F_1 = \pi(0.06 \text{ m})^2\cdot[(14{,}000 \text{ N}) / \pi(0.16 \text{ m})^2 + (750 \text{ kg/m}^3)\cdot(9.8 \text{ m/s}^2)\cdot(1.5 \text{ m})]$

$F_1 = \pi(0.0036 \text{ m}^2)\cdot[(14{,}000 \text{ N}) / \pi(0.0256 \text{ m}^2) + (750 \text{ kg/m}^3)\cdot(9.8 \text{ m/s}^2)\cdot(1.5 \text{ m})]$

$F_1 = (0.0036 \text{ m}^2)\cdot[(14{,}000 \text{ N}) / (0.0256 \text{ m}^2) + (11{,}025 \text{ N})\pi]$

$F_1 = 1{,}969 \text{ N} + 125 \text{ N}$

$F_1 = 2{,}094 \text{ N}$

11. D is correct.

$F_B = \rho Vg$

$F_{net} = F_{object} - F_B$

$F_{net} = (41{,}800 \text{ N}) - (1{,}000 \text{ kg/m}^3)(4.2 \text{ m}^3)(9.8 \text{ m/s}^2)$

$F_{net} = (41{,}800 \text{ N}) - (41{,}160 \text{ N})$

$F_{net} = 640 \text{ N}$

12. C is correct.

$F_{net} = F_{object} - F_B$

$F_{net} = mg - \rho_{fluid} Vg$

$\rho_{fluid} = -(F_{net} - mg) / V_{sphere}g$

$\rho_{fluid} = -[42 \text{ N} - (9.2 \text{ kg})\cdot(9.8 \text{ m/s}^2)] / [(9.2 \text{ kg} / 3{,}650 \text{ kg/m3})\cdot(9.8 \text{ m/s}^2)]$

$\rho_{fluid} = 1{,}950 \text{ kg/m}^3$

13. E is correct.

Tensile strength = F_{max} / A

$F_{max} = \text{A(tensile strength)}$

$F_{max} = (\pi / 4)\cdot(0.0035 \text{ m})^2\cdot(2{,}860 \times 10^6 \text{ N/m}^2)$

$F_{max} = 27{,}516 \text{ N} = 27.5 \text{ kN}$

14. B is correct.

By Poiseuille's Law, the volumetric flow rate of a fluid is given by:

$V = \Delta PAr^2 / 8\eta L$

The volumetric flow rate is the volume of fluid that passes a point per unit time:

$V = Av$

where v is the speed of the fluid.

Therefore:

$Av = \Delta PAr^2 / 8\eta L$

$v = \Delta Pr^2 / 8\eta L$

$v = (225 \times 10^3 \text{ Pa})\cdot(0.0032 \text{ m})^2 / [8\ (0.3 \text{ Ns/m}^2)\cdot(1 \text{ m})]$

$v = 0.96 \text{ m/s}$

15. B is correct.

For most substances, the solid form is denser than the liquid phase. Therefore, a block of most solids sinks in the liquid. For pure water, a block of ice (solid phase) floats in liquid water because ice is less dense than liquid water.

Like other substances, when liquid water is cooled from room temperature, it becomes increasingly dense. However, at approximately 4 °C (39 °F), water reaches its maximum density, and as it is cooled further, it expands the distance between molecules and becomes less dense.

This phenomenon is negative thermal expansion and is attributed to strong intermolecular interactions that are orientation-dependent.

Density of water is about 1 g/cm^3 and depends on the temperature. When frozen, the density of water is decreased by about 9%. This decreased density of water is due to the decrease in intermolecular vibrations, which allows water molecules to form stable hydrogen bonds with other water molecules around.

As hydrogen bonds form, molecules lock into positions as hexagonal structures.

Even though hydrogen bonds are shorter in the crystal than in liquid, this position locking decreases the average coordination number of water molecules as the liquid reaches solid phase.

Notes for active learning

CHAPTER 9

Thermal Physics

Page intentionally left blank

Heat and Temperature

Heat as energy

Heat is energy and exists independently from any medium; *heat is a way to transfer energy*.

Heat and temperature are often used interchangeably in everyday vernacular.

However, this usage of interchanging heat and temperature is incorrect.

Temperature as kinetic motion

Temperature measures *kinetic energy* (KE) in a specific medium and depends on the material.

Temperature and heat are connected principles, but heat is the *total energy* due to molecular motion. In contrast, temperature is the average measure of the *kinetic energy* in a substance.

Some substances respond drastically to a change in thermal energy. This response depends on the material and heat capacity.

Endothermic *vs.* exothermic processes

> *Endothermic* processes experience heat gained by the system from the surroundings, which equals heat lost by the surroundings.
>
> *Exothermic* processes cause the system's heat loss to equal the heat gain of the surroundings.

Thermal energy

The quantity of thermal energy transmitted from one body to another measures heat.

An object does *not* "contain" heat but can have a certain amount of thermal energy.

Heat can be measured in joules (J) and calories (cal).

On a microscopic level, thermal energy is related to atomic and molecular *vibrations*.

Matter is made of billions and trillions of small molecules; even in solids, these molecules and atoms are never entirely still.

The movement of molecules determines how much thermal energy they hold.

Calories

Calories relate heat to changes in temperature, which is often useful when solving problems in thermal physics.

calorie (lowercase *c*) is the heat needed to raise the temperature of one gram of water by one degree Celsius (g/°C).

$$1\ cal = 1\ \frac{g}{°\mathrm{C}} = 4.19\ J$$

Food Calories (uppercase *C*) are not equivalent to heat calories. This resembles the relationship between kilograms and grams (1 Calorie = 1,000 calories).

Another difference is that while nutrition Calories measure stored energy just as calories do, they also measure the energy stored in the chemical bonds of food that are broken down and stored when digested.

Calories measure energy in a substance and can be converted to equivalent conventional energy.

For example, the calories contained in 5 lbs. of spaghetti have enough energy to brew a pot of coffee; calories in one piece of cheesecake can power a 60 W incandescent light bulb for 1.5 hours; 222 Big Macs contain enough energy to drive a vehicle for almost 90 miles.

Energy conversion factors

Conversion relationships:

1 calorie = 4.2 J

1 Calorie (with capital C) = 1,000 calorie

1 Calorie (with capital C) = 1 kilocalorie

1 Calorie = 4,200 J

Conversions for water

1 gram = 1 cubic centimeter

1 cubic centimeter = 1 mL

Temperature Scales

Fahrenheit and Celsius

Three commonly used temperature scales are *Fahrenheit*, *Celsius* and *Kelvin.*

Daniel Fahrenheit (1686-1736) proposed the classic English system of measuring temperature (i.e., °F). in 1724 by dividing the difference between the boiling and freezing points of water into 180 equal degrees (°F).

Anders Celsius (1701-1744) developed the modern metric measuring temperature by dividing water's boiling and freezing points into 100 equal degrees (°C).

Celsius scale was originally the *centigrade scale.*

In the United States, the temperature unit is *Fahrenheit* (°F).

However, the SI unit for temperature is *Celsius* (°C), with water freezing at 0 °C and boiling at 100 °C (instead of 32 °F and 212 °F).

Kelvin scale

William Thomas, 1st Baron Kelvin (1824-1907), developed the Kelvin (K) temperature scale.

Kelvin scale is based on *absolute zero* (i.e., 0 K, –273.15 °C, or –459.67 °F) and is the most important science scale for temperature.

The size of units on the Kelvin scale is the *same* as for degrees on the Celsius scale.

The conversion between the two scales is:

$$°F = \frac{9}{5}°C + 32$$

Kelvin is another unit of temperature often used in scientific calculations. This scale is often used for gases, as low temperatures are common.

Kelvin (K) is the measure of the *absolute temperature.*

Absolute zero is the coldest theoretical temperature any substance can have, equal to 0 K.

0 K is equivalent to –273 °C.

Relative temperature scales

Three thermometers display equal temperatures and standard temperature units are shown.

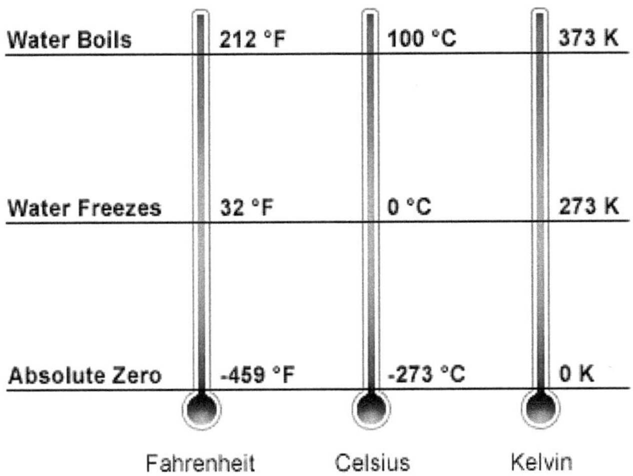

Comparison between the three temperature scales

Converting between scales

Temperature is the average amount of *kinetic energy* of the particles (i.e., molecules and atoms) of a substance.

Absolute zero when these particles theoretically have no KE and no movement.

Kelvin and Celsius use the same step in their degree measurements, making calculations easier.

For example, a temperature difference is the same in Kelvin as in Celsius, so the values need not be converted when asked to calculate a temperature difference in either of these units.

If required to convert between Kelvin and Celsius, the conversion can be performed by adding 273 to the Celsius temperature to get Kelvin:

$$K = °C + 273$$

The temperature of absolute zero may never be attained. Scientists have achieved temperatures close to zero but never fully reached them.

In 2003, the lowest temperature recorded was by a group of scientists from MIT (Massachusetts Institute of Technology). Researchers successfully cooled sodium (Na) gas to a temperature of half a billionth of a degree above absolute zero but not at absolute zero.

Fahrenheit and Celsius temperature scale conversions

Convert between Fahrenheit and Celsius by:

$$°F = (1.8 \times °C) + 32$$

For example, what is the temperature in °F for 28 °C?

$$°F = (1.8 \times °C) + 32$$

$$°F = (1.8 \times 28\ °C) + 32$$

$$°F = (50.4) + 32$$

$$°F = 82.4$$

Therefore, 28 °C equals 82.4 °F

For example, what is the temperature in °C for 70 °F?

$$°F = (1.8 \times °C) + 32$$

Rearrange:

$$(1.8 \times °C) = °F - 32$$

$$(1.8 \times °C) = °F - 32$$

$$°C = (°F - 32) / 1.8$$

$$°C = (70 - 32) / 1.8$$

$$°C = 21$$

Therefore, 70 °F equals 21 °C

Kelvin and Celsius temperature scale conversions

Convert between Kelvin and Celsius by:

$$K = °C + 273$$

For example, what is the temperature in °C for 295 K?

$$K = °C + 273$$

Rearrange:

$$°C = K - 273$$

$$°C = 295 - 273$$

$$°C = 22$$

Therefore, 295 K = 22 °C.

Reference values among temperature scales

	K	**°C**	**°F**
Absolute zero	0	–273	–460
Freezing point of water / melting point of ice	273	0	32
Room temperature	298	25	77
Body temperature	310	37	99
Boiling point of water / condensation of steam	373	100	212

Heat Capacity and Specific Heat

Heat capacity, specific heat and molar heat capacity

	Heat capacity	**Specific heat capacity**	**Molar heat capacity**
Definition	Amount of heat required to raise temperature of a substance by 1 K (1 °C) Ability of a substance to absorb heat	Amount of heat required to raise the temperature of 1 g of a substance by 1 K (1 °C) Amount of heat added (or removed) to change the temperature	Amount of heat required to raise temperature of 1 mole of substance by 1 K (1 °C) At constant pressure unless noted
Equation	$C_H = q / \Delta T$	$C_S = q / m\Delta T$ C_S (specific) equals C_P (constant pressure)	$C_M = q / n\Delta T$
Intrinsicality	Intrinsic property	Non-intrinsic property	
Unit	J / °C	J / g °C or J / Kg K	J / mol °C or J / mol K
Property	Extensive property	Intensive property	Intensive property
Substance quantity	Depends on amount	Independent of amount	
Specificity	Non-specific	Highly specific	
Examples	1 L H_2O = 4182 J/kg °C 100 g Cu = 38.5 J/ °C	H_2O = 1 calorie 1 calorie = 4.18 J/g °C Cu = 0.385 J/g °C	

Heat capacity

Heat capacity (or *thermal capacity*) is a physical property of a substance's ability to hold heat.

Heat capacity (C_H) of a substance is the energy required to raise the temperature of a substance by *one degree Celsius*. It is the absorbed heat energy ratio to the resulting temperature change.

Heat capacity is a product of an *object's mass* (*m*) and a *substance's specific heat capacity (c)*.

When measuring the heat capacity of a substance, a variable must be held constant since the heat capacity of a system depends on the temperature, the pressure, and the volume of the system. These are noted with either a c_p or a c_v, respectively.

SI unit for heat capacity is the joule per Kelvin (J/K).

Isobaric and isochoric processes

Gases and liquids are typically measured at constant volume and subjected to pressure for a constant reading.

> *Isobaric process* (c_p) is when *pressure* is held constant, defining the *heat capacity.*
>
> *Isochoric process* (c_v) is when the *volume* is held constant.

Heat capacity *vs.* specific heat capacity

Heat capacity and specific heat of substances are similar.

> *Specific heat* (C_S) is the ratio of energy absorbed to the temperature rise per unit of mass of the substance.
>
> *Heat capacity* is the ratio of energy absorbed to the rise in temperature.

Specific heat *c* is a characteristic of the substance and thus does not change.

Water has a relatively *high* specific heat capacity; 4.2 J of heat energy is needed to raise one gram of water 1 °C.

High specific heat of the water *helps modulate global temperatures*. Temperature fluctuations during day and night would be much larger if the water's heat capacity were low. This is evident in a desert where water is scarce. Daytime temperatures are high, yet nighttime temperatures can drop to near freezing.

Specific heat

Specific heat capacity (C_S) is the amount of heat required to raise the temperature of 1 gram of a substance by 1 °C.

Specific heat *c* is a characteristic of the substance and thus does not change. Metals tend to have *low* specific heat capacities.

Heat capacity and the specific heat of a substance are similar. *Specific heat capacity* differs from heat capacity because it relates to *the heat energy required to raise one gram of a substance by one degree Celsius.*

> *Specific heat capacity* (or *specific heat*) is the heat capacity of 1 gram of a substance.
>
> *Molar heat capacity* is the heat capacity of 1 mole of a substance.

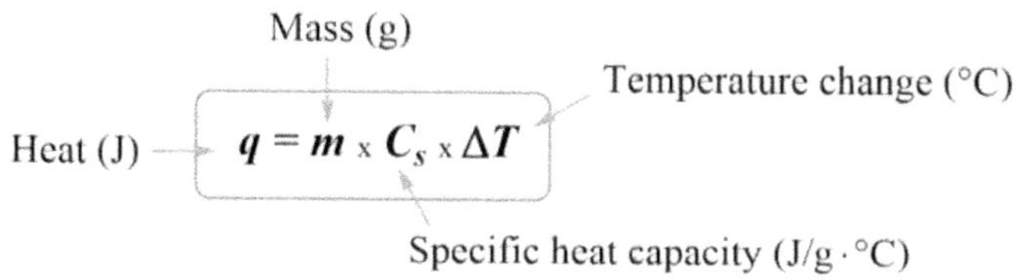

Specific heat capacity

Specific heat relates to heat and temperature

Units for specific heat are calories or joules per gram per °C. For example, the specific heat of water is 1 calorie (or 4.18 joules) per gram per Celsius degree.

The specific heat capacity of water is 4,180 joules per kilogram per degree Celsius (J/kg°C); 4,180 J are needed to raise the temperature of 1 kg of water by 1 °C.

Specific heat is related to heat and temperature by:

$$q = mC_S\Delta T$$

where q is the heat flow, m is mass measured in grams, C_S is the specific heat capacity of the substance, and ΔT is the temperature change.

Specific heat (C_S) of a substance, heat (Q), and temperature are related by:

$$Q = mC_S\Delta T$$

where Q is the heat transferred to the material (J), m is the mass of the object being heated (kg), and ΔT is the temperature change (K).

Calculating specific heat

For example, 3,200 J of heat is added to 1.0 kg of water at 10 °C. (Use specific heat of water C_S = 4,190 J/kg·°C)

Rearrange the specific heat equation to solve ΔT:

$$Q = mC_S\Delta T$$

$$\Delta T = \frac{Q}{mc}$$

$$\Delta T = \frac{3{,}200\,J}{1.0\,kg \cdot 4{,}190\,\frac{J}{kg} \cdot °C}$$

$$\Delta T = 0.76\ °C$$

where Q is the heat transferred to the material (J), m is the mass of the object being heated (kg), and ΔT is the change in temperature (K).

ΔT represents the temperature change. This value for the change (not the absolute value) must be either added to the initial temperature to determine the final temperature or subtracted from the final to find the initial.

A reaction must be contained within a system to observe all the products.

Specific heats of selected materials

Material	***c*(J/kg·K)**
Aluminum	879
Concrete	850
Diamond	509
Glass	840
Helium	5,193
Water	4,181

Low *vs.* high specific heat

Substances such as metals (e.g., copper) have a *low specific heat* because it does not take much energy to transfer the heat and excite the molecules.

Raising the value of their average kinetic energy (raising their temperature).

Materials that are difficult to heat (e.g., rubber) have a much *higher specific heat* because more energy is required to raise the temperature by the same amount. The previous table shows that water has a much higher heat capacity than other common materials found on land.

Specific heat modulates environmental temperatures

Raising an ocean by 1 °C takes more energy than raising city sidewalks by the same amount. Thus, land warms faster during the day than water when subjected to the same amount of thermal energy from the Sun. This is the reason for the sea breeze effect.

Hot land heats the air above it, causing it to rise and create a low-pressure system above it. The cooler water creates a high-pressure system due to air cooling down and descending. The difference in pressures creates airflow from the sea to the land during the day as a sea breeze.

At night, the land cools faster than the water (i.e., lower specific heat capacity), and the effect is reversed as air flows from land to the sea.

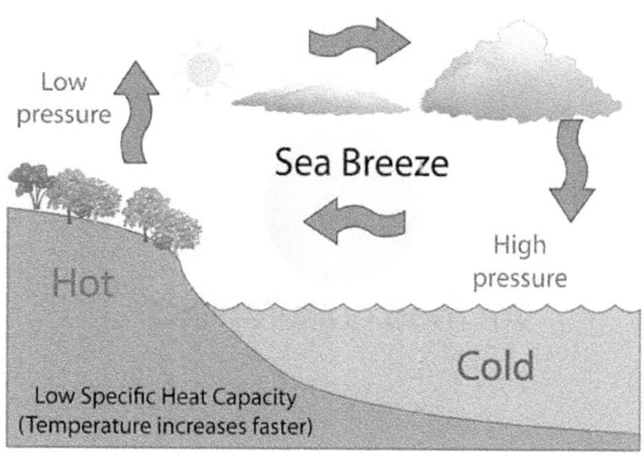

High specific heat of water helps modulate global temperatures

Notes for active learning

Calorimetry

Food samples

Calorimetry is an experimental method to calculate enthalpy changes involving heat flow.

Calorimeters measure the amount of energy in food samples experimentally. Food is placed inside a closed container surrounded by water and burned in the presence of oxygen. Water temperature change is related to the energy released, commonly reported in nutritional Calories (or kilojoules units). Note the term *Calories* in uppercase.

Even though cells within the body combust the molecules differently than a calorimeter, the calorimeter provides an accurate *caloric value* because of the same end products.

Calibration is an experiment in which hot water of known mass and temperature is poured into an insulated cup containing cold water of known mass and temperature. The composite sample equilibrates to a maximum temperature, and a calorimeter constant is determined below.

Nutritional Calories

Nutritional Calories (Cal) are energy units equal to 1,000 calories (cal) and a kilocalorie (kcal).

1 Calorie = 1,000 calories

Nutrient Molecule in Food	Example	Cal/g	kJ/g
Carbohydrate	Table sugar, potatoes, flour	4	17
Protein	Meats, fish, beans	4	17
Fat	Oil, butter	9	38

Foods contain varying amounts of energy, as shown

Calorimeters

Calorimeters are ideally considered an insulated system. The heat liberated by the reaction process is transferred to the other substances within the calorimeter or the calorimeter itself. No heat is lost to the surroundings.

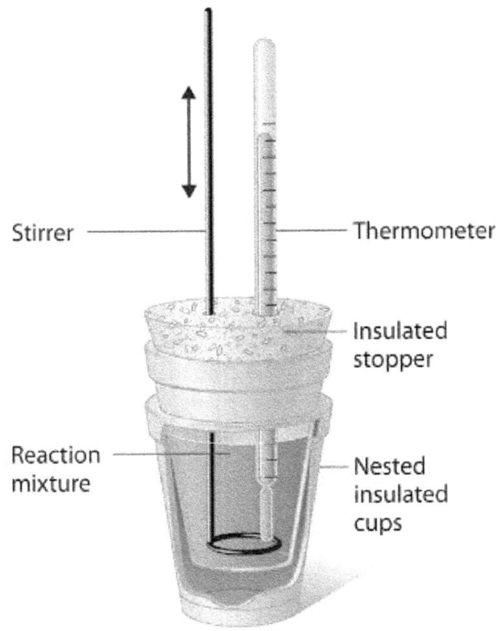

Components of a simple calorimeter

Since no thermal energy is lost in a calorimeter, the heat of a reaction can be measured by inserting a thermometer into a calorimeter and recording the temperature readings before and after the reaction. A stirrer ensures that all calorimeter contents are mixed and react uniformly.

Constant-pressure calorimeters

Constant-pressure and *constant-volume calorimeters* are two types of calorimeters.

Constant-pressure calorimeter is where heat flow is measured at constant atmospheric pressure.

A perfect calorimeter does not exist; no calorimeter completely insulates heat flow. The calorimeter itself, as part of the surroundings, absorbs some heat. Therefore, the calorimeter must be calibrated to obtain accurate results. Styrofoam cups are simple, constant-pressure calorimeters often used in introductory chemistry classes. Although a Styrofoam cup's temperature-insulating properties are reasonably good, the cup absorbs heat.

Calorimeter constant K_{cal}

Calorimeter constant K_{cal} is determined by a composite sample equilibrating at max. temp.

Basic principle of calorimetry:

heat lost by hot water = heat gained by cool water + heat gained by calorimeter

Determine the mathematical equations:

$$-m_{hot}C_{water}(T_{final} - T_{intial,\ hot}) = m_{cool}\ c_{water}(T_{final} - T_{initial,\ cool}) + K_{cal}\ (T_{final} - T_{initial,\ cool})$$

Sign is negative for heat lost by the hot water because heat loss is negative. Substitute known masses and temperatures into the expression. Since the cool water was in the cup, the calorimeter's initial and cool temperatures were the same.

For example, suppose 60.1 g of water at 97.6 °C is poured into a cup calorimeter containing 50.3 g of water at 24.7 °C. The final temperature of the combined water samples reaches 62.8 °C. What is the calorimeter constant? (Use c_{water} = 4.184 J/g°C)

$$-(60.1\text{ g})\cdot(4.184\text{ J/g°C})\cdot(62.8\text{ °C} - 97.6\text{ °C})$$
$$= [(50.3\text{ g})\cdot(4.184\text{ J/°C})\cdot(62.8\text{ °C} - 24.7\text{ °C})] + [K_{cal}\ (62.8\text{ °C} - 24.7\text{ °C})]$$

$$-(60.1\text{ g})\cdot(4.184\text{ J/g°C})\cdot(-34.8\text{ °C})$$
$$= [(50.3\text{ g})\cdot(4.184\text{ J/°C})\cdot(38.1\text{ °C})] + [K_{cal}\ (38.1\text{ °C})]$$

$$[8.75 \times 10^3\text{ J}] - [8.02 \times 10^3\text{ J}] = [K_{cal}\ (38.1\text{ °C})]$$

$$0.73 \times 10^3\text{ J} = [K_{cal}\ (38.1\text{ °C})]$$

$$[0.73 \times 10^3\text{ J}] / (38.1\text{ °C}) = K_{cal}$$

$$K_{cal} = 19.2\text{ J/°C}$$

Heat flow

After calibrating the calorimeter, the specific heat capacity will be determined.

A known quantity of metal is heated to a known temperature and placed in an insulated cup calorimeter containing a known quantity and temperature until the temperature equilibrates.

For example, suppose 28.2 g of an unknown metal is heated to 99.8 °C and placed into a cup calorimeter containing 150.0 g of water at 23.5 °C. The temperature of the water equilibrates at a final temperature of 25.0 °C. What is the specific heat capacity of the metal? (Use the calorimeter constant of 19.2 J/°C).

Consider heat flow:

heat lost $_{\text{hot metal}}$ = heat gained $_{\text{water}}$ + heat gained $_{\text{calorimeter}}$

$$-m_{hot}C_{water}(T_{final} - T_{intial,\ hot}) = m_{cool}\ c_{water}(T_{final} - T_{initial,\ cool}) + K_{cal}\ (T_{final} - T_{initial,\ cool})$$

$$-(28.2\ g)\cdot(C_{metal})\cdot(25.0\ °C - 99.8\ °C) =$$

$$= [(150.0\ g\ )\cdot(4.184\ J/g°C)\cdot(25.0\ °C - 23.5\ °C)] + [19.2\ J/°C(25.0\ °C - 23.5\ °C)]$$

$$-(28.2\ g)\cdot(C_{metal})\cdot(-\ 74.8\ °C)$$

$$= [(150.0\ g\ )\cdot(4.184\ J/g°C)\cdot(1.5\ °C)] + [19.2\ J/°C(1.5\ °C)]$$

$$2.11 \times 10^3\ g°C\ (C_{metal}) = (9.4 \times 10^2\ J) + (29\ J)$$

$$2.11 \times 10^3\ J\ (C_{metal}) = 969\ J$$

Solving for specific heat capacity:

$$C_{metal} = 969\ J\ /\ 2.11 \times 10^3\ g°C$$

$$C_{metal} = 0.459\ J/\ g°C$$

Constant-volume calorimeters

Constant-volume calorimeter is where heat flow is measured at constant volume.

Bomb calorimeter (*decomposition vessel*) is the most used constant-volume calorimeter.

Combustion is when oxygen reacts with a substance to produce carbon dioxide (CO_2), water (H_2O), and energy.

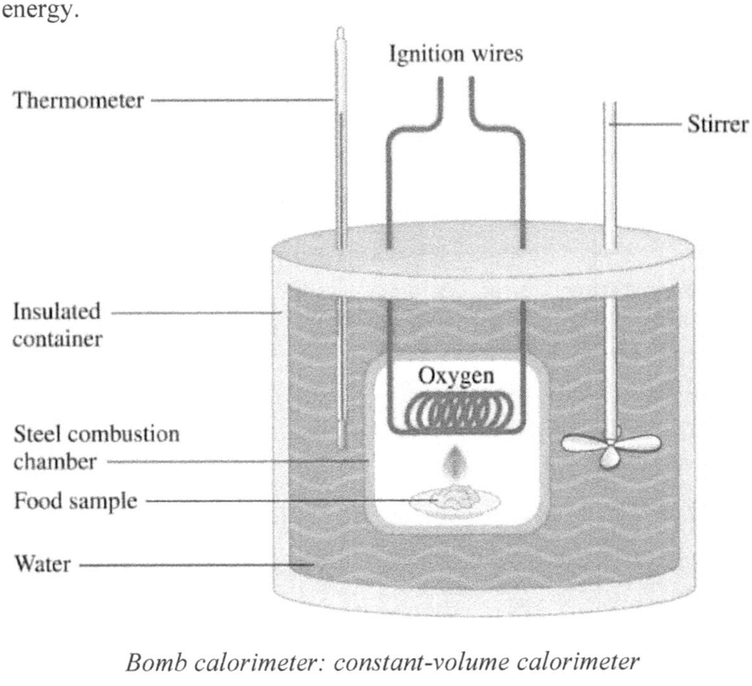

Bomb calorimeter: constant-volume calorimeter

Heat Transfer

Rate of heat transfer

When heat transfers by conduction, the rate of heat transfer (H) can be calculated in joules per second (J/s) or watts (W).

Amount of energy (heat) conducted per unit:

$$H = \frac{kA\Delta T}{t}$$

where ΔT is the temperature difference across the object (K), A is the cross-sectional area (m^2), t is the thickness of the material (m), and k is the thermal conductivity of the material (J/s·m·k).

Heat transfer (H) rate is a material characteristic usually provided in the problem.

Heat transfers between two objects by *conduction*, *convection*, or *radiation*.

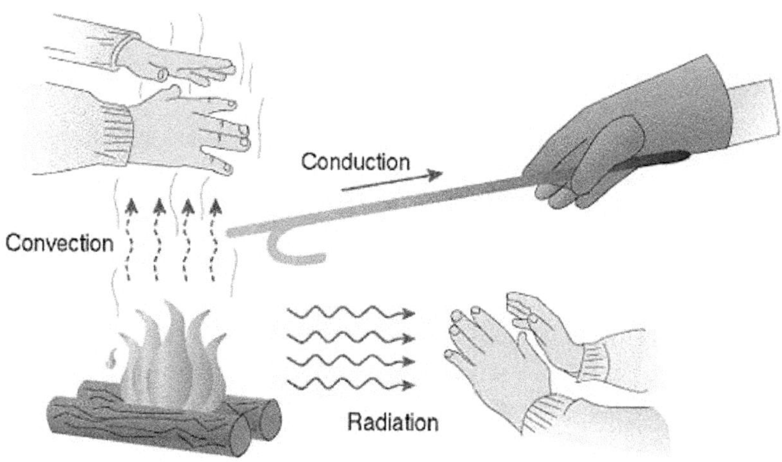

Heat transfer by convection, conduction, and radiation

Heat flows from hotter object

According to the First Law of thermodynamics, *heat always flows from a hotter to a colder system until the two systems are in thermal equilibrium*.

Conduction is the flow of heat between two materials in *direct contact*.

Convection is the flow of energized (e.g., air) molecules that settle and transfer their energy by *direct contact* using conduction.

Radiation is the heat flow (i.e., energy) emitted by matter through *electromagnetic radiation* (e.g., photons or electromagnetic waves).

Thermal conductivities of selected materials

Material	k (J/s·m·K)
Aluminum	237
Concrete	1
Copper	386
Glass	0.9
Stainless Steel	16.5
Water	0.6

Thermal conductivities of common materials

Conduction transfers heat by contact

Conduction is heat transfer between substances in *direct contact* (i.e., must be touching).

When particles of a hotter substance vibrate, these molecules bump into nearby particles and transfer some energy.

Conduction is an essential means of heat transfer for solids, and metals are conductive. It occurs when heat is transferred through *direct contact.*

For example, on a frigid winter night, when a person wraps their hands around a cup of hot cocoa, their fingers become warm because they are in direct contact with the hot mug, which is in direct contact with the hot liquid. The same process is happening at a molecular level.

The liquid in the mug contains heat energy, and the molecules of the hot liquid are in constant motion (high kinetic energy). These molecules continuously collide with the molecules of the mug, increasing their motion. The collisions gradually transfer energy from the hot liquid molecules to the mug molecules.

The mug transfers heat to a person's hands through the same conduction mechanism. The person perceives energy transfer as increasing temperature; thus, their hands feel warm.

Calculating heat conduction

When calculating heat conduction through an object, it is essential to note that the cross-sectional area refers to where the heat is being conducted. For the mug example, the heat is being conducted through its outer curved surface.

When the mug is placed on a table, and the heat conduction to the table is calculated, the cross-sectional area is the area of the bottom of the mug instead of the area of the sidewalls. Therefore, the cross-sectional area is the circumference of the mug times its height. The thickness, in this case, is the thickness of the mug itself.

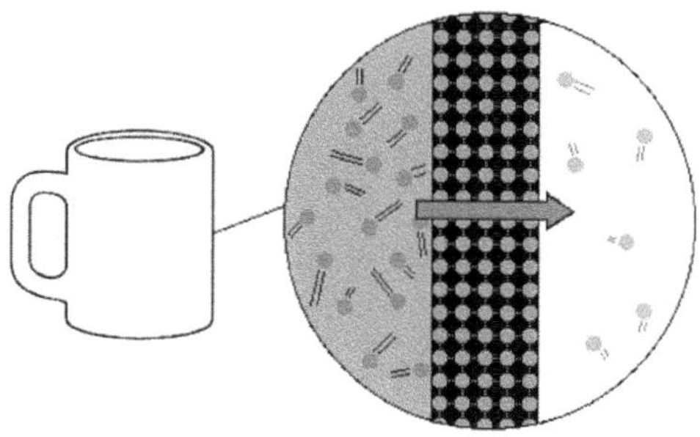

Conduction (heat transfer by direct contact) from the hot liquid to the walls of a mug

Convection transfers heat by motion of particles

Convection refers to the *physical flow of matter*. It is when heat flows energized molecules from one place to another through the movement of fluids.

Convective heat transfer occurs by diffusion (i.e., random Brownian motion of particles) and advection (i.e., larger-scale flowing currents transport matter or heat).

For example, convection occurs when warmer areas of a liquid or gas rise to colder areas, and the cooler liquid or gas replaces the warmer areas.

Gas is fluid, as it can flow and change shape depending on its container.

A pot of water is placed on a hot stove. The water at the bottom of the pot warms first. As the water warms, it rises and is replaced by cooler water. This circulation of the heated water transfers thermal energy from the bottom of the pot to the top.

This movement of molecules is a closed fluid flow pattern of *convection currents*.

Convection ovens work similarly with gas instead of liquid molecules. Air is blown past a heating element to the oven chamber. The hot element energizes the air molecules when passing through the heating element. Energized air molecules are circulated throughout the oven by airflow provided by the fan and continue until all molecules are energized to the same amount.

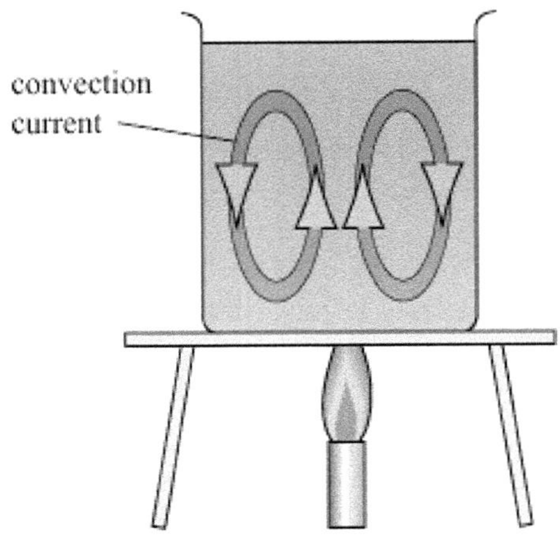

Convention current represented by the circulating water

Radiation transfers heat through electromagnetic waves

Radiation transfers heat *without* contact between the heat source and the heated object. It consists of electromagnetic waves traveling at the speed of light and can occur in a vacuum.

It has *no mass exchange*, and *no medium is required* to transfer heat.

Radiation transfers energy through electromagnetic waves emitted from all objects or surfaces with heat energy and does not require a medium; thus, radiation can occur in a vacuum.

Objects with more energy emit more radiation energy and do not require a medium; thus, radiation can occur in a vacuum.

For example, a hot piece of metal radiates heat as infrared electromagnetic waves. As the metal gets hot, it also radiates heat in the visual spectrum, thus making the metal appear "red hot."

Sun is a radiation source, releasing electromagnetic waves through space and the atmosphere to heat Earth. Similarly, holding hands next to a fire heats hands through radiation.

Radiation can be important even in situations with an intervening medium. An example is the heat transfer between a living entity and its surroundings.

For example, the sun's thermal energy travels through the 93 million miles of space to warm the Earth, even passing through the mesosphere, the coldest part of Earth's atmosphere, which can reach temperatures of -173° C.

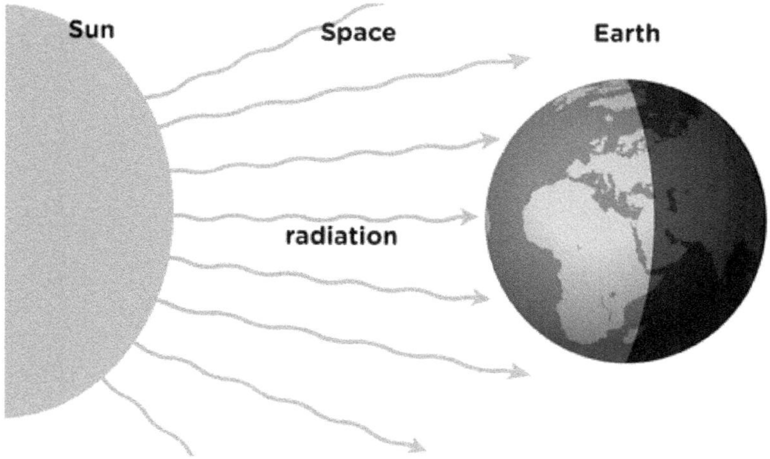

Radiation emitted from the Sun travels to the Earth's surface

Power radiated by an object is related to its temperature:

$$P = e\sigma A(T^4 - T_C^4)$$

where P is the power radiated (W), e is the emissivity of the radiator, σ is Stefan's constant equal to 5.67×10^{-8} W/m^2·K, A is the area of the radiating surface (m^2), T is the temperature of the radiator (K), and T_C is the temperature of the surroundings (K).

Notes for active learning

Thermodynamic Systems

Thermodynamic variables

Thermodynamics defines variables related to energy of physical and chemical changes.

Thermodynamic variables:

ΔE or ΔU	measures the *total internal energy* of a system
q	measures the *heat* of a system
w	measures the *work* of a system
ΔH	measures the *enthalpy change* of a system – whether a process is exothermic or endothermic
ΔG	measures the *Gibbs free energy change* of a system – whether a process is spontaneous or nonspontaneous
ΔS	measures the *entropy change* of the system or surroundings – whether the randomness of the system or surroundings is increasing or decreasing

Steady-state processes occur without a change in the internal energy.

System and surroundings

Thermodynamic system is a body of matter or radiation confined by partitions, with defined permeabilities separating it from its surroundings.

System defines the parameters of the universe being studied, such as the substances involved in a chemical reaction or a phase change.

Surroundings are everything other than the system being studied.

Isolated, closed, and open systems

Thermodynamic system describes processes and matter within a boundary.

Isolated system has no heat transfer, work, or matter with its surroundings.

Closed system may transfer heat and work to its surroundings but *not* matter between the system and surroundings.

Open system can transfer heat, work, or matter between the system and surroundings.

Note: *heat* and *temperature* are different quantities.

Heat is a form of *thermal energy that can leave or enter a system.*

Temperature measures the *average kinetic energy* (KE) of particles in a system.

Characteristics of thermodynamic systems

System	Exchanges with surrounding	Total amount of energy	Example	Illustration
Open	Energy & Matter	Not remain constant	Solution in an open flask	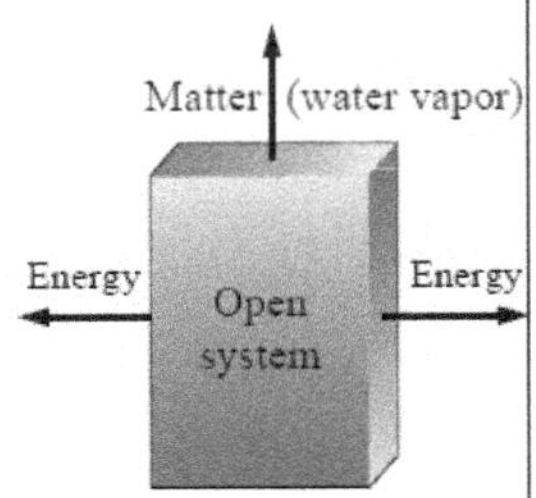
Closed	Only Energy	Not remain constant	Solution in a sealed flask	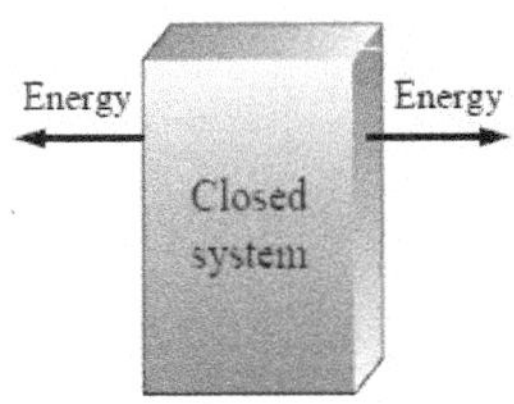
Isolated	Neither energy nor matter	Remains constant	Sealed flask in a thermos flask	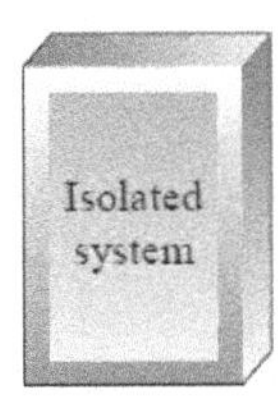

State functions

State function describes the energy required to change the system into another equilibrium state.

State functions (i.e., state quantities) are a group of equations representative of the properties of a system, which depend *only* on the current state of the system, not the path used to achieve the current state.

State functions are quantities with values *not* dependent on the path used to measure the value.

State functions ($\Delta E°$, $\Delta H°$, $\Delta S°$, $\Delta G°$) depend only on the *initial and final states* of the system.

State function variables (e.g., pressure, volume, temp) are a function *of other state variables*.

Thermodynamic state functions include *enthalpy* (H), *entropy* (S), and *Gibbs free energy* (G).

Path functions

Process functions quantitatively describe the transition between equilibrium states of a thermodynamic system.

Path functions depend on the *path taken* to reach one state from another. Therefore, different routes give different quantities (e.g., work, heat).

Quantities, such as work (w) and heat (q), are not state functions, but are *path functions*, represented by *lowercase letters*.

Thermodynamics

Heat, work and temperature

Thermodynamics relates heat, work, and temperature to energy, entropy, and properties of matter and radiation.

Thermodynamics answers several fundamental questions regarding *whether* a reaction occurs, to what *extent*, and whether it *releases or absorbs heat.*

An important concept in thermodynamics is the thermodynamic system, the region under study. Everything else is the surroundings.

Systems are separated from the surroundings by a boundary. Transfers of energy as work, heat, or matter between the system and surroundings occur through the boundary.

Four laws of thermodynamics

Thermodynamics states four laws characterizing heat and energy relationships by describing the behavior of physical quantities such as temperature, heat, work, energy, and entropy.

> *Zeroth Law of thermodynamics* (or *concept of temperature*) states that ***heat flows from the hotter body to the colder*** *when placed in contact until thermal equilibrium.*

> *First Law of thermodynamics* (or *conservation of energy*) states that *the total energy in a system remains constant as* ***energy can neither be created nor destroyed.***

> *Second Law of thermodynamics* (or *concept of entropy*) states that *the universe's natural order* ***favors a net increase*** *in the overall entropy (i.e., disorder).*

> *Third Law of thermodynamics* (or *entropy at absolute zero*) states that *the entropy of a perfectly ordered, pure crystalline solid is* ***zero at the temperature of absolute zero*** *(equal to 0 K, –273.15 °C or –459.67 °F).*

Zeroth Law of Thermodynamics

Thermal equilibrium

Zeroth Law of Thermodynamics is the logical predecessor of the First and Second Laws of Thermodynamics.

Zeroth Law states that *if two systems are in thermal equilibrium with a third system, the two initial systems are also in thermal equilibrium.*

Three systems in thermal equilibrium contain the same amount of heat energy (same temperature) and do *not* exchange heat; a fundamental principle underlying thermodynamics.

For example, if system A is in equilibrium with B, and system C is in equilibrium with B, then system A and C must be in equilibrium, and all three systems must be at the same temperature.

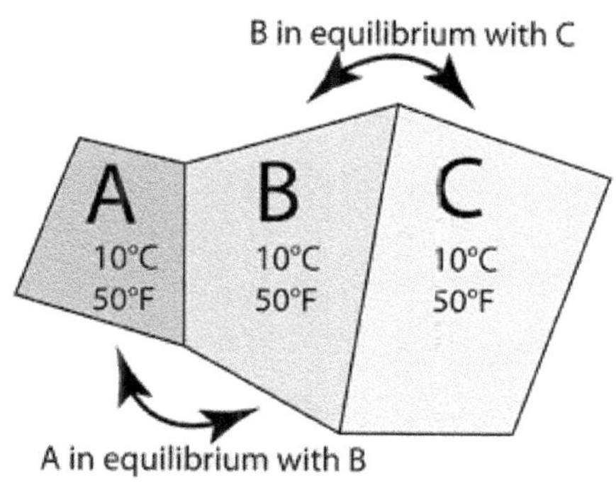

Zeroth Law of Thermodynamics with three systems in thermal equilibrium

However, it is vital to notice that although no net heat transfer occurs between systems in equilibrium, thermal energy is technically transferred between the related systems.

The exchange of thermal energy requires that every unit of energy passed from any system has the same value of energy passed back into the system. This holds even if the two systems contain materials with different specific heats.

This observation of thermal energy transfer means that there must be a property that can be considered the same upon which heat transfer depends, and that property is temperature.

Concept of temperature

Zeroth Law of thermodynamics states that *heat flows from the hotter to the colder body when in contact until thermal equilibrium.*

It establishes that *temperature is a core, measurable property of matter.*

Separated by a permeable barrier, two systems are in thermal equilibrium, and *no heat* transfers.

After thermal equilibrium is achieved, no heat transfers, defining the *concept of temperature.*

Expressing thermal equilibrium

If two systems are in thermal equilibrium with a third system, each is in thermal equilibrium.

Thermal equilibrium is expressed by:

$$T_{\mathrm{A}} = T_{\mathrm{B}}$$

and

$$T_{\mathrm{C}} = T_{\mathrm{B}}$$

Then

$$T_{\mathrm{A}} = T_{\mathrm{C}}$$

Zeroth Law is crucial for calculations of thermodynamics (i.e., heat and energy relationships).

First Law of Thermodynamics

Conservation of energy

Energy, like matter, is always conserved; it can neither be created nor destroyed.

First Law of Thermodynamics states that *energy can neither be created nor destroyed and the* ***total energy*** *in a system remains constant. Energy is conserved in a chemical reaction.*

First Law of thermodynamics relates to the *internal energy* of a substance.

Internal energy increases if heat is added to the system or if work is done on the system.

Disordered motion of particles

Change in total *internal energy* of a system equals contributions by heat (q) and work (w):

$$\Delta U = q - w$$

where ΔU is the change in internal energy, q is the contribution from heat, and w from work.

Internal energy refers to the microscopic energy of the disordered motion of particles and is represented by the symbol (U).

Internal energy cannot be seen and has no apparent effect on the object's motion, but it does play a part in the thermodynamic properties of the substance.

This principle is the foundation for the law of *conservation of energy*.

For example, a glass of water does not look like it contains kinetic or potential energy. Water molecules are whizzing around on the atomic scale, moving hundreds of m/s. If the water is dumped from the glass, its internal energy does not affect its motion.

Thermodynamic properties of the substance include potential and kinetic energy

Sign conventions

$\Delta U = q - w$ uses the following sign conventions:

q is *positive* if heat is *absorbed into* the system (i.e., heating)

q is *negative* if heat is *released from* the system (i.e., cooling)

w is *positive* if work is done *by* the system (i.e., compression)

w is *negative* if work is done *on* the system (i.e., expansion)

First Law of thermodynamics:

$$\Delta U = q + w$$

where q is *positive* if heat is *absorbed into* the system and *negative* if heat is *released from* the system; work (w) is positive if done *on the* system or negative if done *by* the system.

Heat (*q*) and work (*w*) absorbed or released

Use the sign conventions in the chart below:

q	+	Heat ***absorbed*** by the system	E_{system} increases
q	–	Heat ***released*** by the system	E_{system} decreases
w	+	Work done ***on*** the system	E_{system} increases
w	–	Work done ***by*** the system	E_{system} decreases

Heat or energy *flowing out* of system is *negative.*

Heat or energy *flowing into* system is *positive.*

Other natural sciences use the opposite sign convention; note the sign convention used.

Internal energy

The difference between the amount of heat added to the system (Q) and the work done by the system (W) is equal to the total change in internal energy (ΔU):

$$\Delta U = \Delta Q - \Delta W$$

As with other energy, the unit of internal energy is the joule (J), although it is also expressed using calories (cal). By convention, work done on a system is positive, and work done by a system is negative.

Occasionally, the above equation is written as $\Delta U = \Delta Q + \Delta W$, and ΔW is defined as the work done on the system. Be sure to note which form is being used.

Calculating internal energy

The quantities in $\Delta U = \Delta Q + \Delta W$ can determine what a thermodynamic system is experiencing.

For example, in a chemical reaction, if ΔQ is a quantity less than zero ($\Delta Q < 0$), the reaction is losing heat. This reaction is *exothermic,* and heat is a product.

If ΔQ is a quantity greater than zero ($\Delta Q > 0$), the reaction absorbs heat from its surroundings. This is an *endothermic* reaction, and heat is a reactant.

Internal energy is expressed:

$$\Delta U = q + w$$

where ΔU is the change in internal energy, q is the contribution from heat, and w from work.

For example, consider internal energy, heat, and work. For a process in which 3.4 kJ of heat flows out of the system while the system on the surroundings does 4.8 kJ of work, what is the internal energy?

Calculate internal energy:

$$\Delta U = q + w$$

$$\Delta U = -3.4 \text{ kJ} + (-4.8 \text{ kJ})$$

$$\Delta U = -8.2 \text{ kJ}$$

where ΔU is the change in internal energy, q is *negative* as heat is *released from* the system, and work (w) is negative when done *by* the system.

Use the appropriate heat (q) and work (w) signs.

Energy has flowed from the system to the surroundings by –8.2 kJ.

Endothermic and exothermic reactions

Figure below depicts an endothermic (absorbing) and exothermic reaction (producing heat).

Reaction is exothermic on the left because heat is *produced and released* to the surroundings.

Reaction is *endothermic* on the right because heat is *absorbed* from the surroundings.

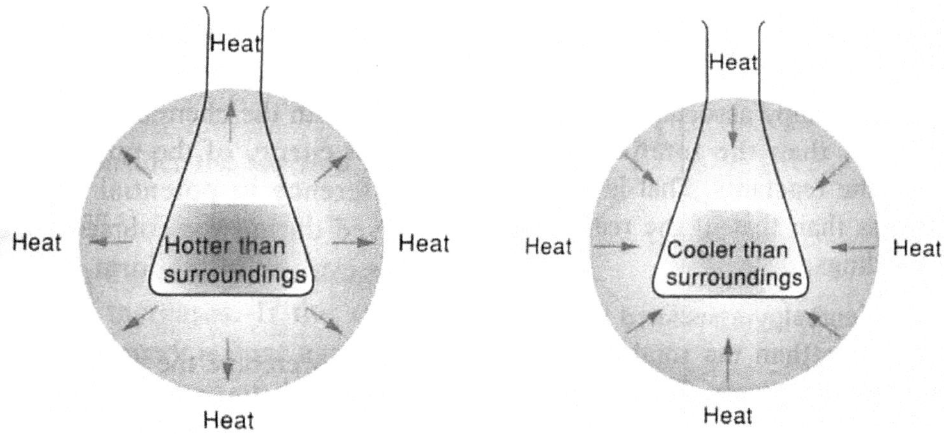

Exothermic process releases heat (left), and the endothermic process absorbs heat (right)

Second Law of Thermodynamics

Concept of entropy (*S*)

Second Law of thermodynamics is that *the universe favors a net increase in the overall entropy.*

Isolated systems increase in entropy and *never decrease.*

Open systems decrease entropy as entropy increases for the surroundings.

Second Law of Thermodynamics states *that in an isolated system (no transfer of heat, work, or matter), a process increases entropy or stays constant, but entropy never decreases.*

Entropy is a measure of the randomness or disorder of a system. Entropy measures the system's thermal energy per unit temperature, which is unavailable for useful work.

Work requires ordered molecular motion.

Open systems transfer energy

An exception to the Second Law of Thermodynamics (i.e., entropy cannot decrease) is an *open system* where heat, work, or matter can be transferred to surroundings; the system's entropy decrease is related to an increase in the entropy of the surroundings.

Second Law states that *heat flows spontaneously* from a hot object to a cold object but never spontaneously flows in the opposite direction. A bowl of ice never gets colder in a warm room; the ice always gets warmer.

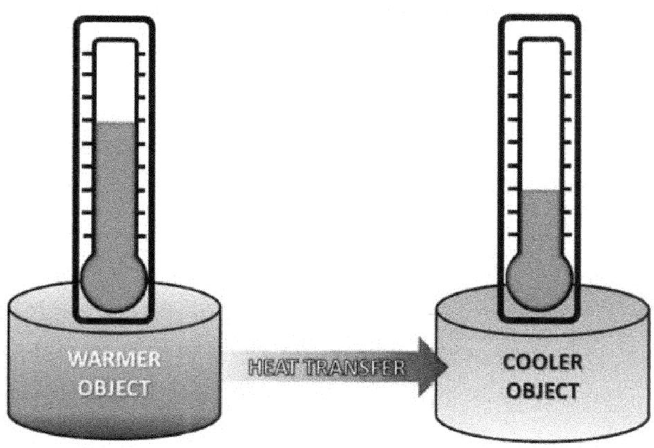

Heat flows spontaneously from a hot object to a cold object

Entropy measures disorder

Entropy measures disorder within a system and likelihood—how likely a system resembles the same state twice or how likely a reaction is to happen.

Entropy of a natural system never decreases, as it is the natural tendency of all things to move towards maximum disorder—maximum entropy.

For example, gas expands to fill space with more degrees of freedom to arrange particles—more *multiplicities*.

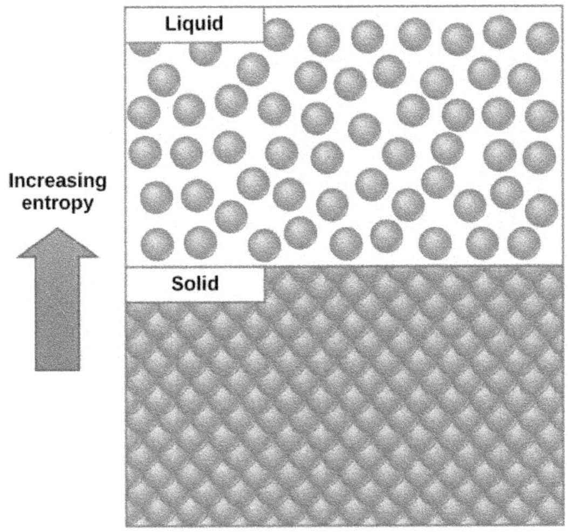

A solid has less entropy than a liquid, which has much less entropy than a gas

According to the Second Law of Thermodynamics (i.e., entropy can stay constant or increase but never decrease), entropy is given by:

$$\Delta S \geq \frac{Q}{T}$$

where ΔS is the change in entropy of a process (J/K), Q is heat transferred (J), and T is the temperature (K).

If a system strives to increase entropy (i.e., disorder), it does not mean that the system cannot become more orderly. This means that to do so, outside energy must be transferred into the system, and it cannot be done on its own.

Nonspontaneous and spontaneous processes

Nonspontaneous processes require energy (or work) for the system to decrease entropy.

Spontaneous processes move towards randomness without energy input to increase entropy.

For example, a valve connects a container with air and another container in a vacuum. When the valve is opened, the air flows from the container to the container in a vacuum. No work or energy is needed for this to happen. The process is *spontaneous*, and entropy *increases*.

For the reverse to occur, energy must be *added to the system* to re-pressurize the original container with air. This process is an example of a *nonspontaneous* reaction.

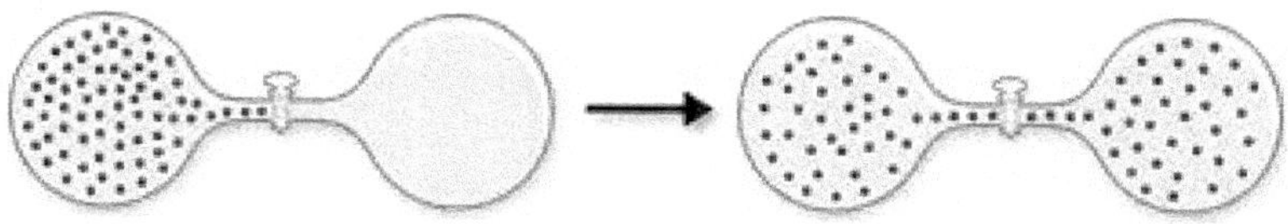

Air flows from the container to the container in a vacuum

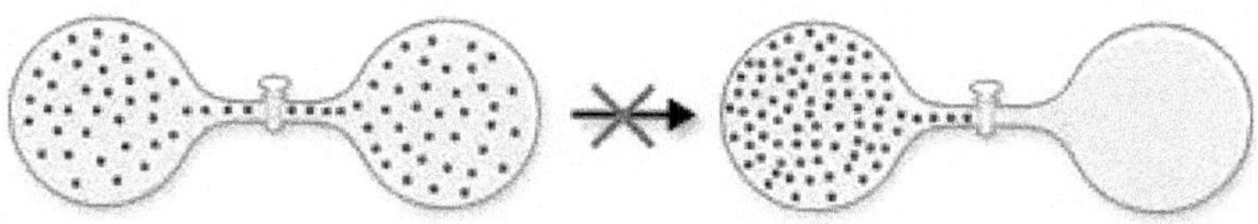

Nonspontaneous reactions do not occur without the input of energy

The universe's entropy always *increases* when a *spontaneous process* occurs.

Entropy of the universe is expressed as:

$$\Delta S_{universe} > 0$$

Reversible and irreversible processes

Thermodynamic processes are *reversible* or *irreversible.*

Reversible processes are idealized, while the irreversible process frequently occurs in nature.

Reversible processes

$$\Delta S = q / T$$

Irreversible processes

$$\Delta S > q / T$$

where ΔS is the change in entropy, q is the contribution from heat, and T is the temperature (K).

Natural processes are *irreversible*, so entropy change is greater than the heat transfer over temperature. Due to the irreversible nature of processes, the universe's entropy is increasing.

Reactions at equilibrium are *reversible* with no net change to the system or surroundings and proceed in both directions.

Reversible processes reverse direction if some change is made to some system property:

$$\Delta S_{\text{total}} = \Delta S_{\text{system}} + \Delta S_{\text{surroundings}}$$

ΔS of the system equals in magnitude, but *opposite in sign*, to the ΔS of the surroundings.

Entropy related to heat and enthalpy

Entropy change of surroundings is related to heat (q) flow, enthalpy (ΔH), and temperature:

$$\Delta S_{\text{surroundings}} = -\,q \,/\, T$$

or (at constant pressure)

$$\Delta S_{\text{surroundings}} = -\,\Delta H \,/\, T$$

Combining these two equations:

$$\Delta S_{\text{surroundings}} = \Delta S_{\text{system}} - (\Delta H_{\text{system}} \,/\, T)$$

where ΔS is the change in entropy, ΔH for enthalpy, q is the contribution from heat, and T is the temperature (K).

ΔS from tabulated values

ΔS is calculated from standard molar entropies as for ΔH and ΔG by heats of formation.

Equation to calculate ΔS from tabulated values:

$$\Delta S^{\circ}_{\text{rxn}} = \Sigma\, n(\Delta S^{\circ}_{\text{f}} \text{ of products}) - \Sigma\, n(\Delta S^{\circ}_{\text{f}} \text{ of reactants})$$

where n represents moles of each reactant (or product) as a balanced chemical equation, and ΔS is the change in entropy.

For example, determine the entropy change (ΔS) for the above chemical equation as (ΔH):

$$16\ H_2S\ (g) + 8\ SO_2\ (g) \rightarrow 16\ H_2O\ (l) + 3\ S_8\ (s)$$

$\Delta S^{\circ}_{\text{rxn}}$ as a measure of the entropy change of a system:

$$\Delta S^{\circ}_{\text{rxn}} = [(16\text{ mol})\cdot(S^{\circ}_{\text{f}} \text{ of } H_2O\ (l)) + (3\text{ mol})\cdot(S^{\circ}_{\text{f}} \text{ of } S_8\ (s))]$$
$$-\ [(16\text{ mol})\cdot(S^{\circ}_{\text{f}} \text{ of } H_2S\ (g)) + (8\text{ mol})\cdot(S^{\circ}_{\text{f}} \text{ of } SO_2\ (g))]$$

From the table of standard *entropies of formation*, substitute:

$$\Delta S^\circ_{rxn} = \Sigma\, n(\Delta S^\circ_f \text{ of products}) - \Sigma\, n(\Delta S^\circ_f \text{ of reactants})$$

$$\Delta S^\circ_{rxn} = [(16 \text{ mol})\cdot(69.91 \text{ J/mol·K}) + (3 \text{ mol})\cdot(31.80 \text{ J/mol·K})] - [(16 \text{ mol})\cdot(205.79 \text{ J/mol·K}) + (8 \text{ mol})\cdot(161.92 \text{ J/mol·K})]$$

$$\Delta S^\circ_{rxn} = [1{,}118 \text{ J/K} + 95.40 \text{ J/K}] - [3{,}292.6 \text{ J/K} - 1{,}295.4 \text{ J/K}]$$

$$\Delta S^\circ_{rxn} = -3{,}375 \text{ J/K}$$

where ΔS°_{rxn} is a measure of the entropy change of a system, and the negative sign accounts for the heat of the surroundings opposite in sign to the heat of the system.

Entropy calculations

For example, the reaction enthalpy (ΔS_{surr}) was –1,876 kJ at a standard temperature of 298 K. The entropy change was calculated as –3,375 J/K (or –3.375 kJ/K). What is the entropy change of the surroundings?

Using the equation:

$$\Delta S_{surroundings} = -\,\Delta H \,/\, T$$

$$\Delta S_{surroundings} = -(-\,1{,}876 \text{ kJ}) \,/\, 298 \text{ K}$$

$$\Delta S_{surroundings} = 6.30 \text{ kJ/K or } \Delta S_{surroundings} = 6.30 \times 10^3 \text{ J/K}$$

where $\Delta S_{surroundings}$ is change in entropy, ΔH is the change in enthalpy, and T is temperature (K).

As a *positive* value, the entropy of the surroundings *increased.*

Total entropy change is the sum of the entropy change of the system and the surroundings:

$$\Delta S_{total} = \Delta S_{system} + \Delta S_{surroundings}$$

$$\Delta S_{total} = -3.375 \times 10^3 \text{ J/K} + 6.30 \times 10^3 \text{ J/K}$$

$$\Delta S_{total} = 2.92 \times 10^3 \text{ J/K}$$

where ΔS_{total} is the total entropy change, ΔS_{system} is the entropy change of the system, and $\Delta S_{surroundings}$ is the entropy change of the surroundings.

Entropy measures degrees of freedom

Entropy (S) measures *randomness* (or *disorder*) from the number of ways (i.e., *degrees of freedom*) a thermodynamic system can be arranged.

Entropy (or disorder) of the universe is *increasing.*

Entropy uses energy divided by temperature; the unit is Joules per Kelvin (J/K).

Entropy is expressed as *entropy per unit mass* (J/K·kg) or *unit amount of substance* (J/K·mol).

For example, imagine taking a piece of paper, tearing it into small pieces, and throwing them in the air. How much energy is needed to restore it to its original form?

Consider tearing paper as *spontaneous* and putting it together as a *nonspontaneous* process.

Temperature and entropy relationship

The higher the entropy and lower the energy (i.e., temperature), the more stable the system.

Higher entropy + lower energy (lower temperature) → more stability

Lower entropy + higher energy (higher temperature) → less stability

As *temperature increases*, *entropy increases*:

$T = 0$ K

particles are in equilibrium lattice positions, and S relatively low

$T > 0$ K

molecules vibrate, S increases

As T increases, more violent vibrations occur, and the entropy (S) increases.

Entropy for gasses, liquids, and crystals

Entropy differences between phases of matter (i.e., solid, liquid, gas) are related to the *degrees of freedom* of kinetic energy (KE).

Solids have low entropy because particles forming a crystal solid are packed tightly.

Particles of a solid are confined to vibrating in fixed positions and limited in *degrees of freedom.*

Liquids are intermediate in entropy as intermolecular forces hold particles as a liquid but have increased *degrees of freedom* to move.

Gases are high in entropy because the gas particles have complete *freedom* to move with a few particles arranged in an orderly fashion.

Low-pressure gas experiences little to no intermolecular forces of attraction or repulsion.

Entropy of phase changes

Entropy changes accompanying phase changes can be estimated, at least qualitatively.

Freezing decreases entropy as the relatively disordered liquid becomes a well-ordered solid.

Boiling results in a substantial increase in entropy as the liquid becomes a more highly disordered gas.

Sublimation (i.e., from a solid to a gas) is the phase transition with the greatest entropy change for a substance.

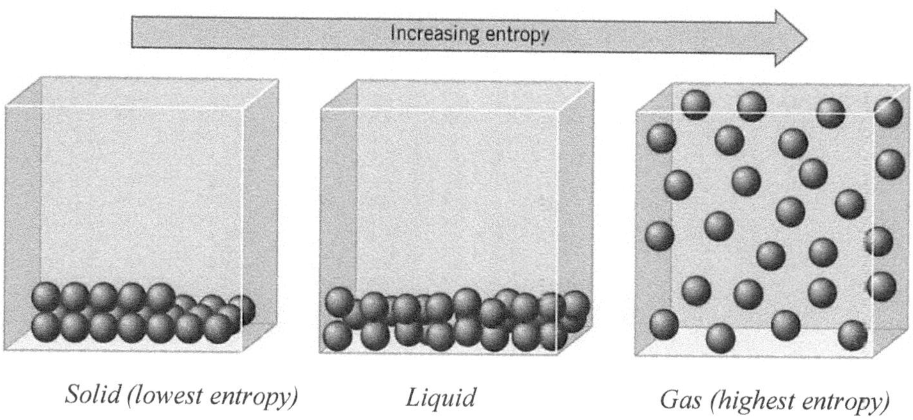

The figure displays a crystalline solid, liquid, and gas with entropy (i.e., randomness) increasing to the right.

Entropy for gases

Entropy *increases* as gas volume *increases* (i.e., decreasing pressure).

Consider a gas separated from a vacuum by an impermeable partition (figure on the left).

Removing the partition (center image) allows more energy distribution.

Gas expands to achieve higher particle distribution (figure on the right).

Gas molecules' configuration is *random* and *more positive S* (more disorder).

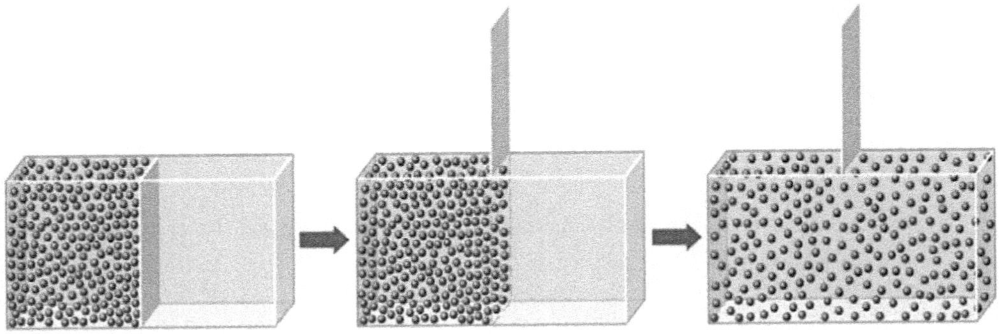

Gas particles are highly ordered on the left (low S) and random on the right (high S)

Number of particles affects entropy reactions, producing more particles as a *positive* ΔS.

Adding particles to a system (below on the right) increases the number of ways energy can be distributed.

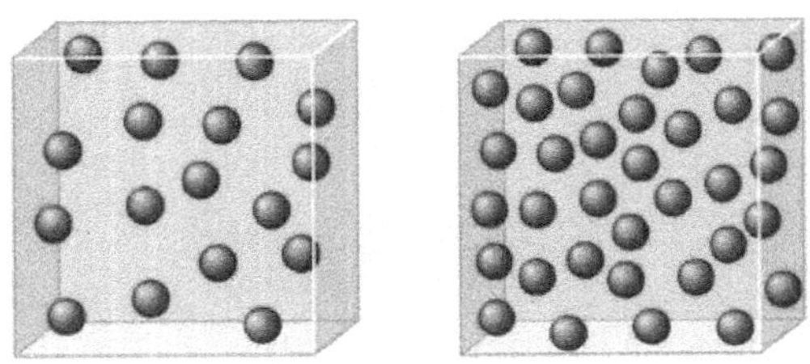

Low entropy *High entropy: more disorder*

Third Law of Thermodynamics

Absolute zero

Third Law of thermodynamics states that *as the temperature of a system approaches absolute zero, the entropy approaches a constant value.*

Typically, this constant value is zero. In this definition, the entropy of a system is related to the number of microstates possible.

Microstates refer to the atomic configurations possible with the equation:

$$S = k \cdot ln\,(\Omega)$$

where S is the entropy (J/K), k is Boltzmann's constant (1.381×10^{-23} J/K), and Ω is the number of microstates.

Only one microstate is possible at absolute zero, and so Ω equals 1:

$$S = k \cdot ln(1) = k \cdot 0 = 0$$

The entropy of a system at absolute zero is zero.

Minimum energy state

However, some systems have more than one minimum energy state as the temperature approaches absolute zero; the entropy levels off at a value other than zero.

Cooling any process to absolute zero in a finite number of steps is impossible.

In left image, $T = 0$ can be reached following infinite steps (step lines between X_1 and X_2).

In the image on the right, $T = 0$ is reached by a finite number of steps, each getting closer to zero but never to zero.

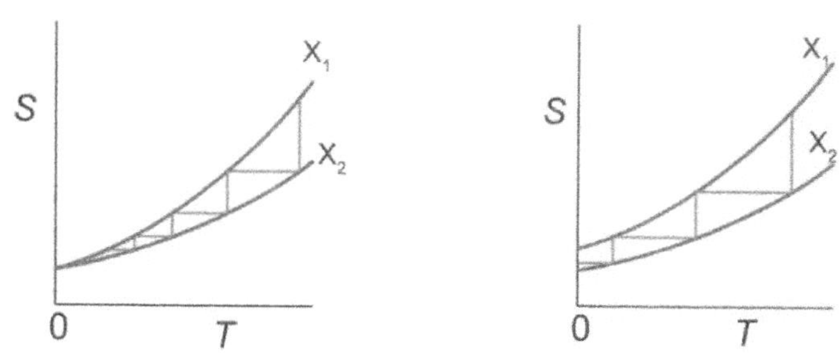

Entropy vs. temperature illustrating the step to approach absolute zero

Entropy at absolute zero

Third Law of Thermodynamics states that *the entropy of a perfectly ordered, pure crystalline solid is zero at absolute zero.*

At *absolute zero* (coldest temperature possible on Kelvin scale equals 0 K, –273.15 °C, or –459.67 °F), the particles' average kinetic energy (KE) equals zero.

Particles' average kinetic energy (KE) increases as entropy increases.

With increasing temperature (i.e., KE), substances transition to liquid and then gaseous state.

For a substance, the gas state has the highest entropy.

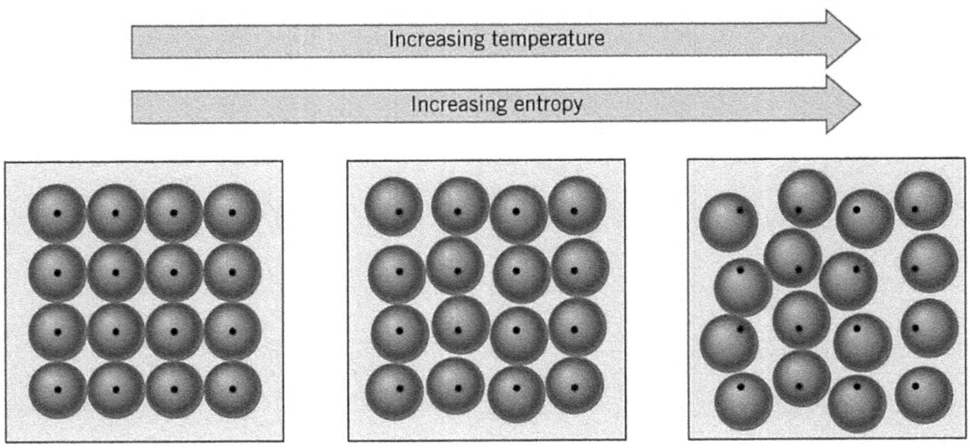

Crystalline solids have low KE with particles arranged orderly.

Liquid particles vibrate from higher KE, and entropy (S) is moderate.

Gas particles have large KE, and entropy (S) is large.

Enthalpy

Energy changes

Enthalpy (ΔH) of a system is the amount of heat used at constant pressure in a system or reaction and is given in SI units of Joules.

Energy changes of the system can be written as:

$$\Delta U = \Delta H - P\Delta V$$

where ΔH is the change in enthalpy of a system (J), P is the pressure (Pa), and ΔV is the change in volume (m^3).

During an isobaric process (pressure is held constant), such as that conducted in an open container, the change in enthalpy equals the amount of heat transferred during the process. Most reactions do not generate many gaseous products, so little work is associated with the reaction.

Energy change of the system equals the change in enthalpy:

$$\Delta E \approx \Delta H$$

Enthalpy is equal to the transfer of thermal energy (heat); it is only valid at constant pressure.

Enthalpy changes can occur at standard temperatures and pressures.

Standard state refers to substances in their most stable form (i.e., the lowest-energy state).

Standard state in thermodynamics refers to one-atmosphere pressure (1 atm) and a temperature of 25 °C (298 K).

For example, oxygen is O_2 (diatomic gas) at the standard state, and carbon is C (solid graphite).

Standard heats of reaction (ΔH)

Standard heat of reaction (*standard enthalpy of reaction* $\Delta H°_{rxn}$) is the change in heat content when one mole of the matter is transformed by a chemical reaction under standard conditions (1 atm and 298 K).

Theoretical and experimental methods calculate the *enthalpy change* of a chemical reaction.

General equation to calculate the heat of a reaction (ΔH) from the difference (delta or Δ) in the energy of the reactants and products is:

$$\Delta H = H_{products} - H_{reactants}$$

Standard heats of formation (ΔH°_f)

Standard heat of formation (*standard enthalpy of formation* ΔH°_f) is the change in heat content when one mole of a compound or molecule forms from constituent elements in standard states.

Standard enthalpy of formation has been determined for many molecules, and the ΔH°_f for the most common molecules are reported and tabulated data.

When choosing ΔH°_f from a table, choose the value corresponding to the appropriate state of matter (solid, liquid, aqueous, gas).

Standard heat of formation for an element in its standard state equals 0 kJ/mol.

For example, calculate the heat of reaction from the heat of formation and determine if the process is exothermic or endothermic for the balanced chemical equation:

$$16\ H_2S\ (g) + 8\ SO_2\ (g) \rightarrow 16\ H_2O\ (l) + 3\ S_8\ (s)$$

From thermodynamic quantities, use values for the heat of formation of each component in the reaction; set the equation to calculate the heat of the reaction:

$$\Delta H = H_{products} - H_{reactants}$$

$$\Delta H^\circ_{rxn} = [16\ mol(\Delta H^\circ_f\ of\ H_2O\ (l)) + 3\ mol(\Delta H^\circ_f\ of\ S_8\ (s))] - [16\ mol(\Delta H^\circ_f\ of\ H_2S\ (g)) + 8\ mol(\Delta H^\circ_f\ of\ SO_2\ (g))]$$

$$\Delta H^\circ_{rxn} = [16\ mol(-285.8\ kJ/mol) + 3\ mol(0\ kJ/mol)] - [16\ mol(-20.2\ kJ/mol) + 8\ mol(-296.8\ kJ/mol)]$$

$$\Delta H^\circ_{rxn} = [-4{,}573\ kJ + 0\ kJ] - [-323\ kJ + (-2{,}374\ kJ)]$$

$$\Delta H^\circ_{rxn} = -4{,}573\ kJ + 323\ kJ + 2374\ kJ$$

$$\Delta H^\circ_{rxn} = -1{,}876\ kJ$$

ΔH°_{rxn} is negative, and the reaction is exothermic; the system releases heat to the surroundings.

Enthalpy for exothermic and endothermic reactions

Exothermic reactions have a change in enthalpy of less than zero.

$$\Delta H < 0$$

Endothermic reactions have a change in enthalpy greater than zero.

$$\Delta H > 0$$

Since enthalpy (H) is a state function, it represents a system property; it can be used to describe a reaction or an equation for a reaction.

Calculating the heat of reaction (ΔH)

Calculate the heat of reaction (ΔH) using these steps:

1) Identify the given and needed quantities.
2) Write an expression using the heat of reaction and molar mass needed.
3) Write the conversion factors, including the heat of the reaction.
4) Set the expression to calculate heat.

Hess' Law of heat summation

In 1840, Swiss-born Russian chemist Germain Hess (1802-1850) published Hess' Law of constant heat summation.

Hess' Law describes an essential relationship in physical chemistry that *total enthalpy change (ΔH) during the complete course of a chemical reaction is independent of the number of steps.*

Hess' Law states that the change of enthalpy in a reaction is equal to the sum of the products' enthalpy minus the sum of the reactants' enthalpy:

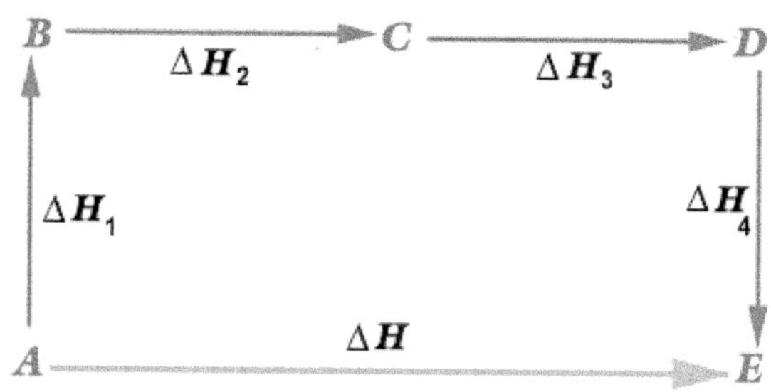

$$\Delta H_{rxn} = \Sigma H_{products} - \Sigma H_{reactants}$$

Enthalpy is a *state function* (i.e., depends only on initial and final states, *not* the reaction's path).

Hess' Law of Heat Summation is: expressed

$$\Delta H^\circ_{rxn} = \Sigma\, n(\Delta H^\circ_f \text{ of products}) - \Sigma\, n(\Delta H^\circ_f \text{ of reactants})$$

where n represents the moles of each reactant or product in the balanced chemical equation, the Greek letter Σ means the sum of the variables.

Hess' Law states that the total energy change throughout a reaction is the same, regardless of whether the reaction occurs in one step or several steps.

Notes for active learning

Gibbs Free Energy (G)

Reaction spontaneity

In 1870, Josiah Willard Gibbs (1839-1903) expressed *Gibbs free energy* (G) as the maximum amount of energy present in molecules available *to do non-expansion work* at constant temperature and pressure.

Free energy change (ΔG) is the difference of free energy in the product and reactant molecules.

The equation for Gibbs free energy ΔG is:

$$\Delta G = \Delta H - T\Delta S$$

where G is Gibbs free energy, measured in Joules (J).

Calculating Gibbs free energy

Because ΔG depends on signs, it can be calculated from estimated values.

If ΔG is equal to *zero*, the reaction is at *equilibrium*.

If ΔG is *negative*, the reaction is *spontaneous*.

If ΔG is positive, the reaction is *nonspontaneous*.

For example, if ΔG is *negative* at a given temperature and pressure, the reaction is *spontaneous*.

Chemical bonds may break in the reactants when molecules react, and others form in products. Breaking bonds requires energy, while making bonds releases energy, free energy exchanges during a reaction. Not all chemical bonds have the same strength or require the same energy.

Gibbs free energy change (ΔG) equals the maximum energy produced during a reaction harnessed as work.

Spontaneous *vs.* nonspontaneous reactions

Table approximates ΔG and whether the reaction is spontaneous or nonspontaneous.

ΔH	**ΔS**	**Result**
-	+	Spontaneous at ***all*** temperatures
+	+	Spontaneous at ***high*** temperatures
-	-	Spontaneous at ***low*** temperatures
+	-	Not spontaneous at ***any*** temperatures

Reaction energy diagrams

A chemical reaction's thermodynamics can be represented on a *reaction energy diagram* showing the energy of the reactants, products, activation energy, and free energy change ΔG.

Energy appears on the y-axis, and the reaction progress (time) from the starting reactants to the final products appears on the x-axis.

Each reaction has an *energy of activation* (E_{ac}) or *energy barrier* before the reactants transform into products.

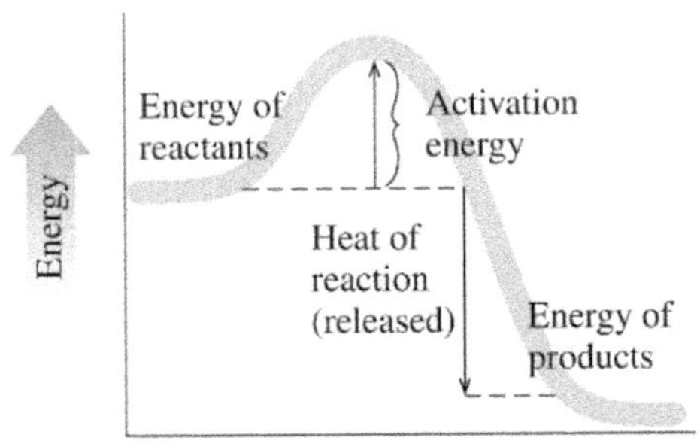

Reaction diagram of energy vs. time (reaction progress)

Gibbs free energy equation

There are several methods to calculate Gibbs free energy ΔG.

Gibbs free energy is a state function, so one method is analogous to the methods for enthalpy and entropy.

Gibbs free energy expression of ΔG°_f products *minus* ΔG°_f reactants:

$$\Delta G^\circ_{rxn} = \Sigma\, n(\Delta G^\circ_f \text{ of products}) - \Sigma\, n(\Delta G^\circ_f \text{ of reactants})$$

where n represents moles of each product or reactant given by the balanced chemical equation coefficients.

Like enthalpy, the standard Gibbs free energy of formation (ΔG°_f) of elements in their standard states is 0 kJ/mol.

Notice the similarity of Gibbs free energy to *Hess' Law of constant heat summation.*

Carnot Cycle

Kelvin-Planck statement

Kelvin-Planck statement of the Second Law of Thermodynamics references heat engines. It states that *it is impossible to extract an amount of heat from a "hot reservoir" (Q_2) and use all of it to do work*. Heat must be exhausted to a "cold reservoir" (Q_1).

A perfect heat engine would use all the energy from the hot reservoir to do work. This is impossible because all real heat engines lose some heat to the outside environment.

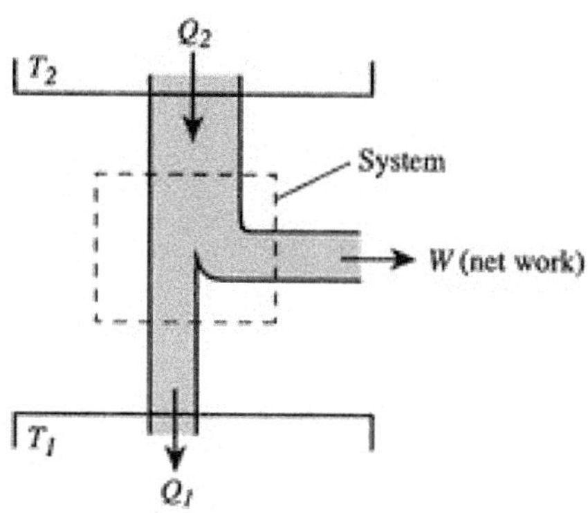

Extracting heat from a "hot reservoir" (Q_2) and using it to do work

Efficient heat engine cycle

Carnot Cycle is the most efficient heat engine cycle, with 2 isothermal and adiabatic processes.

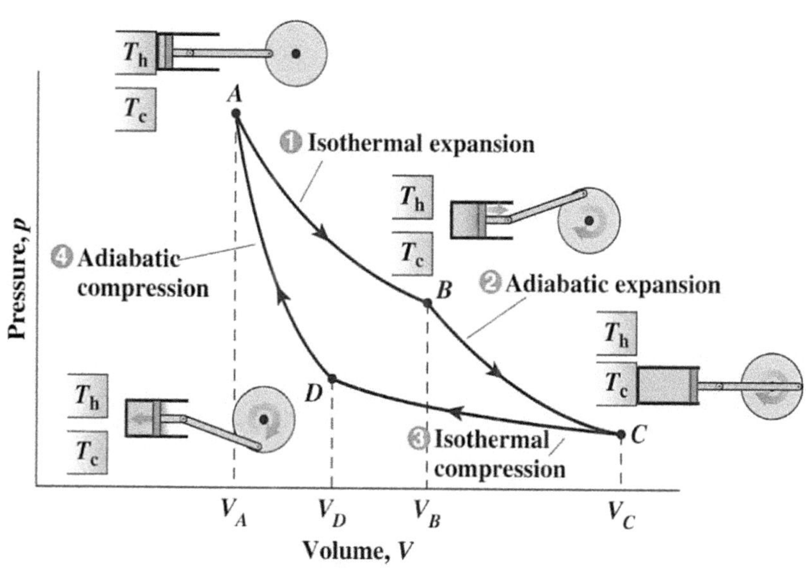

Carnot Cycle is an idealized situation and cannot be replicated in everyday applications

Carnot efficiency

From the Second Law, an equation of the *Carnot efficiency* sets a *limiting value on the fraction of heat taken* from the hot reservoir, which can be used for work:

$$n = \frac{T_H T_C}{T_H} \times 100\%$$

where n is the Carnot efficiency, T_H is the temperature of the hot reservoir (K), and T_C is the temperature of the cold reservoir (K).

Process must be reversible for Carnot efficiency, and the change in entropy must be zero. This is impossible, as no process of a real engine is reversible, and every process increases entropy because maximum entropy is the universe's natural state.

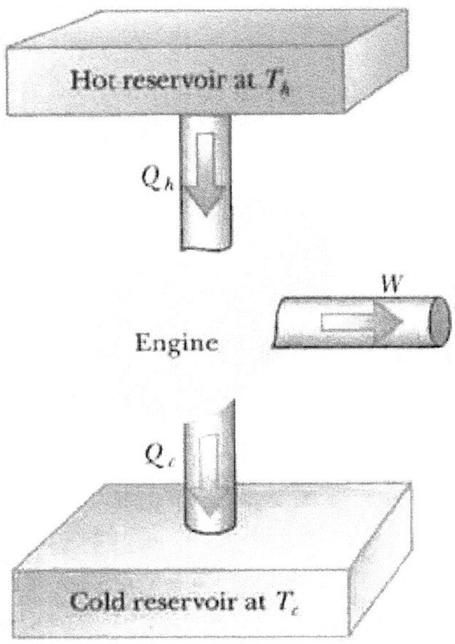

Thermal Expansion

Volume changes related to temperature

Thermal expansion is the tendency of matter to *change volume* from temperature changes.

Thermal expansion occurs through *heat transfer*.

When a substance is heated, the molecules increase kinetic energy and move more; thus, greater average separation is maintained, increasing volume (expansion).

Coefficients of thermal expansion

When a material changes temperature, it shrinks (or expands), depending upon the temperature difference.

As the temperature increases, molecules within the material gain energy and vibrate faster. This vibration increases the distance between the molecules and causes the material to expand.

If the material does not experience a phase change (liquid to gas), this expansion of the material can be related to the change in temperature.

Coefficient of thermal expansion (α) is the degree of expansion divided by temperature change.

Linear expansion: $\Delta L / L_0 = \alpha \Delta T$

Area expansion: $\Delta A / A_0 = 2\alpha \Delta T$

Volume expansion: $\Delta V / V_0 = 3\alpha \Delta T$

The value of the coefficient α depends on the type of expansion.

Coefficient of linear expansion

Coefficient of linear expansion measures the expansion (or contraction) per unit of length of a material when the temperature changes by 1 °C. This term is also known as *expansivity*.

Coefficient of thermal expansion (α) is given as:

$$\alpha_{linear} = \frac{\Delta L}{L_i \cdot \Delta T}$$

where α_{linear} is the coefficient of thermal expansion (K^{-1}), ΔL is the change in length (m), L_i is the initial length (m), and ΔT is the change in temperature (K).

In the figure below, a bar of material is at an initial temperature and has a length denoted as L.

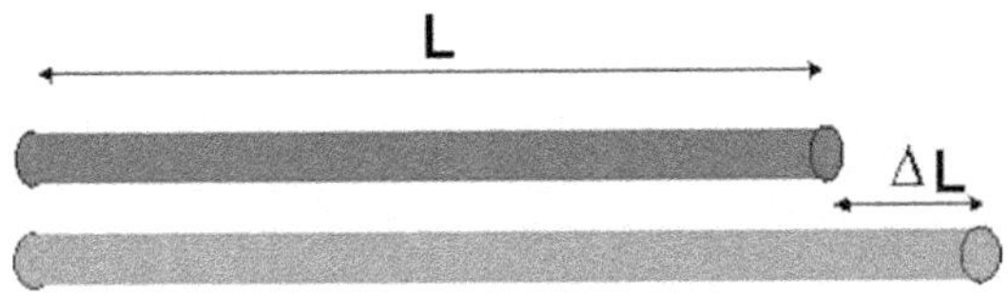

If the bar is heated, its temperature rises and expands linearly by some length ΔL

For example, a metal rod of 7.00 m is heated to 35 °C. If the rod's length expands to 7.12 m after some time, calculate the linear expansion coefficient (Use room temperature = 27 °C).

Calculate the change in length ΔL from the initial L_0= 7.00 m and expanded length L = 7.12 m.

ΔL = 7.12 m – 7.00 m

ΔL = 0.12 m

calculate ΔT from the initial room temperature T_0 = 27 °C and the heated T = 35 °C.

ΔT = 35 °C – 27 °C

ΔT = 8 °C

convert ΔT to Kelvin

ΔT = 8 + 273

ΔT = 281 K

linear expansion expression:

$\Delta L / L_0 = \alpha \Delta T$

Isolate α:

$\alpha = \Delta L / (L_0 \times \Delta T)$

α = 0.12 m / (7 m × 281 K)

$\alpha = 6.1 \times 10^{-5}$

Coefficient of area expansion

Objects can undergo area expansion and volume expansion due to the same principles.

Coefficient of area expansion is expressed as:

$$\alpha_{area} = \frac{\Delta A}{A_i \cdot \Delta T}$$

where α_{area} is the coefficient of thermal expansion (K^{-1}), ΔA is the change in the area (m^2), A_i is the initial area (m^2), and ΔT is the change in temperature (K).

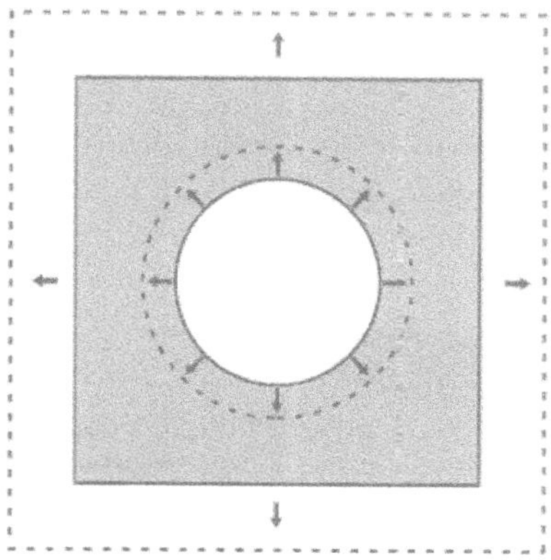

Coefficient of area expansion depends on temperature change

Coefficient of volume expansion

Coefficient of volume expansion is expressed as:

$$\alpha_{volume} = \frac{\Delta V}{V_i \cdot \Delta T}$$

where α_{volume} is the coefficient of thermal expansion (K^{-1}), ΔV is the change in volume (m^3), V_i is the initial volume (m^3), and ΔT is the change in temperature (K).

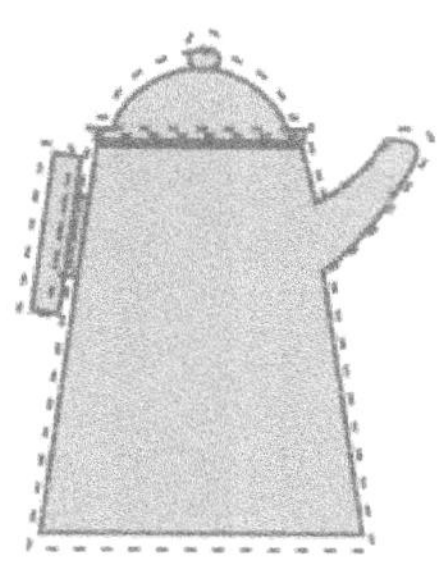

Coefficient of volume expansion

Engineering considerations

Thermal expansion is significant when designing products or structures that experience temperature changes, which makes the design and construction of bridges and aircraft difficult.

Design must allow for varying component expansion and contraction if the product uses materials with different thermal expansion coefficients. If not, the components experience considerable stress when expanding and cannot.

Bridges in regions that get cold have metallic joints where the bridge starts and ends.

In the summer, these expansion joints might be completely closed.

In the cold of winter, the thermal contraction of the bridge materials changes. The frigid air above and below causes the bridge to "*shrink*" and open space between the metal expansion joints. This prevents the bridge from cracking or breaking because of inherent material strain during fluctuations in temperature.

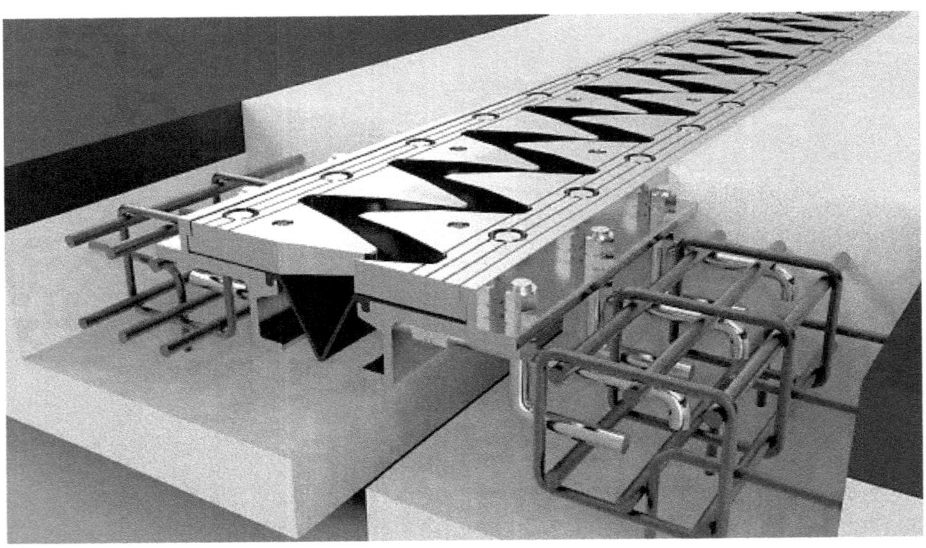

Expansion joints permit variation in material size without failure

Kinetic Theory of Gases

Macroscopic properties of gas

Thermal energy (or internal energy) is the sum of the *kinetic energies* of the particles. Thermal energy of a gas is directly proportional to the molecules' kinetic energy (KE).

The higher the kinetic energy of the molecules, the more thermal energy the system possesses.

The lower the kinetic energy of the molecules, the less thermal energy the system possesses.

The figure below illustrates a closed container of gas where the gas molecules are in random motion throughout the container.

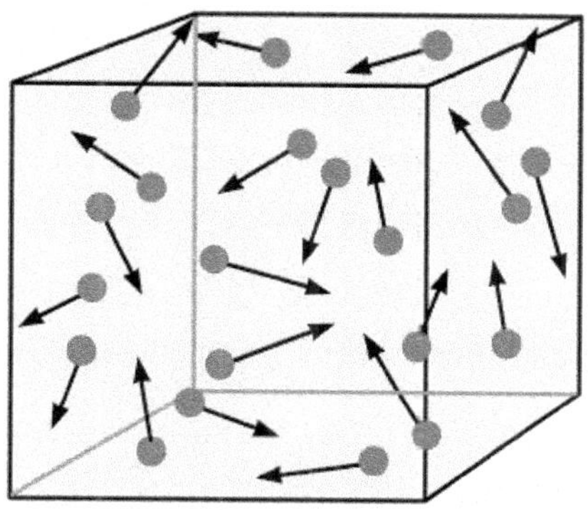

Gas molecules exhibit kinetic energy from random motion in the container

Kinetic Molecular Theory

Kinetic Molecular Theory of Gases relates the microscopic values of atomic kinetic energy to quantities like temperature and pressure.

Kinetic molecular theory (KMT) uses atomic theory to describe the behavior of gases. This theory developed over 100 years, culminating in an 1857 paper by German physicist Rudolf Clausius (1822-1888).

Kinetic molecular theory explains the macroscopic properties of a gas (e.g., pressure and temperature) in terms of its microscopic components (e.g., molecules or atoms). It originated in the ancient idea that matter consists of tiny invisible atoms in rapid motion. The idea was revived and developed between the 17th and 19th centuries to explain gases' properties.

Five assumptions for gases

Five premises for the kinetic theory of gases:

1. Gases are comprised of molecules. Molecules may be treated as perfect spheres of mass (*point masses*), and the space between each of these masses is many times greater than the diameter of the gas molecules.
2. Motion of each gas molecule is *random*, with no order or pattern to either the direction or magnitude of the velocity.
3. Molecules follow *Newton's Laws of Motion*. Each gas molecule moves in a straight line with a constant velocity. If gas molecules collide, the molecules exert an equal but opposite force on the other.
4. Collisions of molecules are *perfectly elastic*; they *lose no kinetic energy*.

   Therefore, since intermolecular attractive and repulsive forces are nonexistent, the gas molecules' total kinetic energy (KE) remains *constant* before and after the collision.
5. Average kinetic energy (*KE*) for gas particles is proportional to the absolute temperature, regardless of the chemical identity or atomic mass.

   At 0 K (absolute zero or –273 °C), the molecules (and orbiting electrons) are not moving and have no volume.

These rules display gases as ideal substances and only approximate their actual behavior. However, their description is remarkably accurate. From these assumptions, laws are derived to describe the behavior of gases.

Boltzmann's constant

Final assumption (#5) for KE is expressed by:

$$\text{KE} = \tfrac{1}{2} mv^2$$

$$\tfrac{1}{2} mv^2 = (3/2)\, k_B T$$

So,

$$\text{KE} = (3/2)\, k_B T$$

where *KE* is kinetic energy, m is mass, v is velocity, T is the temperature (Kelvin), and k_B is *Boltzmann's constant* ($k_B = 1.38 \times 10^{-23}$ J/K).

Boltzmann's constant relates macroscopic and microscopic behavior of gases to their particles.

Root-mean-square speed

KE = (3/2) k_BT equation is significant because it states that a gas particle's average kinetic energy is proportional to its absolute temperature. Increasing the temperature of the gas particles increases the total speed and overall energy.

Gas atoms are infinitesimally small; nearly impossible to measure a particle's speed accurately.

Speed of gas particles is defined by the *root-mean-square speed* (u_{rms}) by the equation:

$$u_{rms} = \sqrt{\frac{3RT}{MM}}$$

where R is ideal gas constant, T is absolute temperature (K), and MM is molar mass of gas.

Root-mean-square speed written with the *Boltzmann constant*, k_B:

$$u_{rms} = \sqrt{\frac{3kT}{m}}$$

where k is the Boltzmann constant, T is the absolute temperature (K), and m is the mass of one gas molecule.

Boltzmann distribution plot

Plotting each gas particle's speed distribution at a given temperature shows a slightly asymmetric curve. This speed-distribution curve is the *Maxwell-Boltzmann distribution curve.*

Boltzmann distribution plot shows that the curve's peak corresponds to the most probable speed.

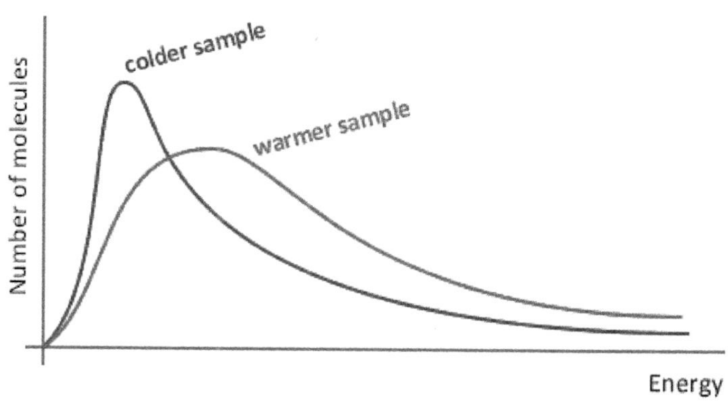

Maxwell-Boltzmann Distribution curves of molecular speeds at two temperatures

If the temperature (K) remains constant, the total kinetic energy remains unchanged.

However, the energy is distributed in several ways, and the gas particles travel at many speeds. This distribution changes continually as the gas atoms collide with each other and with the container walls.

The curve flattens (i.e., the net area under the curve is greater) and shifts to the right at higher temperatures, indicating that more gas molecules move at higher speeds and possess more kinetic energy.

Diffusion and effusion

Diffusion is when a substance (solute or particle) spreads from a high-concentration region to a lower concentration. It is explained using the kinetic theory of motion.

According to the kinetic theory, heavier gases diffuse more slowly than lighter gases because of the difference in their travel speed.

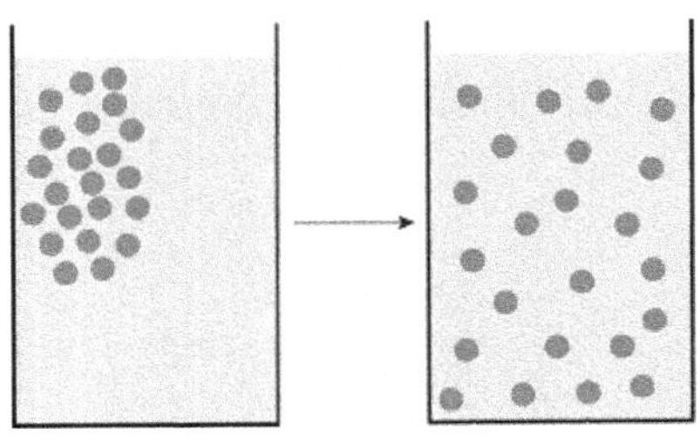

Diffusion of solutes in a solvent

Effusion is the flow of gas particles under pressure from one compartment to another through a small opening, as shown below.

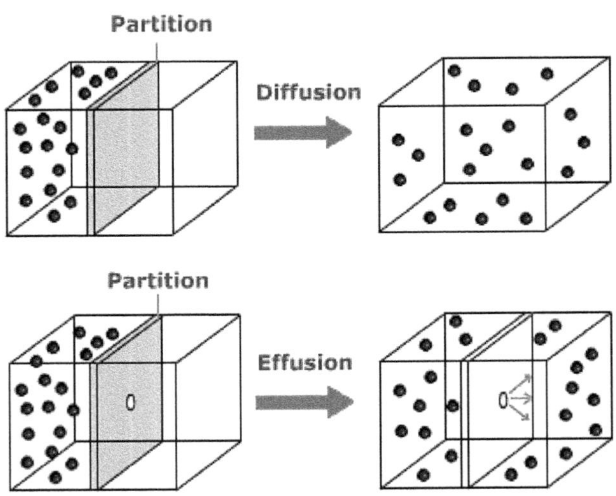

Diffusion and effusion of gas particles

Graham's Law

Scottish chemist Thomas Graham (1805-1869) developed *Graham's Law*, which states that *at constant temperature and pressure conditions, the rates at which two gases diffuse are inversely proportional to the square root of their molar masses.*

Graham's Law is expressed by:

$$\frac{r_1}{r_2} = \sqrt{\frac{MM_2}{MM_1}}$$

where r_1 and r_2 are the diffusion rates of gas 1 and gas 2, respectively, and MM_1 and MM_2 are the molar masses of the gases.

Graham's Law applies to the effusion of gas particles, and the equation for effusion is the same as the equation for diffusion.

Ideal Gas Law

Ideal Gas Law explains the relationship between pressure (P), volume (V), and temperature (T).

Gas law is expressed as:

$$PV = nRT$$

where n is number of moles, and R is universal gas constant with a value of 8.314 J/ mol·K.

One mole is equal to 6.023×10^{23} molecules. Technically, a mole is the number of hydrogen atoms in one gram of hydrogen. Atoms are tiny, so counting them in moles is more practical than counting them individually. Ideal Gas Law applies to the pressure a gas exerts on a cylinder with a moving wall.

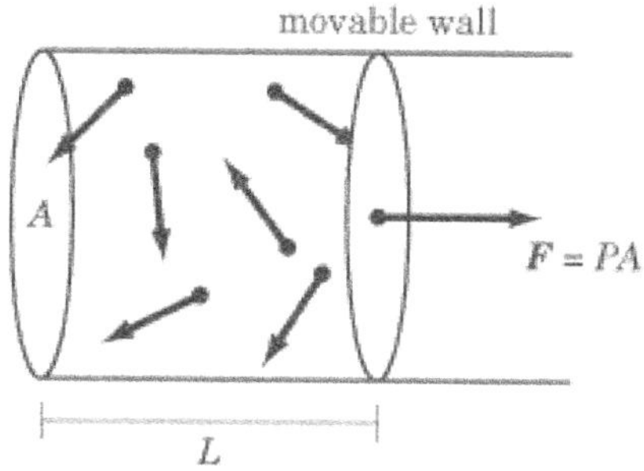

Force that the gas exerts on the wall equals $F = PA$

Since pressure equals $P = F/A$, the force the gas exerts on the wall equals $F = PA$.

If this force moves the wall back a length of L, the cylinder volume increases by $\Delta V = LA$.

By solving for A and substituting this into the equation for force, the result is:

$$F = P\Delta VL \text{ or } P\Delta V = FL$$

Work is *force multiplied by distance traveled.* By pushing the wall at a distance of L with a force of F, the gas has done work equal to FL.

When gas does work, it symbolizes a change in energy.

If a change in PV is equal to a change in energy, then PV is the total energy of the gas.

For the ideal gas law, nRT is the expression for the total kinetic energy of the gas molecules.

Ideal gas law (PV = nRT) can be related to the number of molecules (N) and Boltzmann's constant (k), which has a value of 1.381×10^{-23} J/K:

$$PV = NkT$$

Other gas laws derive from the Ideal Gas Law by holding specific variables constant.

The number of moles (n) and the gas constant (R) are constant, so four other gas laws result from the Ideal Gas Law:

Boyle's Law

Charles' Law

Combined Gas Law

Closed Container Law

Ideal gases follow KMT (isothermal process)

As described previously, *kinetic molecular theory (KMT)* is based on a theoretical ideal gas.

Ideal gases follow *five* assumptions of KMT in which gas molecules:

move in random motion

are point masses with no volume,

there are no intermolecular forces,

collisions are elastic and

kinetic energy of molecules is proportional to temperature.

Compared to an ideal gas, the behavior of a real gas is complicated. By conceptualizing ideal gas behavior, real gas behavior is easier to comprehend.

In the 17th and 18th centuries, scientists realized several relationships between gases' properties. Boyle's, Charles', and Avogadro's Law relate pressure, volume, and temperature properties.

These laws were developed empirically and are individual cases of the ideal gas equation:

$$PV = nRT$$

Boyle's Law (isothermal process)

In 1661, Robert Boyle (1627-1691) studied the compressibility of gases. He observed that *the volume of a fixed amount of gas at a given temperature is inversely proportional to the pressure exerted on the gas.*

Boyle's Law applies when the temperature of a gas is held constant. It states that an increase in pressure causes a decrease in volume or that a decrease in pressure causes an increase in volume.

Boyle's Law is expressed as:

$$P_1V_1 = P_2V_2$$

where subscripts 1 and 2 are same gas sample under two sets of pressure and volume conditions.

A plot of volume *vs.* pressure for a gas illustrates that Boyle's Law is a particular case of the ideal gas law, where n (moles of gas) and T (temperature) are constant (figure below).

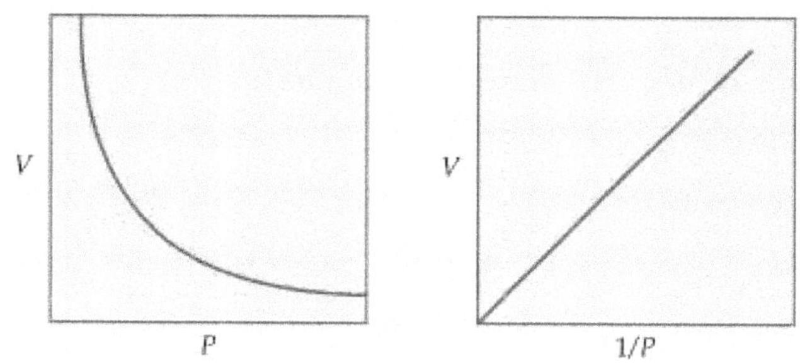

Boyle's Law: as pressure increases, volume decreases

For example, what is the new pressure in a container of gas in a 15.0 L container at a pressure of 5.00 atm when the volume decreases to 0.500 L?

Substitute the known quantities (P_1, V_1, and V_2) into Boyle's Law equation to solve P_2. Note: the volume units need to be consistent. In this example, volumes are expressed in units of liters.

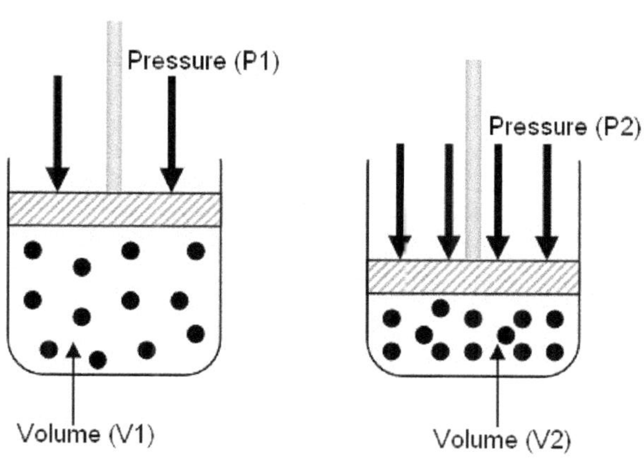

Boyle's Law is where n (moles of gas) and T (temperature) are constant

Boyle's Law:

$$P_1V_1 = P_2V_2$$

Solve for P_2:

$$P_1V_1 / V_2 = P_2$$

$$[(5.00 \text{ atm})\cdot(15.0 \text{ L})] / 0.500 \text{ L} = P_2$$

$$P_2 = 150 \text{ atm}$$

Charles' Law (isobaric process)

French scientist Jacques Charles (1746-1823) developed the relationship between temperature and gas volume.

Charles' Law states that *at constant pressure, the volume of a gas is directly proportional to its absolute temperature (Kelvin)*. It describes the behavior of a gas under constant pressure.

In this relationship, volume and temperature are *directly proportional*; when temperature increases, volume increases; when the temperature decreases, the volume decreases.

Charles' Law is expressed as:

$$\frac{V_1}{T_2} = \frac{V_2}{T_2}$$

where subscripts 1 and 2 refer to the same gas sample at different temperature and volume.

Temperature vs. volume plot for a gas illustrates that Charles' Law is another application of the ideal gas law, where n and P are constant, as shown below.

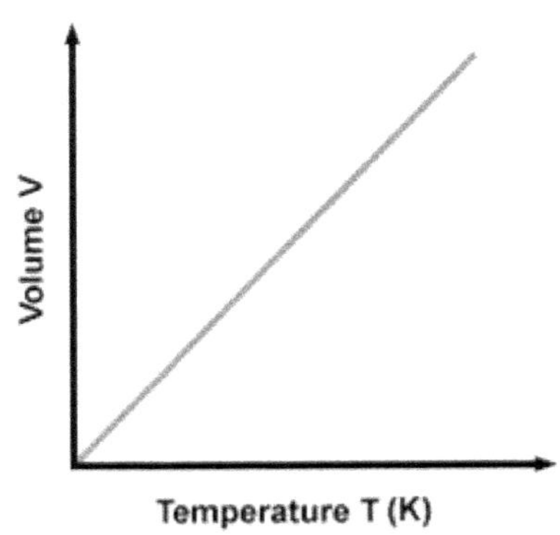

Charles' Law states that as temperature increases, volume increases

For example, if there is 25.0 L of gas at 0 °C, and the temperature is raised to 100 °C, what is the volume of the gas?

First, the temperature is converted to Kelvin.

$T_1 = 0\ °C + 273 = 273\ K$

$T_2 = 100\ °C + 273 = 373\ K$

Substitute the known quantities into the equation for Charles' Law.

$V_1 / T_1 = V_2 / T_2$

Rearrange and solve for V_2:

$V_1\ T_2 / T_1 = V_2$

$(25.0\ L)\cdot(373\ K) / 273\ K = V_2$

$V_2 = 34.2\ L$

Volume and temperature are directly proportional, when temperature increases, volume increases

Gay-Lussac's Law

In 1802, French chemist Louis Gay-Lussac (1778-1850) proposed *Gay-Lussac's Law,* which states that *the pressure of a given sample of gas is directly proportional at a constant volume to its absolute temperature (Kelvin).*

Gay-Lussac's Law is expressed by:

$$\frac{P_1}{T_1} = \frac{P_2}{T_2}$$

where subscripts 1 and 2 refer to the same gas sample with different temperatures and pressure.

Temperature vs. pressure plot for a gas illustrates that Gay-Lussac's Law is another particular case of the ideal gas law, where n and V are constant.

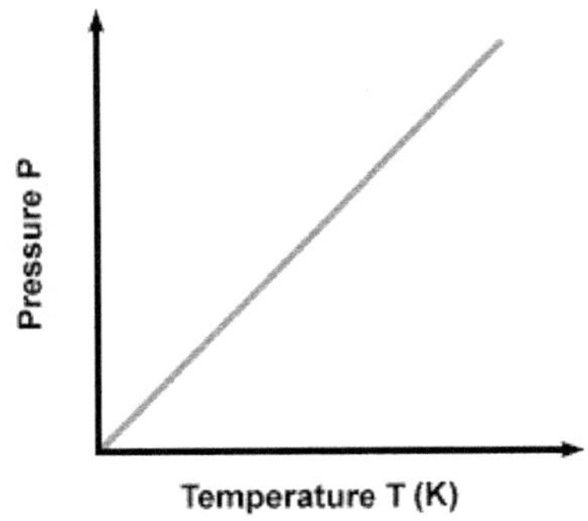

Gay-Lussac's Law states that as temperature increases, pressure increases

For example, suppose a gas is at 30.0 atm pressure and 100 °C, and the temperature changes to 400 °C. What is the new pressure of the gas?

Convert the temperature to units of Kelvin.

$T_1 = 100\ °C + 273 = 373\ K$

$T_2 = 400\ °C + 273 = 673\ K$

Gay-Lussac's Law solves P_2 by substituting the quantities into the equation.

$$\frac{P_1}{T_1} = \frac{P_2}{T_2}$$

$P_2 = [(30.0\ atm)\cdot(673\ K)]\ /\ 373\ K$

$P_2 = 54.1\ atm$

Closed Container Law (constant volume)

Closed Container Law describes the behavior of a gas under constant volume. In such cases, pressure and temperature are directly proportional:

$$\frac{P_i}{T_i} = \frac{P_f}{T_f}$$

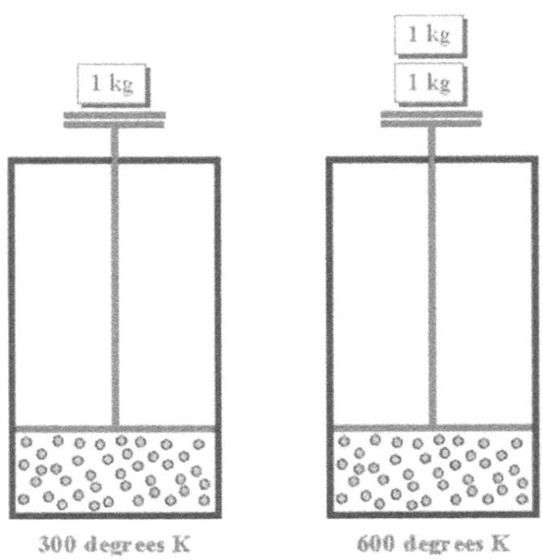

Combined Gas Law

Combined Gas Law amalgamates Boyle's and Charles' Laws. The combined gas law is *not* the ideal gas law (PV = nRT); it is expressed as an equation to relate changes in temperature, volume, and gas pressure.

Combined Gas Law relates the pressure, volume, and temperature of a gas and is expressed as:

$$\frac{P_i V_i}{T_i} = \frac{P_f V_f}{T_f}$$

where subscripts 1 and 2 refer to the same gas sample but under different temperatures, pressure, and volume.

For example, suppose a gas is at 15.0 atm pressure, with a volume of 25.0 L and a temperature of 300.0 K. What would the volume of the gas be at standard temperature and pressure? (Use standard pressure = 1.00 atm and standard temperature = 273.0 K)

Use the *combined gas law* at standard temperature (T_2) and pressure (P_2):

$(P_1V_1) / T_1 = (P_2V_2) / T_2$ solve for V_2:

$(P_1 \times V_1 \times T_2) / (T_1 \times P_2) = V_2$

$V_2 = [(15.0 \text{ atm})\cdot(25.0 \text{ L})\cdot(273.0 \text{ k})] / [(300.0 \text{ K})\cdot(1.00 \text{ atm})]$

$V_2 = 341 \text{ L}$

Avogadro's Law

The volume of a gas in a container is affected by pressure, temperature, and moles of gas.

The relationship between the quantity of gas and its volume was derived by Italian scientist Amadeo Avogadro (1776-1856).

Avogadro's Law states that *gases at a given temperature and pressure occupying a volume are directly proportional to the number of moles of gas present.*

Avogadro's Law is expressed as:

$$\frac{n_1}{V_1} = \frac{n_2}{V_2}$$

where n_1 and n_2 are moles of gas 1 and gas 2, and V_1 and V_2 are the volumes of gas 1 and gas 2.

A plot of *volume vs. moles* of gas (shown below) illustrates that Avogadro's Law is a particular case of the ideal gas law where T and P are held constant.

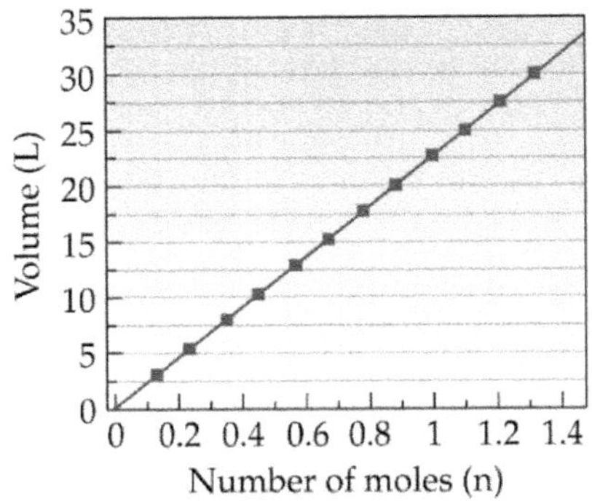

Avogadro's Law states that as moles of a gas increase, volume increases

For example, if 8.00 moles of a gas occupy a volume of 4.00 L at constant pressure and temperature, what volume of gas would 16.0 moles of this gas occupy at the same temperature and pressure?

Use *Avogadro's Law*

$$n_1 / V_1 = n_2 / V_2$$

where n_1 and n_2 are moles of gas 1 and gas 2, and V_1 and V_2 are the volumes of gas 1 and gas 2.

Solve for the V_2

$$V_2 = (V_1 \times n_2) / n_1$$

$$V_2 = [(4.00\text{ L})\cdot(16.0\text{ mol})] / 8.00\text{ mol}$$

$$V_2 = 8\text{ L}$$

Phase Diagrams

Phases of substances

Most substances exist in solid, liquid, or gas form.

Solids have a definite shape and volume (e.g., ice). The molecules in a solid vibrate at a fixed position and a solid cannot be compressed.

Liquids have a definite volume but take their container shape (e.g., water). The molecules move about but are close and bound by intermolecular forces.

Gases have neither a definite volume nor a definite shape; they take the container's volume and shape (e.g., steam or water vapor). The molecules move independently and are not held together by intermolecular forces, making gas easily compressible.

Phases at equilibrium

Phase diagrams indicate the relationship between temperature, pressure, and associated phases, plotting pressure, temperature, or volume for thermodynamically distinct phases (solid, liquid, or gas) at equilibrium.

Phase of a substance (i.e., solid, liquid, or gas) mostly depends on temperature and pressure.

If the pressure is exceptionally high, a substance is less likely to be in a gas phase because the pressure brings the molecules so close that the substance either liquefies or sublimes.

Phase change diagram

When an ice block is placed in the Sun, it slowly absorbs thermal energy through radiation and undergoes phase changes. The ice cube changes from solid to liquid to gas.

> *Melting point* of a substance is the temperature at which it changes from a solid to a liquid, and it has the same temperature range as the *freezing point*; the reverse reaction occurs at the same temperature.
>
> *Boiling point* (or vaporization) of a substance is the temperature at which it changes phase from a liquid to a gas, the same temperature range as the *condensation point*.

From the graph below, once the ice cube absorbs enough heat to melt (i.e., reaches 0 °C), its temperature remains constant.

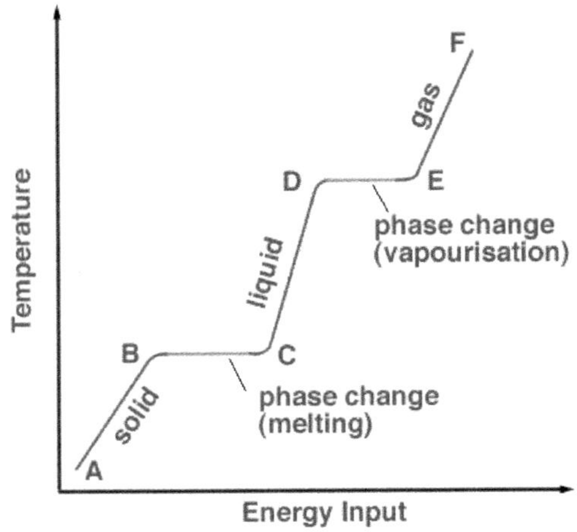

Phase diagram relates temperature changes and states

Melting is points B to C on the graph where all molecules have reached a temperature of 0 °C.

Ice melts only when this occurs, as illustrated by the graph's horizontal plateau from B to C.

When liquid water changes to a gas, the liquid absorbs enough heat to *vaporize* (i.e., at 100 °C).

Liquid temperature remains constant from point D to E until all molecules reach 100 °C.

As illustrated by the *horizontal plateau* from D to E, the *liquid vaporizes* only when this occurs.

Heat during phase changes

Although the energy (heat) input increases with temperature, it does not remain constant as the temperature does during a phase change. Technically, heat converts the potential energy stored within each atom to kinetic energy before changing phase.

As energy is continuously added to the process, the substance uses the added heat to transition each molecule *before increasing* the temperature.

Pressure *vs.* temperature relationship

Phase diagram graphically displays the phases of a substance of pressure *vs.* temperature.

The graph is split into three disproportionate parts, representing the substance's three phases.

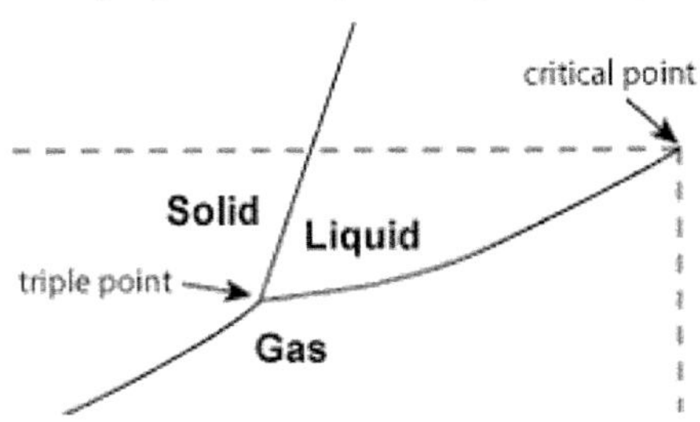

PV phase diagram with triple point and critical point labeled. Dashed horizontal line shows the critical pressure, and vertical dashed line shows the critical temperature.

Triple point, critical point, and critical temperature

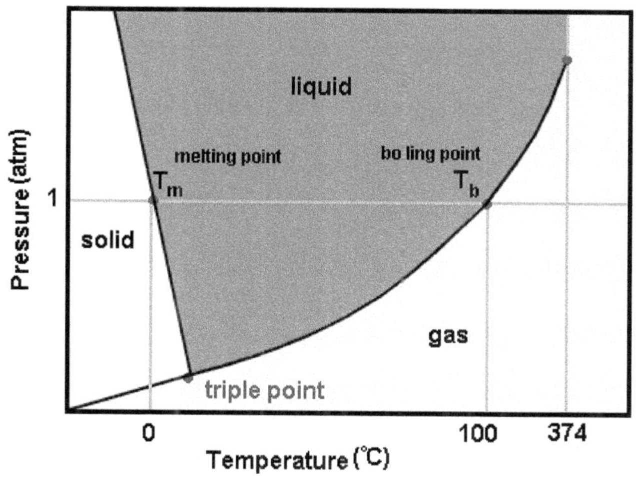

PV phase diagram with triple point and critical point labeled.

Melting point and *boiling point* are found by drawing a horizontal line at 1 atm of atmospheric pressure. The horizontal line intersects with a line on the phase diagram; the substance changes phases at that temperature.

Triple point is temperature and pressure at which three phases of matter coexist in equilibrium.

Critical point is temperature and pressure at which liquids and gases become indistinguishable.

Critical temperature is when a liquid cannot form, no matter the pressure.

Phase line boundaries

Solid-liquid boundary is where the *solid and liquid* exist in equilibrium.

Solid-gas boundary is where *solid and gas* exist in equilibrium.

Liquid-gas boundary is where *liquid and gas* exist in equilibrium.

Positive slope graphs

Solid-liquid phase boundary (or fusion curve) for most substances, has a positive slope (i.e., rise over run), so the *melting point increases with pressure*. This is true when the solid phase is denser than the liquid phase.

The greater the pressure, the closer the molecules of the substance are brought to each other, increasing the effect of intermolecular forces. Thus, the substance requires a higher temperature for its molecules to have enough energy to break the fixed pattern of the solid phase and enter the liquid phase.

A similar concept applies to liquid–gas phase changes.

Negative slope graphs

Water is an exception, which has a solid-liquid boundary with a negative slope so that the melting point decreases with pressure.

Ice (solid water) is less dense than liquid water, as ice floats on water.

At a molecular level, ice is less dense because it has a more extensive network of hydrogen bonding, which requires a greater separation of water molecules.

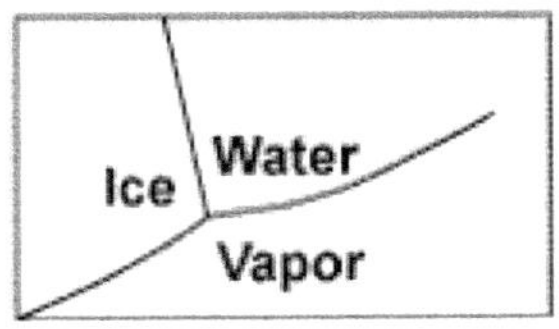

Phase diagram of pressure vs. pressure for water differs because the solid-liquid boundary is slanted to the left. Water (liquid) is denser than ice (solid); if the pressure increases, ice melts into water.

Heat Transformation

Fusion and vaporization

Latent heat (*latent energy* or *heat of transformation*) is the energy released or absorbed by a thermodynamic system during a constant-temperature process — usually a first-order phase transition (i.e., constant temperature during "mixed-phase" states).

Latent heat is the energy supplied or extracted to change the state of a substance without changing its temperature (i.e., first-order phase transition).

Examples of latent heat are *latent heat of fusion* (solid to liquid) and *latent heat of vaporization* (liquid to gas) for phase changes when a substance condenses or vaporizes at a specified temperature and pressure.

Heat and phase changes

Energy is released for:

Condensation: gas changes phase into liquid

Freezing: liquid transforms into a solid

Deposition: gas changes phase directly into solid

Reverse processes use the same amount of energy.

Vaporization (reverse of condensation): liquid changes phase into a gas

Melting (reverse of freezing): solid changes phase into liquid

Sublimation (reverse of deposition): solid changes phase directly into a gas

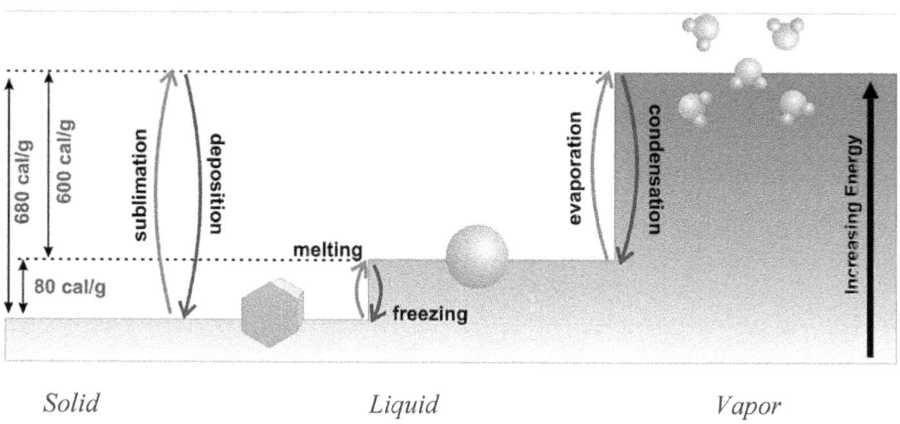

Latent heat

Latent heat is the amount of heat required for a phase change. Latent heat has been reported for most substances. In general, only computing the total heat is required for the reaction.

Heat required to convert a substance from one phase to another is:

$$Q = m \cdot l$$

where l is the latent heat (J/kg), and m is the mass of the substance (kg).

Latent heats are sometimes given in molar heat or joules per mol (J/mol). Divide by the molecular weight of the substance if given the mass in grams.

If any of these processes are reversed, the latent heat is negative (reverse process).

Heat of fusion

Heat of fusion (ΔH_{fusion}) is the energy to melt a given quantity of solid to liquid at a constant temperature; the energy required to melt the solid times the number of moles.

Heat of fusion (ΔH_{fusion}) is the energy to melt a given quantity of solid to liquid at a constant temperature; the energy required to melt the solid times the number of moles.

Heat of fusion ΔH_{fusion} is expressed as:

$$q = n \times \Delta H_{fusion}$$

where q is heat energy, n is the number of moles of the solid, and ΔH_{fusion} is the heat of fusion.

Most substances have reported reference values for the heat of fusion (ΔH_{fusion}).

If the particles become excited (solid to a liquid, liquid to a gas), the change in entropy is always positive because heat is added to the system.

For this process, use latent heat of fusion. The total heat required is the heat of fusion.

If the particles become less excited (gas to a liquid, liquid to solid), the change in entropy is always negative because heat is removed from the system.

A negative change in entropy results in a more ordered molecular structure.

The total heat required for this process is the *heat of vaporization.*

Heat of vaporization

Heat of vaporization (ΔH_{vap}) is the energy to vaporize a given quantity of liquid to gas at a constant temperature.

It is the energy required to vaporize the liquid times the number of moles of that liquid.

Most substances have reported reference values for the heat of vaporization (ΔH_{vap}).

Heat of vaporization ΔH_{vap} is expressed as:

$$q = n \times \Delta H_{vap}$$

where q is heat energy, and n is number of moles of the liquid, and ΔH_{vap} is heat of vaporization.

Sublimation is when a solid bypasses the liquid phase and transitions directly into the gas phase.

The heat required for *sublimation* sums up the *heat of fusion* and the *heat of vaporization.*

Deposition is the reverse reaction of sublimation.

$$Q_{sub} = Q_{fus} + Q_{vap}$$

Total heat of reaction

For a process with multiple phase changes, the total heat of the reaction is the sum of the separate heat needed for each phase change and the addition of heat.

For these problems, consider the First Law of Thermodynamics, which states that heat must be conserved (heat gained = heat loss), and the equation for heat ($Q = mc\Delta T$), where c is the specific heat capacity, and m is the mass.

For example, when calculating the change in heat (change in entropy) when 2 kg of ice at 0 °C is warmed to 120 °C, a few equations are needed using the known values.

- melting point = 0 °C
- boiling point = 100 °C
- latent heat of fusion = 3.33×10^5 J/kg
- latent heat of vaporization = 2.26×10^6 J/kg
- specific heat capacity = 4,186 J/kg °C

Heat required to phase change the *ice into liquid water*

$$Q = m\, l_{fusion}$$

$$Q = (2\text{kg})\cdot(3.33 \times 10^5\ \text{J/kg}) = 6.66 \times 10^5\ \text{J}$$

Heat required to heat the *water to its boiling point*:

$$Q = mc\Delta T$$

$$Q = (2\ \text{kg})\cdot(4186\ \text{J/kg °C})\cdot(100\ \text{°C} - 0\ \text{°C}) = 837{,}200\ \text{J}$$

Heat required to phase change the *liquid into gas:*

$$Q = m\, l_{vap} = (2\text{kg})\cdot(2.26 \times 10^6\ \text{J/kg})$$

$$Q = 4.52 \times 10^6\ \text{J}$$

Heat required to heat the water to 120 °C:

$$Q = mc\Delta T$$

$$Q = (2\ \text{kg})\cdot(4{,}186\ \text{J/kg °C})\cdot(120\ \text{°C} - 100\ \text{°C}) = 167{,}440\ \text{J}$$

Total heat required for the process to occur is the *sum of the separate heats*:

$$Q_{total} = Q_{fusion} + Q_{heat\_1} + Q_{vap} + Q_{heat\_2}$$

$$Q_{total} = 6.66\times10^5 + 837{,}200 + 4.52\times10^6 + 167{,}440$$

$$Q_{total} = 6{,}190{,}640\ \text{J}$$

Calculating heat of fusion and heat of vaporization

Heat of fusion and heat of vaporization use units of kJ/mol.

However, they may be expressed as J/g, where energy is obtained by multiplying the latent heat by the substance's mass rather than moles.

For example, how much kJ is required when 31.5 g of H_2O melts at its melting point of 0 °C? (Use molar mass for H_2O = 18 g/mol and molar heat of fusion for water ΔH_{fusion} = 6.02 kJ/mol).

Use molar mass relationship:

mass (g) = moles (mol) × molar mass (g/mol)

Rearrange to find the number of moles:

moles (g) = mass (g) / molar mass (g/mol)

moles = (31.5 g) / (18 g/mol)

moles = 1.75 mol

Use heat of fusion (ΔH_{fusion}):

$$q = n \times \Delta H_{fusion}$$

$$q = 1.75\ \text{mol} \times 6.02\ \text{kJ/mol}$$

$$q = 10.54\ \text{kJ}$$

10.54 kJ heat (i.e., positive value) is required to melt 31.5 g of water.

Pressure *vs.* Volume Phase Diagrams

PV graphs

Pressure vs. volume diagram (*PV* graph) describes system pressure and volume changes.

PV diagrams show the thermodynamic process by graphing pressure against volume (P *vs.* V).

Work done by the system is equal to the area under the curve.

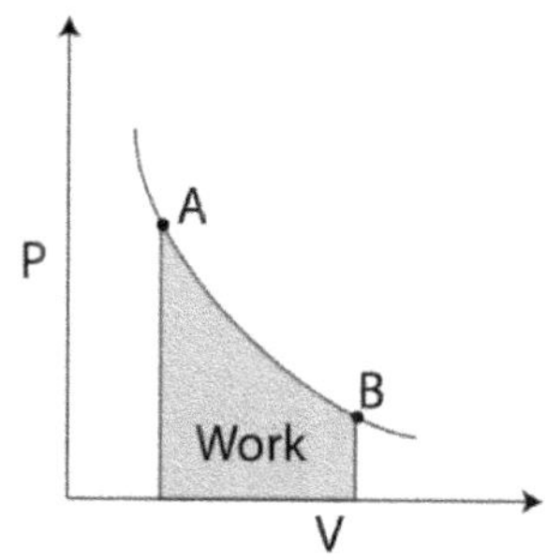

Area under the curve on the above PV diagram equals the work done

Cyclic pressure *vs.* volume reactions

In cyclic reactions, the work done is the *area enclosed* in the graph.

These reactions are often comprised of different processes, as shown below.

work is *done* ***by*** the gas when the volume increases and pressure decreases.

work is *done* ***on*** the gas when the volume decreases and pressure increases.

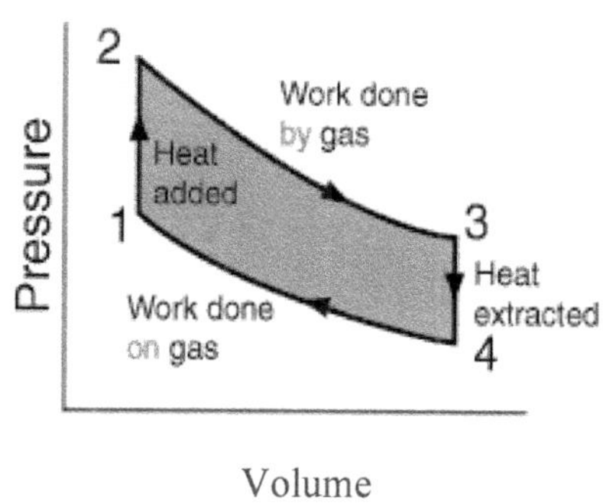

Points 4 to 1 show work done ***on*** *the gas; points 2 to 3 work done* ***by*** *the gas*

Calculating work done by gas

Work can be calculated for the gas phase of a gas and its container.

Gases may *expand* or *compress*; the *pressure* of the gas and *volume changes* determine work:

Figure illustrates:

expansion is work done ***by*** the system ($\Delta E = q - w$)

compression is work done ***on*** the system ($\Delta E = q + w$)

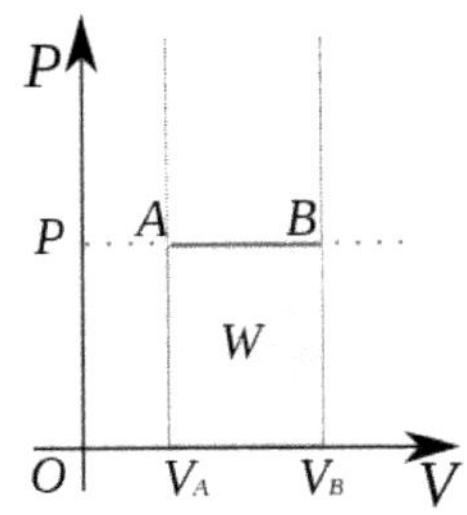

Shaded area (W) represents the work done by the gas

Positive and negative volume changes

Volume change (ΔV) is the final minus initial volume of gas:

$\Delta V = V_{final} - V_{initial}$

Volume is increased to *expand* a gas (ΔV is *positive*).

Gas (of the system) must do work (w) on surroundings; thus, work must be *negative.*

Volume is decreased to compress a gas (ΔV is *negative*).

Surroundings must do work (w) on the system; thus, work must be *positive.*

Thermodynamic work cycles

Processes may labeled *adiabatic*, *isothermal*, *isobaric*, and *isochoric.*

Adiabatic process has no transfer of heat ($Q = 0$); the change in energy equals the work.

Adiabatic processes are often confused with the isothermal process because it is sometimes assumed that no heat transfer means no change in temperature. However, this is not true—in isothermal instances, heat energy must be allowed into (or out of) the reaction to maintain a constant temperature.

Adiabatic processes, unlike an isothermal reaction, occur quickly and have a PV graph with the *same shape* but are *much steeper* than isothermal processes.

Adiabatic processes

Adiabatic process is when energy transfer within the system results in *work only*.

For example, a rapid expansion or contraction of gas is nearly adiabatic.

Adiabatic processes have *no heat* transfer.

Adiabatic processes *cannot decrease entropy*.

temperature is constant

$q = 0$

$\Delta E = w$

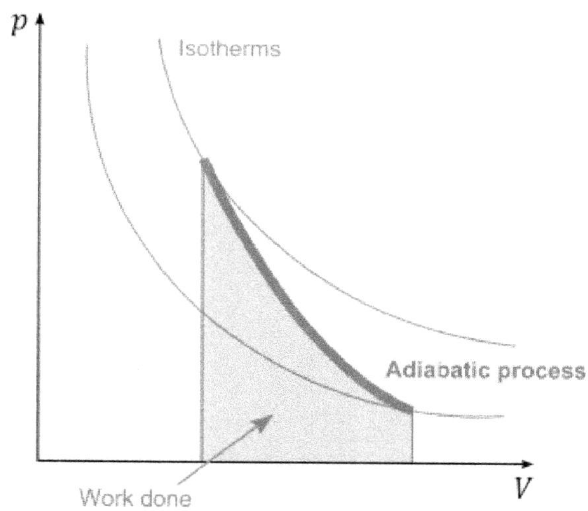

Adiabatic process PV diagrams of work with isothermal lines called isotherms

Adiabatic processes include a tire pump or a sharp expulsion of breath.

Isochoric processes

Isochoric process (or *constant-volume process*, *isovolumetric process*, and *isometric process*) is when the volume of the closed system remains *constant*; these include those inside a closed and rigid container or a constant volume thermometer.

volume is constant

$w = 0$

$\Delta E = q$

Thermodynamic process is the addition or removal of heat; the isolation of the contents of the container establishes the closed system, and the inability of the container to deform imposes the *constant-volume* condition.

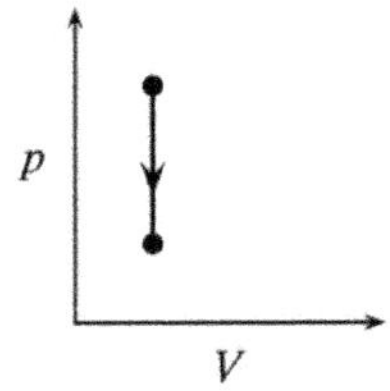

Isochoric process PV diagrams with constant volume show a vertical line

Isochoric processes include heating (or cooling) contents of a sealed, inelastic container.

Isothermal processes

Isothermal reaction keeps temperature constant ($\Delta T = 0$). The change in internal energy must stay the same, as an increase in thermal energy (heat) increases temperature ($\Delta U = 0$).

temperature is constant

$\Delta T = 0$

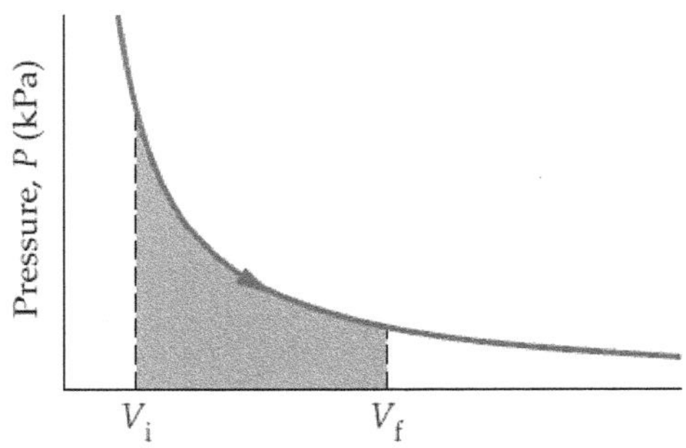

Isothermal process PV diagram at constant temperature

Isothermal process is when the temperature of the system remains constant; typically, when a system is in contact with an outside thermal reservoir. A change in the system allows the system to adjust temperature via the reservoir through heat exchange.

Isobaric processes

Isobaric process is when the pressure of the system remains constant.

Isobaric reactions keep the pressure constant; work equals the pressure times the volume change ($W = P\Delta V$).

pressure is constant

$\Delta P = 0$

$w = P\Delta V$

Examples of isobaric reactions include a piston in an engine or a flexible container open to Earth's atmosphere.

When pressure is constant, the PV graph of an isobaric process is a horizontal line.

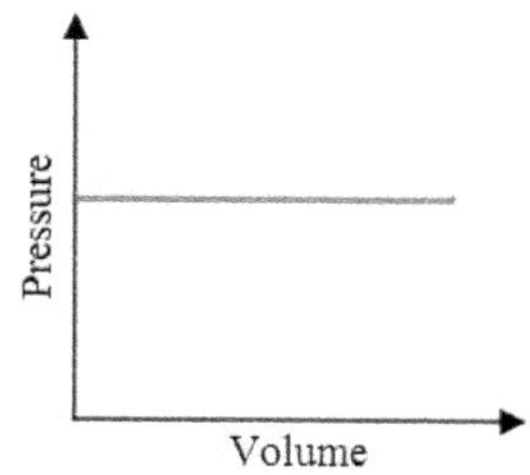

Isochoric process is an isovolumetric process because the volume is constant.

Since there is no movement, the work is zero, and the change in energy is due to the amount of heat added to the system.

$W = 0$

$\Delta E = Q$

Heat transferred to the system does work and *changes* the internal energy (*U*) of the system.

Thermodynamic work processes summary

Process	**Constant**	**PV Diagram**	**Ideal Gas Law**	**First Law of Thermodynamics**
Isobaric	Pressure	Horizontal line	V α T	$\Delta U = Q + W$
Isochoric	Volume	Vertical line	P α T	$\Delta U = Q$
Isothermal	Temperature	Curved line	PV α T	$\Delta U = 0$
Adiabatic	No heat exchanged	Curved line (jumps to different isotherm)	PV = nRT (only "nR" are constant)	$\Delta U = W$

Thermodynamic work processes, with relationship to Ideal Gas Law (PV = nRT)

Notes for active learning

Chapter Summary

Heat and temperature

Heat is the *total internal energy* due to molecular motion.

Temperature is the *average kinetic energy* (KE) for a substance.

Calories are the heat needed to raise the temperature of one gram of water by one degree Celsius (g/°C).

Kelvin (K) is the measure of absolute temperature.

Absolute zero = 0 K

Kelvin and Celsius use the same increments.

Specific heat is ratio of energy absorbed to temperature increase per unit mass.

Calorimeters measure the *heat of a reaction.*

Temperature remains *constant* during phase change.

Phase diagram

Triple point: all three phases exist simultaneously.

Critical point: the substance is always a gas.

$PV = NkT$

Boltzmann's constant 1.381×10^{-23} J/K

Heat transfer

Conduction requires *direct contact.*

Rate of heat transfer (H) by conduction:

$$H = \frac{kA\Delta T}{L}$$

Convection uses *energized particles* from one place to another.

Radiation uses *electromagnetic waves.*

Laws of thermodynamics

Zeroth Law of thermodynamics: two systems in thermal equilibrium with a third system are in *thermal equilibrium.*

First Law of thermodynamics: energy is always *conserved:*

$$\Delta U = Q - W$$

Work was done *by* a system (−)

Work was done *on* a system (+)

$Q < 0$ is exothermic, where Q is heat

$Q > 0$ is endothermic, where Q is heat

Second Law of thermodynamics: *entropy* increases or stays constant, never decreases

Entropy (S) is a measure of disorder within a system. The natural tendency is to move towards maximum disorder:

$$\Delta S \geq \frac{Q}{T}$$

$\Delta S_{\text{universe}} > 0$ spontaneous

$\Delta S_{\text{universe}} = 0$ at equilibrium

$\Delta S_{\text{universe}} < 0$ non-spontaneous (reverse process spontaneous)

Third Law of thermodynamics: as the temperature of a system approaches absolute zero, the entropy (S) approaches a constant value (typically zero).

$$S = k \cdot \ln(\Omega)$$

where S is the entropy (J/K), k is Boltzmann's constant (1.381×10^{-23} J/K), and Ω is the number of microstates

Entropy (S) of a system at absolute zero is zero.

However, some systems have more than one minimum energy state.

Enthalpy *H*

Heat used at constant pressure:

$$\Delta U = \Delta H - P\Delta V$$

$\Delta H < 0$ exothermic

$\Delta H > 0$ endothermic

Hess' Law: $\Delta H_{rxn} = H_{products} - H_{reactants}$

Gibb's Free Energy *G*

$\Delta G = \Delta H - T\Delta S$

$G < 0$: *spontaneous* reaction

$G = 0$: reaction is *at equilibrium* (rate of forward equals reverse reaction)

$G > 0$: *nonspontaneous* reaction

Carnot Cycle

Most efficient heat engine cycle

Two *isothermal* processes and two *adiabatic* processes

Carnot Efficiency:

$$\frac{T_H T_C}{T_H} \times 100\%$$

Thermal expansion

Coefficient of linear expansion (or *expansivity* measures the expansion (or contraction) per unit of length of a material that occurs when the temperature increases or decreases by 1 °C under constant pressure.

Expansivity

$$\alpha_{linear} = \frac{\Delta l}{l_i \cdot \Delta T}$$

where α_{linear} is the coefficient of thermal expansion (K^{-1}), Δl is the change in length (m), l_i is the initial length (m), and ΔT is the change in temperature (K).

Objects undergo area expansion and volume expansion due to the same principles.

***Coefficient of area expansion*:**

$$\alpha_{area} = \frac{\Delta A}{A_i \cdot \Delta T}$$

where α_{area} is the coefficient of thermal expansion (K^{-1}), ΔA is the change in the area (m^2), A_i is the initial area (m^2), and ΔT is the change in temperature (K).

Coefficient of volume expansion:

$$\alpha_{volume} = \frac{\Delta V}{V_i \cdot \Delta T}$$

where α_{volume} is the coefficient of thermal expansion (K^{-1}), ΔV is the change in volume (m^3), V_i is the initial volume (m^3), and ΔT is the change in temperature (K).

Kinetic Theory of Gases

Five critical assumptions for the Kinetic Theory of Gases:

1. Gas molecules move in *random molecular motion* and continue in a straight line until they collide with something.
2. Gas molecule is a "*point mass*," a particle so small that its mass is nearly zero. Therefore, an ideal gas particle has a *negligible volume*.
3. *No attractive intermolecular forces* for gases participate.
4. *Collisions* between gas molecules are *perfectly elastic*. Intermolecular attractive and repulsive forces are nonexistent, so gas molecules' total kinetic energy (KE) remains *constant* before and after collision.
5. *Average kinetic energy* (*KE*) for gas particles is proportional to the absolute temperature, regardless of the chemical identity or atomic mass.

   At 0 K (absolute zero or –273 °C), the molecules (and orbiting electrons) are not moving and have no volume.

Latent heat

Latent heat (l) is the heat (Q) required for a *phase change*:

$$Q = m \cdot l$$

where l is the latent heat (J/kg), and m is the substance mass (kg).

PV diagrams

Work is *area under pressure vs. volume* curve.

Isothermal - *constant temperature*: hyperbola

Adiabatic - *no transfer of heat*: steeper hyperbola

Isobaric - *constant pressure*: horizontal line

Isochoric - *constant volume*: vertical line

Phase diagram: pressure and temperature

Melting point is the temperature at which a substance changes from solid to liquid.

Freezing point is the same temperature as melting point but is the reverse reaction when a substance changes from liquid to solid.

Boiling point (or *vaporization*) is temperature when a substance changes from liquid to gas.

Condensation point is the same temperature as boiling point but is the reverse reaction when a substance changes from gas to liquid.

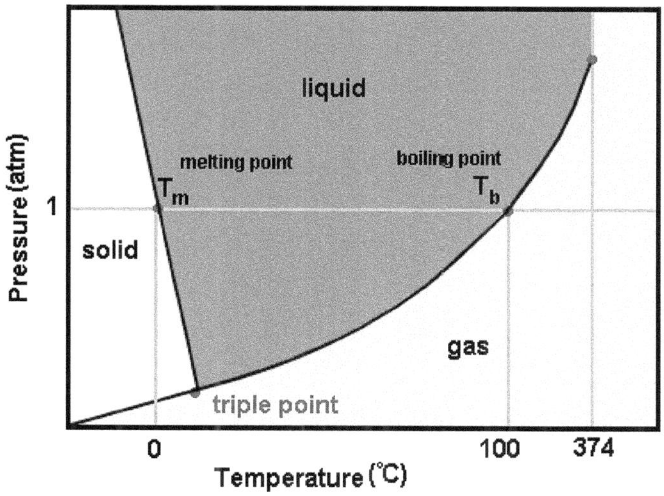

.

Notes for active learning

Practice Problems: Thermal Physics

1. A mass of 0.2 kg ethanol, in the liquid state at its melting point of –114.4 °C, is frozen at atmospheric pressure. What is the change in the entropy of the ethanol as it freezes? (Use the heat of fusion of ethanol $L_f = 1.04 \times 10^5$ J/kg)

A. –360 J/K
B. 54 J/K
C. –131 J/K
D. –220 J/K
E. 285 J/K

2. Isobaric work is

A. $Q - W$
B. $P\Delta V$
C. $P\Delta T$
D. $V\Delta P$
E. $Q + W$

3. An adiabatic process is performed on 9 moles of an ideal gas. The initial temperature is 315 K, and the initial volume is 0.70 m^3. The final volume is 0.30 m^3. What is the amount of heat absorbed by the gas? (Use the adiabatic constant for the gas = 1.44)

A. –18 kJ
B. 32 kJ
C. 9 kJ
D. –32 kJ
E. 0 kJ

4. An 80 g aluminum calorimeter contains 360 g of water at an equilibrium temperature of 20 °C. A 180 g piece of metal, initially at 305 °C, is added to the calorimeter. The final temperature at equilibrium is 35 °C. Assume there is no external heat exchange. What is the specific heat capacity of the metal? (Use the specific heat capacity of aluminum = 910 J/kg·K and the specific heat of water = 4,190 J/kg·K)

A. 260 J/kg·K
B. 324 J/kg·K
C. 488 J/kg·K
D. 410 J/kg·K
E. 535 J/kg·K

5. A chemist uses 120 g of water that is heated using 65 W of power with 100% efficiency. How much time is required to raise the water temperature from 20 °C to 50 °C? (Use the specific heat of water = 4,186 J/g·°C)

A. 136 s
B. 182 s
C. 93 s
D. 232 s
E. 274 s

6. The Second Law of Thermodynamics leads to the following conclusion:

A. the average temperature of the universe is increasing over time
B. it is theoretically possible to convert heat into work with 100% efficiency
C. disorder in the universe is increasing over time
D. total energy of the universe remains constant
E. entropy of the universe remains constant

7. The statement that 'heat energy cannot be completely transformed into work' is a statement of which thermodynamic law?

A. Third
B. Second
C. Zeroth
D. Fourth
E. First

8. What is the change in entropy when 20 g of water at 100 °C is turned into steam at 100 °C? (Use latent heat of vaporization of water $L_v = 22.6 \times 10^5$ J/kg)

A. –346 J/K
B. 346 J/K
C. –80.8 J/K
D. 121 J/K
E. 0 J/K

9. A Carnot-efficiency engine is operated as a heat pump to heat a room in the winter. The heat pump delivers heat to the room at 32 kJ per second and maintains the room at 293 K when the outside temperature is 237 K. Power requirement for the heat pump under these operating conditions is:

A. 6,100 W
B. 3,400 W
C. 7,300 W
D. 14,300 W
E. 11,450 W

10. Which of the relationships is valid for all types of Carnot heat engines?

I. $\eta = 1 - T_C / T_H$
II. $\eta = 1 - |Q_C / Q_H|$
III. $T_C / T_H = Q_C / Q_H$

A. I only
B. II only
C. III only
D. I, II and III
E. I and III only

11. What is the change of entropy associated with 8 kg of water freezing to ice at 0 °C? (Use latent heat of fusion $L_f = 80$ kcal/kg)

A. 1.4 kcal/K
B. 0 kcal/K
C. –2.3 kcal/K
D. –1.4 kcal/K
E. 0.8 kcal/K

12. A Carnot-efficiency engine extracts 515 J of heat from a high-temperature reservoir during each cycle and ejects 340 J to a low-temperature reservoir during the same cycle. What is the efficiency of the engine?

A. 67%
B. 21%
C. 53%
D. 17%
E. 34%

13. A glass beaker of unknown mass contains 65 ml of water. The system absorbs 1,800 cal of heat and rises to 20 °C. What is the mass of the beaker? (Use specific heat of glass = 0.18 cal/g·°C and specific heat of water = 1 cal/g·°C)

A. 342 g
B. 139 g
C. 546 g
D. 268 g
E. 782 g

14. A 0.3 kg ice cube at 0 °C has sufficient heat added to result in total melting, and the resulting water is heated to 60 °C. How much total heat is added? (Use latent heat of fusion for water $L_f = 334$ kJ/kg, latent heat of vaporization $L_v = 2{,}257$ kJ/kg, and specific heat of water = 4,186 kJ/kg·K)

A. 73 kJ
B. 48 kJ
C. 176 kJ
D. 144 kJ
E. 136 kJ

15. The water flowing over a large dam drops a distance of 60 m. If all the gravitational potential energy is converted to thermal energy, by what temperature does the water rise? (Use the acceleration due to gravity $g = 10$ m/s^2 and specific heat of water = 4,186 J/kg·K)

A. 0.34 °C
B. 0.44 °C
C. 0.09 °C
D. 0.14 °C
E. 0.58 °C

Solutions: Thermal Physics

1. C is correct.

Find heat from phase change:

$Q = mL_f$

$Q = (0.2 \text{ kg})\cdot(1.04 \times 10^5 \text{ J/kg})$

$Q = 20{,}800 \text{ J}$

Because the ethanol is freezing, Q should be negative due to heat being released.

$Q = -20{,}800 \text{ J}$

Find change in entropy:

$\Delta S = Q / \text{T}$

$\Delta S = -20{,}800 \text{ J} / (-114.4 \text{ °C} + 273 \text{ K})$

$\Delta S = -131 \text{ J / K}$

2. B is correct.

$\text{W} = \text{P}\Delta\text{V}$

Isobaric means pressure is constant, and the volume is changing.

3. E is correct.

Adiabatic means that no heat enters or leaves the system.

$Q = 0 \text{ kJ}$

4. C is correct.

$Q = mc\Delta\text{T}$

Find the heat added to the aluminum calorimeter:

$Q_A = (0.08 \text{ kg})\cdot(910 \text{ J/kg·K})\cdot(35 \text{ °C} - 20 \text{ °C})$

$Q_A = 1{,}092 \text{ J}$

Find heat added to water:

$Q_W = (0.36 \text{ kg})\cdot(4{,}190 \text{ J/kg·K})\cdot(35 \text{ °C} - 20 \text{ °C})$

$Q_W = 22{,}626 \text{ J}$

continued...

Find total heat added to the system:

$Q_{total} = Q_A + Q_W$

$Q_{total} = 1{,}092\text{ J} + 22{,}626\text{ J}$

$Q_{total} = 23{,}718\text{ J}$

Find specific heat of the metal:

$Q = mc\Delta T$

$c = Q / m\Delta T$

$c = (23{,}718\text{ J}) / [(0.18\text{ kg})\cdot(305\text{ °C} - 35\text{ °C})]$

$c = 488\text{ J/kg}\cdot\text{K}$

5. D is correct.

Watt = 1 J/s

Thermal energy:

$Q = \text{Power} \times \text{time}$

$Q = mc\Delta T$

$P \times t = mc\Delta T$

$t = (mc\Delta T) / P$

$t = [(120\text{ g})\cdot(4.186\text{ J/g}\cdot\text{°C})\cdot(50\text{ °C} - 20\text{ °C})] / (65\text{ W})$

$t = 232\text{ s}$

6. C is correct.

Second Law of Thermodynamics states that entropy is *constant or increases* over time.

Constant entropy process is idealized and does not exist.

Thus, entropy is always increasing over time.

7. B is correct.

Second Law of Thermodynamics states that through thermodynamic processes, there is an increase in the sum of entropies of the system; thus, no engine process is 100% efficient.

8. D is correct.

Find heat from phase change:

$$Q = mL_f$$

$$Q = (0.02 \text{ kg})\cdot(22.6 \times 10^5 \text{ J/kg})$$

$$Q = 45{,}200 \text{ J}$$

Because the water vaporizes, Q should be positive due to absorbed heat.

$$Q = 45{,}200 \text{ J}$$

Find change in entropy:

$$\Delta S = Q / \text{T}$$

$$\Delta S = 45{,}200 \text{ J} / (100\ °\text{C} + 273 \text{ K})$$

$$\Delta S = 121 \text{ J} / \text{K}$$

A positive change in entropy indicates that the disorder of the isolated system has increased.

When water evaporates into steam, entropy is positive because the disorder of steam is higher.

9. A is correct.

Coefficient of performance assuming an ideal Carnot cycle:

$$C_p = Q_H / \text{W}$$

$$C_p = \text{T}_H / (\text{T}_H - \text{T}_C)$$

$$Q_H / \text{W} = \text{T}_H / (\text{T}_H - \text{T}_C)$$

$$\text{W} = [Q_H \cdot (\text{T}_H - \text{T}_C)] / \text{T}_H$$

$$\text{W} = [(32 \times 10^3 \text{ J/s})\cdot(293 \text{ K} - 237 \text{ K})] / (293 \text{ K})$$

$$\text{W} = 6{,}116 \text{ J/s} \approx 6{,}100 \text{ W}$$

10. D is correct.

Carnot efficiency engines can be written as:

$$\eta = 1 - \text{T}_C / \text{T}_H$$

$$\eta = 1 - |\text{Q}_C / \text{Q}_H|$$

Thus:

$$\text{Q}_C / \text{Q}_H = \text{T}_C / \text{T}_H$$

11. C is correct.

Find heat from the phase change:

$$Q = mL_f$$

$$Q = (8 \text{ kg})\cdot(80 \text{ kcal/kg})$$

$$Q = 640 \text{ kcal/kg}$$

Because the water is freezing, Q should be negative due to the released heat.

$$Q = -640 \text{ kcal/kg}$$

Find change in entropy:

$$\Delta S = Q / \text{T}$$

$$\Delta S = -640 \text{ kcal/kg} / (0\ °\text{C} + 273 \text{ K})$$

$$\Delta S = -2.3 \text{ kcal/K}$$

A negative change in entropy indicates that the disorder of the isolated system has decreased. When water freezes, the entropy is negative because water is more disordered than ice. Thus, the disorder has decreased, and entropy is negative.

12. E is correct.

$$Q_H = 515 \text{ J}$$

$$Q_C = 340 \text{ J}$$

Carnot cycle efficiency:

$$\eta = (Q_H - Q_C) / Q_H$$

$$\eta = (515 \text{ J} - 340 \text{ J}) / (515 \text{ J})$$

$$\eta = 0.34 = 34\%$$

13. B is correct.

$$Q = (mc\Delta \text{T})_{\text{water}} + (mc\Delta \text{T})_{\text{beaker}}$$

Change in the temperature is the same for both:

$$Q = \Delta \text{T}[(mc)_{\text{water}} + (mc)_{\text{beaker}}]$$

$$1{,}800 \text{ cal} = (20\ °\text{C})\cdot[(65 \text{ g})\cdot(1 \text{ cal/g}\cdot°\text{C}) + (m_{\text{beaker}})\cdot(0.18 \text{ cal/g}\cdot°\text{C})]$$

$$90 \text{ cal/°C} = 65 \text{ cal/°C} + (m_{\text{beaker}})\cdot(0.18 \text{ cal/g}\cdot°\text{C})$$

$$25 \text{ cal/°C} / (0.18 \text{ cal/g}\cdot°\text{C}) = (m_{\text{beaker}})$$

$$m_{\text{beaker}} = 139 \text{ g}$$

14. C is correct.

Heat to melt ice cube:

$Q_1 = mL_f$

Heat to raise temperature:

$Q_2 = mc\Delta T$

Total heat:

$Q_{total} = Q_1 + Q_2$

$Q_{total} = mL_f + mc\Delta T$

$Q_{total} = [(0.3 \text{ kg})\cdot(334 \text{ kJ/kg})] + [(0.3 \text{ kg})\cdot(4.186 \text{ kJ/kg}\cdot\text{K})\cdot(60\ °\text{C} - 0\ °\text{C})]$

$Q_{total} = (100.2 \text{ kJ}) + [(1.257 \text{ kJ/K})\cdot(60 \text{ K})]$

$Q_{total} = 175.55 \text{ kJ} \approx 176 \text{ kJ}$

15. D is correct.

$PE = Q$

$mgh = mc\Delta T$, cancel m from both sides of the expression

$gh = c\Delta T$

$gh / c = \Delta T$

$\Delta T = [(10 \text{ m/s}^2)\cdot(60 \text{ m})] / 4{,}186 \text{ J/kg}\cdot\text{K}$

$\Delta T = 0.14\ °\text{C}$

Notes for active learning

Notes for active learning

CHAPTER 10

Electrostatics & Electromagnetism

Charge and Conservation of Charge

Conductors and Insulators

Coulomb's Law

Electric Field

Electric Potential and Equipotential Lines

Electric Dipole

Electrostatic Induction

Gauss's Law

Magnetic Fields and Poles

Magnetic Field Force

Faraday's Law

Torque on Current-Carrying Wire

Properties of Electromagnetic Radiation

Electromagnetic Spectrum and Photon Energy

Chapter Summary

Page intentionally left blank

Charge and Conservation of Charge

Charge

Charge, like mass, is an intrinsic property of all matter.

Every particle in the universe has a mass and a charge measured in coulombs (C).

Unlike mass, however, three distinct charges exist: *positive*, *negative,* and *neutral*. (Neutral charge does NOT mean that the particle has no charge; the net charge is neutral.)

Atomic structure

Atom is the basic unit of matter, with electrons orbiting a nucleus of protons and neutrons.

Subatomic particles, specifically electrons and protons, are the basis of all macroscopically observed charge phenomena.

Electron is negatively charged with -1.60×10^{-19} coulombs.

Proton is positively charged with an equal but opposite charge of $+1.60 \times 10^{-19}$ coulombs.

Macroscopic charge

For macroscopic objects with a charge, it is essential to note that only the electrons contribute to the charge because they are mobile.

Protons are part of the core nuclei of the atom and are not mobile.

Charge on an atom is either due to an excess of electrons (if the observed charge is negative) or a shortage of electrons (if the observed charge is positive).

Negative charge: # electrons > # protons

Positive charge: # electrons < # protons

Neutral charge: # electrons = # protons

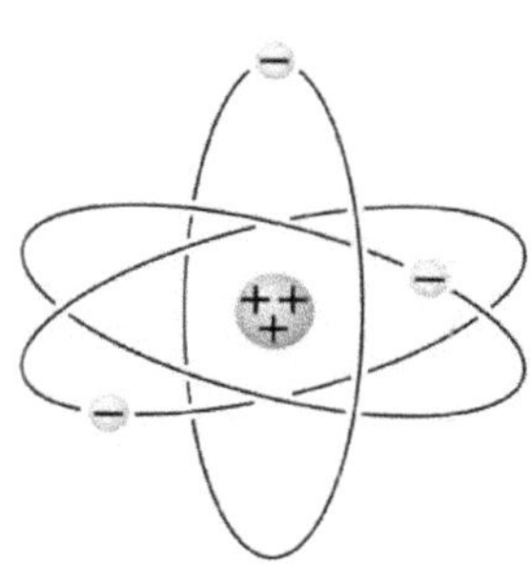

Atom with a core of protons and neutron and orbiting electrons

Properties of charge

Charges have several unique properties that govern the laws of nature.

Most importantly, charges exhibit force upon each other according to the types of charge.

The same charge produces repulsive forces, and opposite charges produce attractive forces.

Figures below show a ruler that is negatively charged (i.e., excess of electrons) and a glass rod that is positively charged (i.e., deficiency of electrons).

In top diagram, two negatively charged rulers are brought close and repel due to like charges.

In middle diagram, two positively charged rulers are brought close and repel due to like charges.

In bottom diagram, the charged ruler and the glass rod attract due to their opposite charges.

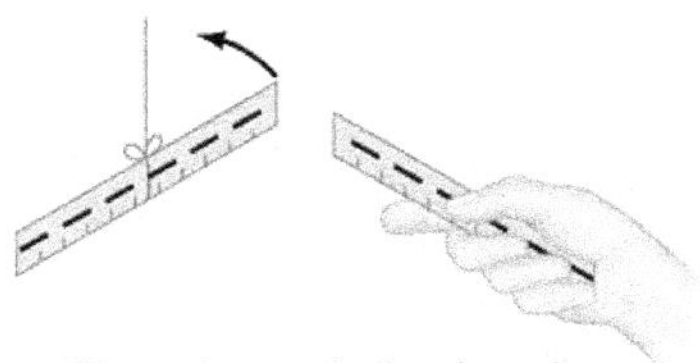

Two charged plastic rulers repel

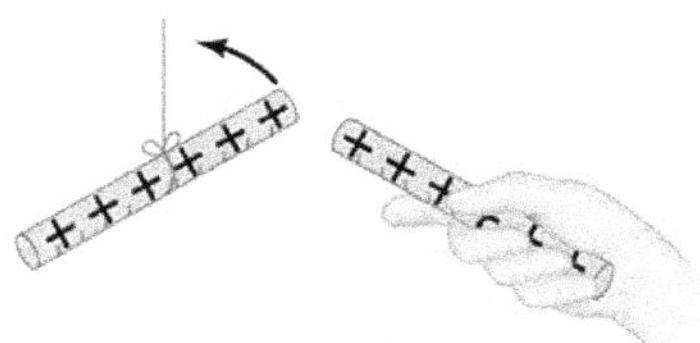

Two charged glass rods repel

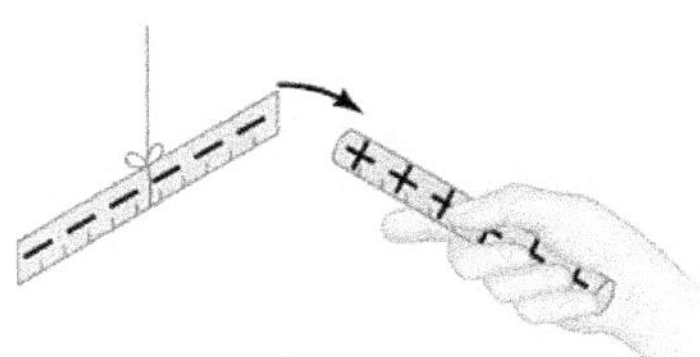

Charged glass rod attracts charged plastic ruler

Like charges repel while opposite charges attract

The charge is always conserved. Like mass and energy, the charge cannot be created nor destroyed. The charge can only be transferred from one source to another.

If a charged object (e.g., the ruler in the above example) is placed on a table for some time and later found neutrally charged, the charge is not destroyed but instead transferred to the surroundings (e.g., air and table).

Fundamental charge

Another unique property of charge is that all charges are quantized.

Magnitude of an electron's (or proton) charge is the *fundamental* and smallest charge unit.

Macroscopically charged objects have a net charge equal to an integer multiple of the fundamental charge.

Fundamental charge can be expressed as:

$$q = ne$$

where q is the net charge (C), n is the integer multiple (i.e., excess electrons or protons), and e is the magnitude of the fundamental charge (1.60×10^{-19} C).

Quantized charge

Quantization of charge enables calculating the excess (or deficiency) of electrons in an object that contributes to the object's net charge.

For example, if the negatively charged ruler from the earlier example is measured to have one Coulomb of charge, then the number of excess electrons in the ruler is:

$$n = \frac{q}{e}$$

$$n = \frac{1.00 \text{ Coulomb}}{1.60 \times 10^{-19} \text{ Coulomb}}$$

$$n = 6.25 \times 10^{18} \text{ electrons}$$

Conductors and Insulators

Conductivity

Conductivity is a measure of a material's *ability to transmit a charge.*

Conductors are materials that have a high conductivity. These materials allow electrical charge (electrons) to "flow" through the material.

Electrons (i.e., negatively charged) move and *not protons* (i.e., positively charged particles).

For example, copper is a good conductor. Electricity is transferred via copper wires in most electrical equipment wires because copper allows electrical charges to flow easily.

Insulators

Insulators are materials that do not transmit charge well and are poor conductors. Conventional insulators are wood, glass, and paper.

In the diagram below, one sphere has been charged positively (i.e., a deficit of electrons), and the other is neutral.

A metal nail is placed atop both spheres in the middle diagram. The metal nail acts as a conductor, allowing electrons to flow through it (from the neutral to the positive sphere) until the charge in each sphere is equal.

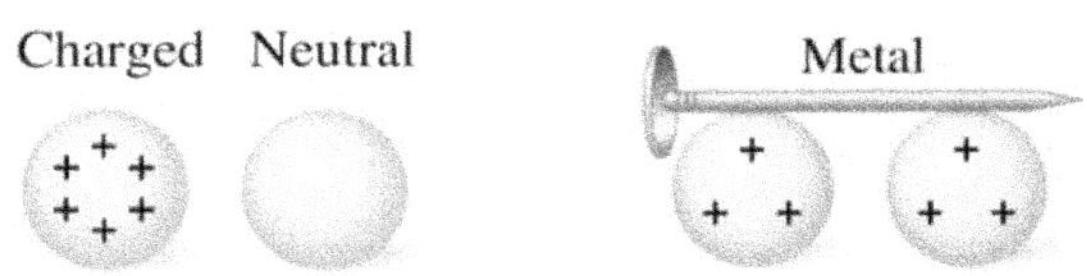

Metal nail acts as a conductor and permits the flow of electrons until charges are equal

A piece of wood is placed across the spheres. Wood is an insulator and does not allow electrons to flow from the neutral sphere to the positive sphere.

Wood is an insulator and does not allow electrons to flow from the neutral to the positive sphere

Coulomb's Law

Attractive or repelling forces

Charges repel or attract, depending on whether the charges are similar (repel) or opposite.

Resulting force is the *electrostatic force* (Coulomb force) and is found using *Coulomb's Law*, which describes the interaction between two charged particles.

Coulomb's force:

$$F = \frac{1}{4\pi \epsilon_0} \frac{q_1 q_2}{r^2}$$

which simplifies to:

$$F = k\frac{q_1 q_2}{r^2}$$

where F is the electrostatic force between the two charges (N), q is the magnitudes of the charges (C), r is the distance between them (m), ϵ_0 is the permittivity of free space (8.854×10^{-12} $C^2/N{\cdot}m^2$), and k is Coulomb's constant with a value of 9×10^9 $N{\cdot}m^2/C^2$.

Electrostatic force

Coulomb's Law strictly applies only to *point charges*.

It refers to the distance from the source, which is only defined for a point, not a distribution.

Coulomb's Law is *not valid* for moving charges.

If two objects have a net charge, then the objects must be approximated as point charges to calculate the force each exerts on the other.

Electrostatic force is always *along the line connecting the charges*.

Charges with the *same sign* are *repulsive*.

Charges with *opposite signs* are *attractive*.

Superposition principle

To find the electrostatic force on point charges, use the *superposition principle* and sum the force from each charge in all *axial directions.*

The superposition principle states that every charge in a medium creates an electric field at point independent of the presence of other charges in that medium and the resultant electric field is a vector sum of the electric fields of individual charges.

For example, what is the net force on Q_3 in the example below?

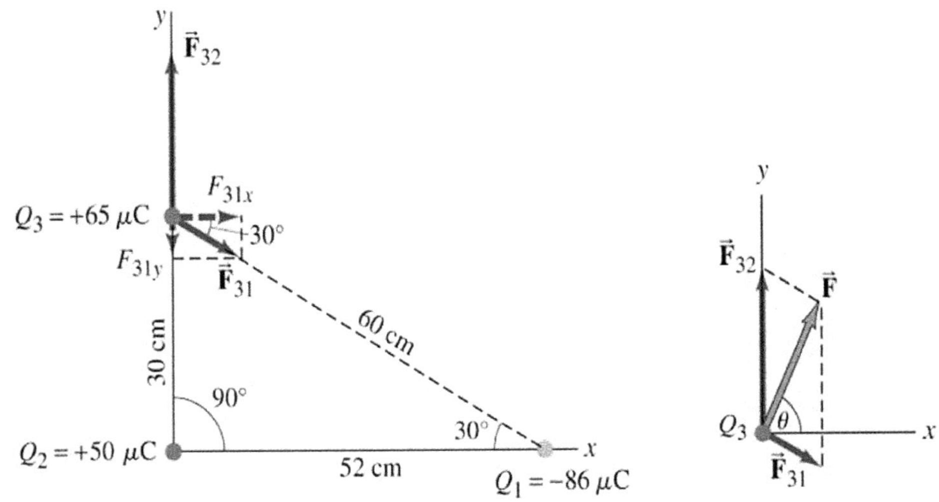

Superposition principle sums the force for each charge in axial directions

$$F_{32} = k\frac{q_1 q_2}{r^2}$$

$$F_{32} = \frac{(9 \times 10^9 \text{Nm}^2)\cdot(65 \times 10^{-6}\text{C})\cdot(50 \times 10^{-6}\text{C})}{(0.3\ \text{m})^2} = 325\ \text{N}$$

$$F_{31y} = \sin\theta\, k\frac{q_1 q_2}{r^2} = \left|\frac{\sin 30\,(9 \times 10^9 \text{Nm}^2)\cdot(65 \times 10^{-6}\text{C})\cdot(-86 \times 10^{-6}\text{C})}{(0.6\ \text{m})^2}\right| = 70\ \text{N}$$

$$F_{31x} = \cos\theta\, k\frac{q_1 q_2}{r^2} = \left|\frac{\cos 30\,(9 \times 10^9 \text{Nm}^2)\cdot(65 \times 10^{-6}\text{C})\cdot(-86 \times 10^{-6}\text{C})}{(0.6\ \text{m})^2}\right| = 121\ \text{N}$$

$$F_{net} = \sqrt{(121\ \text{N})^2 + (325\ N - 70\ \text{N})^2} = 282\ \text{N}$$

Electric Field

Directional vector fields

Every charge is surrounded by *electric fields* (E), which emanate in all directions around the charge. These fields are *directional vector fields*.

Field orientation is in the direction that a positive charge is pushed when placed in the field.

Electric field points *outward* in every direction for *positive charges*.

Electric field points *inward* in every direction for negative charges.

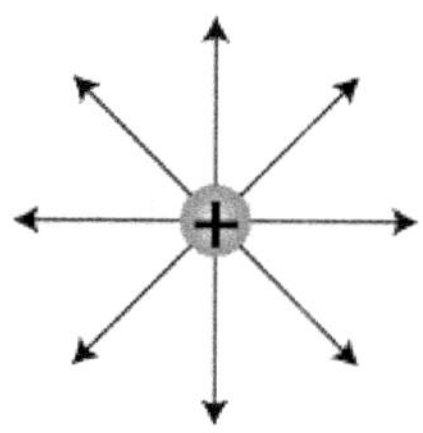

Electric field for (+) charge

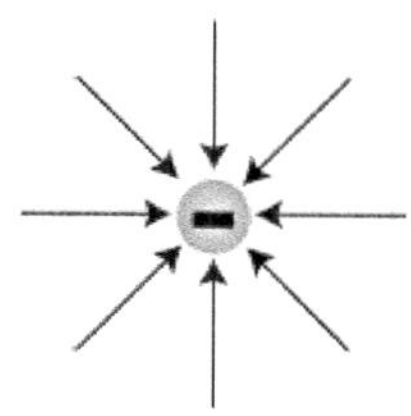

Electric field for (–) charge

Electric field expression

Electric field (E) can be expressed as the force (F) on a charge, divided by the magnitude of charge (q):

$$\vec{E} = \frac{\vec{F}}{q}$$

where (E) is the electric field measured in Newton/Coulombs (N/C).

Substituting force in the electrostatic equation, electric field from a single point charge (Q) is:

$$\vec{E} = \frac{\vec{F}}{q} = \frac{k\,q\,Q\,/\,r^2}{q} = k\frac{Q}{r^2}$$

Equations are rearranged to solve for the force (F) on a point charge (q) in an electric field (E):

$$\vec{F} = q\vec{E}$$

Force on charge q in an electric field

In the figure below, an electric field E points arbitrarily.

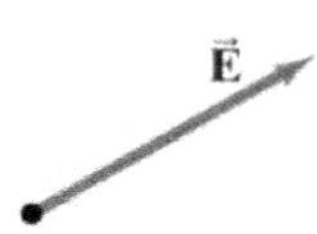

Arbitrary direction of electric field E

If a positive charge q is placed in the same electric field, then the force F on the charge is in the *same direction* as the electric field E.

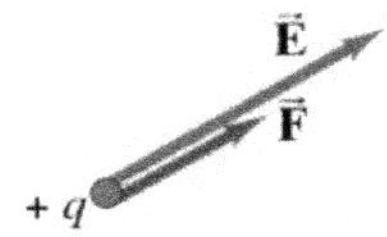

Force F on the positive charge q is in the direction of the electric field E

If a negative charge q is placed in the same electric field, the force F on the charge is in the *opposite direction* of the electric field E.

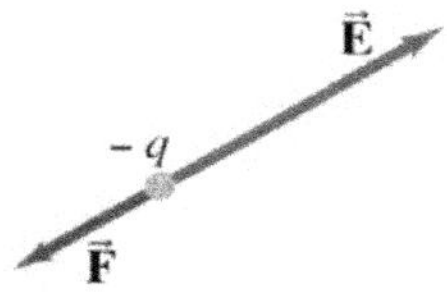

Force F on the negative charge q is in the opposite direction of the electric field E

Solving electrostatics problems

Solve electrostatics problems with electric forces and electric fields by:

Draw a diagram of the entire situation.

Show all charges with signs, electric fields, and force.

Include the direction of electric fields and force.

Use Coulomb's Law and add the forces as vectors.

Like for all vectors, the superposition principle is used to calculate the net strength of multiple electric fields on a point charge:

$$\vec{E} = \vec{E_1} + \vec{E_2} + \ ...$$

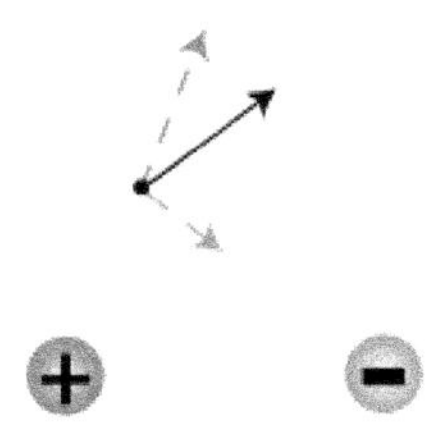

Resultant vector (E) is calculated from the component vectors (E_1, E_2, ...)

Field lines

Like all field lines, the *density of electric field lines* denotes the *strength* of the field.

Electric field is stronger in places where the field lines are close together. If the electric field lines are spread out, the electric field is weaker.

Density of electric field lines determines the magnitude of the charge emitting from the field.

Field due to charge distribution

If the density of the electric field lines is *high*, the charge has a *higher magnitude*.

If the density of the electric field lines is *low*, the charge has a *lower magnitude*.

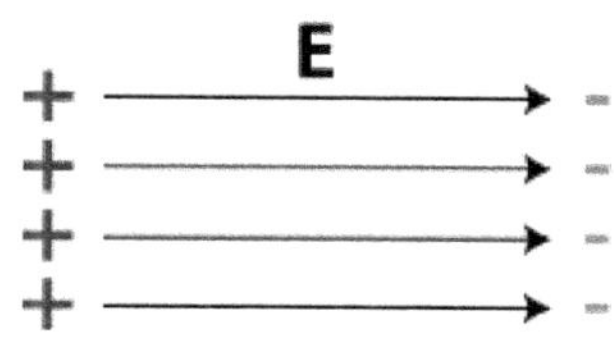

Electric field E points from the positive terminus to the negative terminus

Field due to similar charges

Electric dipole is created when two opposite charges are brought together.

When two charges are close, their electric fields interact depending on the types of charges.

If two positive charges are brought close, the field lines repel.

If two negative charges are brought close, the field lines are the same as the two positive charges, except that the direction of the field lines is reversed.

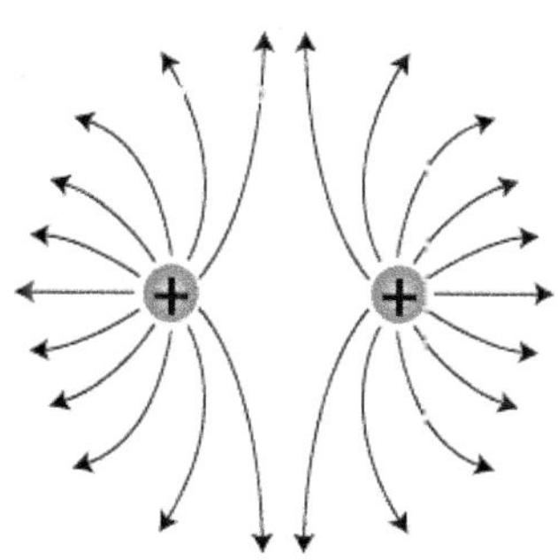

Electric field line E repels upon approach for two positive point charges

Field due to opposite charges

In the diagram below, an electric dipole has the electric field lines originating from the positive charge and moving to the negative charge.

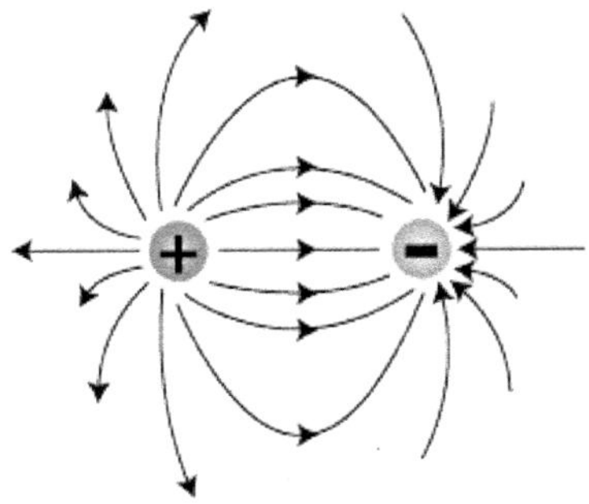

Electric field lines originate at the positive charge

Net charge on a conductor

Net charge on a conductor is on its surface, which radiates electric field lines.

For a charged cylinder, the electric field runs radially perpendicular to the cylinder and is zero (nonexistent) inside the cylinder.

This effect is for all conducting objects; the net electric field is always from the surface and nonexistent within the object.

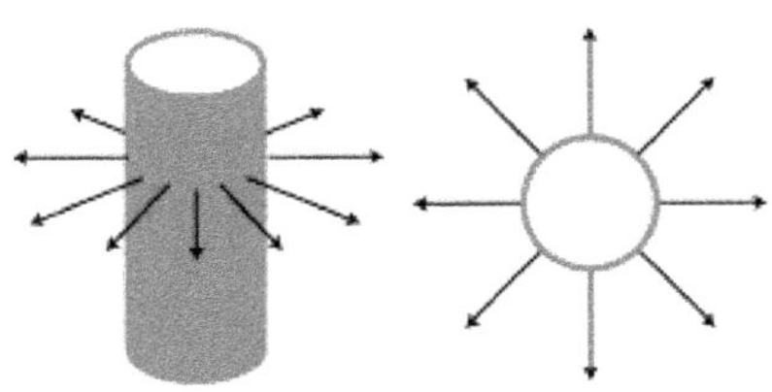

Electric field lines for a positive cylinder; side view and cross-sectional

Electric field lines summary

1. Field lines indicate the direction of the field; the electric field is tangent to the line.
2. Magnitude of the field is proportional to the density of the lines.
3. Field lines start on positive charges and end on negative charges; the number is proportional to the magnitude of the charge.
4. Net electric field within a conductor is zero.

Electric Potential and Equipotential Lines

Electric potential energy

Charges exert a force on each other in relation to the distance between the charges, so charges must have potential energy (PE).

Electric potential energy has units of joules (J):

$$U = \frac{k \cdot Q \cdot q}{r}$$

where U is electric potential energy (J), Q and q are the point charges (C), r is distance between the point charges (m), and k is Coulomb's constant with a value of 9×10^9 N·m²/C².

Electric PE of a single point charge is related to, *but different* from, its electric potential.

Electric potential

Electric potential is the amount of energy that something possesses per charge.

Unit for electric potential is joules per Coulomb (J/C) or volts (V) and is a scalar field associated with potential energy.

Electric potential is found using either of the following two equations.

$$V = \frac{U}{q}$$

or

$$V = \frac{kQ}{r}$$

where V is the electric potential (V), Q is the charge that is causing the potential (C), q is the charge experiencing the potential (the magnitude of q is small), U is the electrical potential energy possessed by q (J), k is Coulomb's constant and r is the distance between the two charges (always positive).

Total electric potential

If multiple charges contribute to the electric potential, then the total electric potential is the sum of the potentials caused by the individual components.

Positive charges produce positive potentials.

Negative charges produce negative potentials.

Plotting total electric potential

Plot below shows the potential due to a *positive* charge.

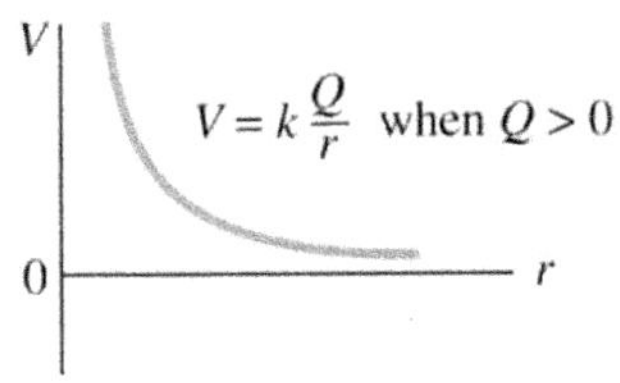

Potential due to positive charge

Plot below shows the potential due to a *negative* charge.

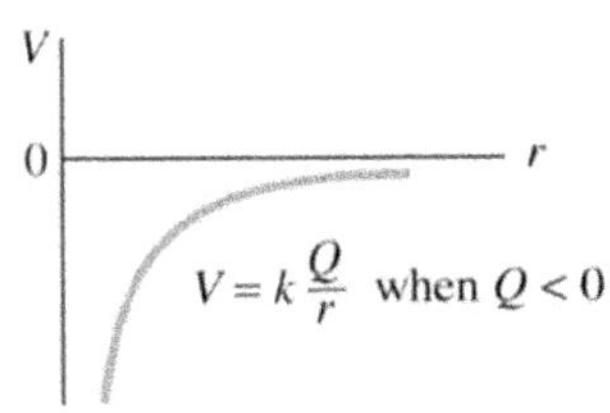

Potential due to negative charge

Determining potential difference

Potential difference is vital in producing forces and moving charges. The process is analogous to moving masses in gravitational fields.

ΔV gives the potential as the difference between the two potentials.

$$\Delta V = V_B - V_A$$

The potential difference is used for several applications.

For example, to determine the difference in potential between two plates of a capacitor or the positive and negative terminals of a battery.

Using potentials instead of fields can make solving problems easier because the potential is a scalar quantity, whereas a field is a vector.

Conservative forces

Coulomb force is conservative. Therefore, the work in an electric field is conservative.

Work required to move a point charge in an electric field depends only on the displacement of the charge, not the path taken by the point charge.

A straightforward method to calculate this work is to use electric potential differences, which can be written to relate to work involved in positioning charges.

$$\Delta V = \frac{W}{q}$$

where ΔV is the electric potential difference (V), W is the work needed to move the point charge (J), and q is the magnitude of the moving charge (C).

Equipotential lines

Equipotential lines are placed around a charge and mark where the electric potential is the same. Equipotential lines are always perpendicular to electric field lines.

In 3D, equipotential lines form a "surface" called an *equipotential surface.*

Movement along an equipotential surface requires no work because the movement is perpendicular to the electric field.

Figures below depict equipotential lines under different circumstances.

The image below is a single positive charge, along with its electric field lines and corresponding equipotential lines. Equipotential lines are perpendicular to the electric field lines at every point.

The image below is a dipole with curved electric field lines and mirrored equipotential lines.

Equipotential lines are perpendicular to the electric field lines at every point.

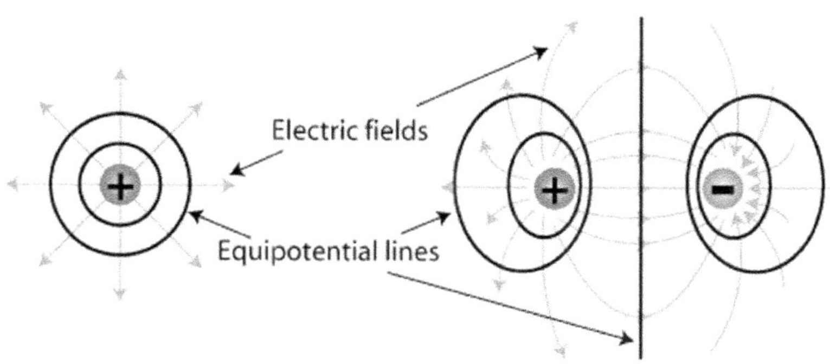

Equipotential lines are perpendicular to the electric field lines

Notes for active learning

Electric Dipole

Dipoles

Dipole is an interaction between a *positive and a negative charge separated* by some distance.

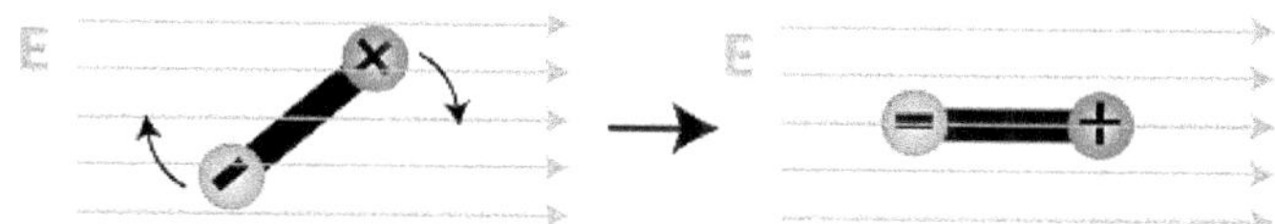

Dipole in an electric field aligns with the electric field

The dipole moment is the product of the charge and the distance between them.

Dipole moment represents the strength of the dipole. The electric field generated by the dipole is proportional to the dipole moment.

A dipole in an electric field aligns itself with the electric field. The positive end of the dipole is in the direction of the electric field (furthest from the source of the electric field). The dipole spins or rotates until it reaches such a position.

Polar molecules

Many molecules exhibit electric dipoles and are *polar molecules* because of the separation of electric charge within the molecule.

A polar molecule has one or more polar bonds, and one end is slightly positive while the other is slightly negative, resulting in an electric dipole moment. The two electrically charged regions are called poles, similar to a magnet's North and South Pole.

Polar molecules possess dipoles, but not all molecules with dipoles are polar. A molecule is polar if the net dipole moment is nonzero.

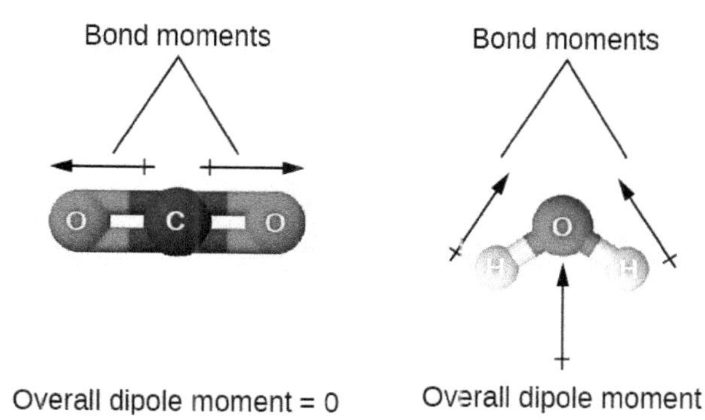

Polar molecules have dipoles, but molecular shape determines polarity. The molecule on the left cancels dipoles by symmetry, while the molecule on the right has a net dipole.

Asymmetric charge distribution

Water (H_2O) molecules are polar. In water molecules, a dipole moment arises from the negative charge near the electronegative O atom and two positive charges near each H atom.

Asymmetric charge distribution creates electric dipole vital to physical and chemical properties.

Microwaves heat food by heating the water in the food. Heating occurs because the oscillating electric field of a microwave (see electromagnetic waves) rotates the water molecule such that the electric dipoles align with the electric field.

Water molecules' movement (kinetic energy) creates heat, increasing the food's temperature.

Temperature is a measure of kinetic energy.

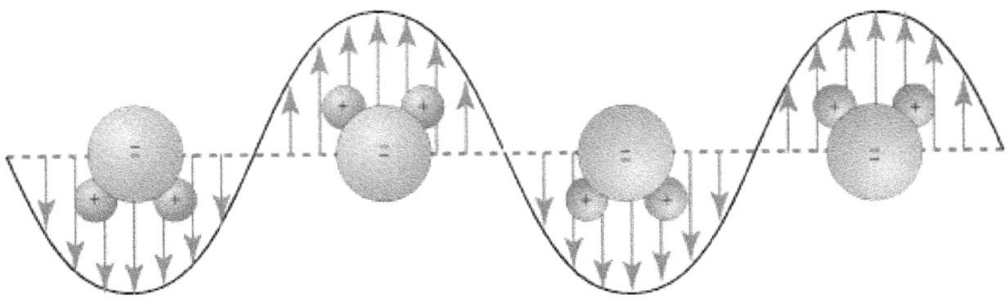

Water molecules have dipoles that align with the electric field

Electrostatic Induction

Electrostatic energy

Electrostatic energy is a charge more or less fixed in a single place.

However, this charge can be transferred to other objects in two ways.

Conduction is the transfer of energy where there is direct contact between the two objects.

Induction is the transfer of energy where there is *no direct contact* between the two objects, and the transfer of energy is across *open space* between two objects.

Induction

Electrostatic induction is where a charged object induces the movement (or redistribution) of charges of another object.

Electrostatic induction occurs when an object is placed in (or near) an electric field.

Metal objects can be charged by:

conduction (i.e., direct contact) or

induction (i.e., without contact).

In the diagram, metal rod B is initially neutral. When the neutral rod B touches a charged rod A, negatively charged electrons flow from the neutral rod B to positively charged rod A.

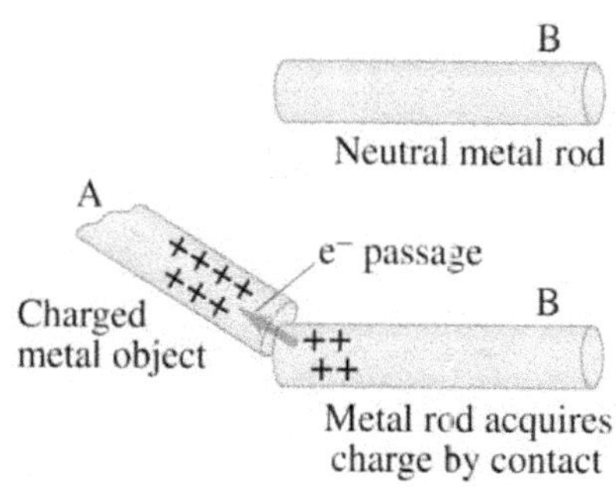

Induction in a neutral metal rod by a charged metal object

Electron transfer

Electrons from the neutral rod are transferred to the ground and leave the surface of their object.

In the middle diagram, the electric field attracts a positive charge when the negatively charged rod is brought near the neutral rod. It causes the negatively charged electrons to be repelled and travel into the ground.

In the bottom diagram, if the ground connection is cut (such that electrons cannot flow back into the system) and the negatively charged rod is removed, the neutral rod has a *positive charge* evenly distributed across its surface.

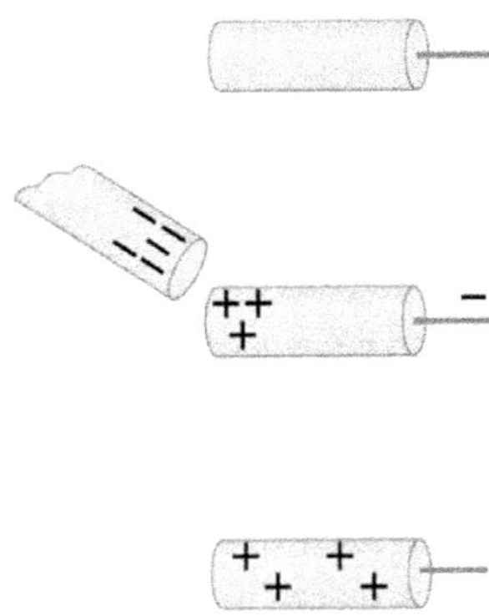

Two rods never touch, and the neutral rod is grounded

Insulators

Insulators are not charged by conduction (or induction) but experience a charge separation as *polarization.*

Unlike conductors, insulators do not have electrons that can move freely about the material.

When a strong charge is brought near the insulator (or nonconductor), the molecules become polarized and oriented such that charges on the molecule align with the electric field (like electric dipoles in an electric field), as shown.

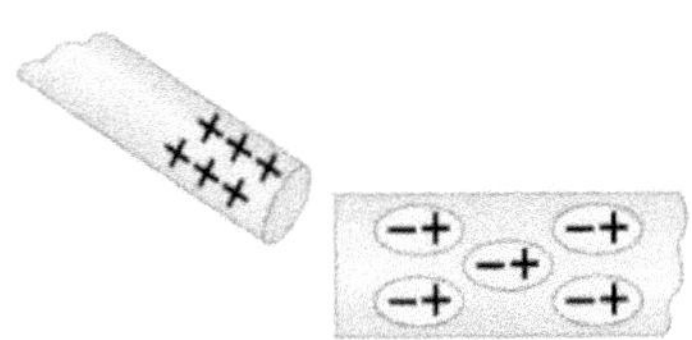

Insulators experience charge separation by polarization

Electroscopes

Electroscopes detect charge in an object. Below is a schematic of a gold-leaf electroscope. The gold arms in the middle hang together if there is no charge, and the arms spread apart when a charge is present.

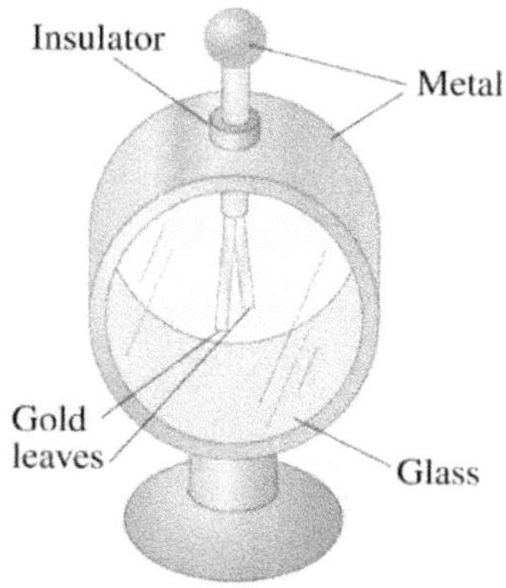

Electroscopes and the leaves determine the sign of an unknown charge

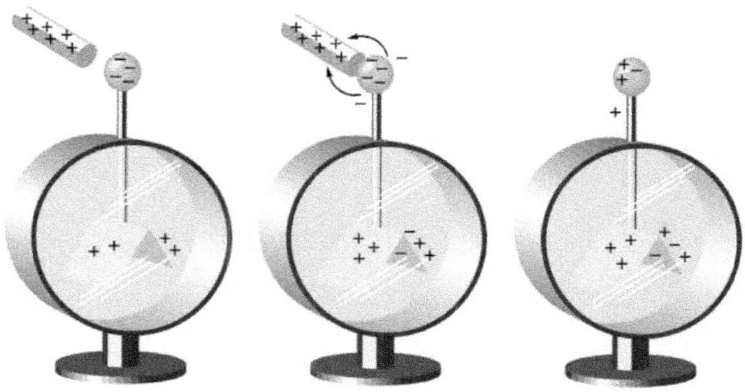

Electroscopes detect charge in an object

In the left diagram below, an object is held near (i.e., induction) the metal knob on top.

In the right diagram below, an object touches (i.e., conduction) the metal knob on top.

If electrons flow due to induction (no contact) or conduction (contact between objects), the gold leaves become similarly charged and repel.

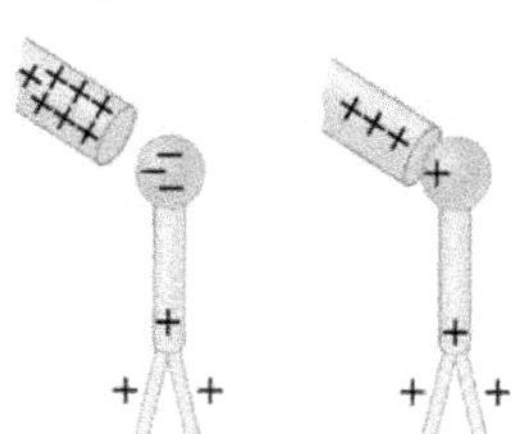

Similarly charged electroscope gold leaves repel each other

Determining sign of charge

Charged electroscopes determine the sign of an unknown charge.

In the middle diagram below, a negatively charged object is held near the knob, and the negatively charged leaves expand. Additional electrons try to escape the like-charged object, causing a stronger negative charge in both leaves.

In the right diagram below, a positively charged object is held near the knob, and the leaves move toward each other. Electrons move into the knob to be close to the positive rod. After electron migration, the leaves are less negatively charged, and there is less force driving the leaves apart.

Electroscope determines the sign of an unknown charge

Gauss's Law

Vector field

Gauss's Law relates the flow of a vector field through a surface to the behavior of the vector field inside the surface.

Vector field is an electric field and is expressed as:

$$\Phi_E = EA \cos(\theta)$$

where Φ_E is the electric flux (V·m), E is the electric field (N/C), A is the area that the field covers, and θ is the angle between the field and plane that runs perpendicular to the direction of field flow.

If the electric field and the surface are parallel, there is zero flux.

If the electric field and the surface are perpendicular, there is full flux.

For an enclosed surface, the electric flux Φ_E equals the charge inside the enclosure (q) over the permittivity of free space (ε_0).

$$\Phi_E = \frac{q}{\varepsilon_0}$$

Electric flux through enclosed space

The net electric flux through an enclosed surface depends on the charge inside. If there is no charge inside, then the net electric flux through the enclosure is zero.

Electric flux through an area is proportional to the number of field lines crossing the area.

In the diagram below, an electric flux passes through area A at an angle with respect to the electric field.

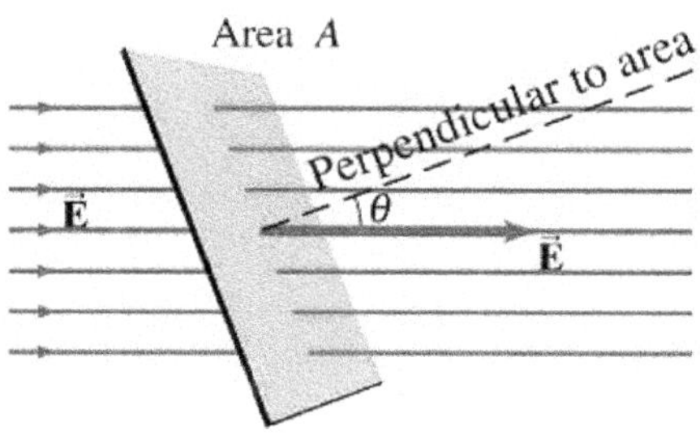

Electric flux passes through an area A at an angle θ to the electric field

Electric flux is proportional to the *electric field perpendicular* to that area.

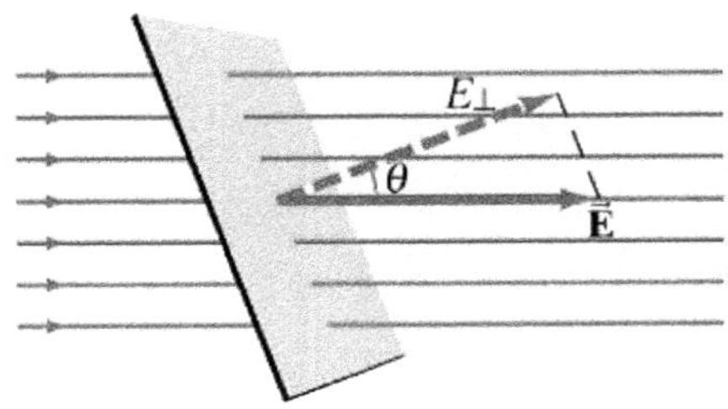

Electric flux passing through the projection of the area is *perpendicular to the electric field.*

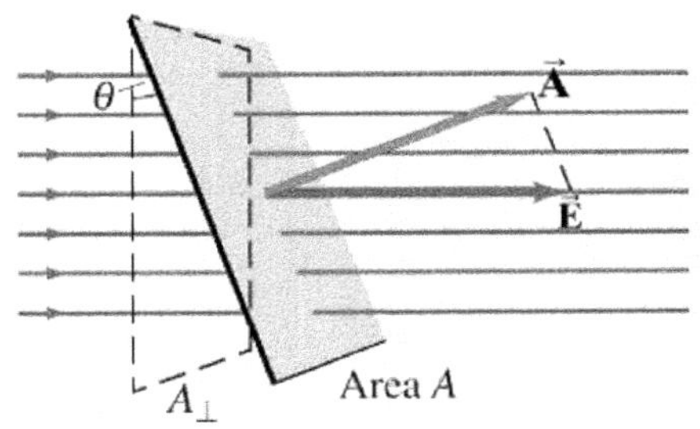

Electric flux through closed surface

Flux through a closed surface of multiple electrical fields is found by:

$$\Phi_E = E_1\Delta A_1 \cos(\theta_1) + E_2\Delta A_2 \cos(\theta_2) + \ldots = \sum E\Delta A \cos(\theta) = \sum E\perp\Delta A$$

where $\sum$ represents the closed surface.

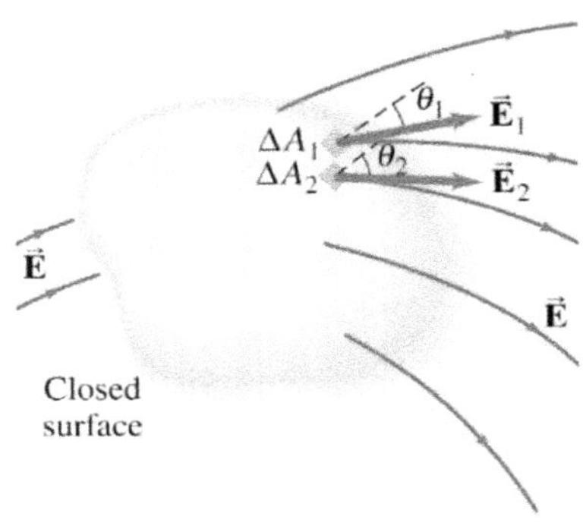

Flux through a closed surface

Net number of field lines through the surface is proportional to the charge enclosed and flux.

Gauss's Law (or *Gauss's Flux Theorem*) can be used to find the electric field when there is a high degree of symmetry in the system:

$$\sum E_{\perp}\Delta A = \frac{q_{encl}}{\epsilon_0}$$

Magnetic Fields and Poles

Magnetism

Magnetism is a natural phenomenon that has essential applications in the modern world. For example, navigation with a compass, memory storage on a computer, and high-speed trains all require magnetism and an understanding of the properties of magnets.

The earliest assumption about magnetism was that it was associated with naturally occurring magnetic materials such as iron, cobalt, or nickel.

There are many ways in which magnetism can be produced.

On the atomic level, the electrons of the material produce magnetism due to their motion (i.e., electrons spin and revolve).

Electric currents can create magnets, the basis of electromagnetism.

Maxwell's equations describe the fundamental laws of electricity and magnetism. They describe how electric and magnetic fields are generated and altered by each other, as well as charges and currents.

In 1864, Maxwell discovered that electric and magnetic fields travel through space as waves with the same speed as light. He proposed that light is a propagating electric and magnetic field wave. The theory describes the interaction between the electric field and the magnetic field.

Magnetic field lines

Magnetic objects emit magnetic fields, visualized by *magnetic field lines*.

Magnetic lines are like *electric field lines* and always start from the *north pole* (i.e., positive charge) and go towards the *south pole* (i.e., negative charge).

Unlike the electric field, magnetic fields always form *closed loops* because all magnets have a north pole and a south pole. The figure shows a bar magnet with a closed loop that magnetic fields created from the north pole to the south pole.

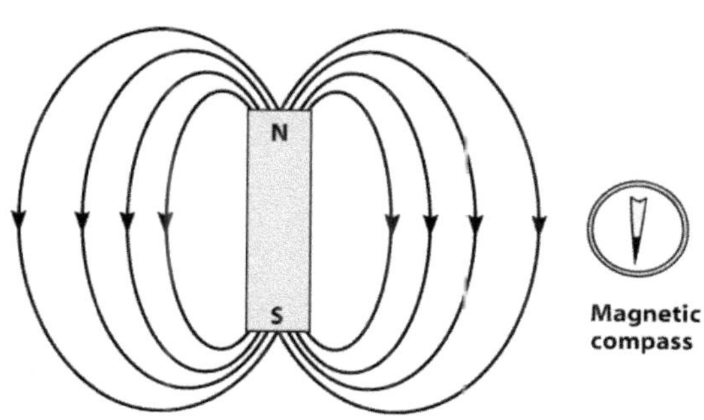

Bar magnet with closed-loop magnetic fields

Earth's magnetic field

Earth's magnetic field originates deep beneath the surface layer and is caused by currents of conductive elements in the Earth's molten core.

Magnetic field lines originate at the geographic south pole of Antarctica (magnetic north pole) and terminate at the geographic north pole (magnetic south pole).

A compass operates by aligning itself with Earth's magnetic field (magnetic fields always form complete loops), allowing navigation with respect to a known axis.

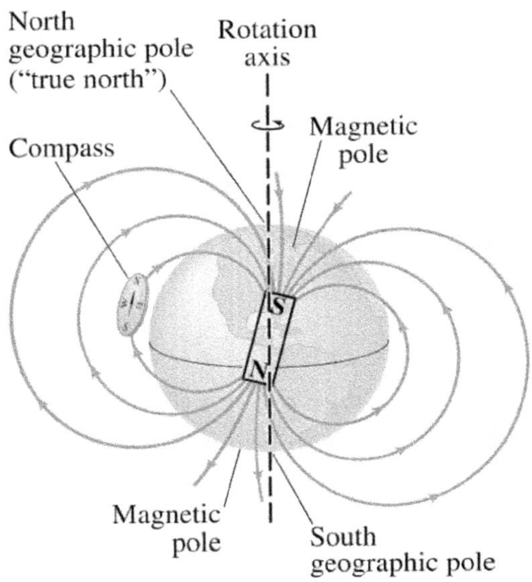

Magnetic field lines originate at the geographic south pole of Antarctica

North and south poles

Magnets never have only one pole; all magnets must have a north and south pole.

If a magnet is broken into two halves, it will not produce two single polarity magnets.

Instead, the two fragments each have a north and south pole. If these pieces are broken again, the process repeats.

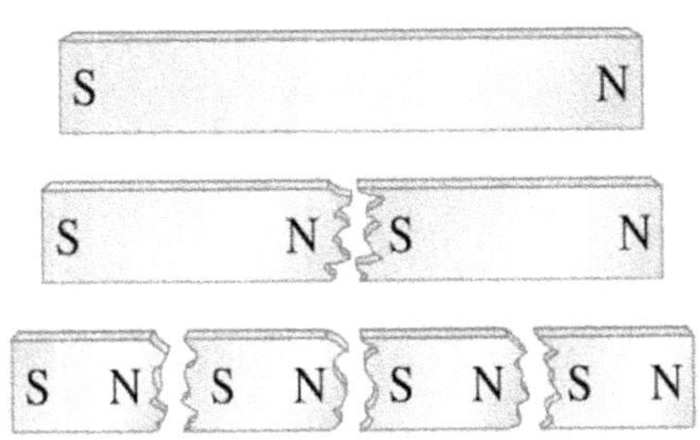

Magnets have a north and south pole during divisions

Magnets have other similar behaviors to electrically charged particles.

Poles of two different magnets exert a force on each according to their polarity.

Opposite poles attract, while like poles repel.

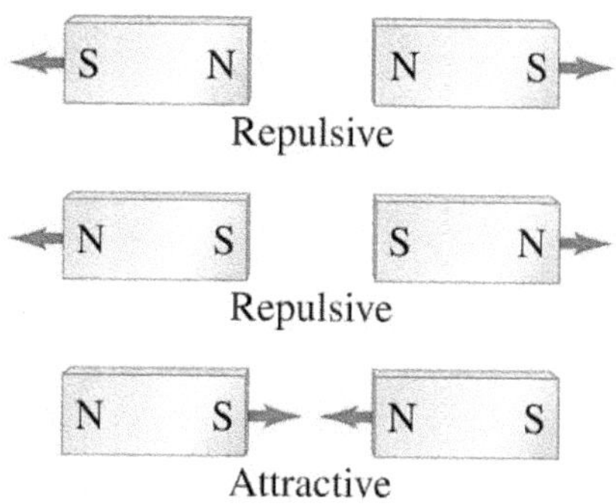

Magnet polarity determines attractive (opposite) or repulsive (similar) forces

Magnetic fields around a current-carrying wire

Charge can travel across a conductor if an electric potential difference exists.

When this occurs, the traveling charge is an *electric current* and is the rate of flow of charge through a conductor over time:

$$I = \frac{\Delta q}{\Delta t}$$

where I is the electric current given in amperes (A), and Δt is time (s).

Current is considered to be the movement of positive charge through a conductor. Although this is technically incorrect (protons are immobile, only electrons can "flow"), diagrams and calculations of current follow this convention.

Movement of electrons is opposite to that shown in the illustrations.

Right-hand rule

When current flows through a wire, a magnetic field is formed that circulates the wire.

The direction of the magnetic field can be found via the *right-hand rule.*

Observe the diagram below. According to the right-hand rule, the thumb goes in the direction of a positive charge (i.e., the direction of current by convention). The direction the fingers curl around in the direction of the magnetic field.

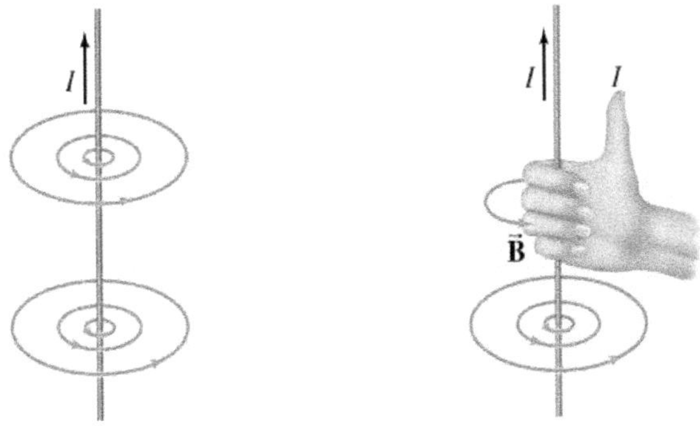

Right-hand rule determines the direction of the magnetic field

Right-hand rule (RHR) has the thumb of the right hand in the direction of a positive charge, the middle finger is the direction of the magnetic field, and the palm faces the direction of the force.

From the right-hand rule, if the charge is negative, then the direction of the force is in the opposite direction.

Physical Situation	Example	How to Orient Right Hand	Result
1. Magnetic field produced by current (RHR-1)	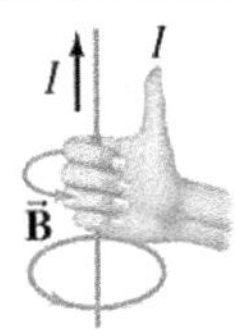	Wrap fingers around wire with thumb pointing in direction of current I	Fingers curl in direction of $\vec{\mathbf{B}}$
2. Force on electric current I due to magnetic field (RHR-2)	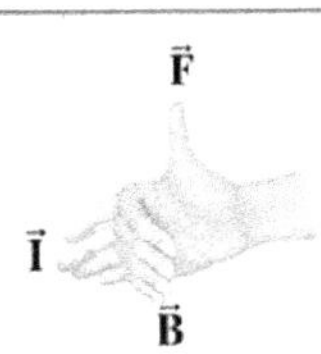	Fingers first point straight along current I, then bend along magnetic field $\vec{\mathbf{B}}$	Thumb points in direction of the force $\vec{\mathbf{F}}$
3. Force on electric charge $+q$ due to magnetic field (RHR-3)	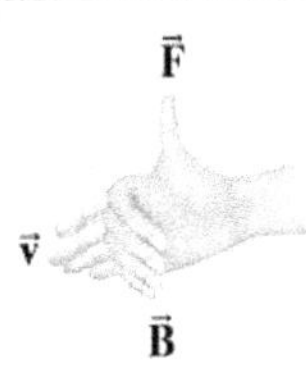	Fingers point along particle's velocity $\vec{\mathbf{v}}$, then along $\vec{\mathbf{B}}$	Thumb points in direction of the force $\vec{\mathbf{F}}$

Current induces magnetic field

To understand the flow of electrons through the conductor, see the figure below.

Notice that the direction of the magnetic field is clockwise because current flows down the wire from top to bottom. However, electrons flow opposite to the conventional current flow. Thus, electrons flow from bottom to top.

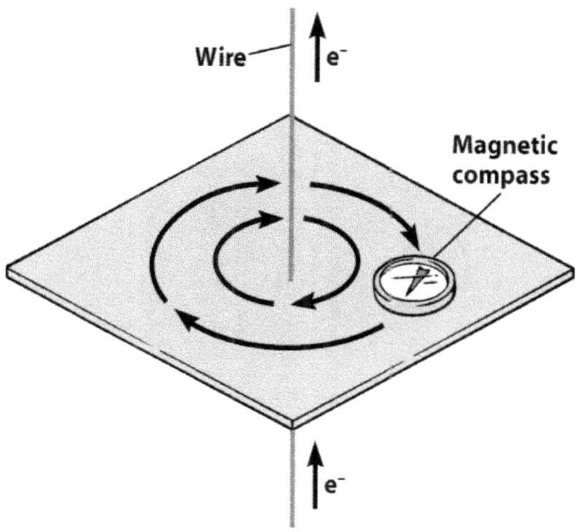

Flowing electrons induce a magnetic current

When a current (I) induces a magnetic field in a straight wire, the magnetic field (B) strength is calculated by *Ampère's Law*:

$$B = \frac{\mu_0 I}{2\pi r}$$

where μ_0 is the constant of permittivity of free space ($4\pi \times 10^{-7}$ T·m/A), I is the current (A), and r is the distance away from the wire (m).

Magnetic fields in solenoids

When a current-carrying wire is wrapped around, such that it forms a loop, it is a *solenoid.*

Solenoids have many applications and are the basis for common electromagnets. The primary importance of a solenoid is that the magnetic field can be manipulated.

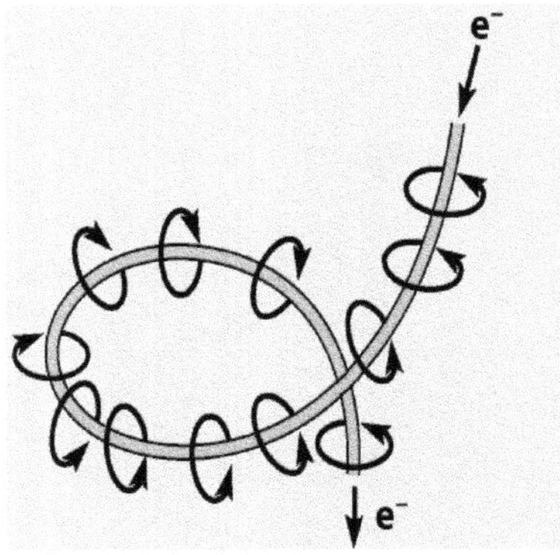

Magnetic field from current flowing through the solenoid

Net magnetic field created by a loop

In the configuration below, the net magnetic field created by the loop is:

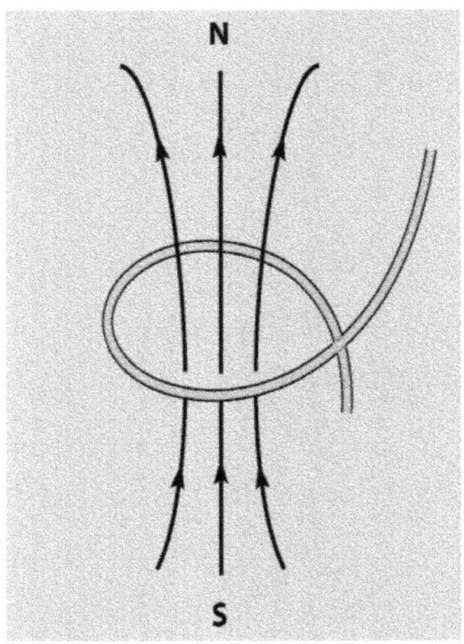

Magnetic field increases proportionally with the number of loops

If the number of loops in the solenoid increases, the magnetic field increases proportionally with the number of loops.

Magnetic field dramatically increases if a piece of iron is inserted into the solenoid.

Magnetic strength

From the equation for the magnetic field strength of a wire, the magnetic field strength through a solenoid (but not outside of it) can be expressed as:

$$B = \frac{\mu_0 \mu_r n I}{l}$$

where n is the number of loops, μ_r is the relative permittivity of core material ($\mu_r = 1$ in air), and l is the length of the solenoid (m).

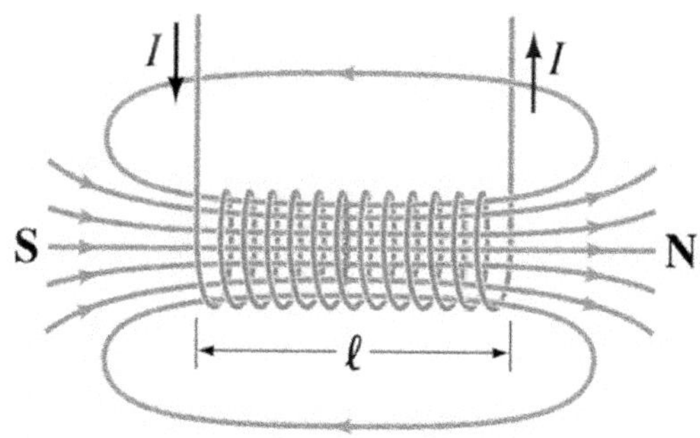

Magnetic field strength through a solenoid is proportional to the number of loops

Notes for active learning

Magnetic Field Force

Particles in a magnetic field

When a charged particle passes through a magnetic field with a specific velocity (which must be perpendicular to the magnetic field), it experiences a force from the magnetic field. This only occurs if the charge has a *velocity perpendicular* to the magnetic field.

If its speed is zero or its direction is parallel to the field, then no force will be exerted:

$$F = qvB \sin(\theta)$$

where q is the electric charge (C), v is the velocity of the charge (m/s), B is the strength of the magnetic field given in the Tesla (T), and θ is the angle between the charge velocity and the magnetic field (degrees). $Sin(\theta)$ may be omitted, as θ is assumed to be 90°.

Magnetic force is always perpendicular to both the magnetic field and the velocity of the charge:

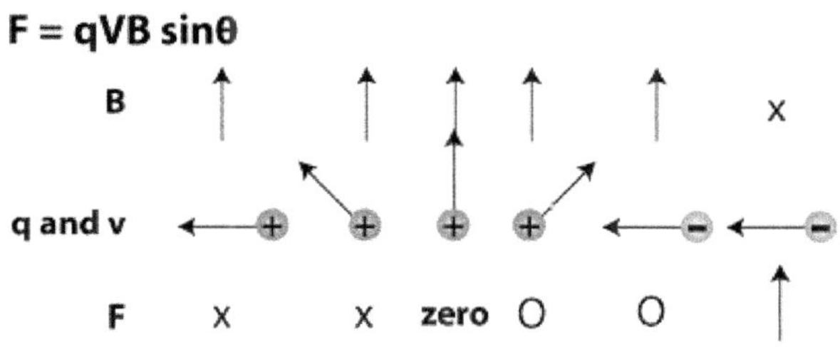

X represents into the page, and O represents out of the page

Motion of charged particles in magnetic fields

When a charged particle travels through a magnetic field, it is deflected according to the magnitude and direction of the field.

The deflection of the particle has important applications (e.g., mass spectrometers) because it can be related to the centripetal force to solve for the mass of the particle.

Equations for centripetal force:

$$F_C = \frac{mv^2}{r}$$

Equation for magnetic field force on a charged particle:

$$F_M = qvB \sin(\theta)$$

If the particle is assumed to be traveling *perpendicular* to the magnetic field, then the magnetic field force becomes:

$$F_M = qvB$$

If a particle with some velocity *perpendicular* to a magnetic field travels in a *circular orbit*, the mass can be solved for by:

$$\frac{mv^2}{r} = qvB$$

$$m = \frac{qBr}{v}$$

Charged particles in a magnetic field

Charged particles as curved line trajectories based on the magnetic field.

When $qvB = mv^2 / r$, there is sufficient centripetal force, and the charged particle orbits.

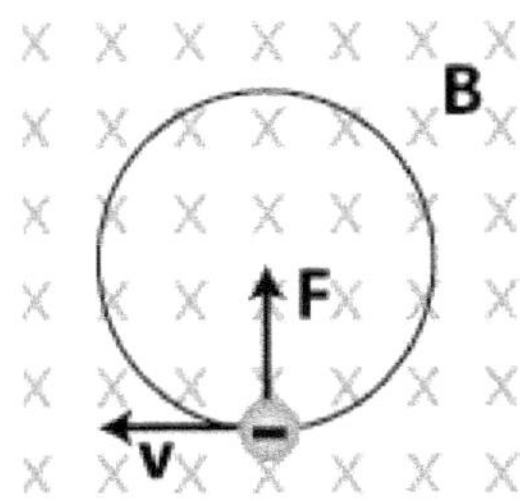

Charged particles as curved line trajectories with $qvB = mv^2/r$

When $qvB > mv^2 / r$, there is too much centripetal force, and the charged particle spirals inward.

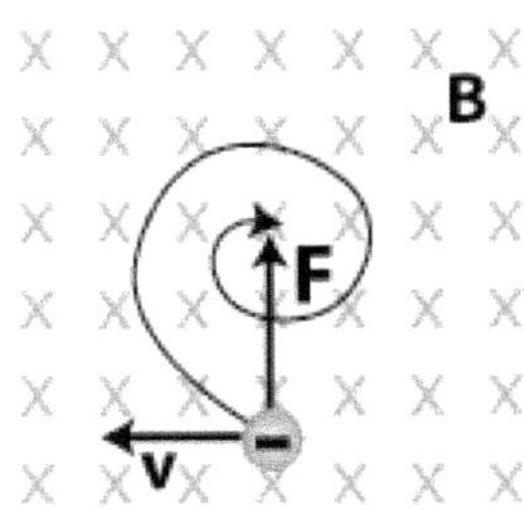

Charged particles as curved line trajectories with $qvB > mv^2/r$

When $qvB < mv^2 / r$, there is not enough centripetal force, and the charged particle escapes orbit.

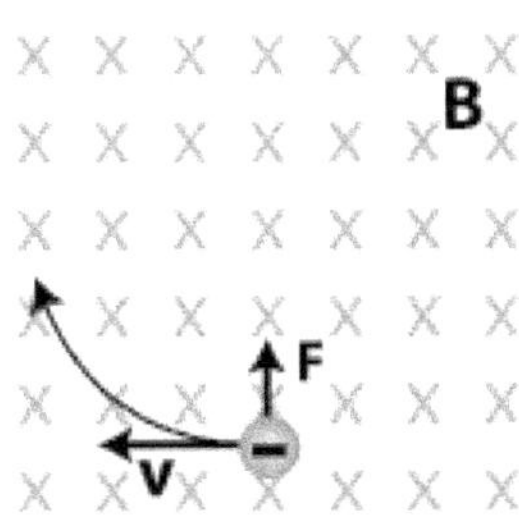

Charged particles as curved line trajectories with $qvB < mv^2/r$

Current-carrying wires

Wires carrying current have charged particles in motion through them.

Magnetic field exerts a calculable force on a current-carrying wire:

$$F = I\ell B \sin(\theta)$$

where I is current through the wire (A), and ℓ is wire length exposed to magnetic field (m).

Again, the right-hand rule (RHR) gives the direction of the force.

In the diagram on the left, current flow has the magnetic field force exerting a *downward force*.

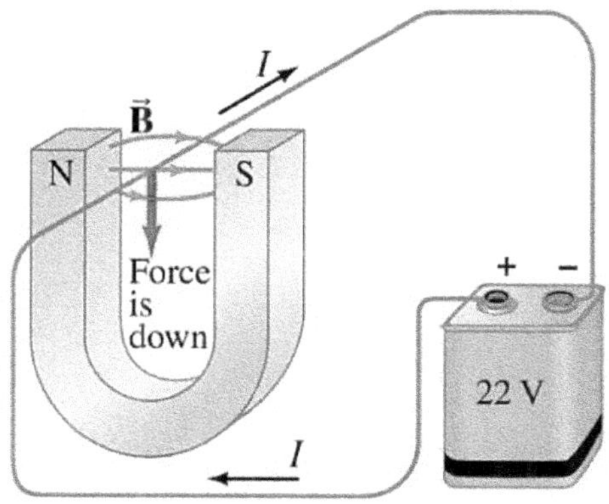

Current-carrying wire exposed to a magnetic field exerts a downward force

In the diagram on the right, the current is reversed, and the magnetic field force is opposite its original direction and *points upwards*.

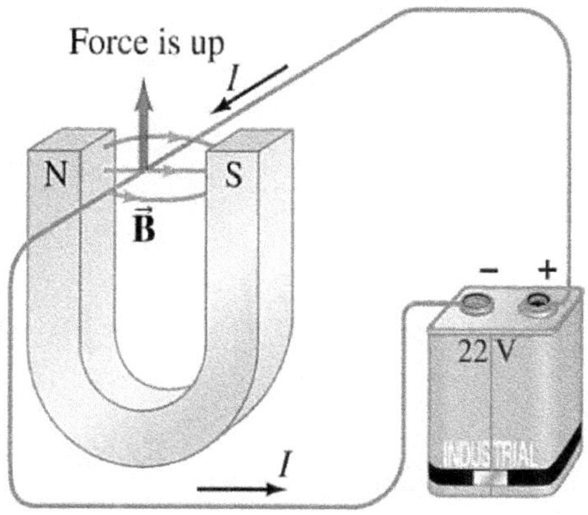

Current-carrying wire exposed to a magnetic field exerts an upward force

Parallel current-carrying wires

When two current-carrying wires, shown below, are parallel and close enough so that their magnetic fields interact, they exert a force on each other according to the magnitude and direction of their respective currents.

Magnitude of the magnetic field force per unit length of the parallel wire section is given by:

$$\frac{F}{l} = \frac{\mu_0}{2\pi}\frac{I_1 I_2}{d}$$

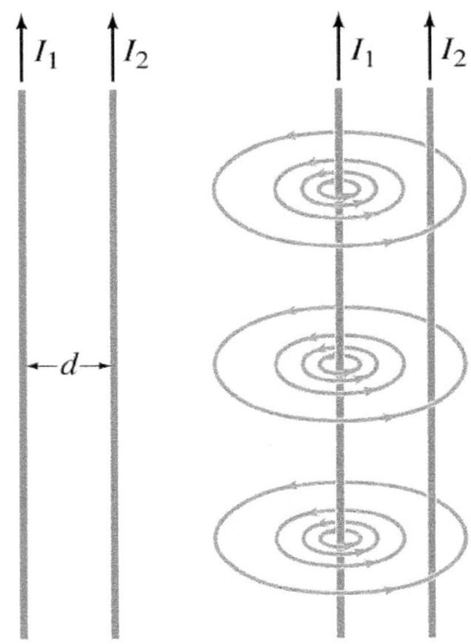

Two nearby parallel current-carrying wires with interacting magnetic fields exert a force

Direction of the force between the wires depends upon the direction of the currents in the wires with respect to the other.

Parallel currents (i.e., moving in the same direction) create an *attractive force.*

Antiparallel currents (i.e., moving in opposite directions) create a *repulsive force.*

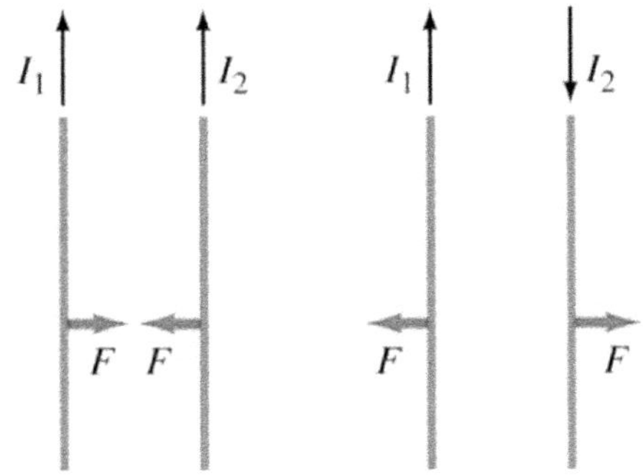

Force F exerted by current I flow in parallel (left) or antiparallel (right)

Faraday's Law

Law of induction

Michael Faraday (1791-1867) expressed the relationship that links the induced electromotive force (EMF) in a wire loop proportional to the magnetic flux change rate through the loop.

Faraday derived the *Law of induction.*

An understanding of magnetic flux is required to discuss the law of induction.

Magnetic flux

Magnetic flux (Φ_B) is similar to electric flux and is the amount of magnetic field passing through an area of a surface.

Weber (Wb) is the SI unit of magnetic flux, equal to volt seconds.

The Weber may also be given in tesla-meters-squared ($T \cdot m^2$).

Magnetic flux (Φ_B) equation is:

$$\Phi_B = B_{\perp}A = BA \cos(\theta)$$

where $B_{\perp}$ is the magnitude of the magnetic field perpendicular to the area (A).

If the field lines are not perpendicular to the area, the right-hand equation must be used where θ is the angle between the field lines and the *vector normal to the surface* (i.e., *area vector*).

Magnetic flux indicates the magnitude of the magnetic field that pierces through the coil.

Rotating coiled wire

The diagram below demonstrates the varying amounts of magnetic flux through the coiled wire as it is rotated with respect to the magnetic field.

Left panel has magnetic flux equal to zero because the coil *area vector* (i.e., the *coil area vector* points normal to the coil's cross-sectional area) is perpendicular to the applied field, $\theta = 90°$.

Middle panel has some magnetic flux as *area vector* of the coil is at $\theta = 45°$ angle to the field.

Right panel has maximum flux because the *area vector* is parallel to the magnetic field, $\theta = 0°$.

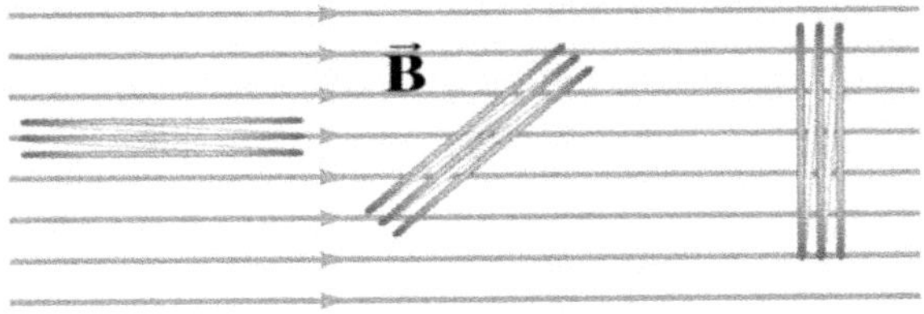

Magnetic flux in the left panel equals zero, the middle panel has some flux, and the right panel has maximum flux with the area vector parallel to the electric field.

In the left panel above, $\theta = 90°$, $\Phi_B = 0$

In the center panel $\theta = 45°$, $\Phi_B = BA\ cos\ 45°$

In the right panel $\theta = 0°$, $\Phi_B = BA$

Faraday's Law can be applied to the voltage induced in a coiled wire under certain situations using an equation for magnetic flux.

Faraday's Law states that the induced voltage in a coil is proportional to the number of loops it contains, multiplied by the rate at which the magnetic flux changes within those loops.

Magnitude of the current is related to resistance of the coil, its circuit, and the voltage induced.

The more coils a wire has, the greater the induced voltage. However, the higher the voltage, the greater the current that pushes back against the magnet.

Rapid motion of the magnet also induces a higher voltage.

Lenz's Law

This relationship is explained in the equation of Faraday's Law of Induction (i.e., Lenz's Law, named after the scientist who added the negative sign in 1834):

$$EMF = -N\frac{\Delta\Phi_B}{\Delta t}$$

Negative sign in Lenz's Law indicates that the induced electromagnetic field (EMF) is oriented, so the current it creates makes a magnetic current *opposing* the change in flux from the original ("external") field.

Diagram below show a loop of wire exposed to a magnetic field. On the right, the loop is stretched such that the area exposed to the flux changes

From Faraday's Law, an EMF is induced and current results through the loop.

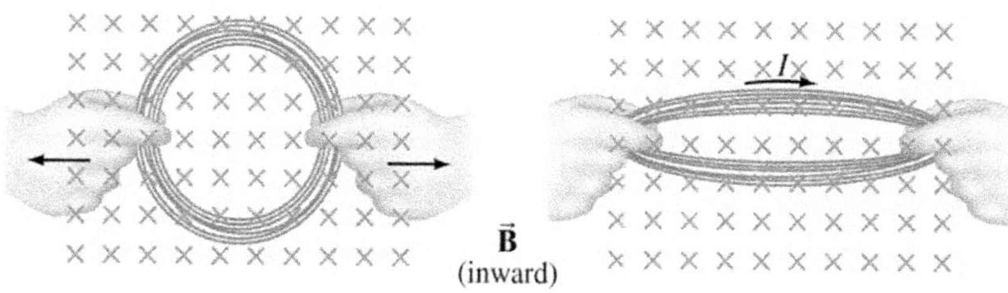

On the right, the flux through the coil decreases because the area decreased

Moving conductor

A moving conductor can form an electromagnetic field (EMF). Whenever a conductor is moved in a uniform magnetic field, emf is induced across the conductor because of the change in the area of the conductor that leaves and enters the magnetic field.

The rectangle represents a conductor with a movable slide. If the slide is moved outward, then the magnetic flux changes.

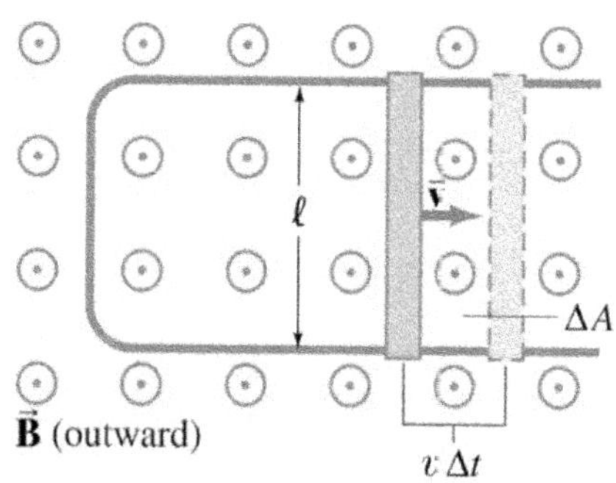

Moving the slider changes the magnetic flux

Induced current from the sliding bar causes a current to form, as seen in the following diagram.

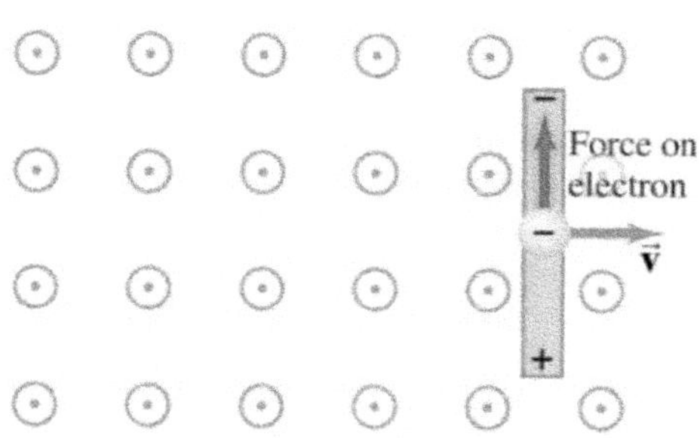

Induced current from the sliding bar causes a current to form

Induced EMF in such a situation has a magnitude of the form:

$$EMF = \frac{\Delta\Phi_B}{\Delta t} = \frac{B\Delta A}{\Delta t} = \frac{B\ell v\Delta t}{\Delta t} = B\ell v$$

Solving problems on Faraday's Law

Solving problems with Faraday's Law:

1. Determine if magnetic flux is increasing, decreasing, or unchanged.

   Due to the induced current points, the magnetic field is opposite to the original field if the flux increases.

   Magnetic field points in the same direction if it is decreasing.

   Magnetic field is zero if the flux is not changing (i.e., meaning of the negative sign in Lenz's Law).

2 Right-hand rule (RHR) determines the direction of the resulting current.

   Point the thumb in the direction of a vector normal to the area vector on the side that the flux exits.

   Direction curled by the fingers as they close toward the palm in the direction of the current.

3. The external field and the field due to the induced current are different.

Torque on Current-Carrying Wire

Galvanometer

When a wire carrying a current forms a loop, the current reverses direction (x- and y-axes) as it travels in and out of the loop.

When the loop is exposed to a magnetic field, a torque is generated about the axis of the loop due to the opposing forces on each side of the loop.

This concept has many applications and is used in galvanometers, motors, and loudspeakers.

Galvanometers use torque on a current loop to measure current.

Galvanometers detect a small electric current and measures its magnitude.

A magnetic needle's movement indicates current and its intensity or that of a coil in a magnetic field, which is an integral part of a galvanometer.

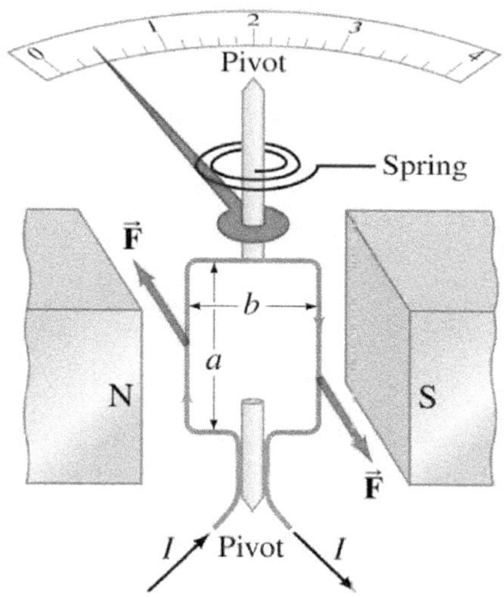

Galvanometer to measure current

Torque

In physics and mechanics, torque (or *moment of force*) is the rotational analog of linear force.

Torque is the rate of change of angular momentum imparted to an isolated body.

If the current increases in the loop, the force from the magnetic field increases proportionally. The dial twists more because of increased torque.

Magnitude of the produced torque is calculated by:

$$\tau = nIAB \sin(\theta)$$

where the quantity nIA is the magnetic dipole moment, n is the number of loops (this equation also applies to a solenoid), I is the current (A), A is the area of the enclosed space (m^2), and θ is the angle of the loop area with respect to the magnetic field (degrees).

Properties of Electromagnetic Radiation

Electromagnetic radiation

Electromagnetic radiation is a form of energy that propagates through space as an oscillating electric and magnetic field. It is considered a wave; however, in quantum mechanics, electromagnetic radiation is considered packets of energy known as *photons*.

Wave nature of electromagnetic radiation is illustrated below.

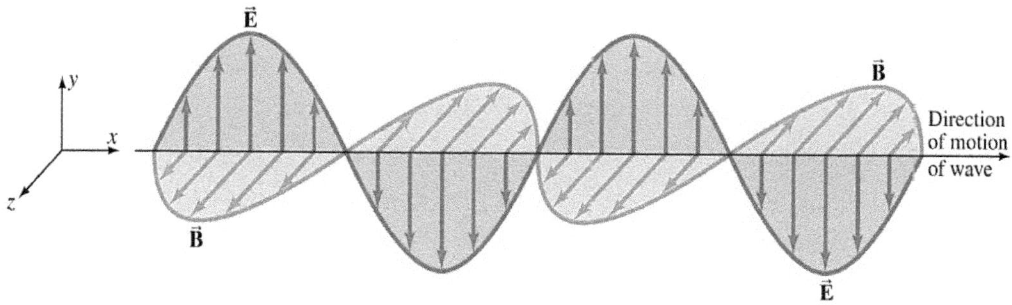

Electromagnetic radiation propagates as an oscillating electric and magnetic field

Velocity equals *c* in vacuum

Electromagnetic radiation does not need a medium to propagate, unlike sound and mechanical waves.

In a vacuum, electromagnetic radiation has a constant speed of:

$$c = 2.99792458 \times 10^8 \text{ m/s} \approx 3.0 \times 10^8 \text{ m/s}$$

c is the standard value for the *speed of light*.

Nothing in the known universe travels faster than light in a vacuum.

Medium's refractive index

Electromagnetic radiation travels slower than this speed in other mediums, according to the value of the mediums' refractive index.

Speed of light is calculated by:

$$v = \frac{c}{n}$$

where v is the speed of light in the medium, and n is the refractive index of the medium.

Electromagnetic waves

Electromagnetic waves are produced by *accelerating charges*.

Acceleration of charges produces oscillations of electric and magnetic fields (electromagnetic radiation), which propagate through space indefinitely until absorbed.

Like all waves, electromagnetic waves can be related by speed, frequency, and wavelength:

$$c = \lambda f$$

where c is the speed of light, λ is wavelength (m), and f is the frequency of the electromagnetic radiation (Hz).

When considering electromagnetic radiation from the quantum mechanical viewpoint, the photons have discrete, quantifiable energies related to their frequency by:

$$E = hf = \frac{hc}{\lambda}$$

where E is the energy per photon (J or eV), h is Planck's constant (6.626×10^{-34} J·s or 4.135×10^{-15} eV), c is the speed of light, λ is the wavelength (m), and f is the frequency of the electromagnetic radiation (Hz).

Electromagnetic momentum

Electromagnetic radiation carries momentum.

Although the wave has no mass, it exerts a force as *radiation pressure*.

When radiation is fully absorbed, the radiation pressure is at a *minimum*:

$$P_{minimum} = \frac{\bar{I}}{c}$$

When radiation is fully reflected, the radiation pressure is at a *maximum*:

$$P_{maximum} = \frac{2\bar{I}}{c}$$

where P is radiation pressure (N/m^2), and I is intensity of the electromagnetic radiation (W/m^2).

Electromagnetic Spectrum and Photon Energy

Electromagnetic radiation

Electromagnetic radiation varies significantly in properties and energies, depending on the frequency of the radiation.

The spectrum of electromagnetic radiation by wavelength and frequency is below.

As *frequency increases*, the *wavelength decreases* (inversely proportional).

High-frequency radiation

The higher frequency forms of electromagnetic radiation (e.g., ultraviolet, X-rays, gamma rays) are known carcinogens (i.e., cancer-causing agents).

High-frequency radiation has high energy, which can cause damage to cells and DNA.

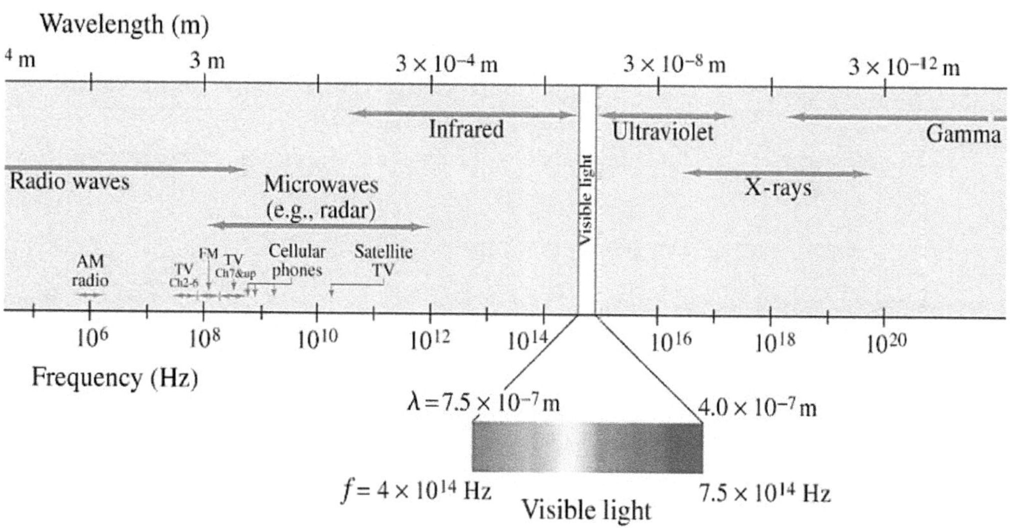

Electromagnetic radiation with associated wavelength and frequency

Frequency, wavelength, and energy relationship

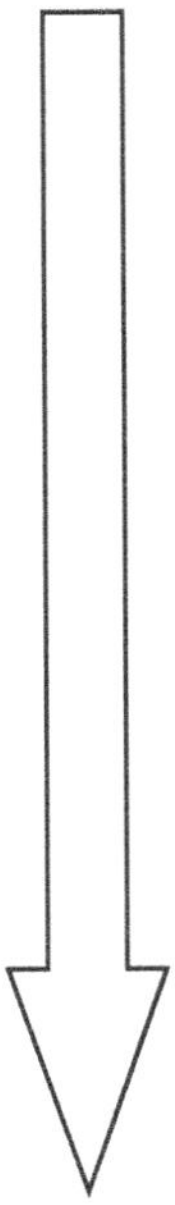

Lower frequency = longer wavelength = less energy	
Radio	Causes electronic oscillations in the antenna.
Microwave	Causes molecular rotation.
Infrared	Causes molecular vibration.
Visible	Excites electrons to higher energy orbitals. Visible light ranges from 400-700 nm. 400 is violet, and 700 is red.
Ultraviolet	Breaks bonds and excites electrons to eject them. UV is considered ionizing radiation.
X-rays	Ionizing radiation, the photoelectric effect.
Gamma rays	Dangerous; more energetic than X-rays.
Higher frequency = shorter wavelength = more energy	

Lower frequency = longer wavelength = *less energy*

Higher frequency = shorter wavelength = *more energy*

Chapter Summary

Charge

Electric charges are *positive* or *negative.*

Charge is *always conserved.*

Charge on an electron: $e = 1.602 \times 10^{-19}$ C

Charge is quantized units of e (i.e., how many times greater than the charge on an electron).

Conductors are materials in which *electrons* are free to move.

Insulators are nonconductors and *do not allow electrons* to move freely.

Objects can be charged by *conduction* or *induction.*

Electrostatic force

Coulomb's Law gives the magnitude of the *electrostatic force*:

$$F = k \times \frac{Q_1 Q_2}{r^2}$$

Electric fields

Electric field is a *force per unit charge*:

$$\vec{E} = \frac{\vec{F}}{q}$$

Electric field is given by a *single-point charge*

$$E = \frac{F}{q} = \frac{kqQ/_{r^2}}{q} = k \times \frac{Q}{r^2}$$

Electric field lines can represent electric fields.

Static electric field inside a conductor equals *zero*; the surface field is *perpendicular* to the surface.

Flux

Electric flux (flow of a field through a closed surface):

$$\Phi_E = E_\perp A = EA_\perp = EA \cos(\theta)$$

Gauss's Law (electric flux through a closed surface):

$$\sum_{\substack{closed \\ surface}} E_\perp \Delta A = \frac{Q_{encl}}{\epsilon_0}$$

Electromagnetic waves

Maxwell's equations are fundamental principles of *electromagnetism*.

Moving or spinning electric charge create a magnetic field.

Electromagnetic waves are produced by *accelerating charges*.

Propagation speed of electromagnetic waves:

$$c = \frac{1}{\sqrt{\epsilon_0 \mu_0}}$$

Fields are *perpendicular* to the direction of propagation.

Wavelength and frequency of EM waves are related: $c = \lambda f$

Electromagnetic spectrum includes all wavelengths (e.g., radio, gamma rays).

Magnets

Magnets have *north* and *south poles*.

Like poles *repel*, unlike *attract*.

Electric currents produce *magnetic fields*.

Magnetic fields

Magnetic fields are in units of Tesla (T).

Parallel currents attract, and *antiparallel currents repel*.

Magnetic field exerts a force on *electric current*:

$$F = I\ell B \sin(\theta)$$

Magnetic field exerts a force on a *moving charge*:

$$F = qvB \sin(\theta)$$

Magnitude of the field of a long, straight current-carrying wire:

$$B = \frac{\mu_0}{2\pi} \times \frac{I}{r}$$

Ampère's Law (magnitude of a field):

$$\sum B_{\|} \Delta\ell = \mu_0 I_{encl}$$

Magnetic field inside a *solenoid*:

$$B = \frac{\mu_0 N I}{\ell}$$

Torque

Torque on a current loop:

$$\tau = NIAB \sin(\theta)$$

Practice Problems: Electrostatics & Electromagnetism

1. In the figure, $Q = 5.1$ nC. What is the magnitude of the electrical force on the charge Q? (Use Coulomb's constant $k = 9 \times 10^9$ N·m²/C²)

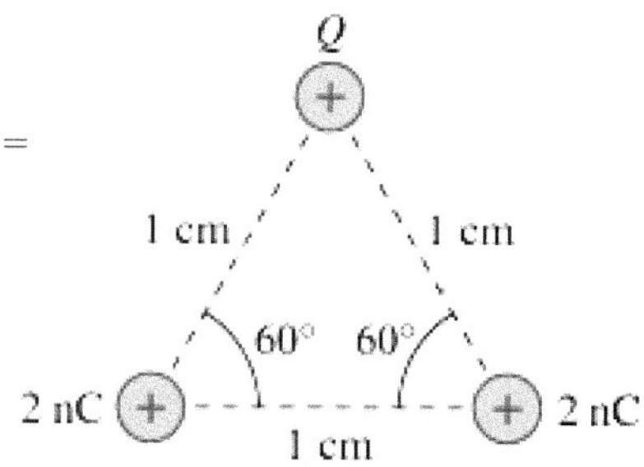

A. 4.2×10^{-3} N
B. 0.4×10^{-3} N
C. 1.6×10^{-3} N
D. 3.2×10^{-3} N
E. 7.1×10^{-3} N

2. Which one of the statements is correct?

A. The north pole of a magnet points towards Earth's geographic North Pole
B. The north pole of a magnet points towards Earth's geographic South Pole
C. Earth's geographic North Pole is the north pole of Earth's magnetic field
D. Earth's geographic South Pole is the south pole of Earth's magnetic field
E. None of the above

3. Two uncharged metal spheres, A and B, are mounted on insulating support rods. A third metal sphere, C, carrying a positive charge, is placed near B. A copper wire is momentarily connected between A and B and removed. Finally, sphere C is removed. In this final state:

A. spheres A and B both carry equal positive charges
B. sphere A carries a negative charge and B carries a positive charge
C. sphere A carries a positive charge and B carries a negative charge
D. spheres A and B both carry positive charges, but B's charge is greater
E. sphere A remains uncharged, and B carries a positive charge

4. Two charges separated by 1 m exert a 1 N force on each other. When pulled to a separation distance of 3 m, what is the force on each charge?

A. 3 N
B. 0 N
C. 9 N
D. 0.33 N
E. 0.11 N

5. A balloon, after being rubbed on a wool rug, can stick to a wall. This illustrates that the balloon has:

I. magnetism   II. net charge   III. capacitance

A. I only
B. II only
C. III only
D. I, II and III
E. I and III only

6. The diagram shows two unequal charges $+q$ and $-Q$, of opposite sign. Charge Q has a greater magnitude than charge q. Point X is midway between the charges.

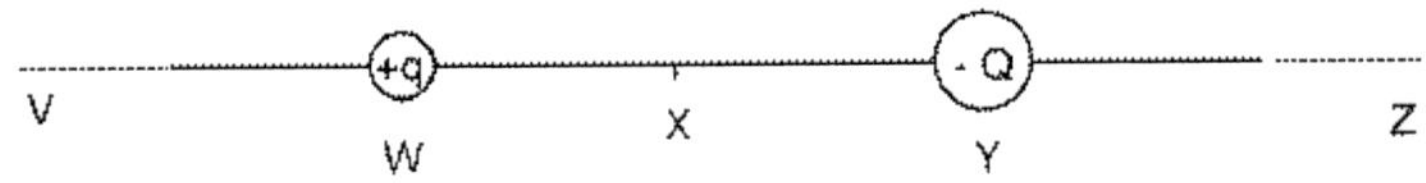

In what section of the line is the point where the resultant electric field could equal zero?

A. VW
B. WX
C. XY
D. YZ
E. VX

7. One coulomb of charge passes through a 6 V battery. Which is the correct value for increasing some battery property?

A. 6 watts
B. 6 ohms
C. 6 amps
D. 6 J
E. 6 N

8. What travels through a conductor near the speed of light when a current is established?

A. Protons
B. Photons
C. An electric field
D. Electrons
E. Neutrons

9. Which statement is accurate for a proton that moves in a direction perpendicular to the electric field lines?

A. it is moving from high potential to low potential and gaining electric potential energy
B. it is moving from high potential to low potential and losing electric potential energy
C. it is moving from low potential to high potential and gaining electric potential energy
D. it is moving from low potential to high potential and losing electric potential energy
E. both its electric potential and electric potential energy remain constant

10. Light having a frequency in a vacuum of 3×10^{14} Hz enters a liquid with a refractive index of 2. What is the frequency of the light in this liquid?

A. 6×10^{14} Hz
B. 1.5×10^{14} Hz
C. 12×10^{14} Hz
D. 3×10^{14} Hz
E. 9×10^{14} Hz

11. With all other factors remaining constant, how does doubling the number of loops of wire in a coil affect the induced emf?

A. The induced emf increases by a factor of $\sqrt{2}$
B. The induced emf doubles
C. There is no change in the induced emf
D. The induced emf quadruples
E. The induced emf increases by a factor of 3

12. A proton travels to the right and encounters a region S, which contains an electric field, magnetic field, or both. The proton is observed to bend up the page. Which of the statements regarding region S is correct?

I. There is a magnetic field pointing into the page
II. There is a magnetic field pointing out of the page
III. There is an electric field pointing up the page
IV. There is an electric field pointing down the page.

A. I only
B. II only
C. I and III
D. II and IV
E. II and III

13. An object with a 6 μC charge accelerates at 0.006 m/s^2 due to an electric field. If the object has a mass of 2 μg, what is the magnitude of the electric field?

A. 0.002 N/C
B. –0.005 N/C
C. 2 N/C
D. –2 N/C
E. –0.07 N/C

14. Two Gaussian surfaces, A and B, enclose the same positive charge $+Q$. The Gaussian surface A has an area two times greater than surface B. Compared to the flux of the electric field through Gaussian surface B, the flux of the electric field through surface A is:

A. two times smaller
B. equal
C. two times larger
D. four times larger
E. four times smaller

15. Which of these electromagnetic waves has the shortest wavelength?

A. γ rays
B. Visible light
C. Radio waves
D. Infrared
E. Microwaves

Solutions: Electrostatics & Electromagnetism

1. C is correct.

$F_e = kQ_1Q_2 / r^2$

$F_e = [(9 \times 10^9 \text{ N·m}^2/\text{C}^2)·(5.1 \times 10^{-9} \text{ C})(2 \times 10^{-9} \text{ C})] / (0.1 \text{ m})^2$

$F_e = 9.18 \times 10^{-4} \text{ N}$

$F_e \sin (60°)$ represents the force from one of the positive 2 nC charges.

Double to find the total force:

$F_{total} = 2F_e \sin (60°)$

$F_{total} = 2(9.18 \times 10^{-4} \text{ N}) \sin (60°)$

$F_{total} = 1.6 \times 10^{-3} \text{ N}$

Sine of the angle is used since only the vertical forces are added because the horizontal forces are equal and opposite, and therefore they cancel.

2. E is correct.

None of the above statements are true.

3. C is correct.

When the positively charged sphere C is near sphere B, it polarizes the sphere, causing its negative charge to migrate towards C and a positive charge to build on the other side of sphere B.

The wire between sphere A and sphere B allows a negative charge to flow to B, creating a net positive charge on sphere A. Once the wire is removed and sphere C is removed, sphere A will have a net positive charge, and B will have a net negative charge.

4. E is correct.

Coulomb's Law: $F_e = kQ_1Q_2 / r^2$

If r is increased by a factor of 3:

$F_{new} = kQ_1Q_2 / (3r)^2$

$F_{new} = kQ_1Q_2 / (9r^2)$

$F_{new} = (1/9)kQ_1Q_2 / r^2$

$F_{new} = F_{original} (1/9)$

continued...

$F_{new} = (1\ N)\cdot(1/9)$

$F_{new} = 0.11\ N$

5. B is correct.

The balloon sticks to the wall because the rubbing on the wool transfers charges to the balloon, leading to an electrostatic force.

6. A is correct.

If charge Q is of a greater magnitude than charge q, then the electric field points toward Q (because it is negative) in section W to Z.

In the VW section, the electric field is the difference in magnitude between q and Q.

If Q has a large enough charge, the difference could equal zero, and no electric field exists.

7. D is correct.

Volt is the potential difference that causes 1 C of charge to increase potential energy by 1 J. Therefore, moving 1 C through 6 V causes the potential energy of the battery to increase by 6 J.

8. C is correct.

Current is caused by a voltage (potential difference) across a conductor.

Whenever a voltage exists, an electric field exists, and this travels near the speed of light.

Electrons do not travel quickly when a current is established and only travel at their drift speed, which is proportional to voltage.

9. E is correct.

Protons moving perpendicular to electric field lines do not get close to the charges creating the electric field. Thus, its electric potential and potential energy remain constant because these values are related to distance from other charges.

Electric Potential Energy:

$$U = kQq / r$$

Electric Potential:

$$V = kQ / r$$

10. D is correct.

Only the wavelength of light changes in different mediums.

Frequency does not change in different mediums.

11. B is correct.

Faraday's Law states that the electromotive force (emf) in a coil is:

$$\text{emf} = N\Delta B A \cos\theta / \Delta t$$

where N = number of loops of wire.

If N doubles, the EMF also doubles.

12. C is correct.

There is a force on the proton up to the page. The electric field points in the direction of the force on positive particles; therefore, it is pointed upwards.

The right-hand rule determines the direction of the magnetic field.

$F = qvB$, where q is a charge, v is velocity, and B is a magnetic field

When F is oriented upwards, the curling fingers from the velocity direction give B into the page.

13. A is correct.

Convert all units to their correct form:

$$F = ma$$

$$F = (2 \times 10^{-6}\ \text{kg})\cdot(0.006\ \text{m/s}^2)$$

$$F = 1.2 \times 10^{-8}\ \text{N}$$

Substituting into the equation for electric field:

$$E = F / q$$

$$E = (1.2 \times 10^{-8}\ \text{N}) / (6 \times 10^{-6}\ \text{C})$$

$$E = 0.002\ \text{N/C}$$

Note: 1 N = 1 kg·m/s^2, not 1 g·m/s^2

14. B is correct.

$$\Phi = Q / E_0$$

For an enclosed charge, the area of the surface does not affect the flux.

15. A is correct.

γ rays are the electromagnetic waves listed that have the shortest wavelength.

Notes for active learning

Notes for active learning

CHAPTER 11

Electric Circuits

Circuits, Electromotive Force, Batteries and Voltage

Current

Resistance and Power

Kirchhoff's Laws

Capacitors

Inductors

DC and LR Circuits

Alternating Currents and Reactive Circuits

RLC Circuits

Transformers and Transmission of Power

Measuring Devices

Chapter Summary

Page intentionally left blank

Circuits, Electromotive Force, Batteries and Voltage

Electric circuits

Electric circuit is a series of connections between a voltage source, circuit elements, and conducting wires, resulting in current flow.

Voltage source is the energy input and is necessary for continuing the flow of energy.

Circuit elements can use some of the energy from the voltage source as heat or light.

Current in a circuit always flows from *high to low potential* through the external circuit.

A complete circuit has current flowing in a complete loop.

Circuit with battery provides a potential difference, causing current to flow from its positive terminal through the light bulb to its negative terminal, completing the loop.

Schematic circuit drawing:

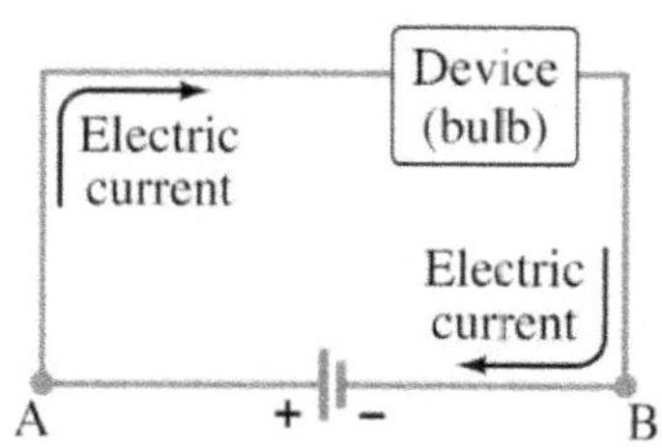

Schematic circuit drawing of the circuit shown above. Note that the schematic drawing looks different than the photograph of the physical circuit that it represents.

Electromotive force

The electric potential difference is vital for solving circuit problems.

Electromotive force (EMF) measures the energy that causes the current flow in a circuit.

EMF is the potential difference between two points in a circuit; the energy per unit of charge given by an energy source. EMF value is the amount of work done on one unit of electric charge, given in volts (V).

Batteries

Batteries are physical devices that produce an electric potential difference and are often used to provide the EMF in electric circuits.

Batteries operate by transforming *chemical energy* into *electrical energy*. Chemical reactions in the battery cell create a potential difference between the two terminals.

Potential difference can be maintained, even if a current is flowing, until the chemical reaction has exhausted itself, and the battery can no longer maintain a potential difference.

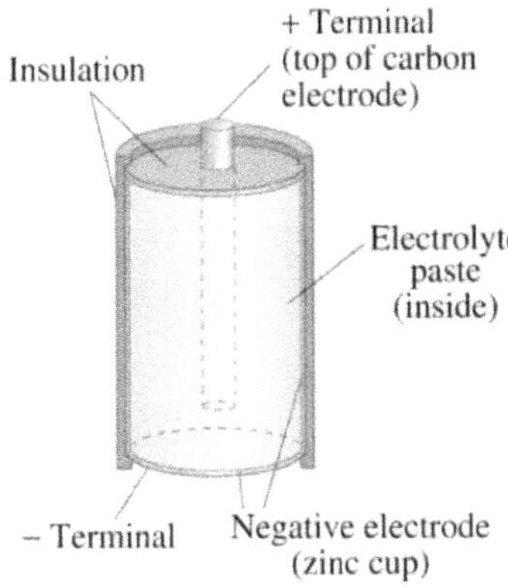

Component of a battery for stored voltage

Internal resistance of a battery

Although batteries are often depicted as having no resistance, all non-ideal batteries will have some internal resistance. The internal resistance of a battery is like a resistor right next to the battery, connected in series.

If an ideal battery in a circuit has no internal resistance, the potential difference across the battery equals the EMF. If the battery does have an internal resistance, the potential difference across the battery equals the EMF minus the voltage drop due to internal resistance.

Terminal voltage is given by:

$$V_{ab} = V_0 - IR_{battery}$$

where V_{ab} is the terminal voltage of the battery (V), V_0 is the voltage before internal resistance (V), I is current through the battery (A), and $R_{battery}$ is the internal resistance of the battery (Ω).

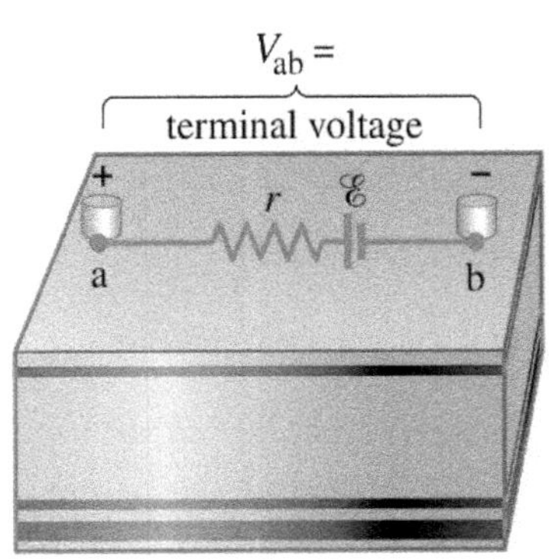

Non-ideal battery with internal resistance and terminal voltage

Ampere-hour

An ampere-hour combines the current with the time a battery takes to discharge completely.

Ampere-hour: 1 ampere of current flowing for one hour. During the hour, the charge transferred in is 3,600 coulombs (ampere-seconds). The ampere-hour rating is displayed on the battery.

Batteries in series

When batteries are in series, the total potential difference provided is the sum of the EMF provided by each battery. This is only the case if the terminals are aligned such that the batteries' terminals go from positive to negative or vice versa.

The current is equal at all points in a series circuit, so the current in any of the series-connected batteries must be the same. Each battery must have the same *amp-hour rating* or some batteries will become depleted sooner than others, compromising the capacity of the series bank.

Total amp-hour capacity of a series battery bank is unaffected by the number of batteries.

Figure below shows two batteries in series (note that these batteries have internal resistance, and the voltage shown is the terminal voltage). Total voltage provided by the two batteries is:

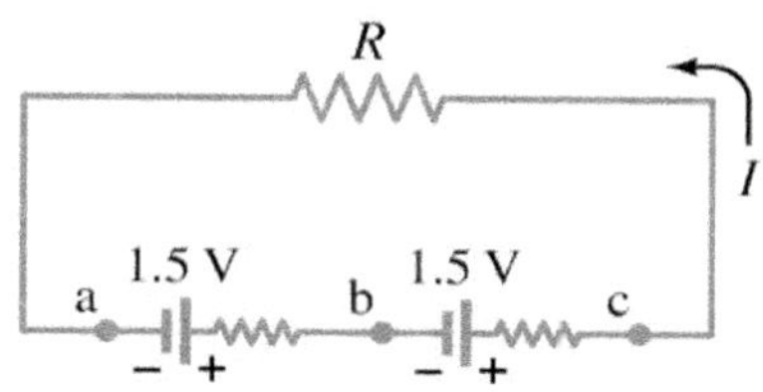

Two 1.5V batteries in series provide 3V

$V_{total} = 1.5V + 1.5V = 3V$

When the batteries are in series but aligned such that, the positive terminal goes to positive (or negative to negative), the total voltage provided is the difference between the two batteries.

In this scenario, the battery with less voltage is being charged:

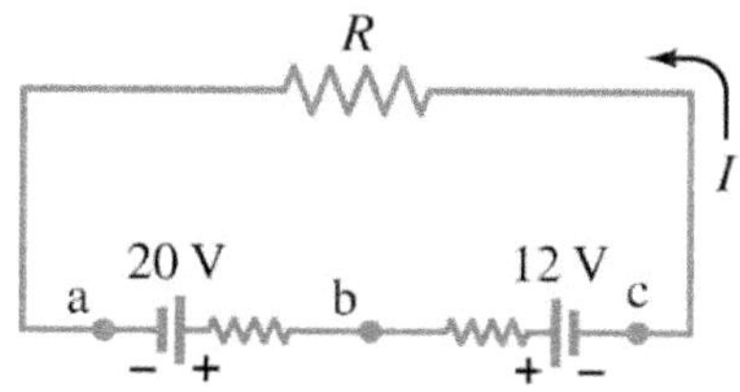

Two batteries in series oriented with similar terminals equal the difference in voltage

$V_{total} = 20\ V - 12\ V = 8\ V$

Batteries in parallel

Batteries in parallel (shown below) produce the same voltage but can support larger current draws depending on the resistance of the circuit.

Connecting batteries in parallel increases total current capacity by decreasing total resistance and increasing amp-hour capacity. Batteries in parallel must have the same voltage rating. For example, a 12V battery should not be connected with a 6V battery.

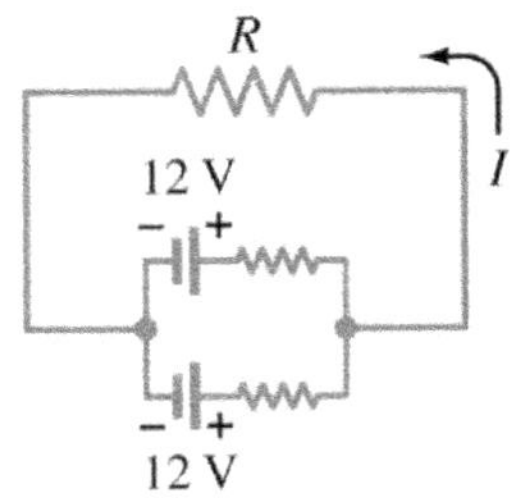

Two batteries in parallel produce the same voltage

$V_{total} = 12\ V$

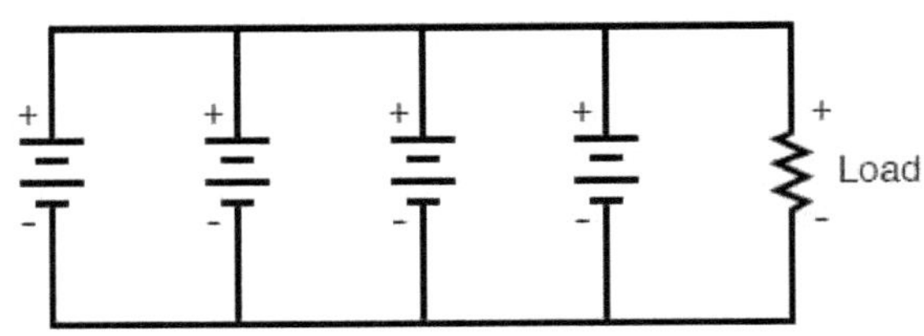

Capacity increases with a parallel battery connection; however, the voltage remains the same

Current in parallel circuits

Each battery can pump a set number of electrons per second for a given circuit, so if batteries are connected in parallel, the number of electrons they push out each second and energy supplied is added; the total current in the circuit increases.

A battery with the same voltage but higher amps is possible when the voltage is the same. The ah rating measures how long that cell can maintain its rated voltage under load.

Batteries of different amp-hour ratings would degrade under load at different rates.

Capacity increases with a parallel battery connection; however, the voltage will be the same.

Each branch has the same voltage across it since each branch has its connection to the power source, and the current from the source divides among branches in the multiple pathway circuit.

As the number of branches increases, the total current supplied by the source increases.

Current

Electrons flow through a conductor

Current results from the flow of electrons through a conductor.

Electrons in a conductor are loosely bound to the nuclei and have large, random speeds dependent upon the temperature of the conductor.

Metals are well-known conductors because one electron from each atom is loosely bound and free to move through the metal lattice.

When an electric potential difference is applied across a metal or any conducting material, it creates an electrical field that passes through the conductor near the speed of light.

Free electrons in the conductor are attracted against the line direction of the electric field due to Coulomb attraction forces.

Drift velocity

As electrons pass through the conduction material, these electrons acquire an average drift velocity, which, although considerably smaller than the thermal velocity of the electrons, creates the electric current through the conductor.

Net velocity for drift velocity: the order of magnitude of drift velocity is 10^{-4} m/s.

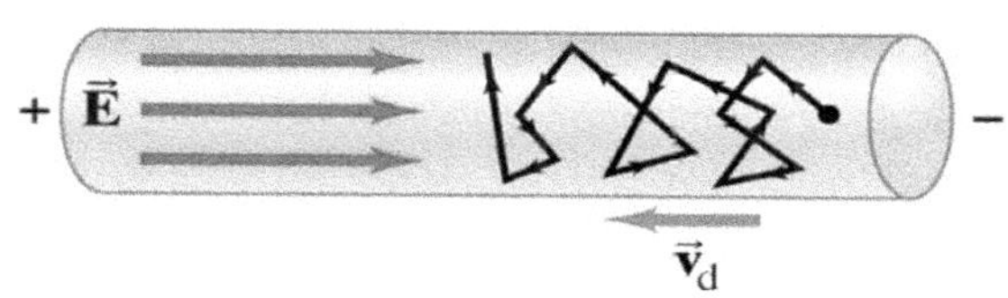

Conductor with an applied potential difference and the resulting electric field

Current through the conductor is calculated as the number of electrons drifting through a unit's volume at the *drift velocity*.

Drift velocity:

$$I = nq_eAv_d$$

where I is current (A), n is the number of electrons per unit volume (electrons/m^3), A is area of the conductor (m^2), v_d is drift velocity (m/s), and q_e is charge of an electron (1.6×10^{-19} C).

Direct and alternating current

Electric current (I) is the rate at which charge flows through a conductor, expressed as:

$$I = \frac{\Delta q}{\Delta t}$$

Current types are *direct current* (DC) and *alternating current* (AC).

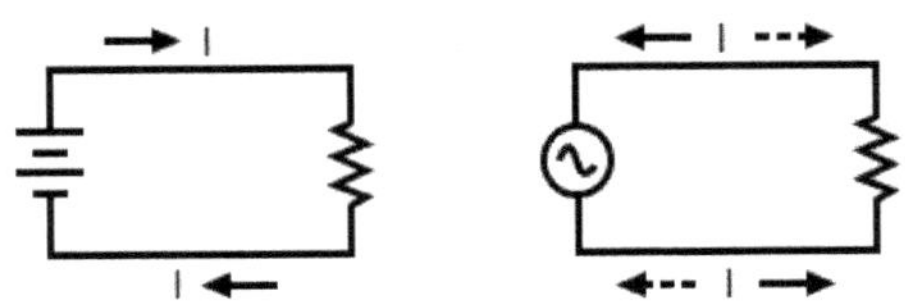

Direct current (D) on the left and alternating current (AC) on the right

Direct current (DC) applies a constant potential difference to the circuit, and current moves only in one direction. For example, direct current is characteristic of electronic devices, batteries, and solar cells.

Alternating current (AC) is when the potential difference is cycled between high and low, creating an oscillatory current flow.

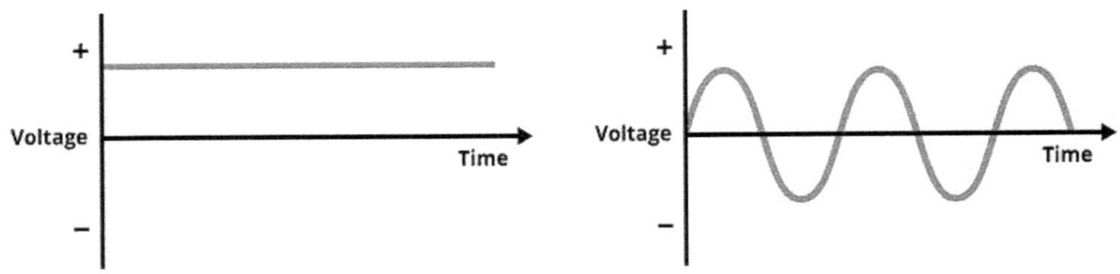

Direct current (D) on the left and alternating current (AC) on the right

Alternating current (AC) is transmitted from the power source to the consumer over high voltage lines and "stepped down" for use in homes and businesses.

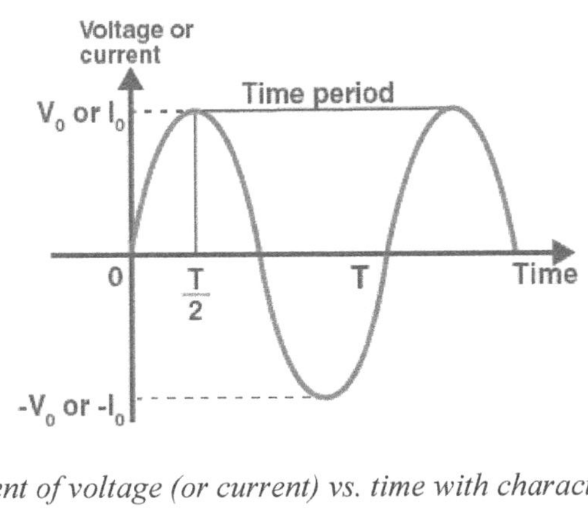

Alternating current of voltage (or current) vs. time with characteristic wave profile

Resistance and Power

Resistivity

Electrical resistance is the loss of current energy (i.e., electron flow) through a material.

Two sources of *electrical resistance*:

collisions with other *electrons* in current, or

collisions with other *charges* in the material.

Collisions are factors that account for resistance.

Resistance is determined by the resistivity of a material (conductance measure) and factors.

In most electronic circuit applications, wires transmit current.

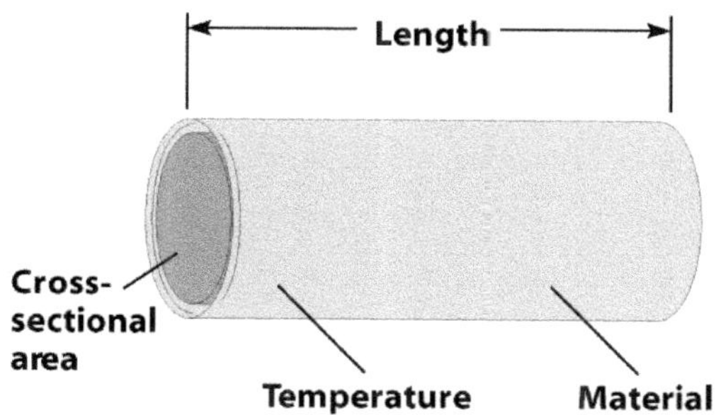

Four variables define resistance in a wire

Resistance of a wire depends upon length, cross-sectional area, temperature, and the material:

$$R = \rho\frac{\ell}{A}$$

where R is the resistance (Ohms or Ω), ρ is the resistivity ($\Omega{\cdot}$m), ℓ is the length of the wire (m), and A is the cross-sectional area of the wire (m^2).

Resistivity proportional to resistance

Resistivity is directly proportional to resistance. Therefore, the greater the resistivity, the higher the resistance of the material.

Resistance is inversely proportional to area (inverse square to radius and diameter). The greater the diameter of the wire, the lower the resistance; there is a larger area for electrons to flow.

From the resistance equation above, a low-resistance wire will be short, have a large diameter, and be made from a low-resistivity material. Extension cords are made of wires with thicker diameters to keep the resistance low.

Resistance depends on temperature

Resistance of a wire is also dependent upon the *temperature* of the material.

Resistivity increases with temperature for any given material:

$$\rho_T = \rho_0[1 + \alpha(T - T_0)]$$

where ρ_T is resistivity at a temperature T (Ω·m), ρ_0 is standard resistivity at standard temperature of T_0 (Ω·m), α temperature coefficient of resistivity (K^{-1}), and T is the temperature (K)

Ohm's Law ($I = V/R$)

Ohm's Law gives the relationship between potential differences across resistance and current.

Ohm's Law states that, in electrical circuits, the current through a conductor between two points is directly proportional to the potential difference (or voltage) across the two points.

Current inversely proportional to resistance

Current is *inversely proportional* to the *resistance* between them:

$$I = \frac{V}{R}$$

or

$$V = IR$$

where V is the potential difference (V), I is the current (A), and R is the resistance (Ω).

Resistors in series

Resistors in a circuit are arranged in *series* or *parallel.*

For a series circuit (shown below), the resistors have current pass through the first resistor, then into the second, into the third, and so on:

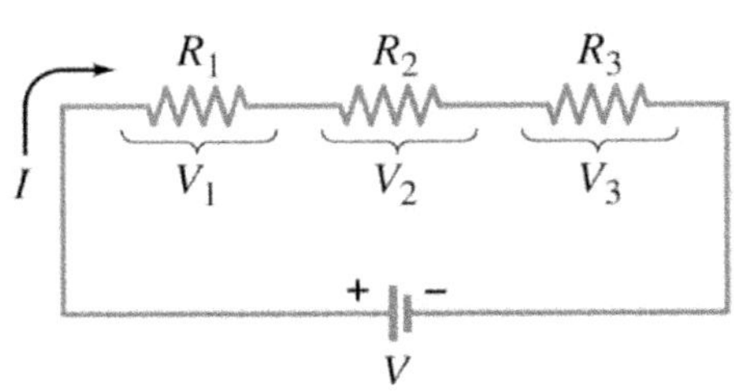

Series circuit current passes sequentially through all resisters

For series circuits, the circuit's total resistance is the sum of the resistances of the components.

Adding elements to the circuit increases the total resistance and decreases the total current.

Equivalent resistance of resistors in series is:

$$R_{\text{equivalent}} = R_1 + R_2 + R_3 + \ldots$$

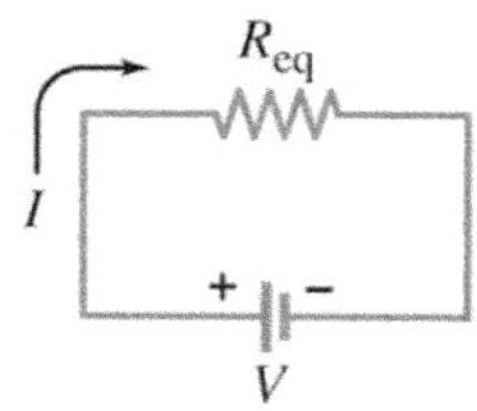

Equivalent resistance of resistors in series

In a series circuit, the current through each resistor is the same ($I_{\text{series}} = I_1 = I_2 = I_3$); however, the voltage drops across each resistor.

Voltage drop among resistors in series is split according to the resistance—a greater resistance equals a greater voltage drop (V = IR).

This can also be expanded for more than one resistor by splitting it into components:

$$V = V_1 + V_2 + V_3 = IR_1 + IR_2 + IR_3$$

Resistors in parallel

Parallel resistors are arranged such that all resistors experience the same voltage drop across them but have different currents according to their resistances.

In a parallel circuit, the current has more than one available path.

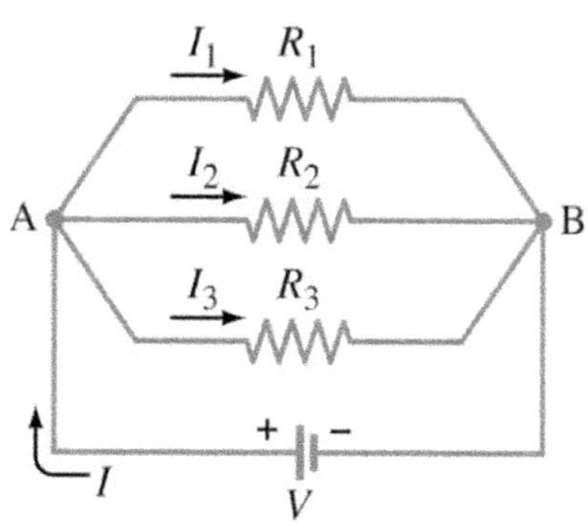

Parallel circuit offers more than one route for the current to flow

Equivalent resistance of resistors in parallel:

$$\frac{1}{R_{equivalent}} = \frac{1}{R_1} + \frac{1}{R_2} + \frac{1}{R_3} + \cdots$$

As stated, for parallel resistors, the voltage across each resistor is the same ($V_{parallel} = V_1 = V_2 = V_3$); however, the current across each resistor is not.

For parallel resistors, the total current is the *sum of currents across each resistor*:

$$\frac{V}{R_{eq}} = \frac{V}{R_1} + \frac{V}{R_2} + \frac{V}{R_3}$$

$$I = \frac{V}{R_{eq}}$$

Most typical circuits (e.g., household circuits) combine series and parallel circuits.

Generally, electrical outlets in a room are in series, but the electrical outlets in separate rooms are in parallel. This is why a blown fuse may cut power to a room but not the entire house.

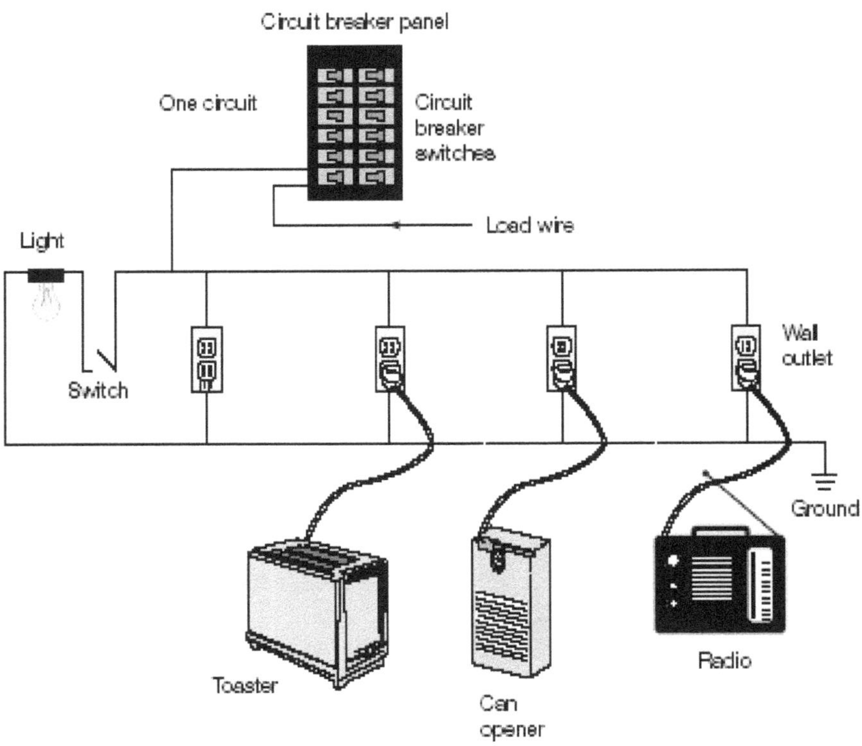

House wiring is typically in parallel between the series circuits in individual rooms

Power in DC circuits

Power (P) in a circuit is the rate at which energy converts from electrical energy into some other form, such as heat or mechanical energy.

SI unit for power is the watt (W), which is equal to joules per second (J/s).

For a component in a DC circuit, power is given by the equation:

$$P = IV$$

where P is the power (W), I is the current (A or ampere), and V is the voltage (V).

Using Ohm's Law, power in an electrical circuit is expressed as:

$$P = I^2R$$

Electrical energy relationships

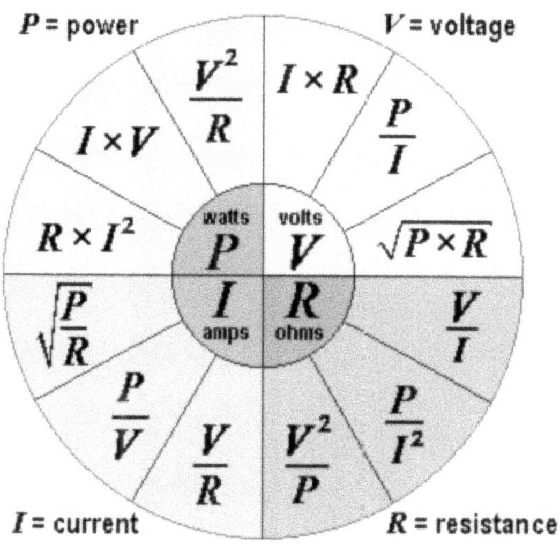

Electrical energy relationships and associated units

Ohms Law relationships

Known Values	Resistance (R)	Current (I)	Voltage (V)	Power (P)
Current & Resistance	---	---	$V = I \times R$	$P = I^2 \times R$
Voltage & Current	$R = \frac{V}{I}$	---	---	$P = V \times I$
Power & Current	$R = \frac{P}{I^2}$	---	$V = \frac{P}{I}$	---
Voltage & Resistance	---	$I = \frac{V}{R}$	---	$P = \frac{V^2}{R}$
Power & Resistance	---	$I = \sqrt{\frac{P}{R}}$	$V = \sqrt{P \times R}$	---
Voltage & Power	$R = \frac{V^2}{P}$	$I = \frac{P}{V}$	---	---

Ohms law formulas solve for unknowns

Kirchhoff's Laws

Kirchhoff's Current Law

Kirchhoff's Current Law expresses conservation of charge; charge is neither created nor destroyed.

Kirchhoff's Laws solve the unknown components within a circuit.

In a circuit, currents entering a junction must equal the sum of all currents exiting the junction.

$$I_{entering,total} = I_{exiting,total}$$

Kirchhoff's Voltage Law

Kirchhoff's Voltage Law (or Kirchhoff's Loop Rule) expresses energy conservation.

The voltage law states that the sum of all voltage drops across any resistive elements within a closed-circuit loop must equal the EMF in the same loop.

$$\sum V_{drop} = EMF_{total}$$

For any complete loop (i.e., traveling around a circuit loop and returning to the point of origin), the total change in voltage is 0.

Evaluating boosts and drops in circuit

Choose a starting point on the circuit and proceed around the circuit in a complete loop (returning to the point of origin).

Evaluate the boosts and drops in voltage as each circuit element is passed.

> Moving across a battery is the same as "up" a battery when approaching the direction of the EMF (from the – to + terminal).
>
> Moving across a battery is the same as "down" a battery if approaching against the EMF (from + to – terminal).
>
> Moving across a resistor is the same as "down" a resistor by the amount IR, where I is the current through the resistor, and R is the resistance in Ω when approaching in the same direction as the current.
>
> Moving across a resistor is the same as "up" a resistor by the amount IR, where I is the current through the resistor, and R is the resistance in Ω when approaching in the opposite direction as the current.

Solving Kirchoff problems

When solving problems that use Kirchhoff's rules, follow these steps:

1. Label each current.
2. Identify unknowns.
3. Apply junction and loop rules; as many independent equations as unknowns will be needed.
4. Solve the equations by being careful with signs.

For example, apply Kirchhoff's Circuit Laws to the circuit below:

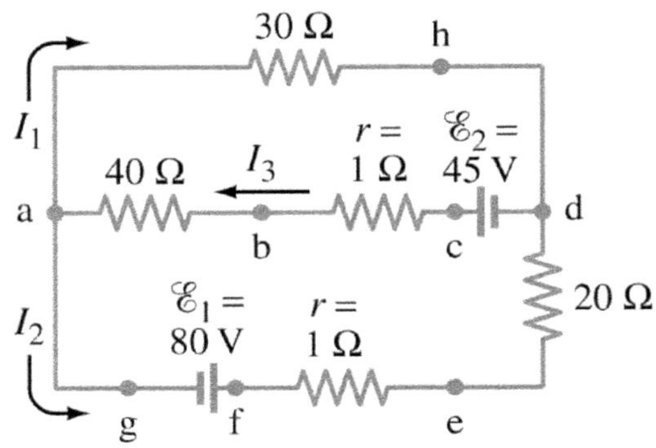

Solve using Kirchhoff's Circuit Laws

Apply Kirchhoff's Current Law:

$I_1 + I_2 = I_3$

Apply Kirchhoff's Voltage Law:

Loop CBAHD: $0\text{ V} = -I_3(1\ \Omega) - I_3(40\ \Omega) - I_1(30\ \Omega) + 45\text{ V}$

Loop FEDCBAG: $0\text{ V} = -I_2(1\ \Omega) - I_2(20\ \Omega) + 45V - I_3(1\ \Omega) - I_3(40\ \Omega) + 80\text{ V}$

Loop FEDHAG: $0\text{ V} = -I_2(1\ \Omega) - I_2(20\ \Omega) + I_1(30\ \Omega) + 80\text{ V}$

Capacitors

Capacitors

Capacitor is a device used to store electrical charge.

A basic capacitor usually consists of two conductors close to each other but *not* touching.

An insulator of a dielectric is between the two conducting plates, which are separated by air.

Left diagram below shows a parallel plate capacitor with no insulator; air is between the plates.

Right diagram depicts a variation of the parallel plate capacitor in which an insulating material is between the coiled plates.

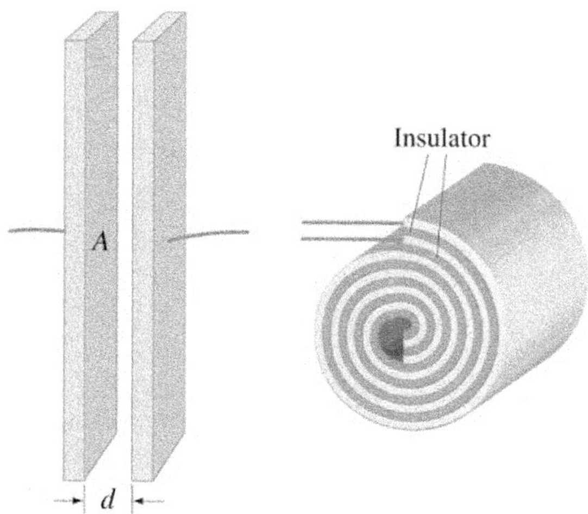

Capacitor without insulation (on the left) and with insulation between the plates

Determining capacitance

Capacitor's ability to store charge is *capacitance*, measured in Farads (F).

Capacitance does not depend on voltage; it is a function of geometry and capacitor material.

Capacitance is expressed by:

$$C = \frac{Q}{V}$$

where C is the capacitance (F), Q is the charge (J or joules), and V is the voltage (V).

Parallel-plate capacitor

Parallel-plate capacitor is the simplest form of a capacitor. As its name implies, it consists of two conductors in the form of plates parallel and separated by air or a dielectric material.

In the diagrams below, a parallel-plate capacitor is connected to a battery. The equivalent circuit diagram is on the right.

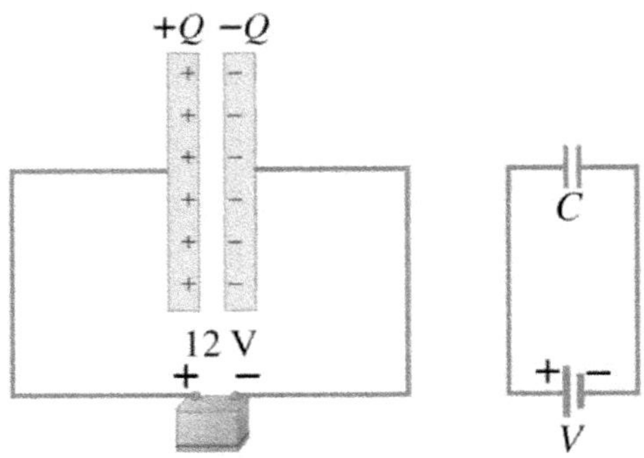

Parallel-plate capacitor is connected to a battery (on the left) and an equivalent circuit diagram (on the right)

Capacitance of a parallel plate capacitor is expressed as:

$$C = \frac{Q}{V} = \frac{\epsilon_0 A}{d}$$

where ϵ_0 is permittivity of space (8.854×10^{-12} F/m), k is relative permittivity of the dielectric material, A is area of plates, not the sum of the areas (m^2), and d is plate separation (m).

Voltage across capacitor

Voltage (V) across the capacitor is the product of the electric field between the plates and the distance of plate separation:

$$V = E \times d$$

where E is the electric field between the capacitor (V·m^{-1}).

Dielectrics

Dielectric is a non-conductive material that does not readily allow a current to pass through it.

Inserting a dielectric between the plates of a capacitor increases the capacitance. The dielectric allows more charge to be stored on the plates of the capacitor.

Diagram below represents an air-filled parallel plate capacitor (left and center) and a dielectric-filled parallel plate capacitor (right).

Left diagram shows an air-filled parallel plate capacitor. The strength of the electric field between the plates limits the amount of charge Q stored.

Center diagram shows that a strong electric field develops if too much charge is stored between the plates. Stored charge ionizes the air and causes an arc of charge to jump across the plates.

Right diagram shows a dielectric placed between the plates. The dielectric material becomes polarized (insulator in an electric field) and inhibits the formation of a strong electric field between the plates.

Accordingly, more charge (and thus a higher capacitance) can be stored on the plates before the dielectric material breaks down, allowing current to pass through.

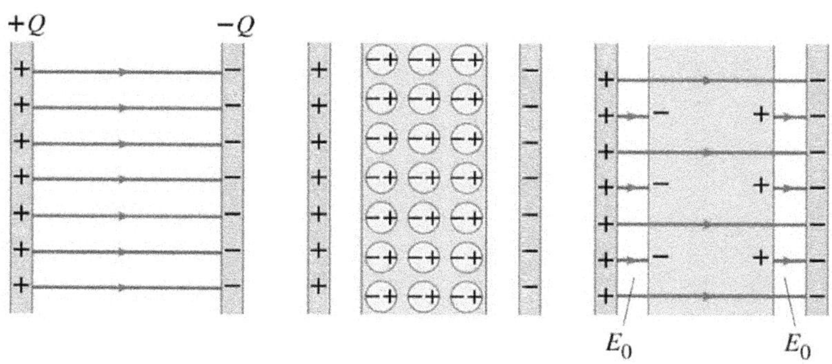

Air-filled parallel plate capacitor (on the left), a strong electric field develops if too much charge is stored between the plates (in the center) and a dielectric is placed between the plates, which inhibits the formation of a strong electric field between the plates (on the right)

Parallel plate capacitor with a dielectric

Capacitance of a parallel plate capacitor with a dielectric:

$$C = \frac{Q}{V} = \frac{k\,\epsilon_0\,A}{d}$$

where C is the capacitance of the capacitor, A is the area of the capacitor, k is the dielectric constant, and ϵ_0 is the permittivity of free space.

Energy of charged capacitor

Charged capacitors store electric energy equal to the work done to charge the capacitor.

$$U = \frac{1}{2}QV = \frac{1}{2}CV^2 = \frac{1}{2}\frac{Q^2}{C}$$

where U is the potential energy of the charged capacitor (J), Q is the charge stored (magnitude of either $+Q$ or $-Q$ on one plate) (C or coulombs), and C is capacitance (F).

Energy density of capacitors

Energy density of the capacitor is defined as the energy of the electric field per unit volume.

Energy density can be expressed as:

$$\text{energy density} = \frac{PE}{volume} = \frac{1}{2}k\,\epsilon_0\,E^2$$

Capacitors in series

Like resistors, capacitors can be connected in *series* or *parallel* in a circuit.

For capacitors in series, the equivalent capacitance is found by the sum of the reciprocals of the individual capacitance:

$$\frac{1}{C_{eq}} = \frac{1}{C_1} + \frac{1}{C_2} + \frac{1}{C_3}$$

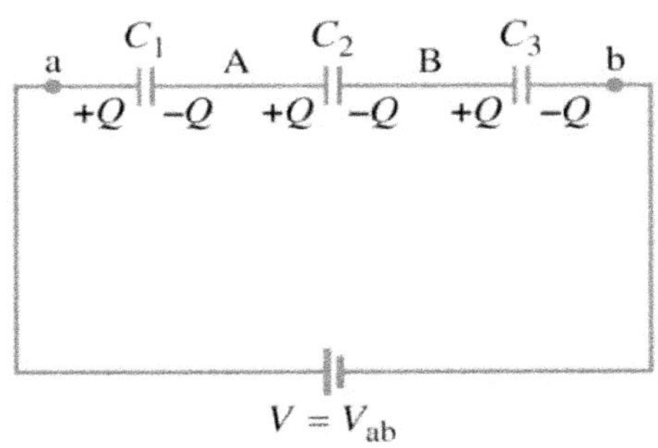

Equivalent capacitance is the sum of the reciprocals of the individual capacitance

Capacitors in series have the same charge Q, but the voltage V across each capacitor is different.

$$V_1 \neq V_2 \neq V_3$$

$$Q_1 = Q_2 = Q_3$$

Capacitors in parallel

Capacitors may be connected in parallel, as depicted below:

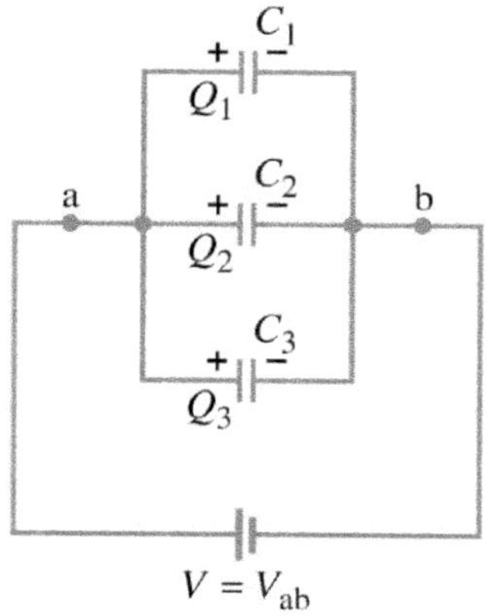

Capacitors connected in parallel have the same voltage

In this example above, all the capacitors have the same voltage.

Equivalent capacitance is the sum of the individual capacitances.

$$V_1 = V_2 = V_3$$

$$C_{equivalent} = C_1 + C_2 + C_3$$

Capacitors in parallel hold different charges:

$$Q_1 \neq Q_2 \neq Q_3$$

Inductors

Inductors

Inductors are circuit devices that resist changes according to Faraday's Law.

Typical inductors consist of coils of wire wrapped around a different core material.

Inductors operate by producing a back electromotive force (EMF) against a change in a current (according to Lenz's Law and Faraday's Law).

Electromotive force

Electromotive force (EMF) equals the *terminal potential difference* when no current flows.

Electromotive force and terminal potential difference (V) are measured in volts; however, they are not the same.

EMF (ϵ) is the amount of energy (E) provided by the battery to each coulomb of charge (Q) passing through.

For example, initially, more current flows to the light bulb because as the magnetic field grows in the coil of wire, it induces its voltage through Faraday's Law. Inductors create a back EMF, opposing the current that caused it.

EMF units

Units for the inductor is the *Henry* (H), and the back EMF produced is:

$$EMF = -L\frac{\Delta I}{\Delta t}$$

where *EMF* is the opposing voltage produced (V), L is the inductance of the inductor (H), and $\Delta I/\Delta t$ is the change in current over the change in time (A/s).

Magnetic field stores energy

Magnetic field produced by an inductor stores energy; similar to the electric field of a capacitor.

Energy stored in this magnetic field is:

$$U = \frac{1}{2} \times \frac{B^2}{\mu_0}$$

where U is the stored energy (J), B is the magnetic field (T or Tesla), and μ_0 is the permittivity of free space (1.2566×10^{-6} m kg/s^2 A^2).

DC and LR Circuits

Resistor-capacitance (RC) circuits

Resistor-capacitance (RC) circuit consists of a resistor and capacitor in series, connected to a potential difference (e.g., battery).

The capacitor begins to charge when the switch is closed, and the circuit forms a complete loop.

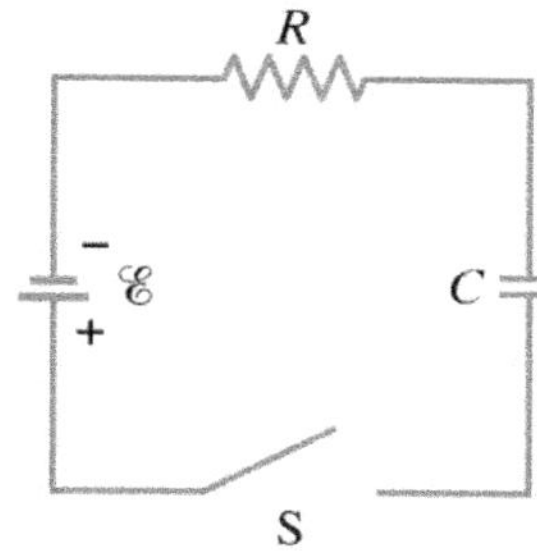

Resistor-capacitance (RC) circuit with the switch open

When the switch is closed, the capacitor begins to charge over time.

Voltage across capacitor

Voltage across the capacitor increases with time and is expressed as:

$$V_C = V(1 - e^{-t/RC})$$

where V_C is the voltage across the capacitor (V), V is the potential difference across the circuit (V), t is time (s), R is resistance (Ω), and C is the capacitance (Farad).

Charge across capacitor

The charge follows a similar curve:

$$Q_C = CV_b(1 - e^{-t/RC})$$

where $V_b = V_R + V_C$. V_R is voltage across resistor (V), and Q_C is charge across the capacitor (C).

Current through circuit

Current through the circuit is the same everywhere for a series circuit:

$$I = \frac{V_b}{R}(e)^{-t/RC}$$

where I is the current through the circuit (A).

Time constant curve

Characteristic time constant curve:

$$\tau = RC$$

where τ is the time constant (s).

Voltage and current

Voltage across the charging capacitor and current across the circuit is shown below.

Diagram below illustrates the *growth curve of voltage over time.*

Notice that the capacitor reaches 63% voltage in one time constant and about 100% voltage after three-time constants.

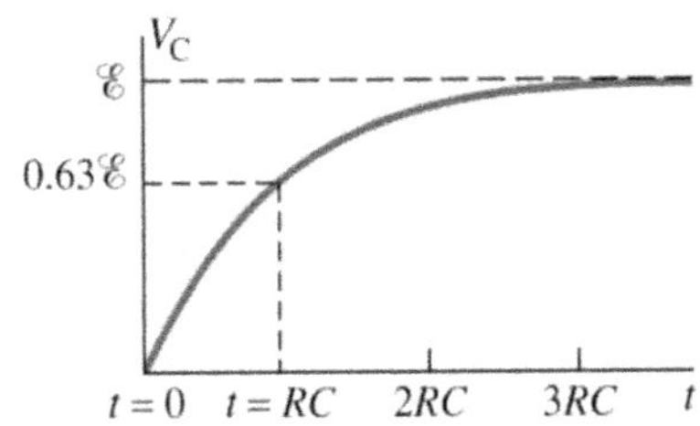

Current vs. time graph plotting voltage

Current decay for discharging capacitor

Diagram below displays the decay in current over time.

Notice that after one time constant, the current is 37% of its original value (t = 0 seconds), and after three time constants, the current is nearly zero amps.

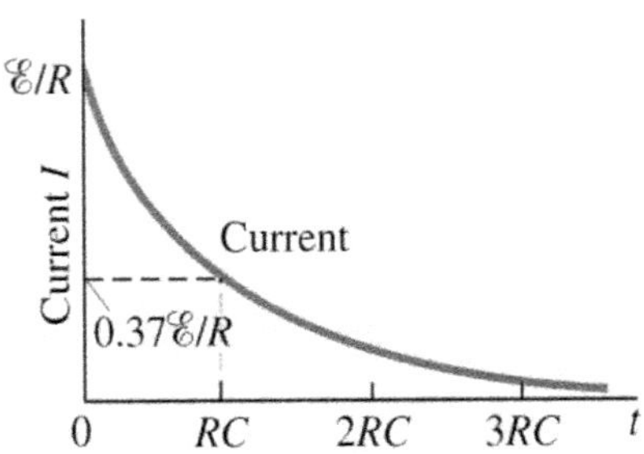

Current vs. time graph plotting current decay

During the discharge of a capacitor, the capacitor acts as a battery and drives the current flow.

Current flow decreases with time as the capacitor discharges.

Capacitor voltage decay

Capacitor voltage decay is expressed as:

$$V_C = V_0(e)^{-t\,/\,RC}$$

where V_0 is the charge voltage across the capacitor (V).

Charge decay and current

Charge through the discharging capacitor follows a similar decay:

$$CV_0\ Q\ (e)^{-t\,/\,RC}$$

Current in the circuit when the capacitor is discharging is expressed as:

$$I = \frac{V_0}{R}(e)^{-t\,/\,RC}$$

When the capacitor discharges, the voltage, charge, and current all follow a similar decay curve.

Voltage decay curve of discharging capacitor

Figure below shows the voltage decay curve of a discharging capacitor; the charge and current graphs have exponential decay curves with this pattern.

Voltage, charge, and current would be 37% of max value (t = 0 seconds) for one time constant.

By three-time constants, the voltage, charge, and current would decay to nearly zero.

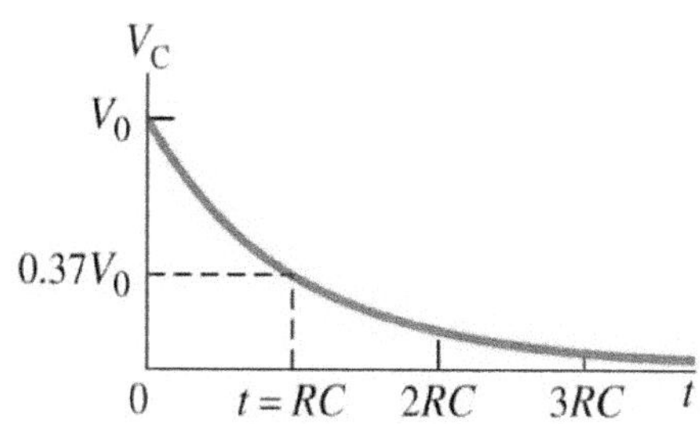

Charge voltage vs. time graph of decay across a capacitor

LR series circuits

LR circuit has an inductor in series with a resistor connected to a potential difference.

The inductor resists change in current through the circuit. Therefore, a gradual rather than abrupt rise in current occurs when the loop is closed.

After a period, the current remains steady and is equal to that if the circuit were a resistor alone (assuming an ideal inductor with no resistance).

LR circuit below has a switch to connect (or disconnect) the circuit to a potential difference (e.g., battery).

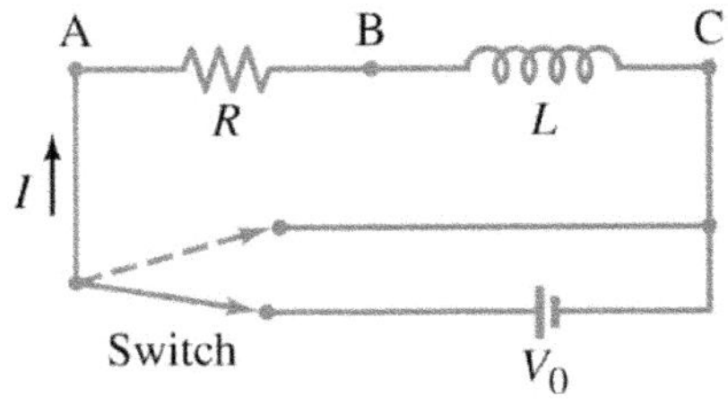

Schematic of a LR circuit with inductor L, resistor R, and circuit switch

Switch is initially open, so there is no potential difference, and no current flows in the circuit.

Voltage and current in LR circuit

When the switch is closed, the voltage across the capacitor is:

$$V_L = V_b e^{-tR} / _L$$

where V_L is the voltage across the inductor (V), R is the resistance of the resistor (Ω), and L is the inductance of the inductor (H or Henry).

Voltage (V) across the resistor is:

$$V_b = V_L + V_R \text{ and } V_R$$

Current (I) across the circuit is:

$$I = \left(\frac{V_b}{R}\right)(1 - e^{-tR/L})$$

where I is the current across the circuit (A).

Time constants for LR circuit

Time constant of an LR circuit is:

$$\tau = \frac{L}{R}$$

where τ is the time constant (s), L is the value of the inductor (Henries), and R is the value of the resistor (Ohms).

Current growth curve

Growth curve of the current through the circuit is below. After one time constant, the circuit has reached 63% of its maximum current.

Maximum current is reached in approximately three-time constants (like the RC circuit).

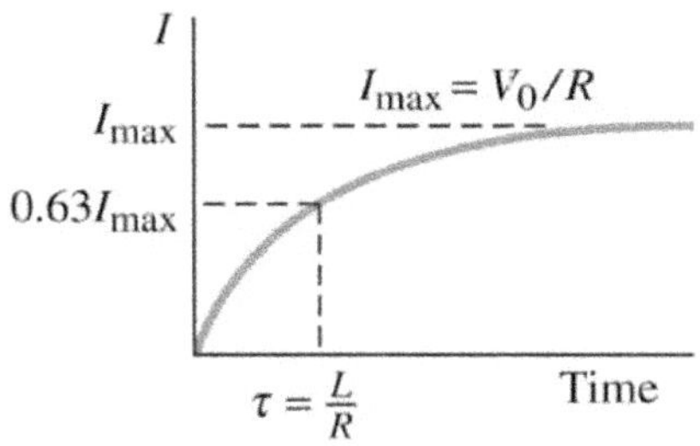

Growth curve and maximum current

Current decay for LR circuit

If the switch is opened, such that the potential difference is no longer across the circuit, but the circuit remains closed, the current decays according to:

$$I = I_{max}e^{-t/\tau}$$

In the graph below, the current decays from its maximum value.

After one time constant, the current is at 37% of its maximum.

After three-time constants, the circuit has approximately zero current.

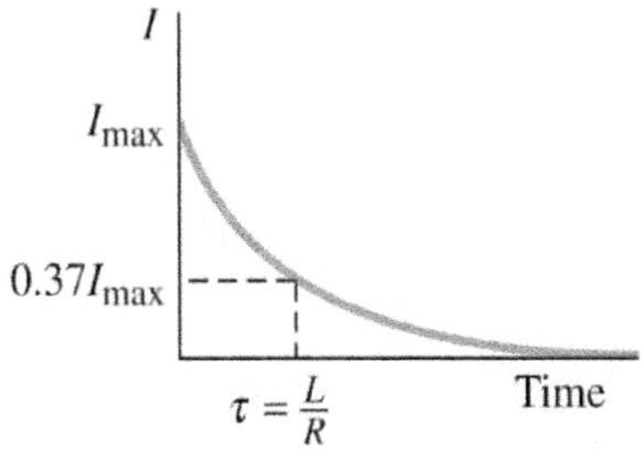

Current vs. time graph with current decay plotted as a line

Notes for active learning

Alternating Currents and Reactive Circuits

Alternating current

As mentioned, a battery's current flows steadily in one direction, which is how *direct current* (DC) is named.

Diagram shows a DC system in which the current is constant over time.

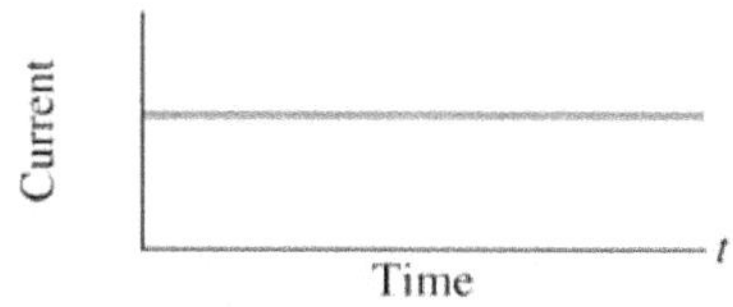

Current vs. time graph for DC circuits

Power supplied by an electrical generation plant is delivered in alternating current (AC).

Diagram shows that alternating current varies sinusoidally as a function of a square or triangular wave.

In most typical applications, the current is supplied as a sinusoidal function.

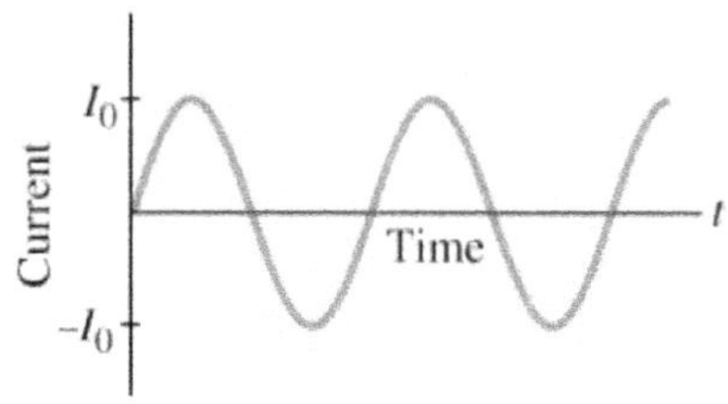

Alternating current is supplied as a sinusoidal function

Sinusoidal variation in current

The sinusoidal variation in current is a function of the frequency:

$$I = I_0 \sin \omega t = I_0 \sin 2\pi f t$$

where I_0 is the max current (amplitude) given in (A), ω is the frequency (rad/s), and f is the frequency (Hz).

Sinusoidal variation in voltage

Like current, voltage varies in a sinusoidal wave when graphed against time.

$$V = V_0 \sin 2\pi f t = V_0 \sin \omega t$$

where V_0 is the max voltage (amplitude) given in (V).

Calculating power

Power delivered is calculated by multiplying the current squared by voltage:

$$P = IV = I(IR) = I^2R = I_0^2R \sin^2 \omega t$$

Usually, the average power is of most interest:

$$\bar{P} = \frac{1}{2}I_0^2R = \frac{1}{2}\frac{V_0^2}{R}$$

where $\bar{P}$ is the average power (W or watts), and R is the resistance (Ω or ohms).

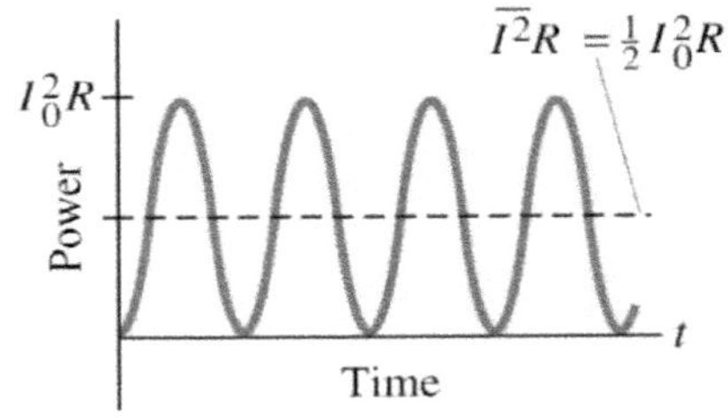

Average power for a power vs. time graph

Root-mean-square for current and voltage

Current and voltage have average values of zero and are expressed as *root-mean-square values*:

$$I_{rms} = \sqrt{\overline{I^2}} = \frac{I_0}{\sqrt{2}}$$

$$I_{rms} = 0.707\ I_0$$

$$V_{rms} = \sqrt{\overline{V^2}} = \frac{V_0}{\sqrt{2}}$$

$$V_{rms} = 0.707\ V_0$$

where I_{rms} is the root-mean-square current (A), and V_{rms} is the root-mean-square voltage (V).

Root-mean-square for Ohm's Law and power

Ohm's Law is written using root-mean-square values:

$$V_{\text{rms}} = I_{\text{rms}}R$$

Average power can be found using root-mean-square values:

$$P_{\text{avg}} = I_{\text{rms}}V_{\text{rms}}$$

$$P_{\text{avg}} = I^2_{\text{rms}}R$$

RLC Circuits

Resistor, inducer, and capacitor circuits

An RLC circuit is an electrical circuit consisting of a resistor (R), an inductor (L), and a capacitor (C), connected in series or parallel.

Circuit's name is derived from the letters used to denote the constituent components of this circuit, where the sequence of the components may vary from RLC.

Resistors, inducers, and capacitors have different phase relationships between current and voltage when placed in an AC circuit than a DC circuit.

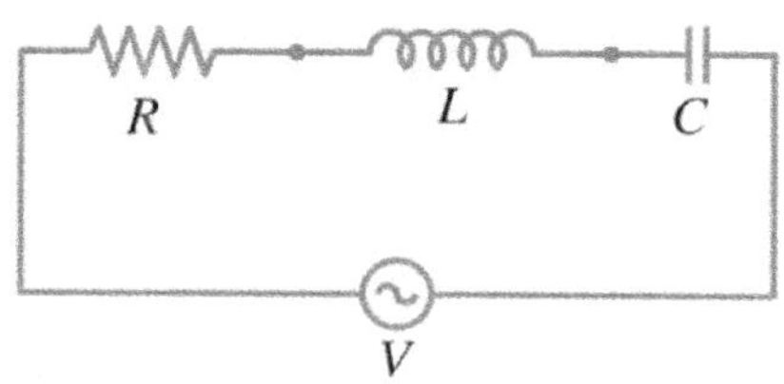

AC circuit with an inductor, resistor, and capacitor

Current and voltage through resistors

Unlike DC circuits, current *vs.* voltage through RLC circuit elements is time-dependent.

Unique characteristics of each element must be known before the entire circuit can be analyzed.

Current through a resistor in an AC circuit is in phase with the voltage, meaning both change signs simultaneously.

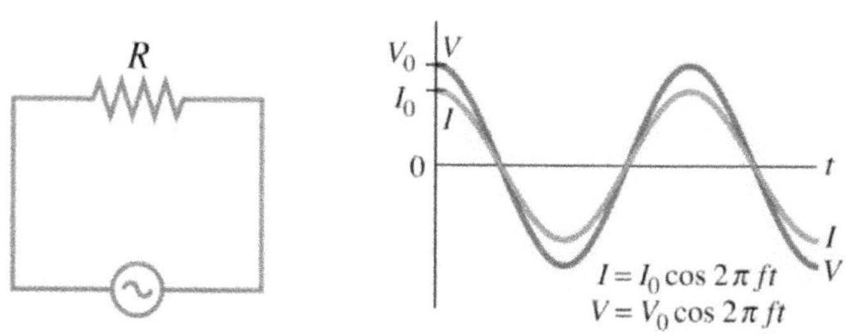

Current through a resistor in an AC circuit is in phase with voltage

Current and voltage through inductors

Phase relationship between current and voltage in inductors is different.

Inductors resist changes in current, so the current lags the voltage by 90°.

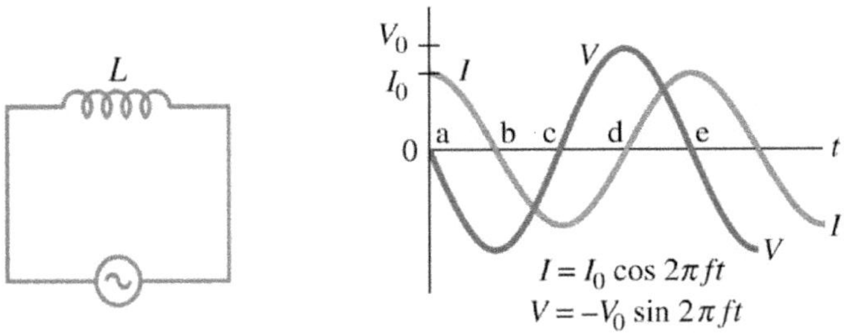

Current through an inducer in an AC circuit lags the voltage by 90°

Current and voltage through capacitors

Conversely, the capacitor's current leads the voltage by 90°.

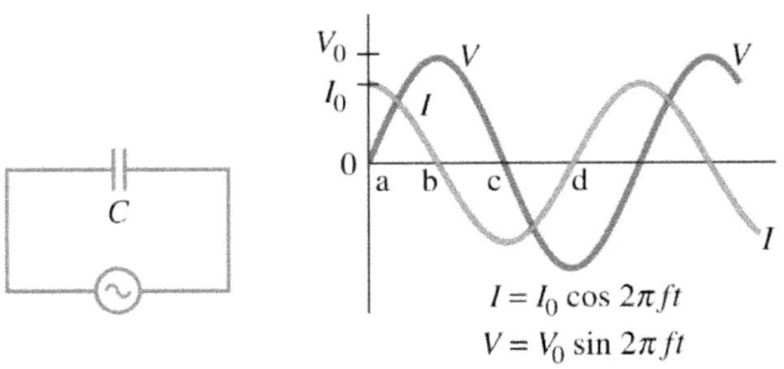

Current through a capacitor in an AC circuit leads voltage by 90°

Reactance of inductors and capacitors

Inductors and capacitors have phase differences with respect to the voltage of their effective resistance (ratio of voltage to current), so the root-mean-square current is calculated differently.

Reactance is the effective resistance of an inductor or capacitor and is given by:

$$X_L = \omega L = 2\pi f L$$

where X_L is the reactance of the inductor (Ω).

Reactance of the capacitor (X_C) is given by:

$$X_c = \frac{1}{\omega C} = \frac{1}{2\pi f C}$$

where X_c is the reactance of the capacitor (Ω).

Effective resistance

Both equations depend on frequency.

From the ratio of voltage to current, the effective resistance (i.e., *impedance*) of the circuit is:

$$Z = \sqrt{R^2 + (X_L - X_C)^2}$$

where Z is the impedance (Ω or ohms).

Root-mean-square current

Root-mean-square current in an AC circuit is:

$$I_{rms} = \frac{V_{rms}}{Z} = \frac{V_{rms}}{\sqrt{R^2 + (2\pi f L - \frac{1}{2\pi f C})^2}}$$

I_{rms} depends on the frequency and is at a maximum when $X_c = X_L$.

Resonant frequency

f_0 is the *resonant frequency*.

Frequency at which I_{rms} is maximum:

$$f_0 = \frac{1}{2\pi}\sqrt{\frac{1}{LC}}$$

where f_0 is the resonant frequency of the circuit (Hz).

Figure below illustrates the current peak at the resonant frequency (f_0) for large resistances and small resistances:

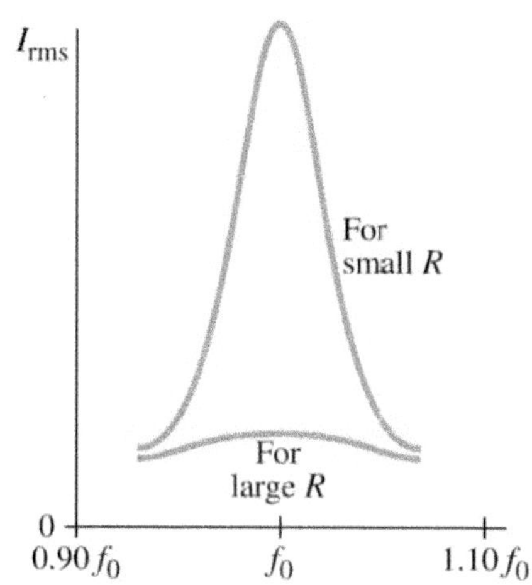

Current I_{rms} peak at the resonant frequency for large vs. small resistance

Notes for active learning

Transformers and Transmission of Power

Power transmission

Power is transmitted in alternating current (AC) for several reasons.

A significant factor is power loss through the power lines when transporting electricity.

If electricity were transported at a high current, the result is a large Ohmic power loss through the power lines.

Power losses are reduced by transporting electricity at high voltages rather than higher currents.

According to the power loss equation:

$$P = I^2R$$

where P is power, I is current, and R is resistance in the line.

Transformers

Transformers are electrical devices that permit electricity to be transported at high voltage and then reduced to lower voltages for the consumers.

Transformers work only if the current changes, so electricity is transmitted as AC.

Transformers operate according to Faraday's Law of electromagnetic induction (1831), which steps-up or steps-down the voltage through a line. This principle permits power companies to reduce power loss during the delivery of electricity.

Transformers have an iron core around which input and output wires are wrapped.

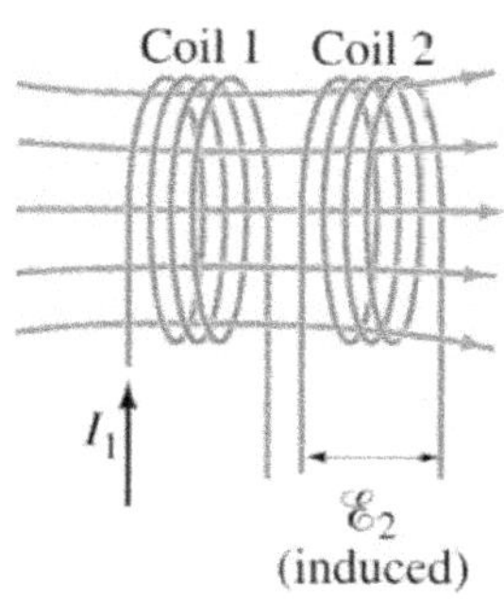

Transformer with an input (coil 1) and output coil (coil 2)

Power delivery system

Power delivery system includes power stations generating electricity, which is sent through a step-up transformer to reduce the power loss during transmission.

Commercial AC electricity runs a high-voltage transmission line to a step-down transformer near the consumer. Electricity is then sent to homes on low-voltage lines, where a final step-down transformer lowers the voltage for the end-user.

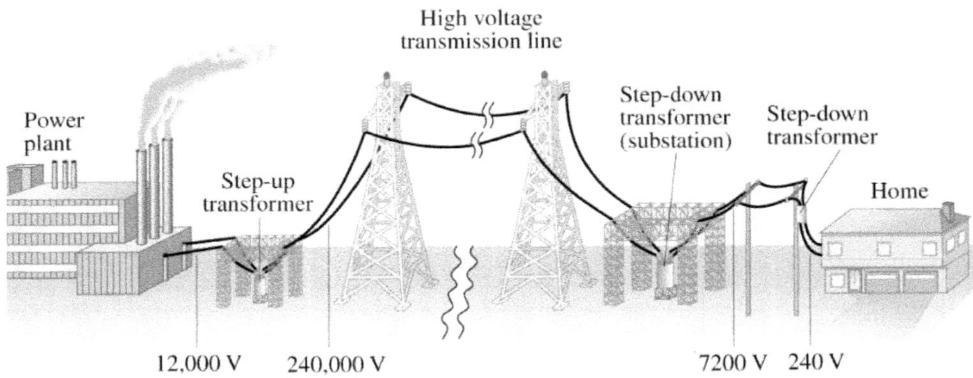

Power delivery system with step-up and step-down transformers

Ratio of voltage in primary and secondary coils

The ratio of the voltages in the primary and secondary coils of a transformer is equal to the ratio of the number of turns in each coil:

$$\frac{V_S}{V_P} = \frac{N_S}{N_P}$$

where V_S is voltage through secondary coil (V), V_P is voltage through the primary coil (V), N_S is the number of turns in the secondary coil and N_P is the number of turns in the primary coil.

Voltage and current relationship

Power is determined by calculating the product of voltage (V) and the current (I).

Input power in the primary coil must equal the output power in the secondary coil.

Relationship between voltage and current is:

$$V_p I_p = V_s I_s$$

where I_p is current through the primary coil (A) and I_s is the current through secondary coil (A).

Energy must be conserved. Therefore, in an ideal system, the ratio of the currents must be the inverse of the ratio of turns:

$$\frac{I_P}{I_S} = \frac{N_S}{N_P}$$

Step-up transformers

In a step-up transformer (i.e., increase voltage), the input wire has a few primary coils around the iron core. The output wire has many secondary coils.

When alternating current travels through the input coil, the changing current (sinusoidal variation in current) induces a magnetic field in the iron core.

A magnetic field loops through the iron core and induces a voltage in the output wire wrapped around the core. Output voltage is proportional to the number of coils wrapped around the core.

Below is a step-up transformer used at power-generating stations to increase the voltage of the power lines and reduce Ohmic losses in the lines.

Notice that the primary coil has far fewer turns than the secondary coil.

Accordingly, the voltage through the secondary coil is much higher, and the electricity can be sent down power lines at a higher voltage with less power loss.

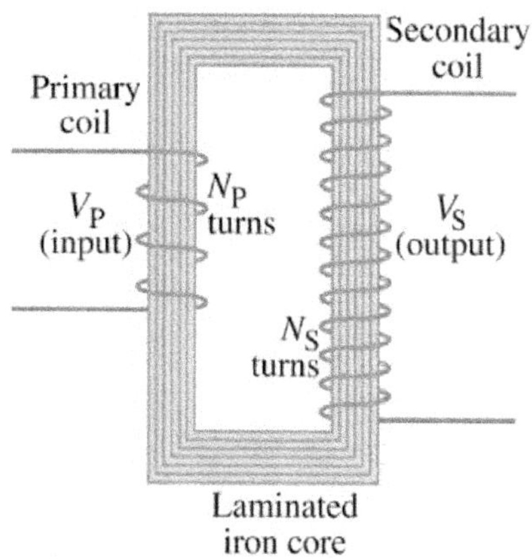

Step-up transformer has the primary coil with fewer turns than the secondary coil

Notes for active learning

Measuring Devices

Ammeter

Electric meters measure various aspects of an electric circuit. There are many types of electric meters, with the *ammeter* and *voltmeter* being the most common.

Ammeter is a device used to measure the *current* through a circuit.

Ammeters usually have two connecting electrodes placed alongside the point in the circuit where the current is to be measured.

Current passes through the ammeter, and the result is displayed.

An ammeter must have almost no internal resistance. The current in the circuit passes through the ammeter, so the ammeter needs low resistance so as not to affect the current reading.

Galvanometer

Galvanometer is an older but more accurate type of ammeter.

Two electrodes are placed at a point in the circuit, and the current flows through them. As the current passes through the coil, it creates a magnetic field whose direction depends upon the direction of the current flow.

The permanent magnets on the side of the coil produce a torque about the coil, which then rotates to display the measured current against a calibrated dial.

Symbol for an ammeter in a circuit diagram is shown below:

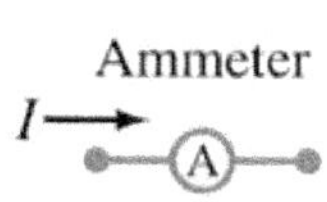

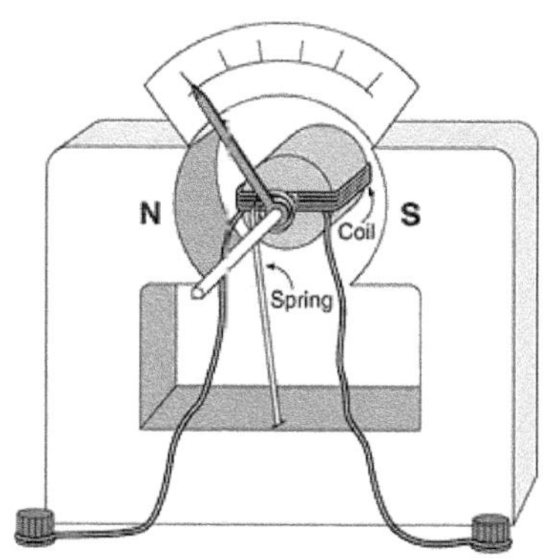

Galvanometer displays measured current

Voltmeter

Voltmeters measure the voltage through two points within a circuit. Like ammeters, voltmeters have two connecting electrodes placed on either side of the element to be measured.

Voltmeters are designed to have infinitely high resistance, theoretically pushing all the current through the circuit element and not letting any current flow through the voltmeter.

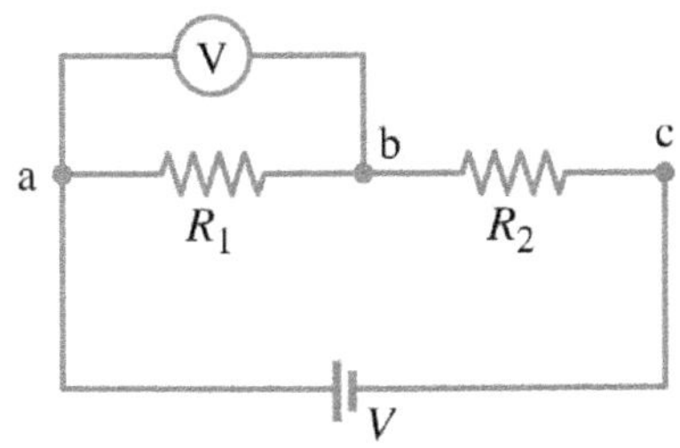

Voltmeter connected across opposite sides of a resistor to measure the voltage through it

Ohmmeter

Ohmmeter is a device used to measure *electrical resistance.*

Unlike ammeters and voltmeters, the ohmmeter has a battery that passes current through resistive elements to measure their resistance.

Figure below depicts an ohmmeter symbol:

Symbol indicating an ohmmeter

Chapter Summary

Circuits

Electric current is the *rate of flow* of electric charge.

Conventional current is in the direction that *positive charge* would flow.

Direct current (DC) is *constant.*

Alternating current (AC) varies *sinusoidally:*

$$I = \frac{V}{R} = \frac{V_0}{R}\sin\omega t = I_0 \sin\omega t$$

Power in an electric circuit:

$$P = IV$$

RC circuits have a characteristic *time constant*:

$$\tau = RC$$

Electric potential

Electric potential is potential *energy per unit charge*:

$$V_a = \frac{\text{PE}_a}{q}$$

A source of an electromotive force (*EMF*) transforms energy from some other form of electrical energy.

Batteries are a source of *constant potential difference.*

Electric potential of a *point charge*:

$$V = k\frac{Q}{r} \quad \text{[single point charge V = 0 at r = } \infty\text{]} \quad = \frac{1}{4\pi\epsilon_0}\frac{Q}{r}$$

Kirchhoff's rules:

Sum of currents entering a junction equals sum of the currents leaving.

Total potential difference around a closed loop is *zero.*

Resistance

Resistance is the ratio of *voltage to current*:

$V = IR$

Ohmic materials have *constant resistance, independent of voltage.*

Resistance is determined by *shape and material*:

$R = \rho\frac{\ell}{A}$ (ρ is the resistivity)

Battery is a source of EMF in parallel with internal resistance.

Resistors in *series*:

$R_{eq} = R_1 + R_2 + R_3$

Resistors in *parallel:*

$$\frac{1}{R_{eq}} = \frac{1}{R_1} + \frac{1}{R_2} + \frac{1}{R_3}$$

Capacitance

Capacitors are non-touching conductors with equal and opposite charges.

Capacitance is given by:

$Q = CV$

Capacitance of a *parallel-plate capacitor* is:

$$C = \epsilon_0 \frac{A}{d}$$

Dielectric is an insulator (usually between plates of parallel-plate capacitor)

Dielectric constant gives a ratio of the total field to an external field.

Energy density in an electric field:

$$\text{Energy density} = \frac{PE}{volume} = \frac{1}{2} \epsilon_0 E^2$$

Capacitors in *parallel:*

$V = C_{eq} V = C_1V + C_2V + C_3V = (C_1 + C_2 + C_3)V$

$C_{eq} = C_1 + C_2 + C_3$

Capacitors in *series*:

$$\frac{Q}{C_{eq}} = \frac{Q}{C_1} + \frac{Q}{C_2} + \frac{Q}{C_3} = Q\left(\frac{1}{C_1} + \frac{1}{C_2} + \frac{1}{C_3}\right)$$

$$\frac{1}{C_{eq}} = \frac{1}{C_1} + \frac{1}{C_2} + \frac{1}{C_3}$$

Average RMS

Current:

$$I_{rms} = \sqrt{\overline{I^2}} = \frac{I_0}{\sqrt{2}} = 0.707\ I_0$$

Voltage:

$$V_{rms} = \sqrt{\overline{V^2}} = \frac{V_0}{\sqrt{2}} = 0.707\ V_0$$

$$I = \frac{\Delta Q}{\Delta t} = neAv_d$$

LRC series circuit:

$$z = \sqrt{R^2 + (X_L - X_c)^2}$$

Magnetic fields and flux

Magnetic flux:

$$\Phi_B = B_{\perp}A = BA\cos(\theta)$$

Changing magnetic flux induces EMF:

$$\mathcal{E} = -N\frac{\Delta\Phi_B}{\Delta t}$$

Induced EMF produces a current that opposes the original flux change.

Changing magnetic fields produce an *electric field.*

Energy density stored in a magnetic field:

$$u = \text{energy density} = \frac{1}{2} \times \frac{B^2}{\mu_0}$$

Transformers

Transformers use induction to change voltage:

$$\frac{V_S}{V_P} = \frac{N_S}{N_P}$$

Mutual inductance:

$$\mathcal{E}_2 = -M\frac{\Delta I_1}{\Delta t}, \quad \text{or} \quad \mathcal{E}_1 = -M\frac{\Delta I_2}{\Delta t}$$

Notes for active learning

Practice Problems: Electric Circuits

1. Two parallel circular plates with radii 7 mm carrying equal-magnitude surface charge densities of ±3 μC/m^2 are separated by a distance of 1 mm. How much stored energy do the plates have? (Use dielectric constant $k = 1$ and electric permittivity $\mathcal{E}_0 = 8.854 \times 10^{-12}$ F/m)

A. 226 nJ
B. 17 nJ
C. 127 nJ
D. 78 nJ
E. 33 nJ

2. The potential difference between the plates of a parallel plate capacitor is 75 V. The magnitude of the charge on each plate is 3.5 μC. What is the capacitance of the capacitor?

A. 116×10^{-6} F
B. 37×10^{-6} F
C. 0.7×10^{-6} F
D. 478×10^{-6} F
E. 4.7×10^{-8} F

3. An electron is released from rest at a distance of 5 cm from a proton. How fast will the electron move 2 cm from the proton? (Use Coulomb's constant $k = 9 \times 10^9$ N·m^2/C^2, the mass of an electron = 9.1×10^{-31} kg, the charge of an electron = -1.6×10^{-19} C and the charge of a proton = 1.6×10^{-19} C)

A. 92 m/s
B. 123 m/s
C. 147 m/s
D. 1.3×10^3 m/s
E. 4.8×10^5 m/s

4. A current of 5 A flows through an electrical device for 12 seconds. How many electrons flow through this device during this time? (Use the charge of an electron = -1.6×10^{-19} C)

A. 5.2×10^{18} electrons
B. 3.8×10^{20} electrons
C. 6.3×10^{19} electrons
D. 1.2×10^8 electrons
E. 9.1×10^{14} electrons

5. 1 mm separates two parallel plates. If the potential difference between them is 3 V, what is the magnitude of their surface charge densities? (Use electric permittivity $\mathcal{E}_0 = 8.854 \times 10^{-12}$ F/m)

A. 64×10^{-9} C/m^2
B. 33×10^{-9} C/m^2
C. 27×10^{-9} C/m^2
D. 16×10^{-9} C/m^2
E. 76×10^{-9} C/m^2

6. A 3 Ω resistor is connected in parallel with a 6 Ω resistor; both resistors are connected in series with a 4 Ω resistor, and all three resistors are connected to an 18 V battery, as shown. If a 3 Ω resistor burnt out and exhibits infinite resistance, which of the following is correct?

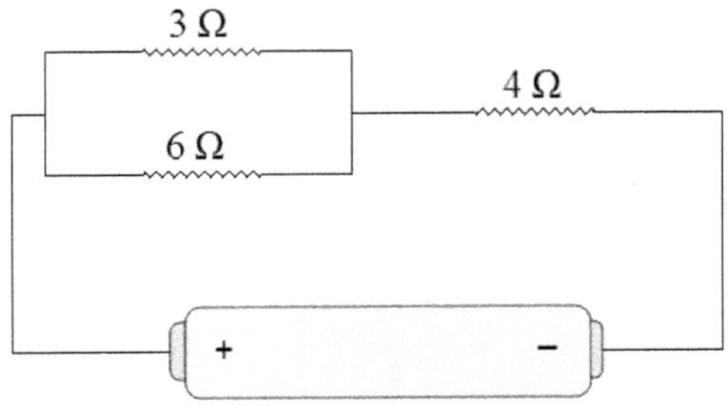

A. The power dissipated in the circuit increases
B. The current provided by the battery remains the same
C. The current in the 6 Ω resistor decreases
D. The current in the 4 Ω resistor decreases to zero
E. The current in the 6 Ω resistor increases

7. Each plate of a parallel-plate air capacitor has an area of 0.005 m^2, and the separation of the plates is 0.08 mm. What is the potential difference across the capacitor when an electric field of 5.6×10^6 V/m is present between the plates?

A. 367 V
B. 578 V
C. 448 V
D. 227 V
E. 834 V

8. What is the current in a wire if a total of 2.3×10^{13} electrons pass a given point in a wire in 15 s? (Use the charge of an electron = -1.6×10^{-19} C)

A. 0.25 μA
B. 3.2 μA
C. 7.1 μA
D. 1.3 μA
E. 0.88 μA

9. The resistivity of gold is 2.44×10^{-8} Ω·m at a temperature of 20 °C. A gold wire, 0.6 mm in diameter and 48 cm long, carries a current of 340 mA. What is the number of electrons per second passing a given wire cross-section? (Use the charge of an electron = -1.6×10^{-19} C)

A. 2.1×10^{18} electrons
B. 2.8×10^{14} electrons
C. 1.2×10^{22} electrons
D. 2.4×10^{17} electrons
E. 6.3×10^{15} electrons

10. When the frequency of the AC voltage across a capacitor is doubled, the capacitive reactance of that capacitor will:

A. become zero
B. decrease to ½ its original value
C. increase to 4 times its original value
D. decrease to ¼ its original values
E. increase to 2 times its original value

11. A charged particle of mass 0.006 kg is subjected to a 6 T magnetic field, which acts at a right angle to its motion. The particle moves in a circle of radius 0.1 m at a speed of 3 m/s. What is the magnitude of the charge on the particle?

A. 3 C
B. 30 C
C. 3.6 C
D. 0.03 C
E. 1.8 C

12. When a current flows through a metal wire, the moving charges are:

I. protons
II. neutrons
III. electrons

A. I only
B. II only
C. III only
D. I and II only
E. I and III only

13. A kilowatt-hour is equivalent to:

A. 3.6×10^6 J/s
B. 3.6×10^6 J
C. 3.6×10^3 W
D. 3.6×10^3 J
E. 3.6×10^3 J/s

14. For the graph shown, what physical quantity does the slope of the graph represent?

A. 1 / Voltage
B. Resistance
C. Power
D. Charge
E. 1 / Resistance

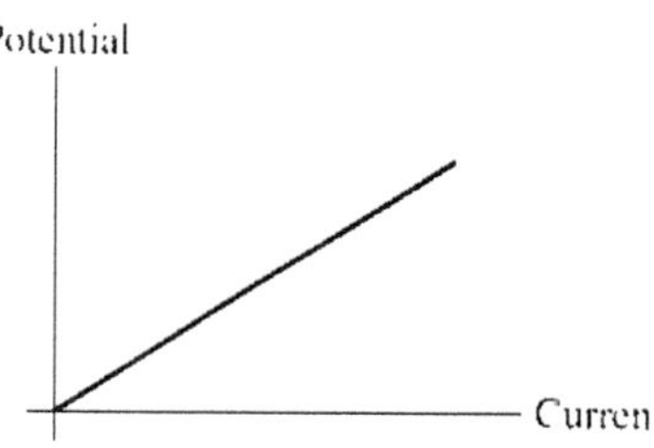

15. What is the voltage across a 15 Ω resistor with 5 A current passing through it?

A. 3 V
B. 5 V
C. 15 V
D. 30 V
E. 75 V

Solutions: Electric Circuits

1. D is correct.

Find area:

$A = \pi r^2$

$A = \pi(7 \times 10^{-3}\ \text{m})^2$

$A = 1.54 \times 10^{-4}\ \text{m}^2$

Find capacitance:

$C = A\mathcal{E}_0 k / d$

$C = [(1.54 \times 10^{-4}\ \text{m}^2)\cdot(8.854 \times 10^{-12}\ \text{F/m})\cdot(1)] / (1 \times 10^{-3}\ \text{m})$

$C = 1.36 \times 10^{-12}\ \text{F}$

Find charge:

$\sigma = 3 \times 10^{-6}\ \text{C/m}^2$

$\sigma = Q / A$

$Q = \sigma A$

$Q = (3 \times 10^{-6}\ \text{C/m}^2)\cdot(1.54 \times 10^{-4}\ \text{m}^2)$

$Q = 4.6 \times 10^{-10}\ \text{C}$

Find potential energy:

$U = ½Q^2 / C$

$U = ½(4.6 \times 10^{-10}\ \text{C})^2 / (1.36 \times 10^{-12}\ \text{F})$

$U = 78 \times 10^{-9}\ \text{J}$

2. E is correct.

Use the relationship:

$Q = CV$

$C = Q / V$

$C = (3.5 \times 10^{-6}\ \text{C}) / (75\ \text{V})$

$C = 4.7 \times 10^{-8}\ \text{F}$

3. B is correct.

Use the energy relationship:

$PE_{before} = PE_{after} + KE$

Electrostatic Potential Energy:

$PE = kQq / r$

Solve:

$kQq / r_1 = kQq / r_2 + ½mv^2$

$½mv^2 = kQq(1 / r_1 - 1 / r_2)$

$v^2 = (2kQq / m)·(1 / r_1 - 1 / r_2)$

$v = \sqrt{[(2kQq / m)·(1 / r_1 - 1 / r_2)]}$

$v = \sqrt{[(2)·(9 \times 10^9 N·m^2/C^2)·(1.6 \times 10^{-19} C)·(1.6 \times 10^{-19} C) / (9.1 \times 10^{-31} kg)]·[(1 / 0.02 m) - (1 / 0.05 m)]}$

$v = 123 m/s$

Note: only the magnitude of the charge is used.

4. B is correct.

1 amp = 1 C/s

\# of electrons = (5 C/s)·(12 s / 1)·(1 electron / 1.6×10^{-19} C)

\# of electrons = 3.8×10^{20} electrons

5. C is correct.

$Q = A\varepsilon_o V / d$

$Q / A = \varepsilon_o V / d$

$Q / A = (8.854 \times 10^{-12} F/m)·(3 V) / (1 \times 10^3 m)$

$Q / A = 27 \times 10^{-9} C/m^2$

6. E is correct.

Equivalent resistance of the 3 Ω and 6 Ω resistors is:

$1 / R_{eq} = 1 / (3 Ω) + 1 / (6 Ω)$

$R_{eq} = 2 Ω$

continued...

Voltage across equivalent resistors (i.e., the 6 Ω resistor) is given by voltage divider relationship:

$$V_6 = 18\text{ V }(2\ \Omega) / (2\ \Omega + 4\ \Omega) = 6\text{ V}$$

Current through the 6 Ω resistor is:

$$I_6 = V_6 / 6\ \Omega = 1\text{ A}$$

After the 3 Ω resistor burns out, the voltage across the 6 Ω resistor is found using the voltage divider relationship:

$$V_6' = 18\text{ V }(6\ \Omega) / (6\ \Omega + 4\ \Omega) = 10.8\text{ V}$$

Now the current through the 6 Ω resistor is:

$$I_6' = V_6' / 6\ \Omega = 1.8\text{ A}$$

The current has increased.

7. C is correct.

Parallel-plate capacitor voltage:

$$V = Ed$$

$$V = (5.6 \times 10^6\text{ V/m})\cdot(0.08 \times 10^{-3}\text{ m})$$

$$V = 448\text{ V}$$

8. A is correct.

$$1\text{ amp} = 1\text{ C/s}$$

$$I = (2.3 \times 10^{13}\text{ electrons} / 1)\cdot(1.6 \times 10^{-19}\text{ C} / 1\text{ electron})\cdot(1 / 15\text{ s})$$

$$I = 0.25 \times 10^{-6}\text{ amps}$$

$$I = 0.25\ \mu\text{A}$$

9. A is correct.

Use the definition of one ampere:

$$1\text{ amp} = 1\text{ C/s}$$

Let I_e denote the current expressed in electrons per second.

Solve:

$$I_e = (340 \times 10^{-3}\text{ C} / 1\text{ s})\cdot(1\text{ electron} / 1.6 \times 10^{-19}\text{ C})$$

$$I_e = 2.1 \times 10^{18}\text{ electrons/s}$$

10. B is correct.

Capacitive Reactance formula:

$$X_c = 1 / (2\pi Cf)$$

If f is doubled:

$$X_{c2} = 1 / [2\pi C(2f)]$$

$$X_{c2} = ½[1 / (2\pi Cf)]$$

$$X_{c2} = ½X_c$$

Capacitive reactance is halved.

11. D is correct.

Find the forces acting on the particle:

$$F_{\text{centripetal}} = F_{\text{magnetic}}$$

$$ma_{\text{centripetal}} = qvB$$

$$mv^2 / r = qvB$$

$$q = mv / Br$$

$$q = [(0.006 \text{ kg})\cdot(3 \text{ m/s})] / [(6 \text{ T})\cdot(0.1 \text{ m})]$$

$$q = 0.03 \text{ C}$$

12. C is correct.

In a solid, the nuclei locations are fixed; only the electrons move.

13. B is correct.

$$1 \text{ kW} = 1{,}000 \text{ W}$$

$$1{,}000 \text{ W} = 1{,}000 \text{ J/s}$$

$$1 \text{ hr} = (60 \text{ min})\cdot(60 \text{ sec/min})$$

$$1 \text{ hr} = 3{,}600 \text{ s}$$

$$\text{kW}\cdot\text{hr} = (1{,}000 \text{ J/s})\cdot(3{,}600 \text{ s})$$

$$\text{kW}\cdot\text{hr} = 3.6 \times 10^3 \text{ J}$$

14. B is correct.

$$V = IR$$

$$R = V / I$$

15. E is correct.

Voltage = current × resistance

$$V = IR$$

$$V = (5\ \text{A})\cdot(15\ \Omega)$$

$$V = 75\ \text{V}$$

Notes for active learning

Notes for active learning

CHAPTER 12

Atomic & Nuclear Structure

Atomic and Nuclear Spectra

Bohr Model and Emission Spectrum of Hydrogen

Atomic Energy Levels

Quantum Mechanics

Electronic Configuration

Atomic Nucleus

Atomic Number and Atomic Weight

Nuclear Forces

Radioactivity

Decay Types

Radioactive Decay Rates

Decay Series

Measuring Radioactive Decay

Fission Reactions

Fusion Reactions

Mass Deficit, Energy Liberated and Binding Energy

Mass Spectrometry

Radiation Exposure

Medical Applications of Radioactivity

Chapter Summary

Page intentionally left blank

Origins of Modern Atomic Theory

Dalton's model of an atom

John Dalton (1755-1844) proposed the *atomic structure* in the early 1800s. Dalton reintroduced atomic theory to explain chemical reactions.

Dalton's theory centered on five main concepts:

1. All matter is made of indivisible particles of *atoms*.
2. An *element* is made up of identical atoms.
3. Different elements have atoms with different masses.
4. Chemical compounds are made of atoms in specific integer ratios.
5. Atoms are neither created nor destroyed in chemical reactions.

Discovery of the electron

Nobel laureate J. J. Thomson (1856-1940) discovered the *electron* in the late 1800s. By performing cathode ray experiments shown below.

Thomson discovered that the electron is negatively charged by measuring the electron's charge-to-mass ratio and identified electrons as a fundamental particle.

To test the properties of the particles, Thomson placed two oppositely charged electric plates around the cathode ray. The cathode ray was deflected from the negatively charged electric plate and towards the positively charged plate, indicating that the cathode ray was composed of negatively charged particles.

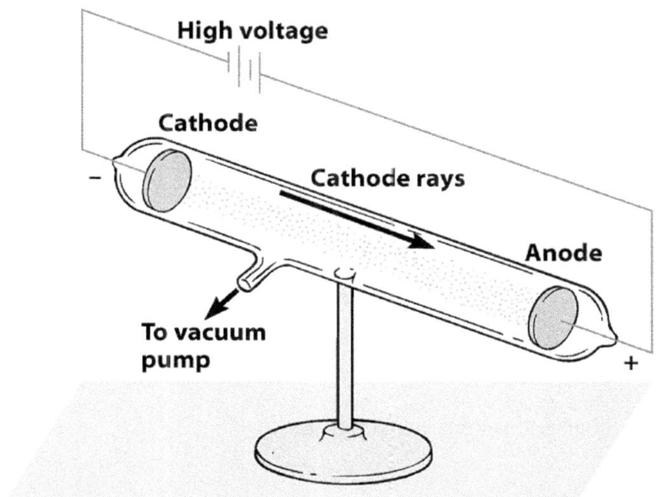

Cathode ray experiment with cathode and anode

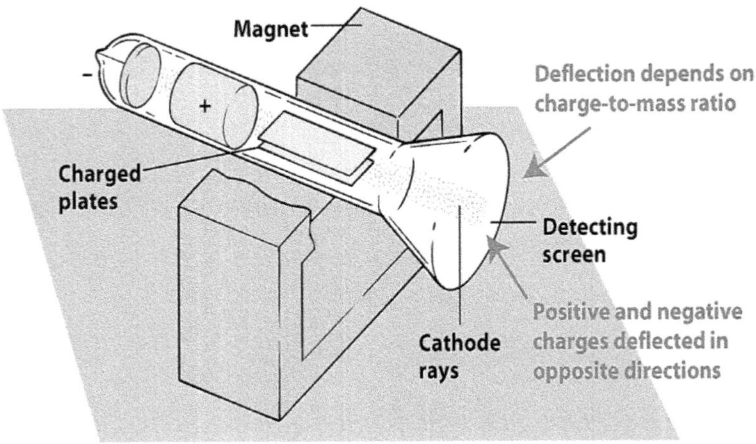

Cathode rays to measure charge-to-mass ratio

Mass of an electron

Around 1906, Nobel laureate Robert Millikan (1868-1953) studied minuscule particles, specifically, charged oil droplets in an electric field.

Millikan observed that the charge on the oil droplets was a multiple of the electron charge. This discovery, along with Thomson's results, was used to calculate the mass of the electron. He observed that the charge changed in steps between one passage of the droplet and the next and were multiples (*quantized*) of a common elementary charge, now known to be 1.602×10^{-19} C.

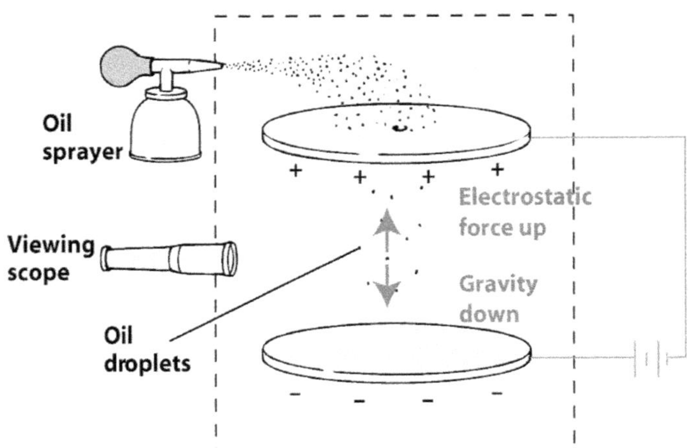

Millikan observed charged oil droplets

Charge-to-mass ratio

Thomson's result (charge-to-mass) was the quantity:

$$\frac{q}{m} = 1.7584 \times 10^{11} \frac{C}{kg}$$

Millikan's result (q = charge) was:

$$q = 1.60 \times 10^{-19} C$$

Combined, the mass of an electron is:

$$m = \frac{q}{1.7584 \times 10^{11}}$$

$$m = 9.11 \times 10^{-31} kg$$

Thomson's "*plum pudding*" model

Dalton claimed that atoms were indivisible and that no smaller quantities were involved. However, Thomson's and Millikan's experiments, along with the discovery of the electron, proved Dalton's statement false.

Thomson's atomic model attempted to reconcile the two properties of atoms known at that time: 1) there are electrons, and 2) atoms have no net electric charge. Based on this knowledge, Thomson theorized that there must be an equal amount of positive charge to cancel the electrons' negative charge. He proposed that the atom was spherical, and this positive charge was everywhere with the negatively charged electrons being distributed evenly throughout it and repelling each other.

This model became known as the "*plum pudding*" model with electrons embedded in a blob of positively charged matter like "*raisins in plum pudding*."

While significant at the time as the first model to describe internal atomic structure, Thomson's model became obsolete following Ernest Rutherford's discovery of the atomic nucleus in 1911. Thomson's model attempted to explain the atom's physical and chemical properties by the number and arrangement of its electrons. While he was generally on the right track, his approach was based on classical mechanics and did not account for quantized energy.

Protons

Nobel laureate Ernest Rutherford (1871-1937), in 1911, conducted an experiment that scattered *alpha particles* by bouncing them off gold foil. Most of the gold atoms are in empty space. Most particles passed through without deflection, but a few scattered at large angles, meaning that a small part of the gold atom must be positively charged.

Alpha particles are *positively charged* particles with *two protons and two neutrons* bound together. They are emitted by an atomic nucleus undergoing radioactive decay and identical to a helium atom nucleus.

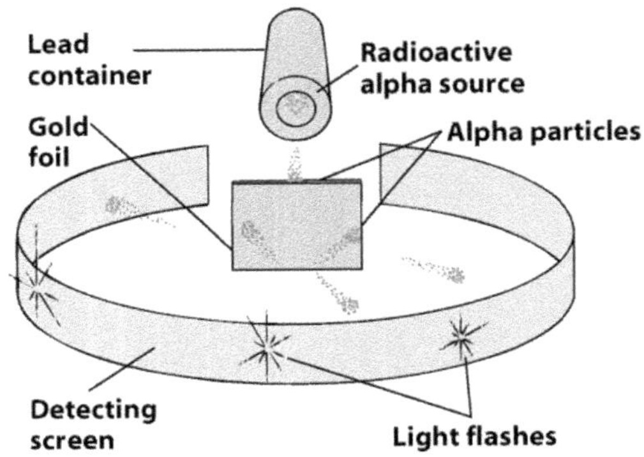

Alpha particles were discovered by being deflected by the dense gold foil

Rutherford concluded that an atom's positive charge resides in a tiny nucleus that contains most of the atom's mass (unlike the spread-out blob of matter containing electrons, as the *plum pudding* model suggested). He named these positively charged subatomic particles *protons.*

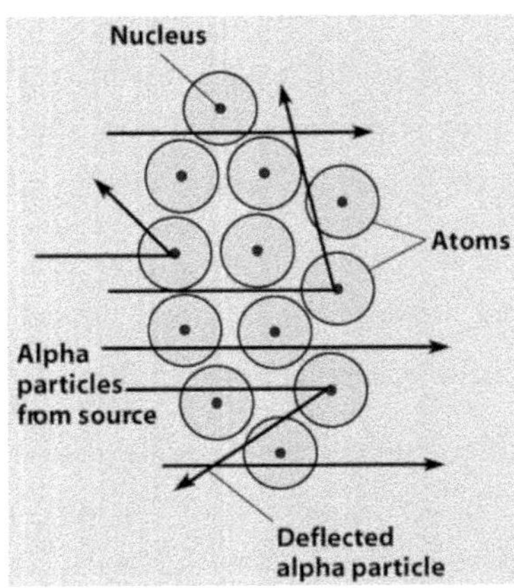

Alpha particle deflection contributed to the current model of the atom's nucleus

Rutherford discovered the atomic nucleus in 1911 and observed the proton in 1919. However, it seemed there must be something in the nucleus besides protons. For instance, helium was known to have an atomic number of 2 but a mass number of 4.

Neutrons

Some scientists thought there were additional protons in the nucleus and an equal number of electrons to cancel out the additional charge. In 1920, Rutherford proposed that an electron and a proton could combine to form a new, *neutral particle.*

In 1932, Nobel laureate James Chadwick (1891-1974) added the existence of neutral *neutrons* in the nucleus to the atomic theory. Neutrons were discovered later than the other subatomic particles because they have no charge; other methods were needed to identify the neutrons other than those used to discover protons and electrons.

Bohr Model and Emission Spectrum of Hydrogen

Bohr model of atoms

In 1913, Nobel laureate Niels Bohr (1885-1962) proposed a model depicting the atom in which the electron orbits the nucleus as the Earth orbits the Sun.

Electrostatic attraction pulls the electron toward the nucleus.

However, the electron orbits at high enough speeds to prevent it from crashing into the nucleus.

Relationship of the electron orbit around the nucleus leads to several important implications:

1. Electrons only exist *in precise allowed orbits.*
2. Within an orbit, the *electron does not radiate.*
3. Radiation is emitted or absorbed *when changing orbits.*

Three implications of Bohr model

1. Electrons *occupy specific orbitals*, which determine their energy.

Orbits are characterized by the orbital number *n,* where $n = 1$ equals the ground state and represents the electron's lowest energy level.

The higher the orbital number *n* of the electron, the higher the energy, and the further the electron is from the nucleus.

2. Electrons *do not radiate* while orbiting the nucleus. As mentioned in earlier chapters, accelerating electric charges emit energy in the form of electromagnetic radiation.

According to classical physics, the electron would radiate away its energy (due to the energy radiated from centripetal acceleration) and crash into the nucleus.

Bohr's model assumes electrons do not radiate energy within energy levels and stay in orbits.

3. Electrons *absorb or release energy when changing energy levels.*

For a ground-state electron to achieve a higher orbital, it must absorb a photon (and thus absorb the photon's energy).

If the electron drops to a lower orbital, then the electron must emit a *photon* equal to the energy difference between orbital levels.

Photons are elementary particles that are a quantum of the electromagnetic field, including electromagnetic radiation such as light and radio waves, and are the force carriers for the electromagnetic force.

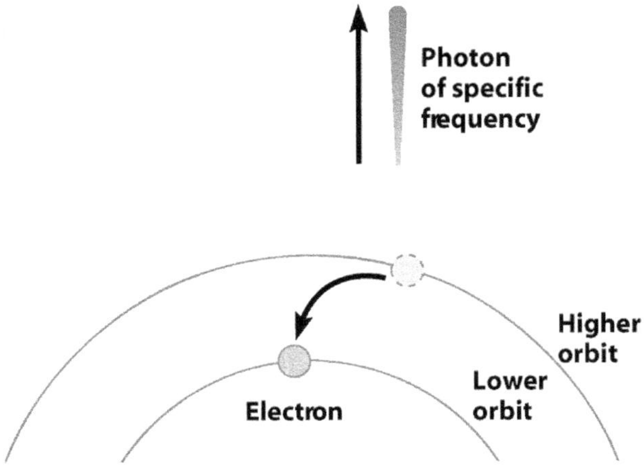

Electrons occupy specific orbitals

Emission line spectrum

An essential application of the Bohr model is identifying elements through their *line spectrum*.

When a high potential difference energizes an element, the electrons are constantly raised to higher energy levels and then dropped back to lower energies.

Emission spectrum of an element is the frequency of electromagnetic radiation emitted due to electrons transitioning from a high-energy state to a lower-energy state. These energies relate the energy differences in the electron orbitals and allow quantization of the specific orbital energies.

Electromagnetic energy of the emitted photons equals the energy difference between the two energy states. Photons (i.e., massless particles) released in the process contain specific energies characterized by frequency.

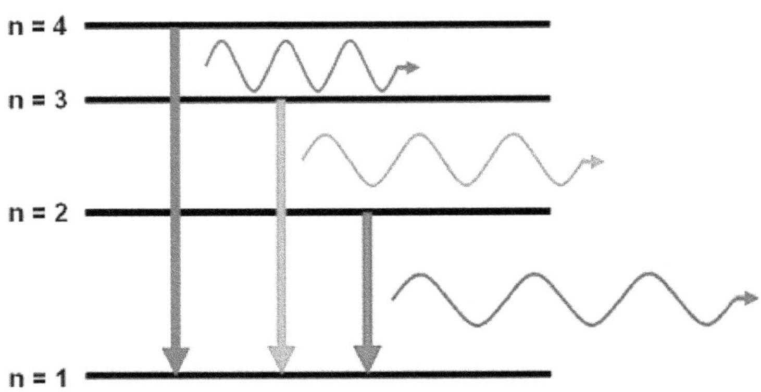

Electromagnetic radiation is emitted due to electrons transitioning from a high-energy to a lower-energy state

Rydberg formula

Emission lines form regular patterns and can be represented by a simple equation.

Emission lines are calculated by combining simple whole numbers using the *Rydberg formula*, which governs the wavelength of the emitted or absorbed radiation:

$$\frac{1}{\lambda} = R\left(\frac{1}{n_f^2} - \frac{1}{n_i^2}\right)$$

where λ is the wavelength, n_f is the final energy level, n_i is the initial energy level, and R is the Rydberg constant of $1.0973 \times 10^7\,\text{m}^{-1}$.

Rydberg formula calculates the wavelength of light emitted by an electron moving between the energy levels of an atom. It works well for transitions between energy levels of a hydrogen atom with only one electron.

However, the Rydberg formula gives incorrect results for atoms with multiple electrons because the amount of screening for inner or outer electron transitions varies, and the equation is too simplistic to compensate for the differences.

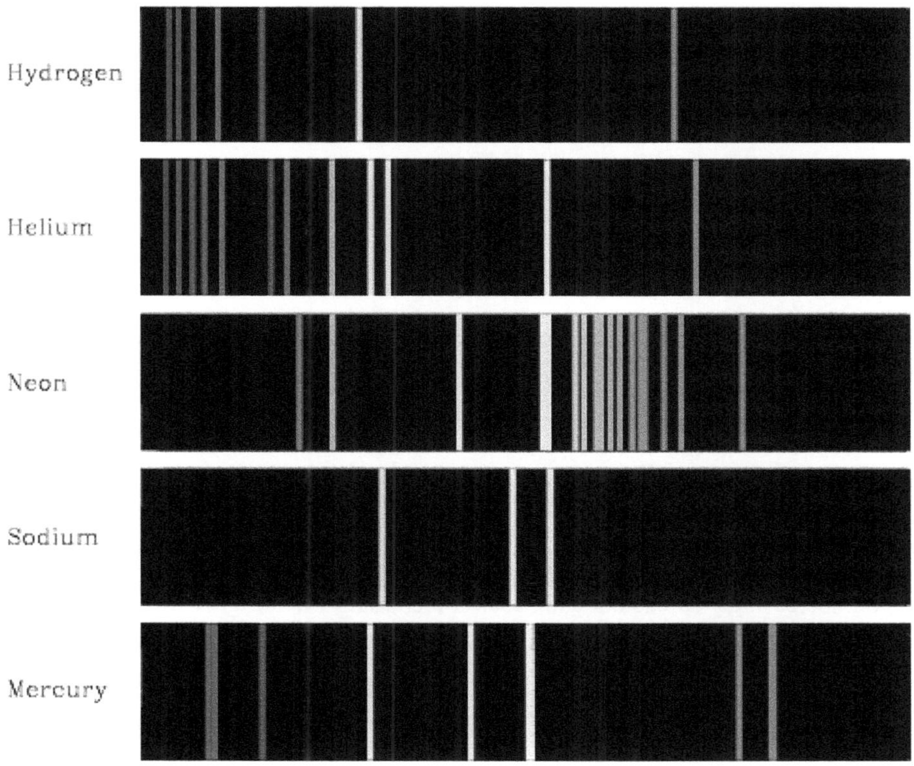

Intensity vs. wavelength emission line spectra for sample elements

Named emission line spectra

Lyman series involve transitions to or from the ground state (n = 1).

Balmer series (in which all lines are in the visible region) corresponds to n = 2.

Paschen series corresponds to n = 3.

Brackett series corresponds to n = 4.

Pfund series corresponds to n = 5.

Humphrey series corresponds to n = 6.

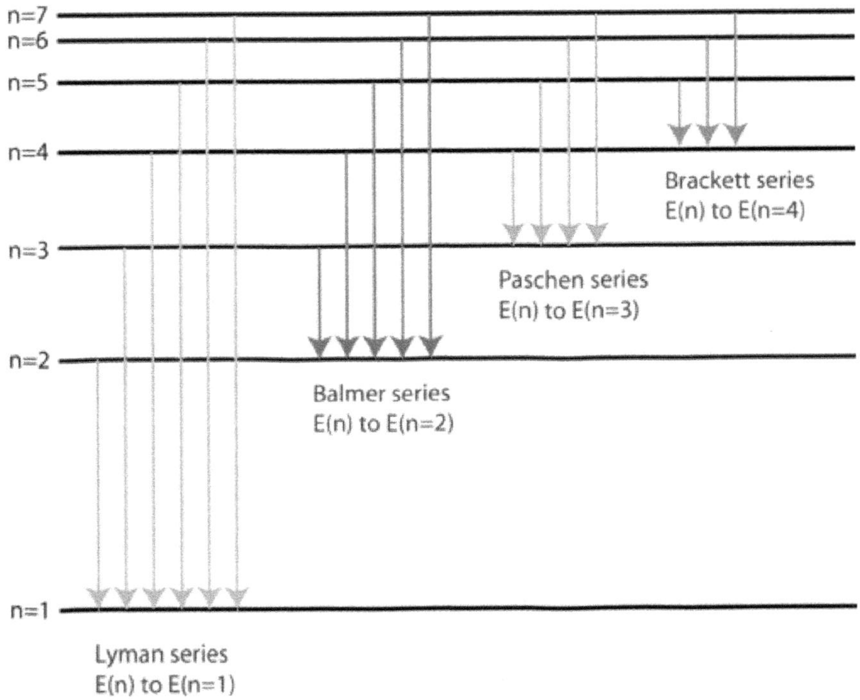

Electron transitions for a hydrogen atom

Further series are unnamed but follow the pattern and equation of the Rydberg equation.

Series are increasingly spread out and occur at increasing wavelengths with lines increasingly faint, corresponding to less rare atomic events. The seventh series of atomic H was demonstrated experimentally at infrared wavelengths in 1972.

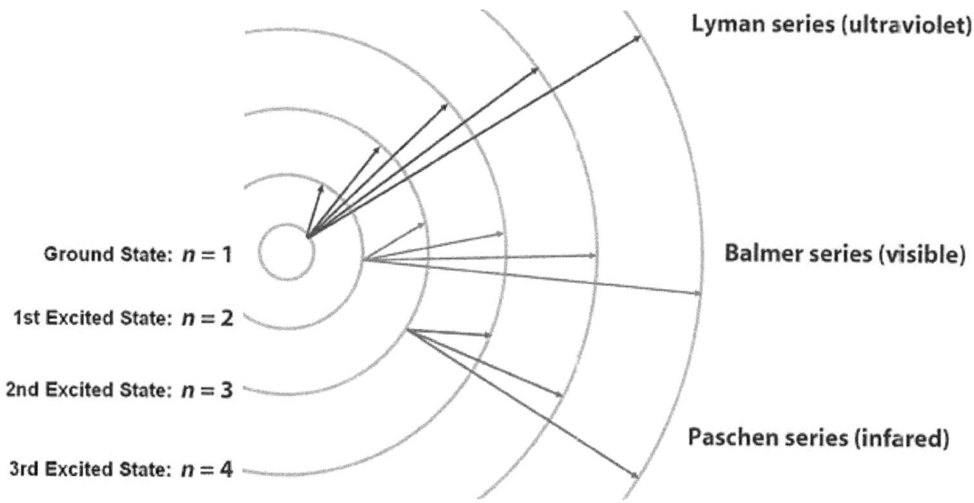

Energy transitions between orbitals for the H atom

Hydrogen line spectra

Emission line spectra of a hydrogen (H) atom have lines representing the changes in quantum energy levels of the atom's single electron.

Hydrogen absorption line spectrum

Hydrogen emission line spectrum

Balmer series

In 1885, Johann Balmer (1825-1898) proposed the Balmer series for the orbital transitions of electrons from any orbital back to n = 2. Wavelength = $1/\lambda = R(1/2^2 - 1/n^2)$

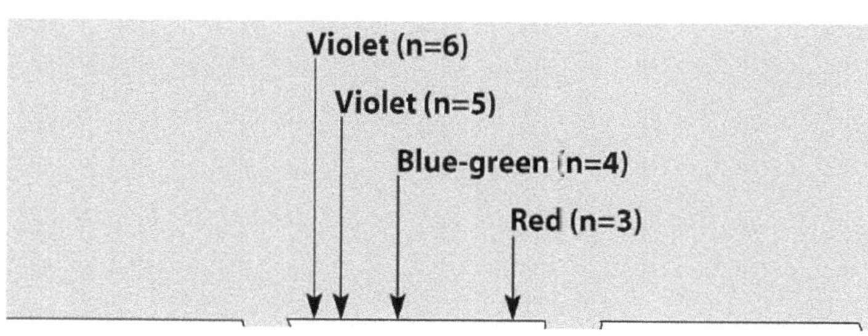

Balmer series to account for electron transitions between orbitals

Notes for active learning

Quantum Theory of Atoms

Atomic energy levels

Quantum theory of the atom describes the series of energy states that an electron can occupy.

Lowest energy state is the "*ground state*," and higher states are "*excited states*."

The energy of the photon emitted (or absorbed) during a transition between the specific energy levels equals the *difference* in state energies.

Violet light represents the higher energy transitions (n = 5 or 6 to n = 2) because violet has the highest visible spectrum energy (mnemonic of ROY G BIV).

Red light has lower energy than violet because of a lower energy transition (n = 3 to n = 2).

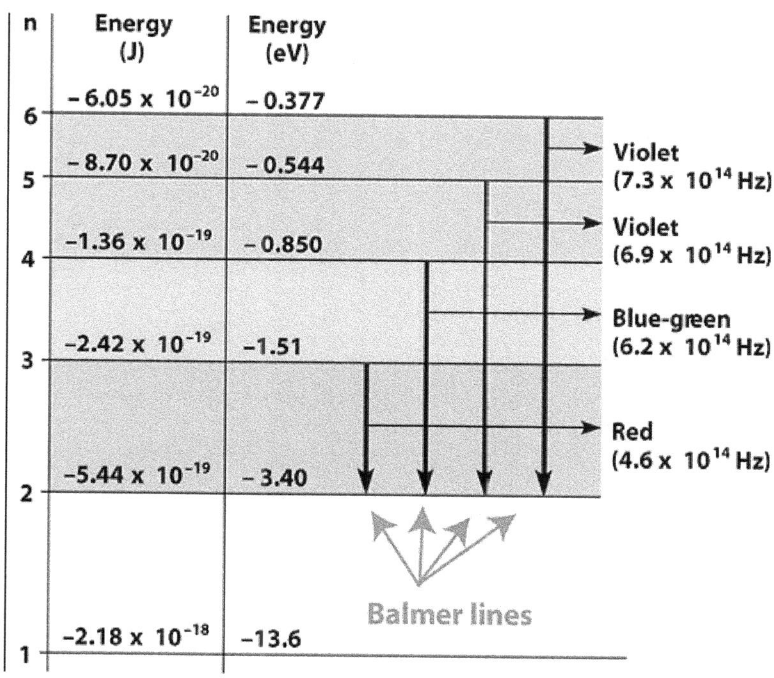

Hydrogen atom and its energy levels

Wave–particle duality

Bohr's theory only holds for the line spectrum of H with only two electrons. When more electrons are added, the theory cannot account for the complications of heavier atoms.

Experiments established the wave-particle duality of light and matter. For example, Young's two-slit experiment (discussed in light wave theory) produced *interference patterns* for photons and electrons.

Wave theory

In 1932, Nobel laureate Louis de Broglie (1892-1987) proposed the existence of *matter waves* regarding the atomic theory.

Matter waves are central to the theory of quantum mechanics as part of wave-particle duality. At scales where measurements have been practical, matter exhibits wave-like behavior. For example, a beam of electrons can be diffracted, like a beam of light or a water wave.

Wavelength is related to *momentum*, and the matter waves in atoms are standing waves.

Louis de Broglie derived the formula for wavelength:

$$\lambda = \frac{h}{mv}$$

where h is Planck's constant 6.63×10^{-34} J·s, λ is particle wavelength, m is mass and v speed.

Wave mechanics

Nobel laureate Erwin Schrödinger (1887-1961) developed the theory of *wave mechanics*.

Wave mechanics treats atoms as three-dimensional systems of waves and incorporates the successful ideas of the Bohr model, along with other foundational principles.

Schrödinger's wave mechanics theory describes the hydrogen atom and other atoms with many electrons, providing a fundamental understanding of chemistry.

Quantum (wave) mechanical theory states that an electron behaves as a standing wave. It describes an electron's possible location in an orbital.

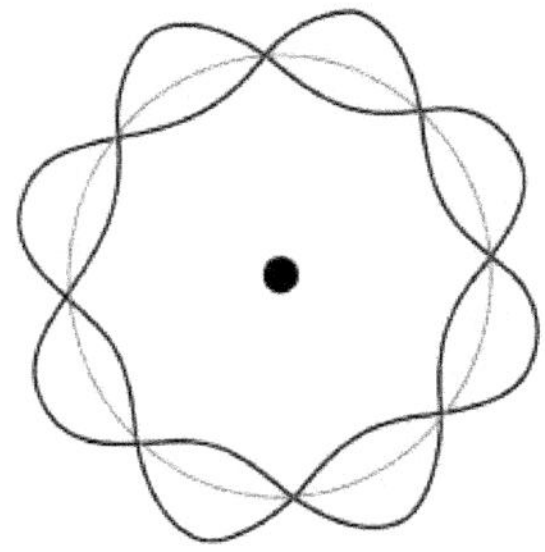

Matter waves around the nucleus with wavelength related to momentum

Quantum Mechanics

Wave functions and probability distributions

Quantum mechanics model is a visualization of *wave functions* and *probability distributions*. It does not represent an electron's motion but indicates the likelihood of finding an electron within a specific orbital within the atom. It is based on probability and not certainty, unlike previous models of the atom.

Quantum numbers specify electronic quantum states and that an electron is delocalized. They define and describe a particular electron in an atom.

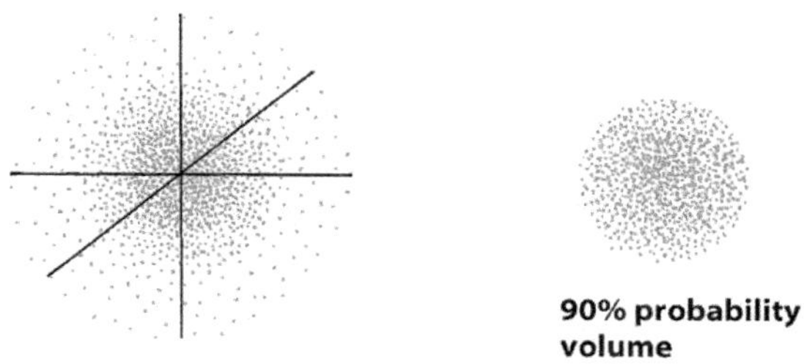

Probability distribution for the location of an orbiting electron

Principal quantum number

Principal quantum number (n) states the energy level (i.e., *shells*) where an electron will most likely be found. n represents the *orbital size* and *average distance from the nucleus*.

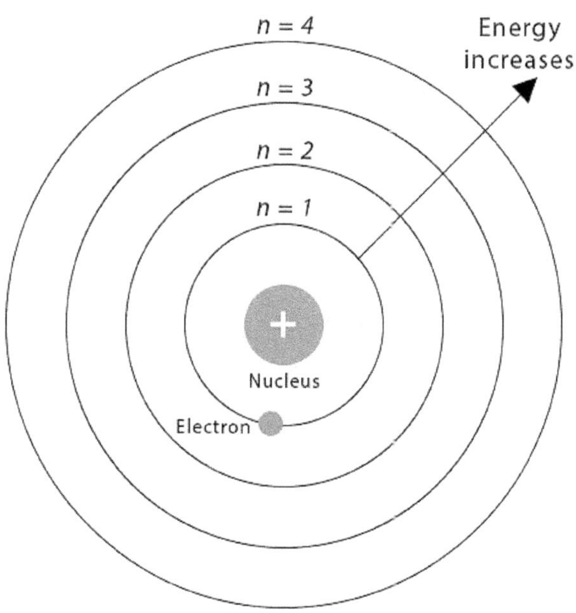

Principal quantum number (n) with relative atomic size and energy

Principal quantum number has positive integer values of $n = 1, 2, 3...$

Electron shells (or principal energy levels) are labeled by principal quantum numbers ($n = 1, 2, 3...$) but can be labeled alphabetically as n = K, L, M. . .

Higher n shells indicate *larger* orbitals *further* from the nucleus and *higher energy* levels.

Angular momentum quantum number

Angular momentum (or *azimuthal*) *quantum number* is the second quantum number designated with the lowercase letter l. It describes the *shape of the orbital* electron occupies.

Lowest value of l is 0, and the highest, depending on the principal quantum number, is $n - 1$.

Angular momentum quantum number describes the spatial distribution (or shape) of the orbits labeled *s*, *p*, *d*, *f*, *g*, *h*, etc. These are *subshells.*

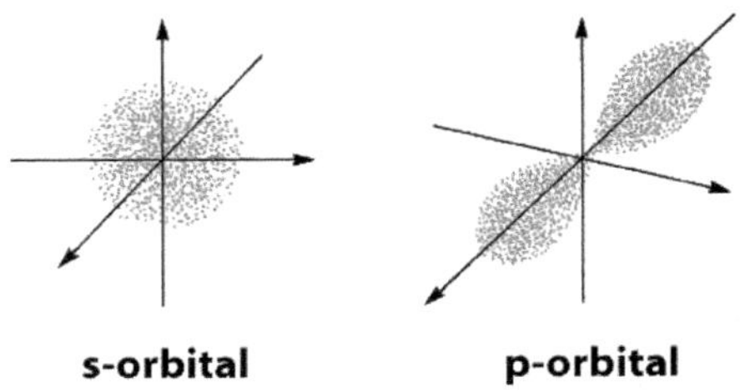

Subshells as spatial distributions of orbitals

Orbital	
s	$l = 0$
p	$l = 1$
d	$l = 2$
f	$l = 3$

Orbitals and associated angular momentum quantum number

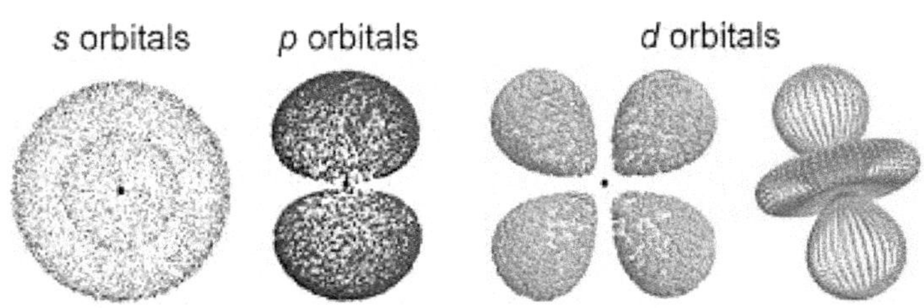

S orbital ($l = 0$), p orbitals ($l = 1$) and d orbitals ($l = 2$)

Magnetic quantum number

Magnetic quantum number (m), the third quantum number, states the spatial orientation of the orbit within the energy (n) and shape (l). Magnetic quantum number (m_l) divides the subshell into specific orbitals that hold the electron.

Number of orbitals is given by $2l + 1$. m_l has values of $-l$ to l, with increments of one, including zero ($-l, .., 0, ..., +l$).

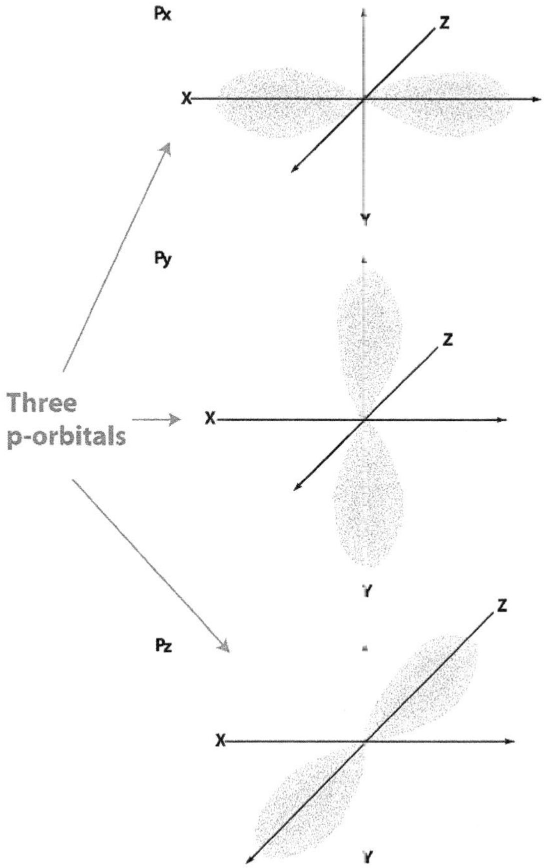

p orbitals orient along the x-, y- and z-axis and can accommodate 6 electrons

m_l range from $-l$ to $+l$, including 0.

- *s* subshell has $l = 0$; the m_l has a single value (0) and one orbital.

  s subshell can hold 2 electrons.

- *p* subshell has $l = 1$; the m_l has values of $-1, 0, +1$.

  p subshell has three orbitals holding up to 6 electrons.

- *d* subshell has $l = 2$; $m_l = -2, -1, 0, +1, +2$.

  d subshell has five orbitals with up to 10 electrons.

- *f* subshell has $l = 3$; $m_l = -3, -2, -1, 0, +1, +2, +3$.

  f subshell has seven orbitals with up to 14 electrons.

Each succeeding subshell holds four more electrons than its predecessor.

Spin quantum number

Spin quantum number, the fourth quantum number, is either +½ or –½ and represents the electron spin orientation. Spin quantum number has only two values because an electron only spins in one of two directions (+½ or –½).

By convention, the first electron to occupy an orbital has a spin number of +½ while the second electron has a spin number of −½.

In orbital diagrams, the +½ electrons are represented by upward arrows, and the −½ electrons are indicated by downward arrows.

Energy level	Number of subshells	Subshells	# of orbitals in subshell	Total number of electrons in a subshell
1	1	*s*	1	2
2	2	*s*	1	2
		p	3	6
3	3	*s*	1	2
		p	3	6
		d	5	10
4	4	*s*	1	2
		p	3	6
		d	5	10
		f	7	14

Quantum numbers represent the locations of the electrons within the atom

Electronic Configuration

Pauli Exclusion Principle

In 1925, Nobel laureate Wolfgang Pauli (1900-1958) proposed the *Pauli Exclusion Principle,* which states that no two electrons in the same atom may have the same four quantum numbers (n, l, m_l, and m_s).

Maximum number of electrons in an orbital is two, and each electron must possess opposite spins (i.e., +½ or –½) because electrons of the same magnetic spin cannot be in the same subshell of an orbital.

Pauli's exclusion principle states that no two electrons in the same atom can have the *same set of quantum numbers.*

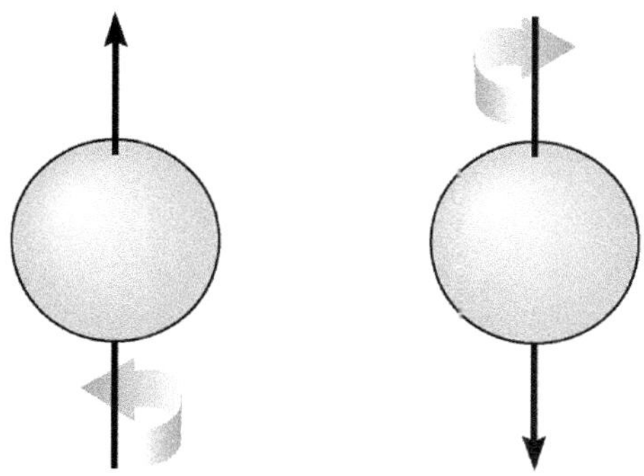

Each orbital contains only two electrons, and the electrons must have opposite spins

Electrons in atomic orbitals

Electron configuration places electrons in atomic orbitals and specifies their quantum state.

Chemical properties of an atom are determined by examining the atom's electronic structure.

When writing electron configurations, following the Pauli exclusion principle is imperative.

Each electron has unique quantum numbers, with a maximum of two electrons per orbital; one spins up, and one spins down.

Electrons fill available orbitals in the order of increasing energy (n).

Each shell only holds a finite number of subshells and the corresponding number of electrons.

Subshells

Electron capacity in each subshell of an orbital is as follows:

s = 2 electrons in 1 subshell

p = 6 electrons in 3 subshells

d = 10 electrons in 5 subshells

f = 14 electrons in 7 subshells

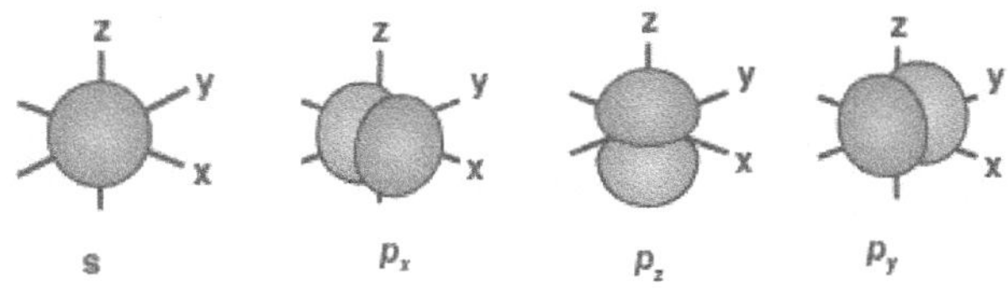

One s orbital (on the left)and three p orbitals (on the right)

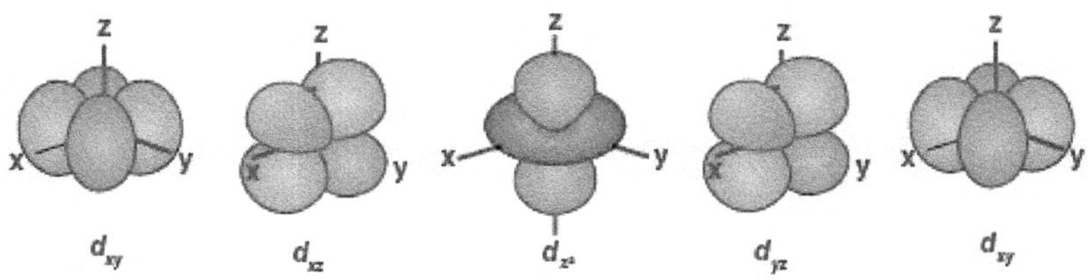

Five d orbitals

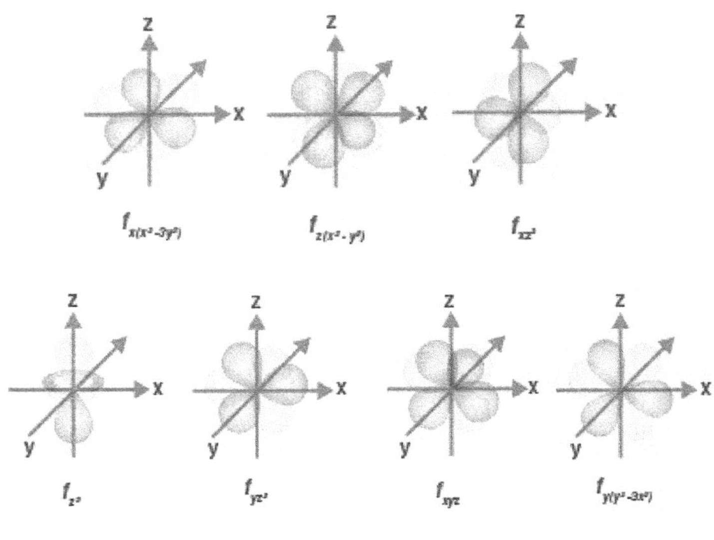

Ten f orbital

Orbitals hold two electrons

Each orbital contains a maximum of two electrons. It is possible to fill half an orbital, but it must be in the outer (valence) shell. For example, strontium has 38 electrons.

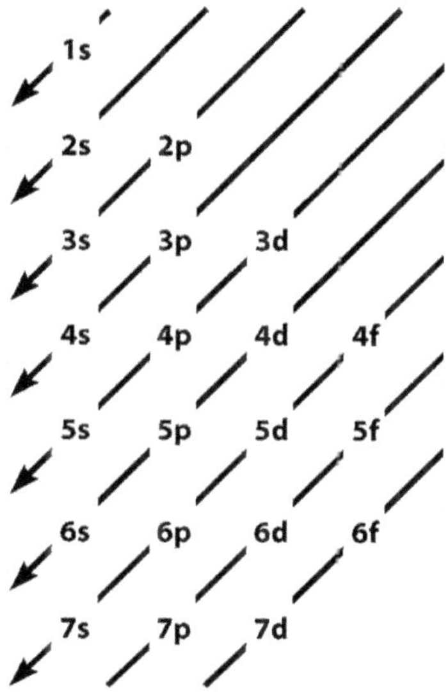

$1s^2\,2s^2\,2p^6\,3s^2\,3p^6\,4s^2\,3d^{10}\,4p^6\,5s^2$

Noble gases (VIIIA) have filled valence shells and are inert (or do not react readily).

Elements with 1 to 3 outer electrons *lose valence electrons* to become positive ions (e.g., metals).

Elements with 5 to 7 outer electrons tend *to gain electrons* and form negative ions (such as nonmetals).

Metalloids are elements that have a mixture of properties of metals and nonmetals. There is no standard definition of a metalloid or agreement on which elements belong to this category.

Six commonly recognized metalloids are boron, silicon, germanium, arsenic, antimony, and tellurium.

Carbon, aluminum, selenium, polonium, and astatine are five less frequently classified metalloid elements.

Periodic Table shows all eleven metalloid elements in a diagonal in the p-block, extending from boron at the upper left to astatine at the lower right.

Periodic Table of the Elements

Atomic Number	Valence
Symbol	
Name	
Atomic Mass	

1 IA 1A	2 IIA 2A	3 IIIB 3B	4 IVB 4B	5 VB 5B	6 VIB 6B	7 VIIB 7B	8	9 VIII 8	10	11 IB 1B	12 IIB 2B	13 IIIA 3A	14 IVA 4A	15 VA 5A	16 VIA 6A	17 VIIA 7A	18 VIIIA 8A
1 -1, +1 **H** Hydrogen 1.008																	2 0 **He** Helium 4.003
3 +1 **Li** Lithium 6.941	4 +2 **Be** Beryllium 9.012											5 +3 **B** Boron 10.811	6 +4,+3,+2,+1 -4,-3 -2,-1 **C** Carbon 12.011	7 +5,+3,-3 **N** Nitrogen 14.007	8 -2 **O** Oxygen 15.999	9 -1 **F** Fluorine 18.998	10 0 **Ne** Neon 20.180
11 +1 **Na** Sodium 22.990	12 +2 **Mg** Magnesium 24.305											13 +3 **Al** Aluminum 26.982	14 +4,-4 **Si** Silicon 28.086	15 +5,+3,-3 **P** Phosphorus 30.974	16 +6,+4,+2,-2 **S** Sulfur 32.066	17 +7,+5,+3,+1 -1 **Cl** Chlorine 35.453	18 0 **Ar** Argon 39.948
19 +1 **K** Potassium 39.098	20 +2 **Ca** Calcium 40.078	21 +3 **Sc** Scandium 44.956	22 +4 **Ti** Titanium 47.88	23 +5 **V** Vanadium 50.942	24 +6,+3 **Cr** Chromium 51.996	25 +7,+4,+2 **Mn** Manganese 54.938	26 +6,+3,+2 **Fe** Iron 55.845	27 +3,+2 **Co** Cobalt 58.933	28 +2 **Ni** Nickel 58.693	29 +2 **Cu** Copper 63.546	30 +2 **Zn** Zinc 65.38	31 +3 **Ga** Gallium 69.723	32 +4,+2,-4 **Ge** Germanium 72.631	33 +5,+3,-3 **As** Arsenic 74.922	34 +6,+4,+2,-2 **Se** Selenium 78.971	35 +5,+3,+1,-1 **Br** Bromine 79.904	36 +2,0 **Kr** Krypton 84.798
37 +1 **Rb** Rubidium 85.468	38 +2 **Sr** Strontium 87.62	39 +3 **Y** Yttrium 88.906	40 +4 **Zr** Zirconium 91.224	41 +5 **Nb** Niobium 92.906	42 +6,+4 **Mo** Molybdenum 95.95	43 +7,+4 **Tc** Technetium 98.907	44 +4,+3 **Ru** Ruthenium 101.07	45 +3 **Rh** Rhodium 102.906	46 +4,+2 **Pd** Palladium 106.42	47 +1 **Ag** Silver 107.868	48 +2 **Cd** Cadmium 112.414	49 +3 **In** Indium 114.818	50 +4,+2,-4 **Sn** Tin 118.711	51 +5,+3,-3 **Sb** Antimony 121.760	52 +6,+4,+2,-2 **Te** Tellurium 127.6	53 +7,+5,+3,+1 -1 **I** Iodine 126.904	54 +6,+4,+2,0 **Xe** Xenon 131.294
55 +1 **Cs** Cesium 132.905	56 +2 **Ba** Barium 137.328	57-71	72 +4 **Hf** Hafnium 178.49	73 +5 **Ta** Tantalum 180.948	74 +6,+4 **W** Tungsten 183.85	75 +4 **Re** Rhenium 186.207	76 +4 **Os** Osmium 190.23	77 +4,+3 **Ir** Iridium 192.22	78 +4,+2 **Pt** Platinum 195.08	79 +3 **Au** Gold 196.967	80 +2,+1 **Hg** Mercury 200.59	81 +3,+1 **Tl** Thallium 204.383	82 +4,+2 **Pb** Lead 207.2	83 +3 **Bi** Bismuth 208.980	84 +4,+2,-2 **Po** Polonium [208.982]	85 +1,-1 **At** Astatine 209.987	86 +2,0 **Rn** Radon 222.018
87 +1 **Fr** Francium 223.020	88 +2 **Ra** Radium 226.025	89-103	104 +4 **Rf** Rutherfordium [261]	105 +5 **Db** Dubnium [262]	106 +6 **Sg** Seaborgium [266]	107 +7 **Bh** Bohrium [264]	108 +8 **Hs** Hassium [269]	109 unknown **Mt** Meitnerium [278]	110 unknown **Ds** Darmstadtium [281]	111 unknown **Rg** Roentgenium [280]	112 +2 **Cn** Copernicium [285]	113 unknown **Nh** Nihonium [286]	114 unknown **Fl** Flerovium [289]	115 unknown **Mc** Moscovium [289]	116 unknown **Lv** Livermorium [293]	117 unknown **Ts** Tennessine [294]	118 unknown **Og** Oganesson [294]

Lanthanide Series	57 +3 **La** Lanthanum 138.905	58 +4,+3 **Ce** Cerium 140.116	59 +3 **Pr** Praseodymium 140.908	60 +3 **Nd** Neodymium 144.243	61 +3 **Pm** Promethium 144.913	62 +3 **Sm** Samarium 150.36	63 +3,+2 **Eu** Europium 151.964	64 +3 **Gd** Gadolinium 157.25	65 +3 **Tb** Terbium 158.925	66 +3 **Dy** Dysprosium 162.500	67 +3 **Ho** Holmium 164.930	68 +3 **Er** Erbium 167.259	69 +3 **Tm** Thulium 168.934	70 +3 **Yb** Ytterbium 173.055	71 +3 **Lu** Lutetium 174.967
Actinide Series	89 +3 **Ac** Actinium 227.028	90 +4 **Th** Thorium 232.038	91 +5 **Pa** Protactinium 231.036	92 +6 **U** Uranium 238.029	93 +5 **Np** Neptunium 237.048	94 +4 **Pu** Plutonium 244.064	95 +3 **Am** Americium 243.061	96 +3 **Cm** Curium 247.070	97 +3 **Bk** Berkelium 247.070	98 +3 **Cf** Californium 251.080	99 +3 **Es** Einsteinium [254]	100 +3 **Fm** Fermium 257.095	101 +3 **Md** Mendelevium 258.1	102 +2 **No** Nobelium 259.101	103 +3 **Lr** Lawrencium [262]

Quantized energy levels for electrons

In 1900, Nobel laureate Max Planck (1858-1947) introduced *quantized energy.*

In 1905, Albert Einstein (1879-1955) discovered that light was made up of quantized photons. Einstein found that higher-frequency photons have more energetic photons. He expressed this with the equation:

$$E = hf$$

where h = Planck's constant = 6.63×10^{-34} J·s, E = photon energy and f = photon frequency.

Distinct emission spectrum lines indicate that electron energy is quantized into discrete energy levels. If electron energy is not quantized, a continuous spectrum would be observed.

Energy of a particular level is calculated using the quantum number and given by the equation:

$$E_n = -\frac{13.6}{n^2}(eV)$$

The equation is negative, so all energies are *negative.*

Electron binding energy

Negative energies mean that energy contributes to the "stability" of the system — the electron's *binding energy.*

More negative (or lower) energy means the compound is more stable in the orbit and more difficult to liberate the electron.

Less negative (or higher) energy means that the compound is less stable in the orbit, and it is easier to remove the electron.

There is no binding energy at the highest energy, 0 eV, so the electron dissociates.

For atoms other than hydrogen, the shape of the energy level curve stays the same. However, the numerator is a constant other than 13.6 eV.

Relationship for atoms other than hydrogen is:

$$E = -\frac{Z^2 R_E}{n^2}$$

where Z is the atomic number.

Higher Z values give more *negative binding energy* (i.e., more stable) because the greater the charge, the more electrostatic attraction.

To compare how tightly bound nuclei are, divide the binding energy by the mass number A to calculate the binding energy per nucleon.

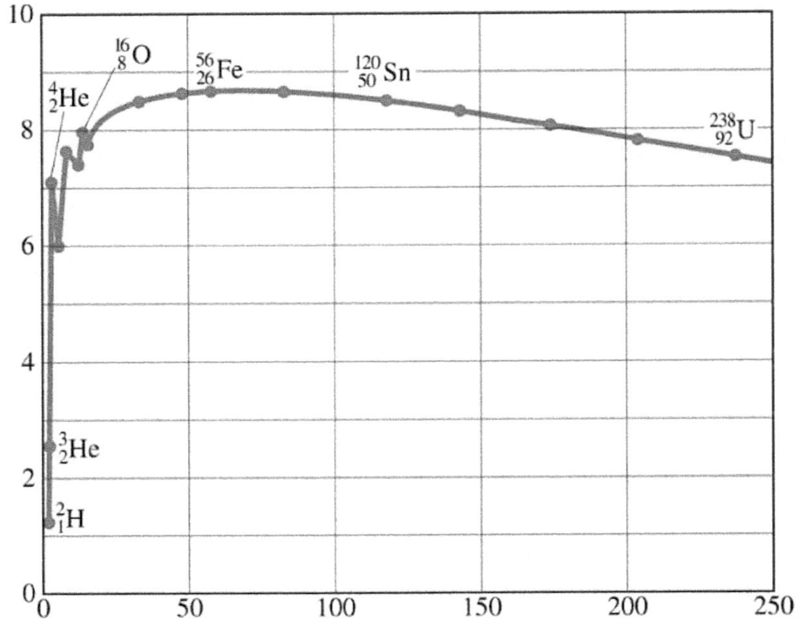

Binding energy per nucleon (MeV) vs. number of nucleons A (or mass number)

Energy emitted or absorbed for energy level transitions

Energy of the emitted or absorbed radiation is:

$$E = hf = hv = h\frac{c}{\lambda}$$

where E is energy, f and v represent frequency, and c is the speed of light.

Energy is *emitted* from transitions to lower energy levels ($n_f < n_i$).

Energy is *absorbed* from transitions to higher energy levels ($n_f > n_i$).

Photon energy is computed using the equation:

$$hf = E_H - E_L$$

Atomic Nucleus

Nuclides

Nuclei are made of protons and neutrons.

Protons have a positive charge and a mass of:

$$m_p = 1.67262 \times 10^{-27}\ \text{kg}$$

Neutrons are electrically neutral and slightly more massive than protons:

$$m_n = 1.67493 \times 10^{-27}\ \text{kg}$$

Neutrons and protons are collectively *nucleons* (as they reside in the nucleus).

Nuclides have different nuclei.

Number of protons defines an element and refers to the neutral atom (having no charge); two atoms with the same number of protons are the same element.

Isotopes

Isotope is an atom with the same atomic number as its counterpart on the periodic table but with a different number of neutrons.

Isotopes often have similar chemical properties but different stabilities. Some isotopes decay and give off radiation particles, while others are stable and do not decay.

For many elements, several different isotopes exist in nature.

Natural abundance is the percentage of an element that consists of a particular isotope in nature.

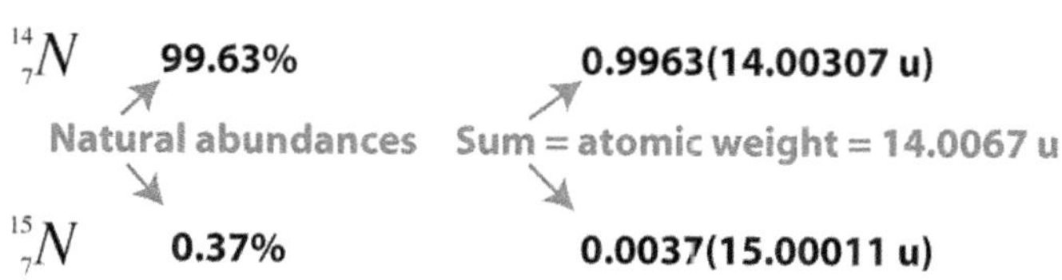

Natural abundance is the percentage of a particular isotope of an element in nature

Atomic particles

Atomic particles			
Name	Mass (amu)	Charge	Location
Proton	1	+1	In the nucleus
Neutron	1	0	In the nucleus
Electron	0	−1	Surrounding the nucleus

Atomic particle, mass, charge, and location of proton, neutron, and electron

Subatomic particles

Subatomic particles are smaller than an atom. According to the Standard Model of particle physics, a subatomic particle can be a:

composite particles are composed of other particles (e.g., a baryon, like a proton or a neutron, composed of three quarks; or a meson, composed of two quarks) or an

elementary particle, which is not composed of other particles (e.g., quarks or electrons, muons, and tau particles).

Proton is a composite particle with two up quarks and one down quark

Particle physics and nuclear physics study these particles and how they interact.

Most force-carrying particles like photons or gluons are called *bosons*, and

1) have discrete quanta of energy, do not have rest mass or discrete diameters (other than pure energy wavelength), and

2) unlike particles with rest mass and cannot overlap or combine.

Quantum field theory

Interactions of particles in quantum field theory are understood as the creation and annihilation of *quanta* of corresponding fundamental interactions, blending particle physics with field theory.

Particle physicists use diverse definitions of a particle, including:

collapsed wave function

quantum excitation of a field

irreducible representation of the Poincare group

an observed phenomenon

Standard Model of Elementary Particles

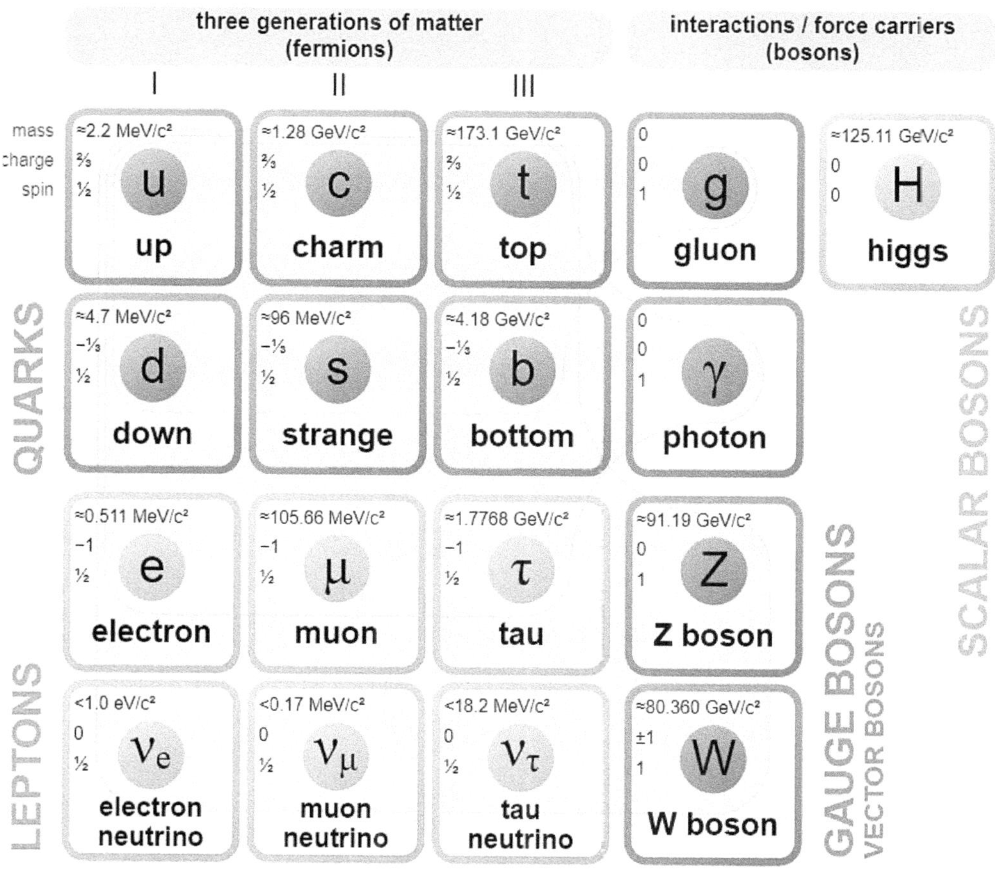

Standard model of elementary particles

Notes for active learning

Atomic Number and Atomic Weight

Atomic number

Number of protons determines an element's *atomic number* (*Z*). As of February 2024, 118 elements have been identified.

Number of nucleons is equal to the *atomic mass* number (*A*).

Atomic mass number is found by adding the number of protons and neutrons.

Neutrons are determined by subtracting the atomic number (*Z*) from atomic mass number (*A*).

Expression:

$N = A - Z$ gives the neutron number.

Atomic weight and mass

Atomic weight is the weighted average of *atomic mass* for isotopes of a given atom and used for an element.

Atomic mass is used for isotopes.

A and *Z* are sufficient to specify a nuclide.

Nuclides are symbolized as follows:

$${}^{A}_{Z}X$$

X is the chemical symbol for the element; it contains the same information as *Z* but in a more recognizable form.

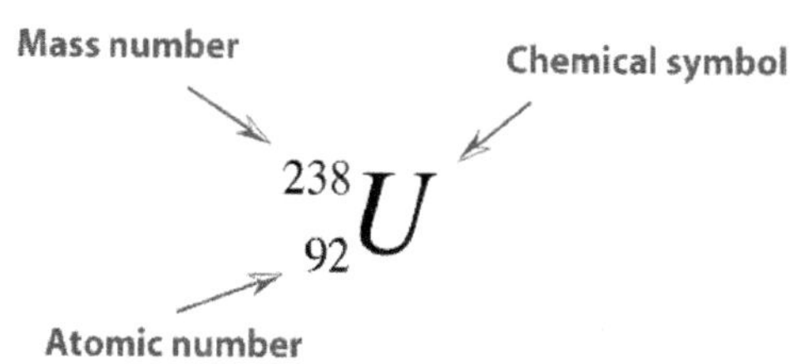

Chemical symbol with atomic number and mass number

Atomic size

Because of wave-particle duality, the size of the nucleus is fuzzy.

Measurements of high-energy electron scattering yield the radius to be:

$$r \approx (1.2 \times 10^{-15}\ m)(A^{\frac{1}{3}})$$

Masses of atoms are measured with reference to the carbon-12 (or ^{12}C) atom, which is assigned a mass of exactly 12 amu (or Daltons).

$$1 \text{ amu} = 1.6605 \times 10^{-27} \text{ kg}$$

$$1.6605 \times 10^{-27} \text{ kg} = 931.5 \text{ MeV}/c^2$$

$$1 \text{ amu} = 931.5 \text{ MeV}/c^2$$

From the table, the electron is considerably less massive than a nucleon.

Rest mass

Rest mass (or *invariant mass*) is the mass of a body as measured when the body is at rest relative to an observer.

Electron mass (or *mass of an electron* and *electron rest mass*) is a fundamental constant defined as the mass of a stationary electron and denoted by m_e.

Electron mass is the invariant mass of an electron.

Rest mass is an *inherent property* of the body.

Rest energy of a particle is given by:

$$E_0 = m_0c^2$$

where m_0 is the particle's rest mass and c is the speed of light.

Object	**kg**	**Atomic mass u**	**MeV/c^2**
Electron	9.1094×10^{-31}	0.00054858	0.51100
Proton	1.67262×10^{-27}	1.007276	938.27
^1_1H atom	1.67353×10^{-27}	1.007825	938.78
Neutron	1.67493×10^{-27}	1.008665	939.57

Rest mass (in kilograms, unified atomic mass units, and MeV/c^2

Nuclear Forces

Strong nuclear and electromagnetic forces

Two forces in the nucleus are the *strong nuclear* and *electromagnetic forces.*

Strong nuclear force binds the nucleons together, contributing to the *binding energy.*

Electromagnetic force is due to the electrostatic repulsion in the group of positively charged protons in the nucleus (i.e., like-charges repel).

Nuclei stay together because the *strong nuclear force exceeds the electromagnetic repulsion.*

Strong force is *short-ranged,* less than 10^{-15} m.

Proton–proton chain reaction

Repulsion between proton particles that it overcomes is the *proton-proton Coulomb repulsion.*

$$^{1}_{1}H + ^{1}_{1}H \rightarrow ^{2}_{1}H + ^{0}_{1}e$$

$$^{2}_{1}H + ^{2}_{1}H \rightarrow ^{3}_{2}He + ^{1}_{0}n$$

$$^{3}_{2}He + ^{3}_{2}He \rightarrow ^{4}_{2}He + 2^{1}_{1}H$$

Coulomb's Law

Coulomb's Law (*F*) relates the repulsion force on two charged particles depending on their charge (*q*) and distance (*r*) from each other.

The electrostatic force between two electric charges can be quantified by Coulomb's law, according to which the force between two electrically charged point particles is directly proportional to the product of the magnitude of the two charges inversely proportional to the square of the distance.

Force (*F*) is significant when the distance is small.

$$F_{Coulomb} = \frac{kq_1q_2}{r^2}$$

Notes for active learning

Radioactivity

Natural radioactivity

Natural radioactivity is a spontaneous emission of particles or energy from an unstable nucleus.

At the end of the 19th century, minerals were found to darken a photographic plate even in the absence of light. This phenomenon is radioactivity.

When an atom is unstable, it will undergo *radioactive decay*.

Maximum stability for a nucleon number is 2, 8, 20, 28, 50, 82, or 126.

The higher the binding energy per nucleon, the more stable the nucleus.

Massive nuclei require extra neutrons to overcome Coulomb repulsion of protons to be stable.

Specific neutron-to-proton ratios

A band of stability depends on the ratio of protons to neutrons, which makes an atom stable.

Coulomb force is long-range; therefore, extra neutrons are needed for stability in high-Z nuclei:

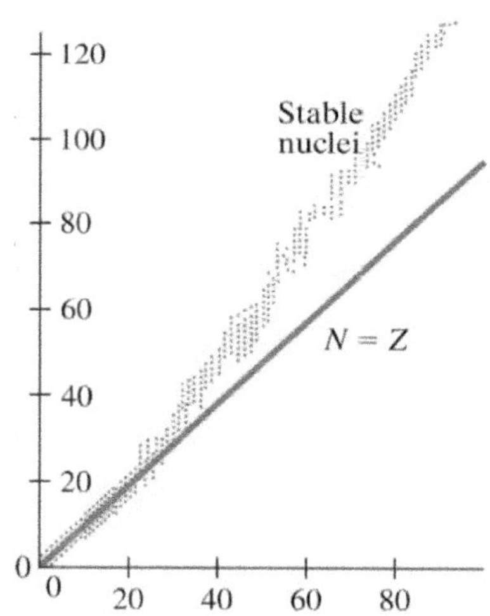

Neutron number ($N = A - Z$) vs. proton number (Z)

Pairs of protons and pairs of neutrons add stability to the atom.

Conversely, an odd number of protons and neutrons makes a less stable atom.

Specific neutron-to-proton ratios are desired for added stability: 1:1 for isotopes with up to 20 protons and (1+1):1 for increasingly heavy isotopes.

Unstable nuclei undergo decay

Unstable nuclei undergo decay; the *weak nuclear force* governs many such decays.

Three types of radioactive decay are *alpha*, *beta*, and *gamma decay*. These are *ionizing radiation* because they ionize material when passing through.

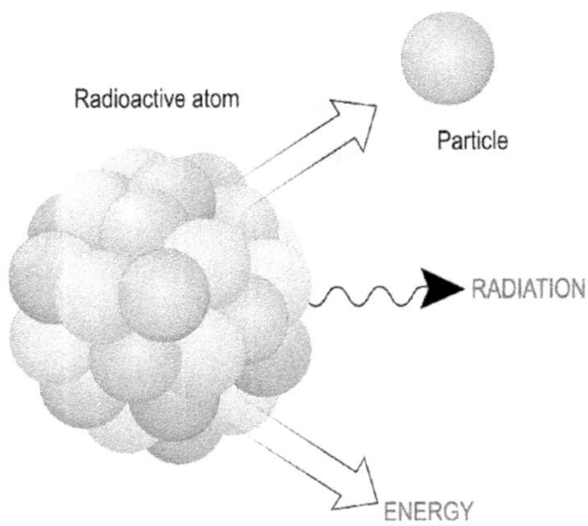

Radioactive decay creates new particles while releasing energy

Alpha emission is the expulsion of a helium nucleus, is the least penetrating, and can be stopped by a piece of paper.

Beta emission is the expulsion of an electron. Beta emission is more penetrating than alpha decay and can be stopped by 1 cm of aluminum.

Gamma decay is an emission of a high-energy photon (electromagnetic radiation).

Gamma decay is the most penetrating emission and requires a thick, dense shield (e.g., 13.5 feet of water, 6.5 feet of concrete, or about 1.3 feet of lead) to block transmission.

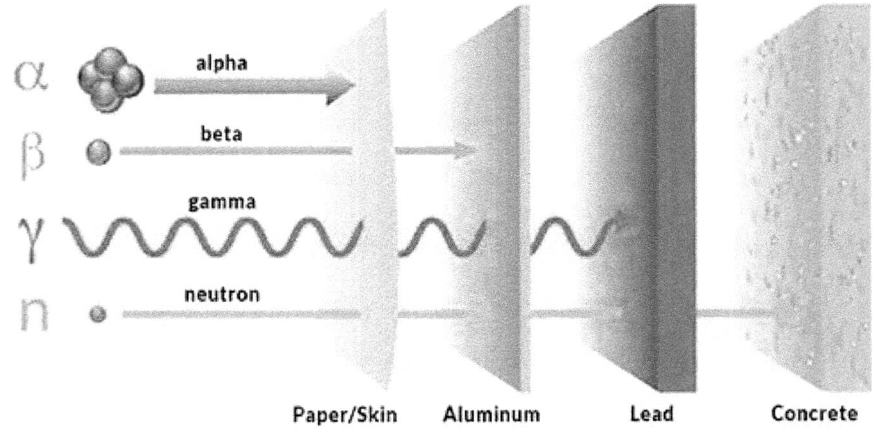

Radioactive particles penetrating paper, aluminum, and lead by alpha, beta, gamma particles, and neutrons

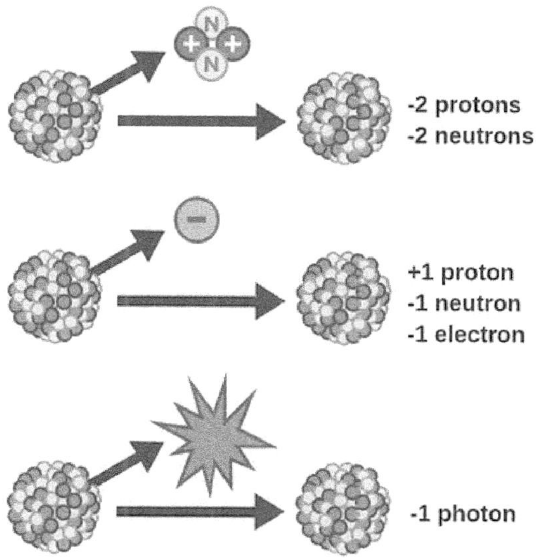

Alpha (top), beta (middle) and gamma (bottom) decay with emitted particles

Unstable condition	Type of decay	Emitted	Product nucleus
More than 83 protons	Alpha emission	${}^{4}_{2}\alpha$ $({}^{4}_{2}He)$	Lost 2 protons and 2 neutrons
Neutron-to-proton ratio too large	Beta emission	${}^{0}_{-1}\beta$ $({}^{0}_{-1}e)$	Gained 1 proton, no mass change
Excited nucleus	Gamma emission	${}^{0}_{0}\gamma$	No change
Neutron-to-proton ratio too small	Other emissions	${}^{0}_{+1}e$	Lost 1 proton, no mass change

Radioactive decay with emitted particles and product nucleus

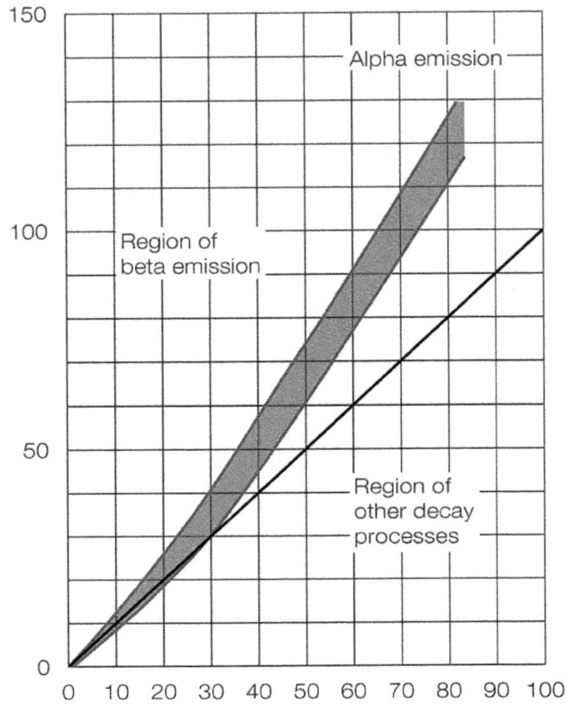

Number of neutrons vs. protons for alpha emission, beta emission, and other decay processes

Magnetic field deflections

Alpha and beta rays are bent in *opposite directions* in a magnetic field, while a magnetic field does *not* bend gamma rays.

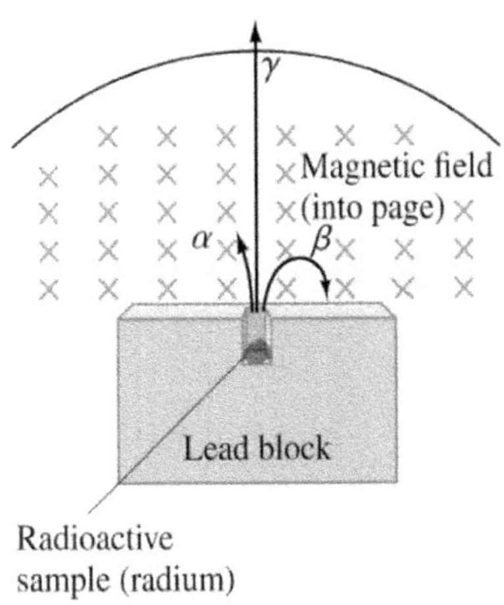

Magnetic field bends charged particles

Decay Types

Conservation of mass and charge

Conservation of mass dictates that the total atomic weight before the decay must equal the total atomic weight after.

Conservation of charge dictates that the total atomic number before the decay must equal the total atomic number after.

Radioactive decay supports the conservation that the total number of nucleons cannot change.

Do not get thrown off by unfamiliar particles. Incorporate these numbers into calculations if they have a weight and a charge. Identifying decay products is just math.

Atomic number (bottom number) determines what element it is.

Alpha decay

For example, the alpha decay of radium-227 (^{227}Ra) into radon-222 (^{222}Rn):

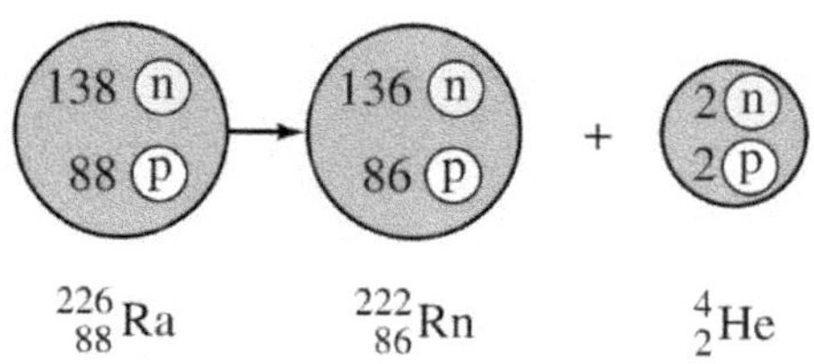

Alpha decay of radium-227 (^{227}Ra) into radon-222 (^{222}Rn)

Alpha decay can be written as:

$$^{A}_{Z}N \rightarrow ^{A-4}_{Z-2}N' + ^{4}_{2}He$$

Alpha decay occurs when the strong nuclear force cannot hold a large nucleus together.

Mass of the parent nucleus is greater than the sum of the masses of the daughter nucleus and the alpha particle; this difference is the *disintegration energy*.

Alpha decay is more likely than other nuclear disintegration because the alpha particle is stable.

Some smoke detectors use alpha radiation—the presence of smoke absorbs the alpha rays and keeps them from striking the collector plate.

Beta decay

Beta-decay occurs when a nucleus emits an e^{-1}. For example, the decay of carbon-14 (^{14}C):

$$^{14}_{6}C \rightarrow {}^{14}_{7}N + e^{-} + \text{neutrino}$$

The nucleus has 14 nucleons, but it has one more proton and one fewer neutron. This decay is an example of an interaction that proceeds via the weak nuclear force.

An electron in beta decay is not orbital; it is *created* in the decay.

Fundamental process is a neutron decaying to a proton, electron, and neutrino:

$$n \rightarrow p + e^{-} + neutrino$$

Neutrinos

Neutrinos were discovered through energy and momentum conservation analysis in beta decay—it could not be a two-particle decay.

Neutrinos are notoriously difficult to detect because they interact weakly, and direct evidence for their existence was unavailable for some time.

Neutrino symbol is the Greek letter nu (v); the beta decay of carbon-14 is written as (the bar over the neutrino means that it is an antineutrino):

$$^{14}_{6}C \rightarrow {}^{14}_{7}N + e^{-} + \overline{v}$$

Beta decay can occur when the nucleus emits a positron rather than an electron:

$$^{19}_{10}Ne \rightarrow {}^{19}_{9}F + e^{+} + \overline{v}$$

Nucleus can capture one of its inner electrons:

$$^{7}_{4}Be + e^{-} \rightarrow {}^{7}_{3}Li + v$$

Gamma decay

Gamma rays are high-energy photons. They are emitted when a nucleus decays from an excited state to a lower state, just as electrons returning to a lower state emit photons.

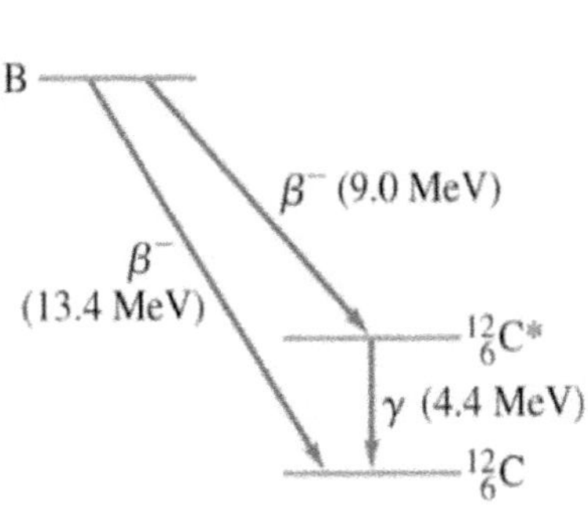

Gamma rays are high-energy photons emitted when a nucleus decays from an excited state

Radioactive Decay Rates

Half-life

Half-life is the time required for ½ a radioactive sample to decay.

For example, a 1 kg sample of an unstable isotope with a one-day half-life:

After 1 day - 500 g remain

After 2 days - 250 g remain

After 3 days - 125 g remain

Isotope	Half-life	Mode of decay
$^{3}_{1}H$ (tritium)	12.26 years	Beta
$^{14}_{6}C$	5,730 years	Beta
$^{90}_{38}Sr$	28 years	Beta
$^{131}_{53}I$	8 days	Beta
$^{133}_{54}Xe$	5.27 days	Beta
$^{238}_{92}U$	4.51×10^{9} years	Alpha
$^{242}_{94}Pu$	3.79×10^{5} years	Alpha
$^{240}_{94}Pu$	6,760 years	Alpha
$^{239}_{94}Pu$	24,360 years	Alpha
$^{40}_{19}K$	1.3×10^{9} years	Alpha

Half-life of sample radioactive isotopes

Exponential decay

Half-life graphs are shown with decreasing exponential:

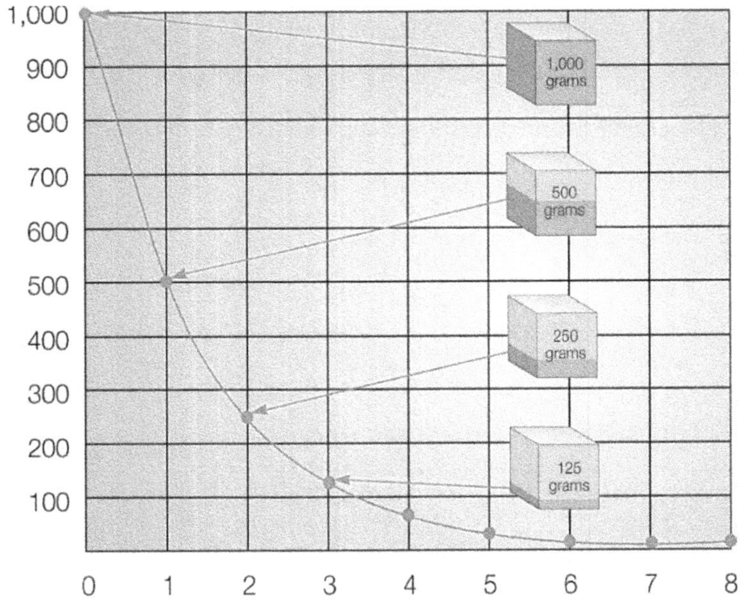

Mass (g) vs. time (days) with decreasing exponential half-life

Decay rate as a function of time

Nuclear decay is a random process; the decay of one nucleus does not influence another.

Decay rate is *unaffected* by temperature, pressure, volume, or other environmental factor.

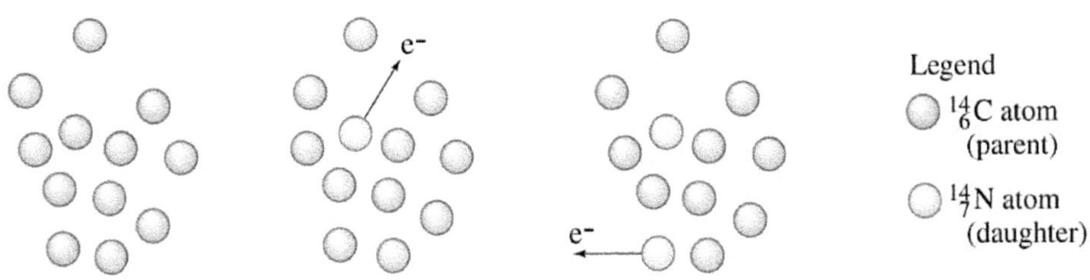

^{14}C with 6 protons, the parent atom decays into ^{14}N with 7 protons daughter atom

Number of decays in a short time interval is proportional to the number of nuclei and time:

$$\Delta N = -\lambda N \Delta t$$

Here, λ is a constant characteristic of that nuclide, the decay constant.

The number of decays equation uses calculus for N as a function of time:

$$N = N_0 e^{-\lambda t}$$

Plotting decay rates

Number of parental nuclei plotted:

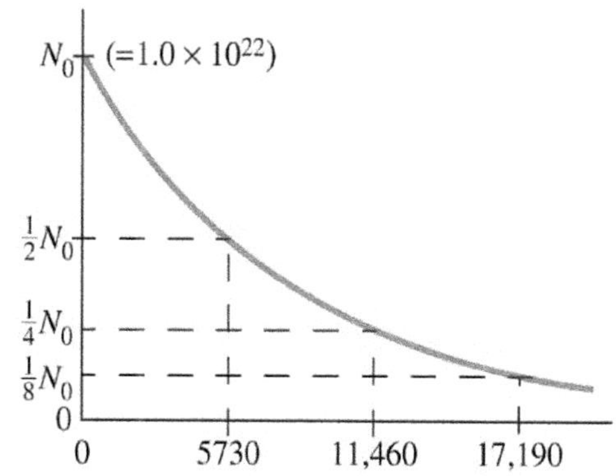

Number of parental nuclei (N) vs. time (years)

Decays per second plotted:

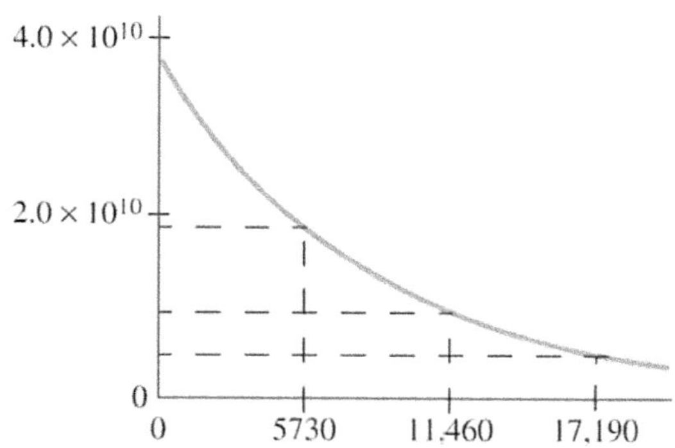

Decays per second (ΔN/Δt) vs. time (years)

Half-life calculations

Half-life is related to the decay constant:

$$T_{\frac{1}{2}} = \frac{\ln 2}{\lambda} = \frac{0.693}{\lambda}$$

It can also be written:

$$N_t = N_{t=0} \cdot \left(\frac{1}{2}\right)^{\#half-lives} = N_{t=0} \cdot \left(\frac{1}{2}\right)^{\frac{t}{half-life}}$$

where $N_{t=0}$ is amount of original starting material, N_t is remaining material, and t is time.

The above is the formal half-life equation, but it may be easier to multiply rather than divide. Therefore, a more user-friendly equation is:

$$N_{t=0} = N_t \cdot 2^{\#half-lives} = N_t \cdot 2^{\frac{t}{half-life}}$$

Semi-log plots

Semi-log plots convert exponential curves into *straight lines*.

- Something that *curves up* becomes a *straight line with a positive slope*.
- Something that *curves down* becomes a *straight line with a negative slope*.
- For exponential decay, a semi-log plot graphs the *log of amount vs. time*.
- For exponential decay, a semi-log plot is a straight line with a *negative slope*.
- Semi-log plot intercepts the x-axis, where the original y-value is 1.

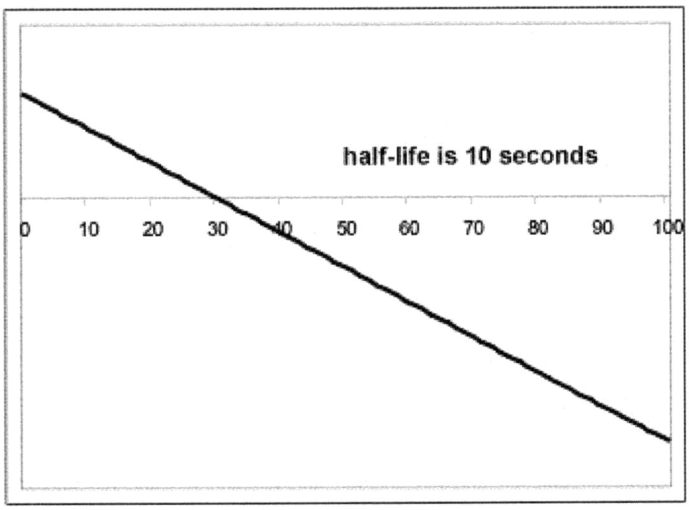

Exponential decay in a semi-log plot of the amount of material (logarithmic scale) vs. time in seconds (plotted in non-logarithmic scale)

Decay Series

Novel isotopes

Decay series occurs when one radioactive isotope decays to another radioactive isotope, which decays to another, and so on.

Decay series allows for the creation of nuclei that otherwise would not exist in nature.

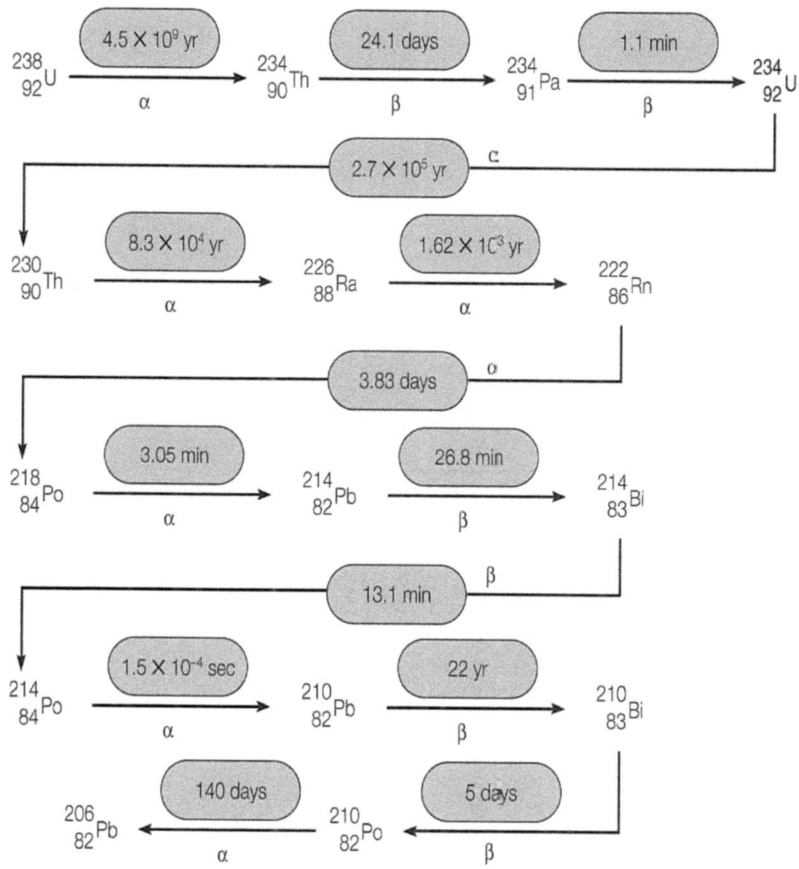

Decay series occurs when one radioactive substance decays into another

Radioactive dating with carbon

Radioactive dating analyzes the fraction of carbon in organic material that is carbon-14. The ratio of ^{14}C to ^{12}C in the atmosphere has been roughly constant over thousands of years.

A living plant or tree constantly exchanges carbon with the atmosphere and will have the same carbon ratio in its tissue. When the plant dies, this exchange stops.

^{14}Carbon has a half-life of about 5,730 years; it decays and becomes a smaller fraction of the plant tissue's total carbon. This fraction can be measured, and the age of the tissue is deduced.

Objects older than about 60,000 years cannot be dated this way because there is little [14]carbon remaining. Other isotopes are helpful for geologic time-scale dating.

Uranium-238 (^{238}U) has a half-life of 4.5×10^9 years and has been used to date the oldest rocks on Earth as about 4 billion years old.

Radioactive decay timelines

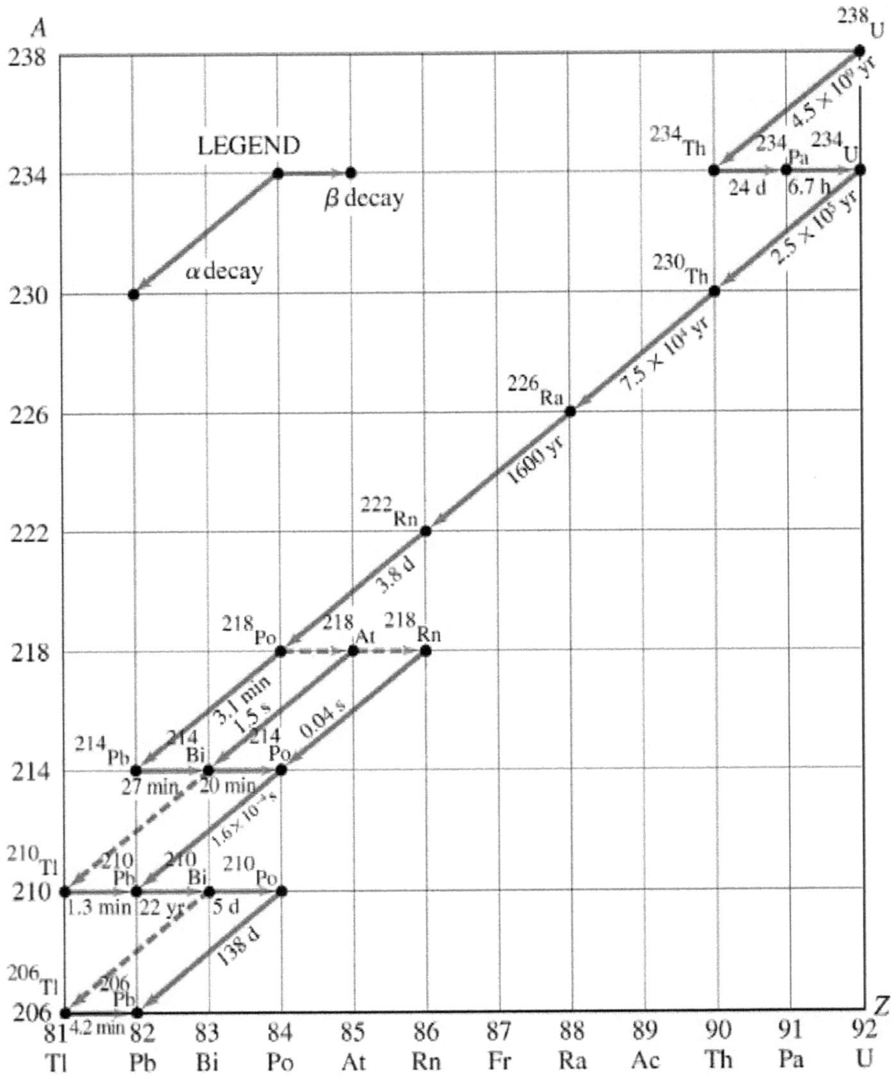

Elements undergo alpha and beta decay with corresponding years noted

Energy barriers

Energy is released when nuclei decay by alpha emission, but nuclei do not decay immediately.

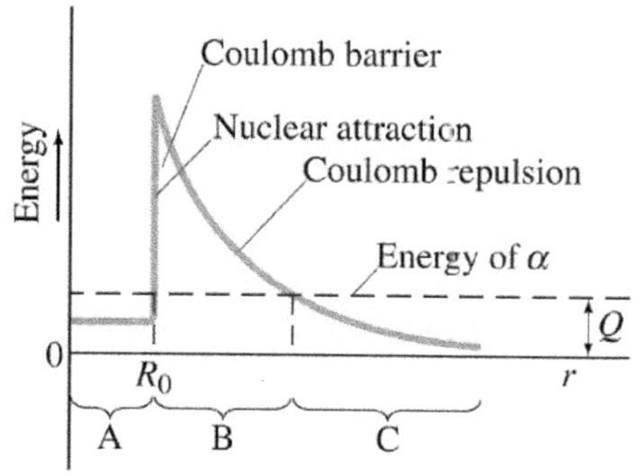

Energy barriers for nuclei decay by alpha emissions

Although energy is released in the decay, there is an energy barrier. The alpha particle escapes through a quantum mechanical phenomenon of *tunneling*.

Heisenberg uncertainty principle, energy conservation differences can be violated temporarily:

$$Q = M_pc^2 - (M_D + m_\alpha)c^2$$

The higher the energy barrier, the less time the alpha particle has to get through it, and the less likely it is to happen; this accounts for the extensive variation in half-lives for alpha decay.

Notes for active learning

Measuring Radioactive Decay

Ionization counters

Ionization counters, such as a *Geiger counter*, detect *radiation-induced ions*. Geiger-Muller tubes (or Gieger counter) operate in a process known as ionization to detect gamma beta and alpha particles and X-rays. It is commonly used in radiation dosimetry (i.e., "absorbed dose" as the amount of radiation energy deposited in tissue divided by the mass of the tissue), radiological protection, experimental physics, and the nuclear industry.

Geiger counters provide "yes" or "no" information, while proportional counters are often used to detect alpha and beta rays since charged particles interact strongly with a gas.

Geiger counter is a gas-filled tube with a wire in the center. The wire is at high voltage; the case is grounded. When a charged particle passes through, it ionizes the gas. The ions cascade onto the wire, producing a pulse.

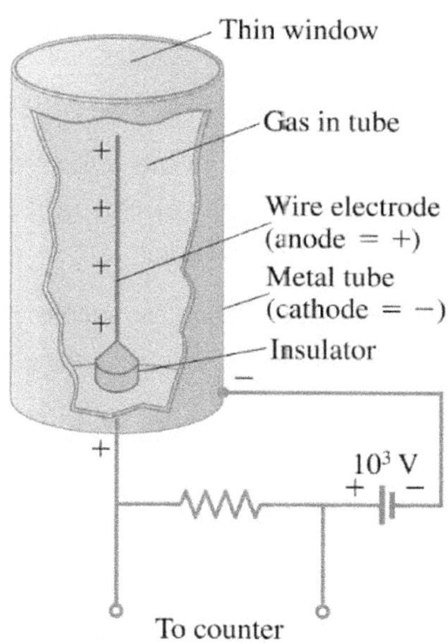

Geiger counter is a gas-filled tube with a wire in the center that detects ions produced

Wire drift chamber

Wire drift chamber is similar but vastly more sophisticated than a Geiger counter.

Many wires are present, some at high voltage and some grounded. In addition to the presence of a signal, the time it takes the pulse to arrive at the wire is measured, allowing precise position measurement. Radiation can be measured at the source.

Activity is the number of *disintegrations per unit of time*.

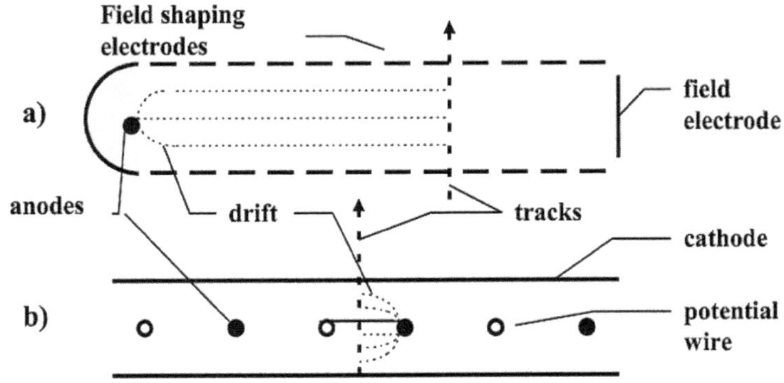

Principle of drift chambers. a) constant drift field b) multi-wire drift chamber

Multi-wire proportional chamber

Wire chamber (or *multi-wire proportional chamber*) is a proportional counter that detects charged particles and photons and provides *positional information* on their trajectory by tracking the trails of gaseous ionization.

Multi-wire chambers use an array of wires at high voltage (anode), which run through a chamber with conductive walls held at ground potential (cathode).

Alternatively, the wires may be at ground potential, and the cathode held at a high negative voltage; the important thing is that a uniform electric field draws extra electrons or negative ions to the anode wires with slight lateral motion.

The chamber is filled with a gas (e.g., argon and methane mixture) such that any ionizing particle that passes through the tube will *ionize surrounding gaseous atoms.*

The electric field across the chamber accelerates the resulting ions and electrons, causing a localized cascade of ionization known as a *Townsend avalanche* (i.e., ionization process for gases) where an electric field accelerates free electrons, collide with gas molecules, and free additional electrons.

Those electrons are accelerated and free additional electrons). This collects on the nearest wire and results in a charge proportional to the ionization effect of the detected particle. The particle trajectory can be found by computing pulses from the wires.

Multi-wire proportional chamber provides positional trajectory during gaseous ionization

Scintillation counter

Scintillation counter uses a scintillator, which is a material that emits light when a charged particle passes through it. The scintillator is light-tight, and the light flashes are viewed with a photomultiplier tube, which has a photocathode emitting an electron when struck by a photon and then a series of amplifiers.

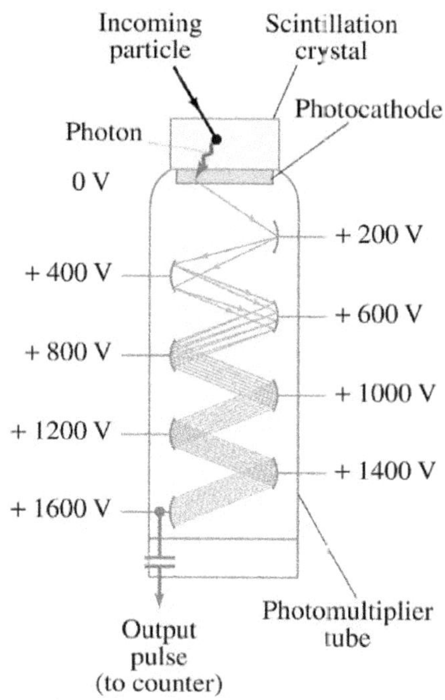

Scintillator emits light when a charged particle passes through it

Cloud chambers

Cloud chambers contain a supercooled gas; droplets form along its track when a charged particle goes through it. A cloud chamber has a sealed environment of a superheated vapor (e.g., water or alcohol).

An energetically charged particle (e.g., alpha or beta particles) interacts with the gaseous mixture by removing electrons from gas molecules via electrostatic forces during collisions. This results in a trail of ionized gas particles.

The resulting ions act as condensation centers around which a mist-like trail of tiny droplets forms if the gas mixture is at the condensation point. Droplets are visible as a *cloud* track with characteristic shapes that persist for several seconds while the droplets fall through the vapor.

For example, an alpha particle track is thick and straight, while a beta particle track is wispy and shows more evidence of deflections by collisions.

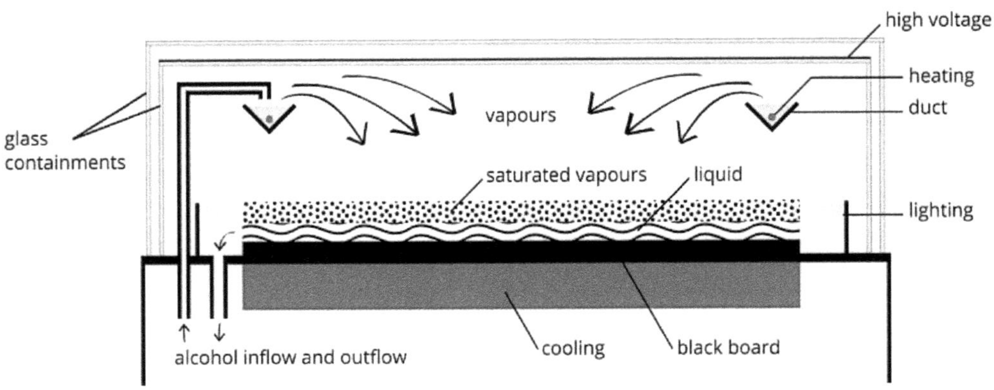

Cloud chamber with supercooled gas

Diffuse cloud chamber

Diffuse cloud chamber differs from the expansion cloud chamber as it is continuously sensitized to radiation, and the bottom must be cooled to a low temperature, colder than −26 °C (−15 °F).

Instead of water vapor, alcohol is used because of its lower freezing point.

Cloud chambers cooled by dry ice or Peltier effect thermoelectric cooling are demonstration and hobbyist devices; the alcohol used is commonly isopropyl.

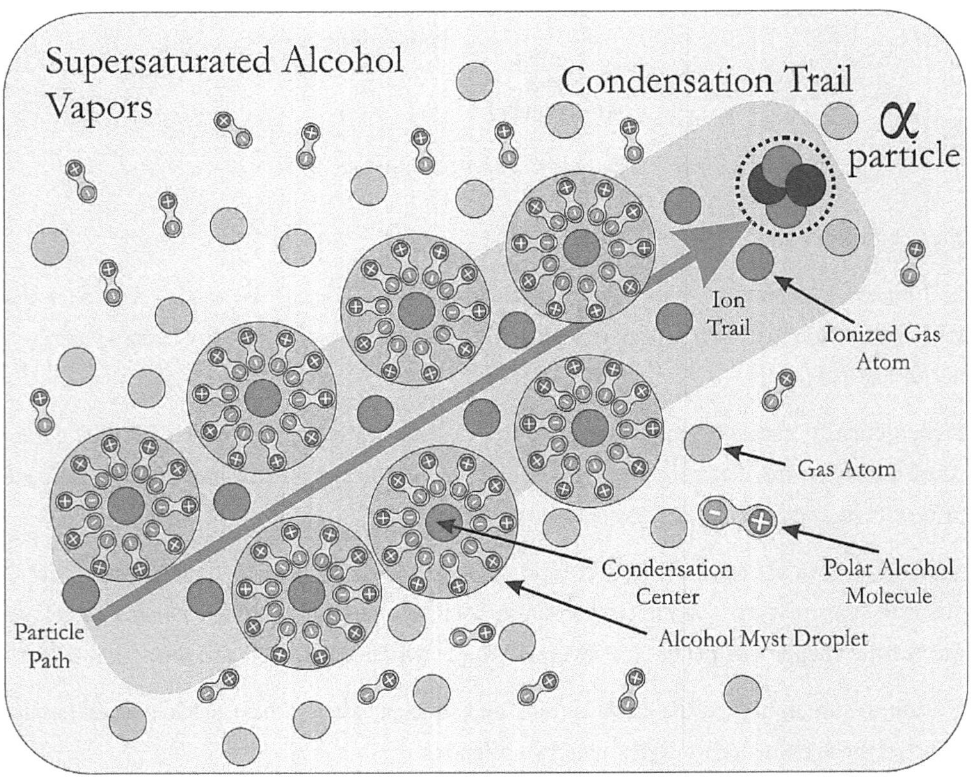

Diffuse cloud chamber

Bubble chamber

Bubble chambers contain a superheated liquid, which forms bubbles. When electrically charged particles pass through a bubble chamber, they ionize the molecules in the chamber medium. The ions trigger a phase transition, and the superheated liquid vaporizes, creating visible tracks as bubbles form along the particle's path.

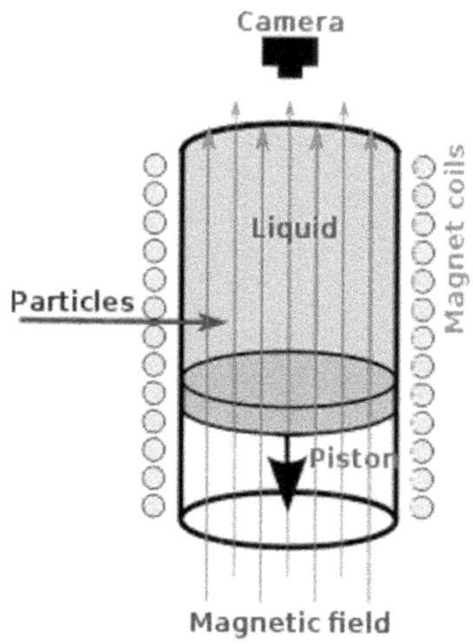

Bubble chamber

Notes for active learning

Fission Reactions

Nuclear reactions

Nuclear reactions take place when another nucleus or particle strikes a nucleus. Two types of nuclear reactions are *fission* and *fusion.*

Since smaller nuclei are stable with fewer neutrons, several neutrons emerge from each fission. These neutrons can be used to induce fission in other nuclei, causing a chain reaction as occurs when Uranium undergoes fission when struck by a free neutron.

Nuclear fission is the process whereby the nucleus of an atom is split into smaller fragments, and the energy released in a fission reaction is tremendous.

Fission products are usually unstable and, therefore, radioactive. Due to being neutron-rich in their atomic number, many quickly undergo beta decay. This releases additional energy through beta particles, antineutrinos, and gamma rays.

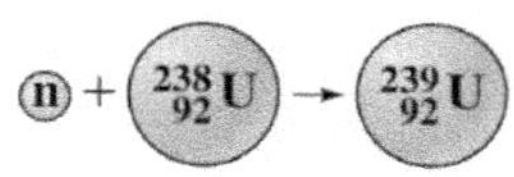

Neutron captured by $^{238}_{92}\text{U}$.

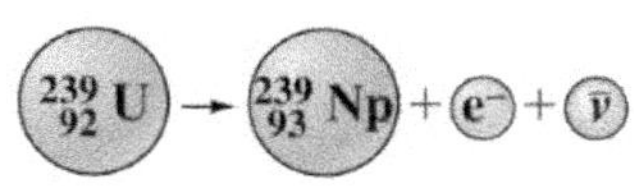

$^{239}_{92}\text{U}$ decays by β decay to neptunium-239.

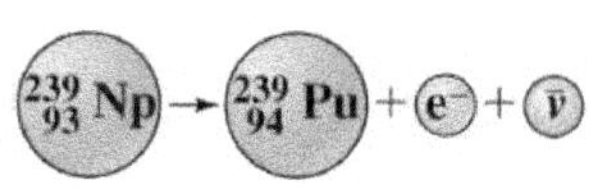

$^{239}_{93}\text{Np}$ itself decays by β decay to produce plutonium-239.

Nuclear fission occurs when a nucleus splits apart

Some problems associated with nuclear reactors include the disposal of radioactive waste and the possibility of an accidental release of radiation (e.g., Chornobyl, Ukraine, in 1986 and Fukushima, Japan, in 2011).

Atomic bombs use fission, but the core is designed to undergo a massive uncontrolled chain reaction when the uranium is formed into a critical mass during detonation.

Transmutation

Transmutation or nuclear transmutation involves a change in the nucleus of af an atom. When the number of protons in an atom's nucleus changes, that atom's identity changes as it is turned into another element or isotope. This transmutation process can be either natural or artificial.

Transmutation is when the original nucleus is transformed into another:

$$^{4}_{2}\text{He} + ^{14}_{7}\text{N} \rightarrow ^{17}_{8}\text{O} + ^{1}_{1}\text{H}$$

Laws of conservation

Nuclear reactions obey the laws of conservation of mass, energy, charge, and momentum.

In a nuclear reaction, total (relativistic) energy is conserved.

In any reaction, linear momentum must also be conserved. Thus, the momentum of the reaction products equals the reactants. Conservation of linear momentum requires that the products have definite amounts of kinetic energy depending on their recoil directions.

Energy and *momentum* must be conserved in nuclear reactions.

$$a + X \rightarrow Y + b$$

Reaction energy

Reaction energy (or Q-value) is the sum of the initial masses less the sum of the final masses, multiplied by c^2:

$$Q = (M_a + M_X - M_b - M_Y)c^2$$

If Q is positive, the reaction is exothermic and occurs regardless of how small the initial kinetic energy is.

If Q is negative, a minimum initial kinetic energy must be available for the reaction to occur.

Neutrons are effective in nuclear reactions, as they have no charge and, therefore, are not repelled by the nucleus.

Critical mass

Critical mass determines if nuclei have sufficient mass and concentration for chain reactions.

In nuclear physics, critical mass is the least material needed to start and maintain a nuclear reaction (e.g., reactors and bombs). At critical mass, the neutrons released as a product of one fission reaction can cause neighboring atoms to fission.

Reactor's critical size is the core's minimum size, and the reflector surrounding it is required to maintain a chain reaction.

Critical mass of a fissionable substance depends upon its density, purity, and design. When a mass of fissionable material increases in density, the mass required to achieve criticality decreases. Furthermore, the density of a substance increases if its volume decreases. Thus, rapidly decreasing the radius of a mass of fissionable material can help achieve *critical mass.*

Compressing the fissile core can lower the critical mass because, at higher densities, emitted neutrons are more likely to strike fissionable nuclei before escaping.

Mass distribution

Mass distribution of fragments for the first two pieces is substantial but usually unequal. Among them, the mass distribution of fragments is the number of fragments in different mass ranges.

For example, the number and mass distribution of fragments formed after an explosion of uncontrolled fragments are related to the projectile's structure, shell materials, and explosives.

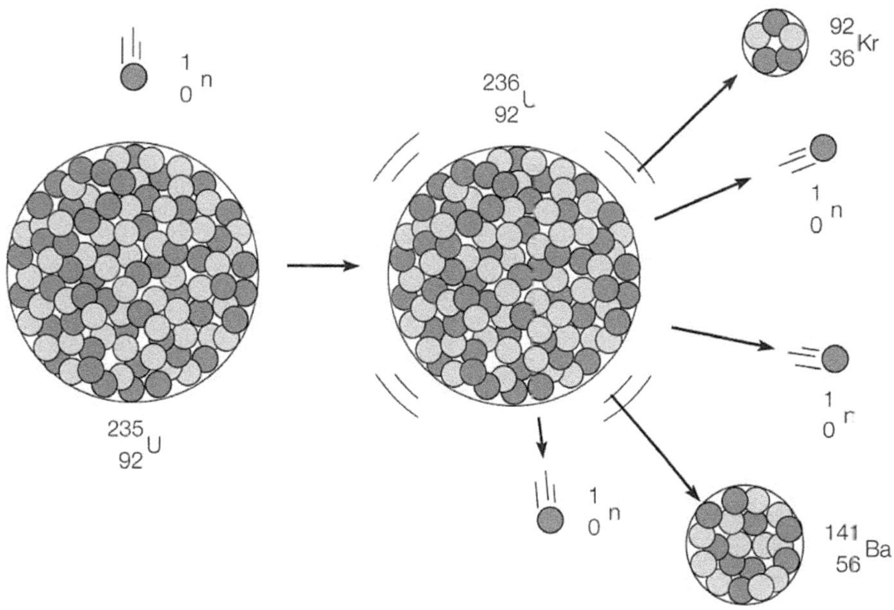

Nuclear reactions involve nuclear decay with daughter nuclei and particle emissions

Self-sustaining reactions

For a nuclear reactor (shown below), the chain reaction needs to be self-sustaining—it continues indefinitely—but is controlled.

Control rods, usually cadmium or boron, absorb neutrons and control the reaction to moderate it and keep it barely critical.

A *moderator* is needed to slow the neutrons; otherwise, their interaction probability is too small. Common moderators are heavy water (D_2O, deuterium) and graphite.

Unless the moderator is heavy water, the fraction of fissionable nuclei in natural uranium (about 0.7%) is too small to sustain a chain reaction. Uranium needs to be enriched to about 2–3%. Neutrons that escape from the uranium do not contribute to fission.

Below critical mass, a chain reaction does not occur because too many neutrons escape.

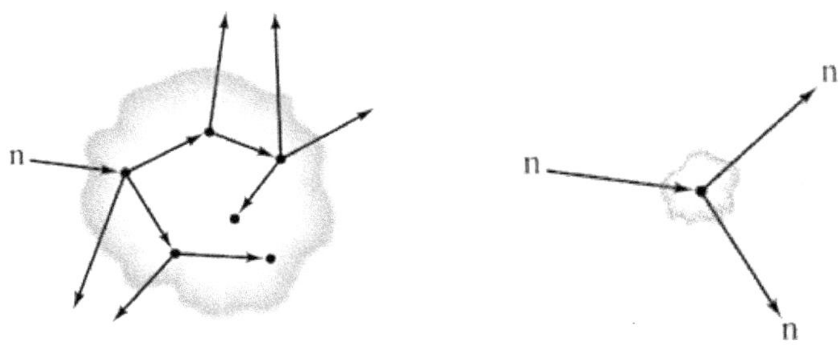

Chain reactions are self-sustaining nuclear reactions

Nuclear fission reactors

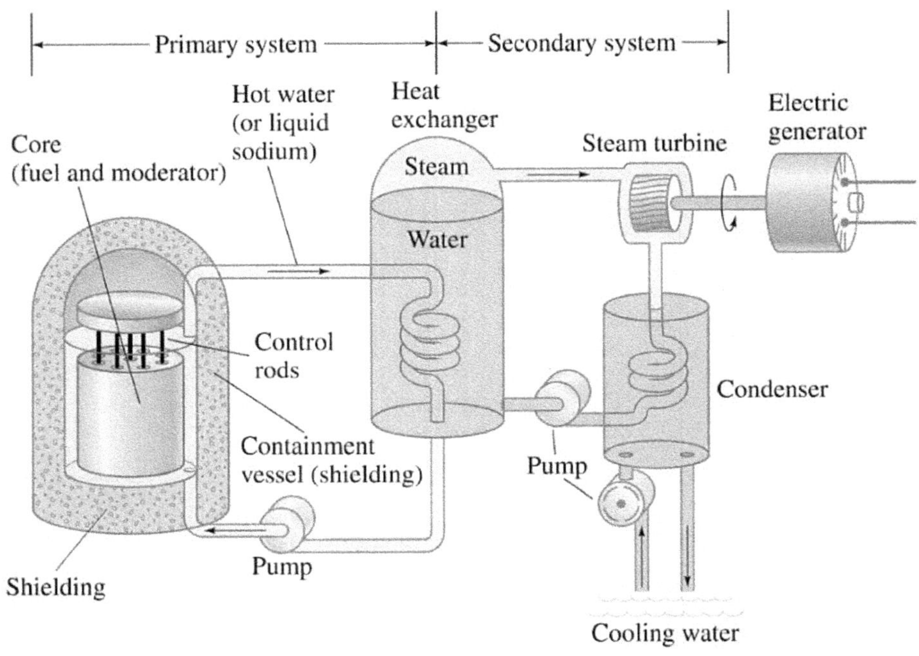

Nuclear reactor is a controlled, self-sustaining fission reaction

Fusion Reactions

Nuclear fusion

Nuclear fusion occurs when less massive nuclei form more massive nuclei. It is the energy source for the Sun and other stars. For example, the fusion process changes hydrogen in the Sun into helium.

Fusion requires a high temperature, high density, and sufficient confinement time.

Controlled fusion is usually studied with magnetic or inertial confinement.

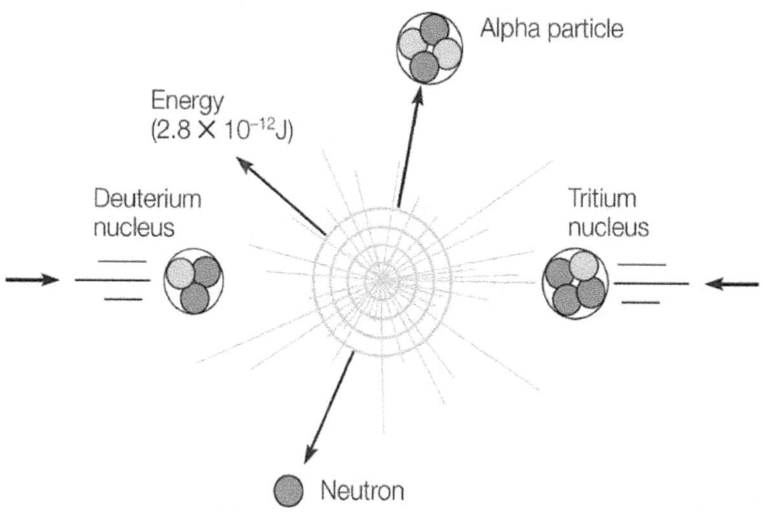

Light nuclei fuse to form heavier nuclei and release energy

Energy released

Fusion reactions are listed with the energy released for each process:

$${}^1_1H + {}^1_1H \rightarrow {}^2_1H + e^+ + v \qquad (0.42\ \text{MeV})$$

$${}^1_1H + {}^2_1H \rightarrow {}^3_2He + \gamma \qquad (5.49\ \text{MeV})$$

$${}^3_2He + {}^3_2He \rightarrow {}^4_2He + {}^1_1H + {}^1_1H \qquad (12.86\ \text{MeV})$$

Fusion products

Net effect of the fusion reaction is to transform four protons into a helium nucleus plus two positrons, two neutrinos, and two gamma rays.

$$4{}^1_1H \rightarrow {}^4_2He + 2e^+ + 2v + 2\gamma$$

More massive stars can fuse heavier elements in their cores, up to iron, the most stable nucleus.

Fusion power reactors

Fusion reactions use standard fuels (i.e., deuterium or tritium) and release much more energy per nucleon than a fission reaction.

A successful fusion reactor has not yet been achieved, but fusion bombs (*thermonuclear bombs*) have been built. Several geometries for containing the incredibly hot plasma that results in a fusion reactor have been developed.

Three fusion reactions are being considered for power reactors:

$$^{2}_{1}H + ^{2}_{1}H \rightarrow ^{3}_{1}H + ^{1}_{1}H \qquad (4.03\ \text{MeV})$$

$$^{2}_{1}H + ^{2}_{1}H \rightarrow ^{3}_{2}He + n \qquad (3.27\ \text{MeV})$$

$$^{2}_{1}H + ^{3}_{1}H \rightarrow ^{4}_{2}He + n \qquad (17.59\ \text{MeV})$$

Mass Deficit, Energy Liberated and Binding Energy

Mass deficit

Mass deficit is the difference between the masses of the *reactants and products*.

Nuclear energy is an interconversion of mass and energy. Nuclear energy is connected to the origins of the universe and the life cycles of stars.

Big Bang theory states that incredibly hot, dense, primordial plasma cooled, creating protons and neutrons. Continued cooling leads to hydrogen atoms, which collapse gravitationally into 1st generation stars.

Nuclei and binding energy

Gravitational attraction is the ultimate source of nuclear energy.

Binding energy is required to break a nucleus into individual protons and neutrons.

Ratio of *binding energy* to the *nucleon number* indicates how stable a nucleus is.

Iron-56 is the most stable nucleus:

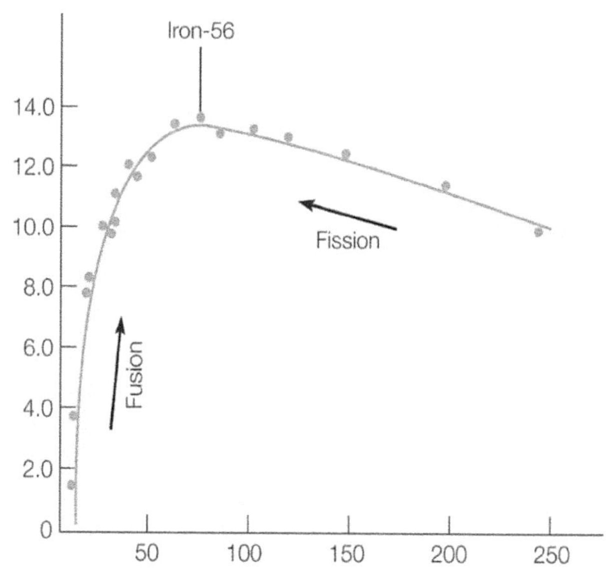

Binding energy per nucleon (10^{-13} J) vs. mass number

Conserving charge and mass number

Balanced equations present nuclear reactions where the charge is conserved, as is mass number.

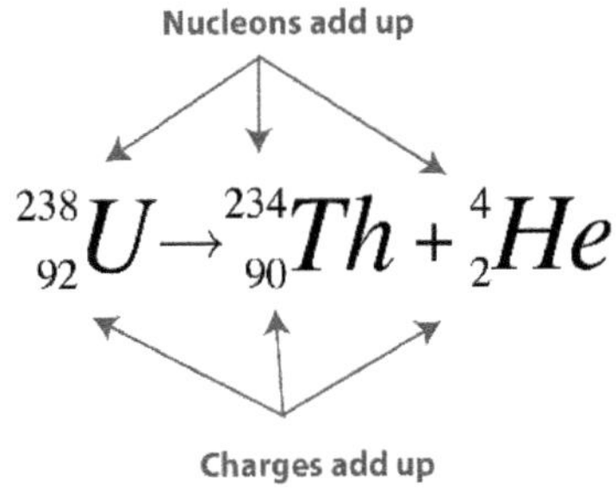

Mass number and charge are conserved

Classifying nuclear particles

Particle	Symbol	Mass number	Charge
Proton	^1_1H (or ^1_1p)	1	1+
Electron	$^{0}_{-1}e$ (or $^{0}_{-1}\beta$)	0	1–
Neutron	^1_0n	1	0
Gamma Photon	$^0_0\gamma$	0	0

Names, symbols, and properties of particles in nuclear equations

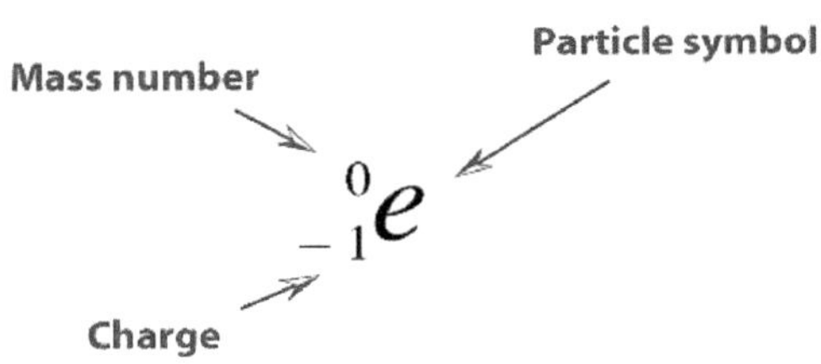

Particles labeled with mass number, charge, and symbol

Mass converts to energy

The difference in mass before and after a reaction is the *mass deficit* (or *mass defect*). If the total mass before the reaction differs after the reaction, then the difference in mass is energy.

Energy is liberated when mass is lost during a reaction because mass and energy are conserved. Total mass and energy before a reaction equals the total mass and energy afterward.

Energy that makes up for the mass deficit is calculated by:

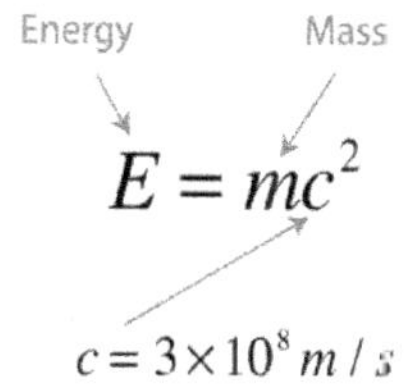

Energy-mass deficit relationship

Mass of the nucleons is higher than the mass of the atom. Mass has become energy (e.g., radiation or kinetic energy) released during the formation of the nucleus. The difference between the mass of the constituents and the nucleus is the *binding energy* of the nucleus.

Energy liberated equals binding energy

Energy liberated equals the *binding energy*, which is used when converting ΔM into its equivalent in energy (ΔMc^2).

$M_{nucleons} = M_{atom} + \text{binding energy}/c^2$

Binding energy commonly refers to nuclear binding energy (energy binding the nucleons).

Binding energy per nucleon is strongest for Iron (^{56}Fe) and weakest for Deuterium (the 2-nucleon isotope of hydrogen).

Electron binding energy is less commonly used because it is called *ionization energy*.

$m_{nucleons} - m_{atom}$ = mass deficit (also mass defect) = Δm

$\Delta m = 4.0330\ u - 4.0026\ u = 0.0304\ u = 28.3$ Mev

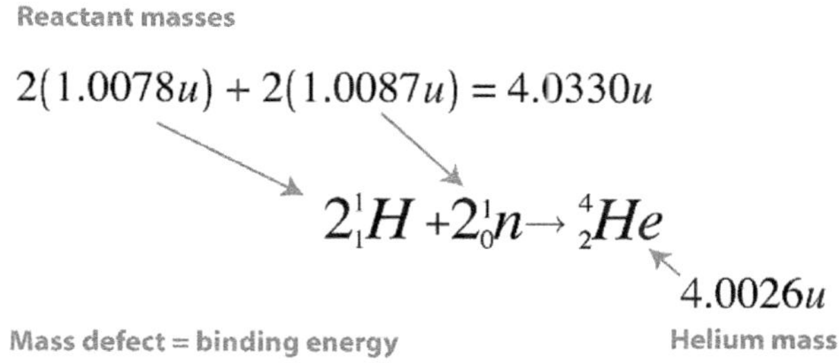

Mass deficit equals binding energy

Notes for active learning

Mass Spectrometry

Identifying chemical composition

Mass spectrometry is a technique that identifies the chemicals in a substance.

Magnetic fields deflect charged particles.

Mass spectrometers convert atoms into ions, increasing their acceleration so they all have the same kinetic energy and deflecting them with a magnetic field.

It creates ions by removing electrons from the atom—even if an atom typically forms a negative ion, a mass spectrometer usually works with positive ions.

Particles are deflected, and their mass can be calculated because the deflection force is known. The curve of their deflection path separates atoms; lighter ions are deflected more.

The higher the positive charge on an ion, the more the magnetic field deflects it.

Mass spectrometer

In a typical mass spectrometer procedure, a sample (i.e., solid, liquid, or gaseous) is ionized by bombarding it with a beam of electrons. This may cause sample molecules to break into positively charged fragments or become positively charged without fragmenting.

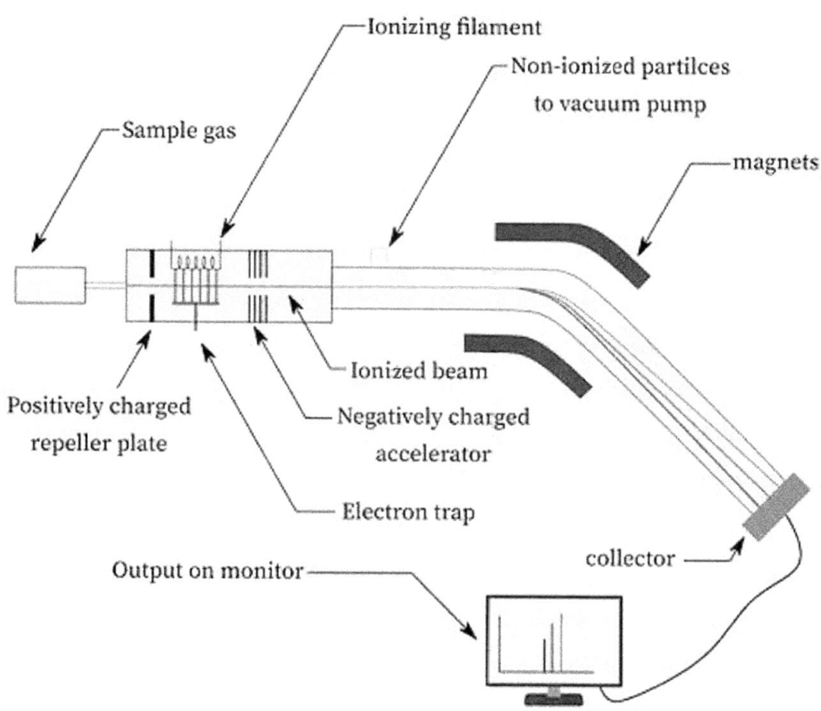

Mass spectrometers measure the mass-to-charge ratio of a sample

Mass-to-charge ratio

Mass spectrometers measure the *mass-to-charge ratio* of a sample and the number of ions in the gas phase, producing a *mass spectrum*.

A mass spectrum is produced, resembling a bar graph (shown below). The mass spectrum gives the concentration of the atom, organized by their mass-to-charge ratio.

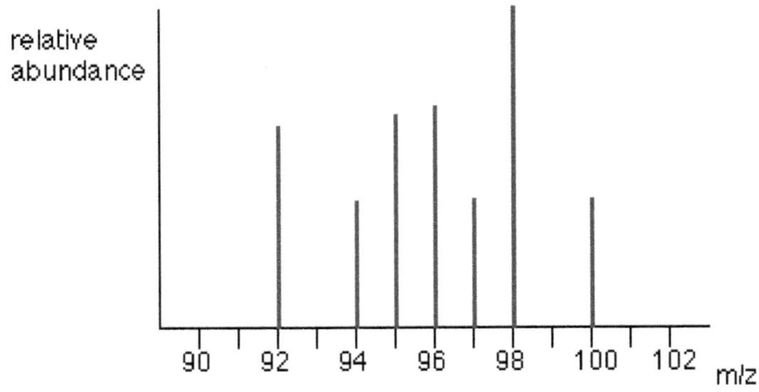

Mass spectrum plotting relative abundance against the mass-charge ratio

Radiation Exposure

Units of radioactivity

Name	Symbol	Units	Multiply by
Becquerel	Bq	Ci	2.7×10^{-11}
gray	Gy	rad	100
sievert	Sv	rem	100
curie	Ci	Bq	3.7×10^{10}
rem	rem	Sv	0.01
rem	rem	millirem	1,000
millirem	mrem	rem	0.001

Names, symbols, units, and conversion factors for radioactivity

Radiation dosimetry

Radiation is measured where absorbed, and dosage is related to its effects on organisms.

Radiation dosimetry measures the "absorbed dose" as the amount of radiation energy deposited in tissue divided by the mass of the tissue.

Absorbed dose measures the effect the radiation has on the absorbing material.

Human exposure is measured in *rem,* but the SI unit is the *millisievert.*

SI unit for dose is the gray (Gy).

Rad (unit of dosage) is the amount of radiation that deposits energy at a rate of 1.00×10^{-2} J/kg.

Radiation damage in tissue

Level	Comment
0.130 rem	Average annual exposure to natural background radiation
0.500 rem	Upper limit of annual exposure to the public
25.0 rem	Threshold for observable effects such as reduced blood cell count
100.0 rem	Fatigue and other symptoms of radiation sickness
200.0 rem	Definite radiation sickness, bone marrow damage, the possibility of developing leukemia
500.0 rem	Lethal dose for 50% of individuals
1,000.0 rem	Lethal dose for all

An approximate single dose, whole-body effects of radiation exposure

Radiobiological effects

High-energy protons and carbon ions exhibit an inverse dose profile, allowing for increased energy deposition with penetration depth.

Heavier ions like carbon beams have increased biological effectiveness characterized by enhanced ionization density in the individual tracks of the heavy particles.

Damage for most *radiobiological effects* is creating double-strand breaks (DSBs) that the cell cannot repair; local ionization density and biological repair capacity are the determining factors.

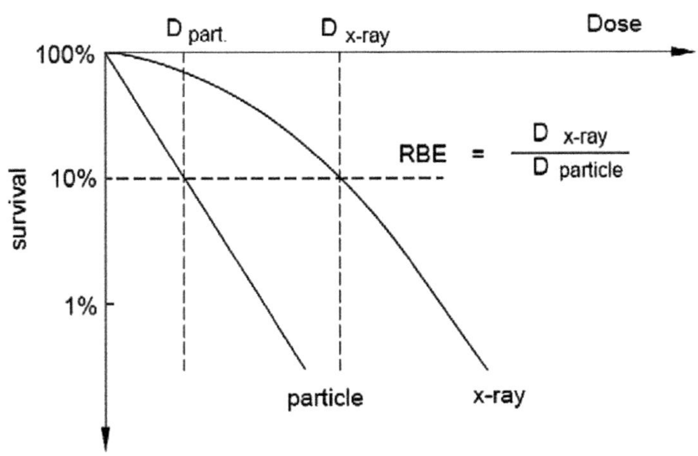

Survival vs. Relative Biological Effectiveness of absorbed particles compared to X-rays

Measuring radiation exposure

Measuring the amount, or dose, of radiation received is crucial.

Relative Biological Effectiveness (RBE) is generally used in experimental radiation biology and related fields. It is related to, but not identical to, the quality factor for dose equivalent.

Compared with higher energy photons (e.g., cobalt-60 gamma rays), lower energy electrons and photons produce more dense clusters of ionizations, leading to more complex damage to the cell's DNA and, thus, a higher RBE.

Type	**Relative Biological Effectiveness (RBE)**
X- and γ rays	1
β (electrons)	1
Protons	2
Slow Neutrons	5
Fast Neutrons	≈ 10
α particles and heavy ions	≈ 20

Relative Biological Effectiveness (RBE)

Notes for active learning

Medical Applications of Radioactivity

Effective dose

Gamma rays are often the most dangerous overall because they penetrate the entire body; however, if ingested, *alpha rays* are the most damaging.

The dose is multiplied by the relative biological effectiveness of an *effective dose.*

Effective dose is expressed in *rem* for a dose measured in *rads.*

For a dose measured in grays (Gy), the effective dose is measured in Sieverts (Sv).

Cancer (i.e., uncontrolled cell growth) is treated with *radiation therapy* to target cancerous cells. For minimal damage to healthy tissue, the radiation source is rotated to penetrate different body parts as it targets the tumor (i.e., a mass of abhorrent cells).

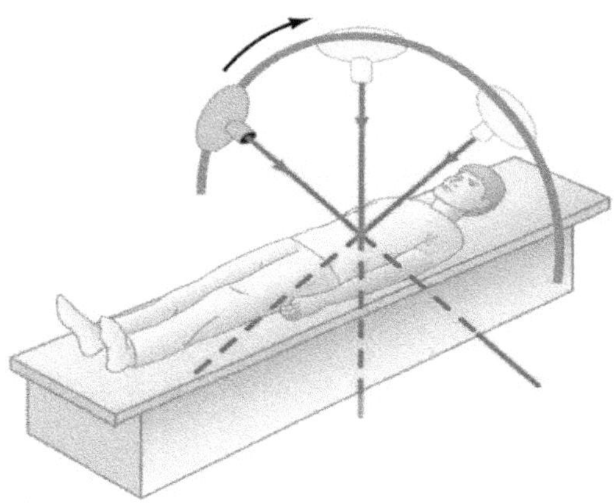

Radiation therapy with a movable emission source to minimize radiation exposure for healthy cells

Medical isotopes

Radioactive isotopes are used in medicine for *diagnostic purposes* as non-invasive scans to detect and evaluate unusual concentrations that could signal a tumor or other problem.

Radionuclides have unstable nuclei, which achieve stability through nucleic changes (fission, emission of alpha particles, or conversion of neutrons to protons or the reverse).

For example, radiopharmaceuticals are linked to specific 3-dimensional (targeting) molecules that recognize cell surface target proteins (i.e., plasma membrane surface proteins)

Isotopes are essential to radiopharmaceuticals, which detect and treat cardiovascular diseases and cancer. Each year, 48 million patients are diagnosed or treated with nuclear medicines.

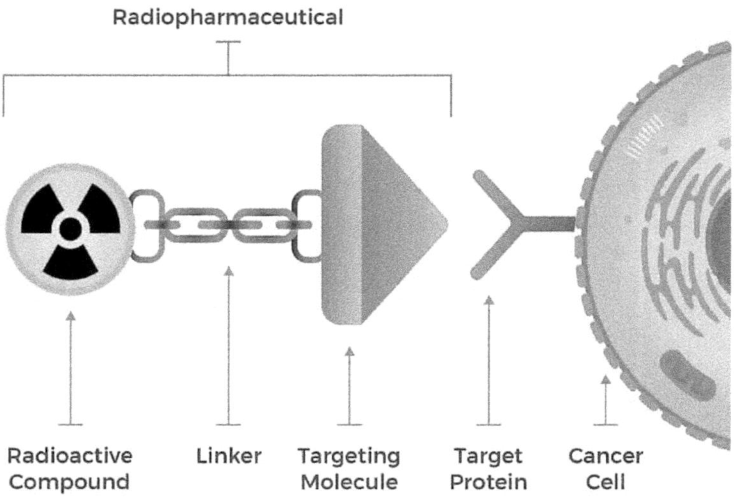

Radionuclides targeting cancerous cells using endogenous cell receptors, Dana Farber Cancer Institute, Harvard Medical School

Tomography

Tomography is imaging by sections or sectioning using any penetrating wave and is used in radiology, atmospheric science, geophysics, and oceanography.

Computed tomography (or CT) refers to a computerized X-ray imaging procedure in which a narrow beam of X-rays is aimed at a patient and quickly rotated around the body, producing signals processed by the machine's computer to generate cross-sectional images, or "slices."

X-ray tomography uses X-ray radiation to penetrate objects. On the way through an object, part of the impinging radiation is absorbed. The longer the radiographic length of the object, the less radiation escapes from the opposite side.

Radioactive tracers can be detected using tomographic techniques, where a three-dimensional image is gradually built up through successive scans.

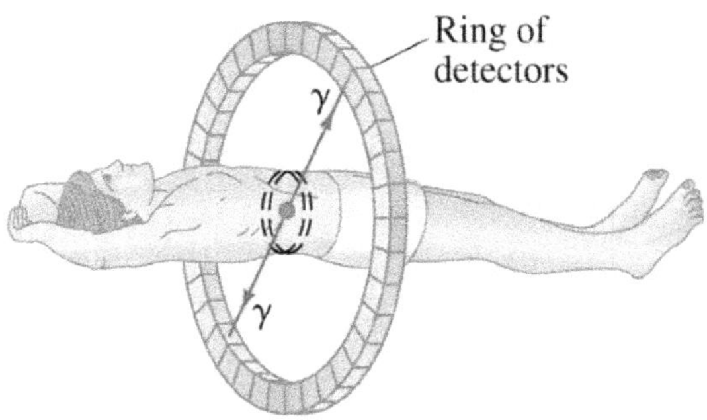

Radioactive tracers are detected using tomographic techniques

Technique	Optimal resolution
Conventional X-ray	½ mm
CT scan, X-ray	½ mm
Nuclear medicine (tracers)	1 cm
SPECT (single photon emission)	1 cm
PET (positron emission)	2 – 5 mm
MRI (NMR)	½ – 1 mm
Ultrasound	0.3 – 2 mm

Scintillation counters in medicine

Scintillation counters are widely used for the assay of radioactive materials and physics research because they can be made inexpensively yet with good quantum efficiency and can measure both the intensity and the energy of incident radiation.

Scintillators are materials that *emit light* when bombarded with high-energy particles or X-rays. In medical or dental X-ray systems, they convert incoming X-ray radiation into visible light captured using film or photosensors.

In scintillation detectors, the material of the detector is excited by luminescence (emission of visible or near-visible light photons) by the absorbed photons or particles.

Thin scintillators are excellent for low-energy gamma rays and high-energy beta particles. The ray or particle will be absorbed within the thin scintillator, and light produced by this interaction passes through the remaining thickness to allow gamma-ray detection.

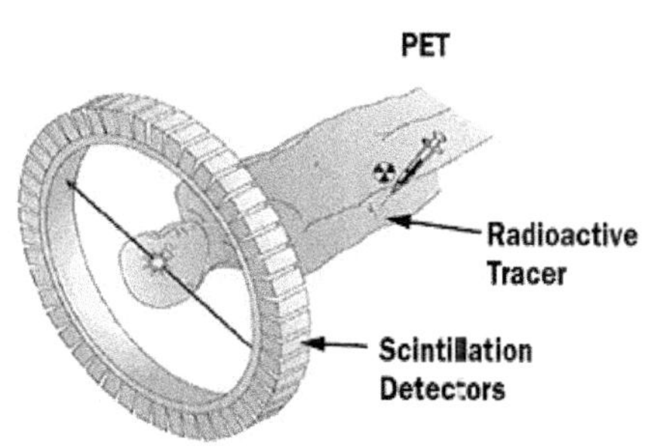

Scintillation detectors detect the high-energy part of the X-ray spectrum

Nuclear magnetic resonance

Nuclear magnetic resonance (NMR) is a technique where the examined object is placed in a static magnetic field, and radio frequency (RF) electromagnetic radiation is applied.

A proton in a magnetic field has its spin either parallel or antiparallel to the field. The field splits the energy levels slightly; the energy difference is proportional to the field magnitude, where the vector B represents the magnetic field.

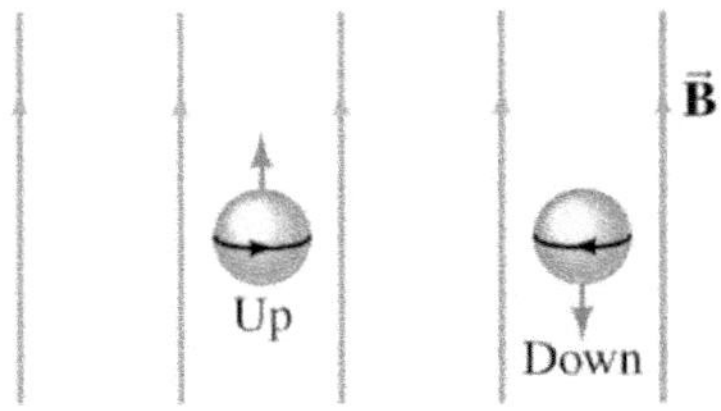

Protons in a magnetic field have their spin either parallel or antiparallel to the field

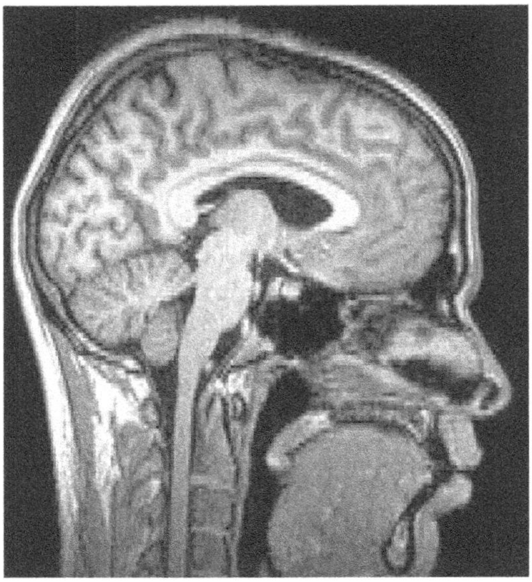

Magnetic resonance imaging (MRI) of the head

Photons are absorbed when the radiation has the proper energy to excite the spin-flip transition. The value of the field depends on the local molecular neighborhood. This value of spin-flip transitions allows information about the structure of the molecules to be determined.

Magnetic resonance imaging

Magnetic resonance imaging works similarly; the transition is excited in hydrogen atoms, the most common in the human body. Giving the field a gradient contributes to image accuracy, as it allows determining the origin of a particular signal.

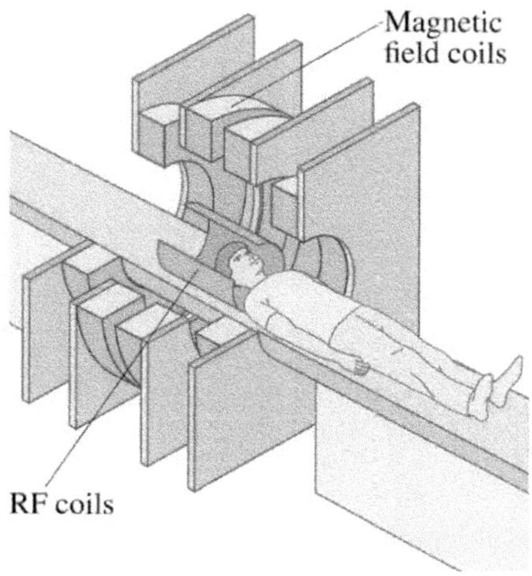

Magnetic resonance imaging is a common and valuable diagnostic tool

Notes for active learning

Chapter Summary

Atoms

- Nuclei contain protons and neutrons—nucleons.
- Total number of nucleons, A, is the atomic mass number.
- Number of protons, Z, is the atomic number.
- Isotope notation: $^{A}_{Z}X$
- Nuclear masses are measured in u; carbon-12 is defined as having a mass of 12 amu.
- 1 amu = 1.6605×10^{-27} kg = 931.5 MeV/c^2
- The difference between the mass of the nucleus and the mass of its constituents is the binding energy.

Decay

- Unstable nuclei decay through alpha, beta, or gamma emission.
- Alpha particle is a helium nucleus; beta particle is an electron or positron; gamma-ray is a highly energetic photon.
- Nuclei are held together by the strong nuclear force; the weak nuclear force is responsible for beta decay.
- Electric charge, linear and angular momentum, mass-energy, and nucleon number are conserved.
- Radioactive decay is a statistical process.
- Number of decays per unit time is proportional to nuclei present:

  $$N = N_0 e^{-\lambda t}$$

- Half-life is the time it takes for half the nuclei to decay.

Nuclear reactions

- Nuclear reaction occurs when nuclei collide, and nuclei are produced.
- Reaction energy or Q-value:

  $$Q = M_P c^2 - (M_D + m_a)c^2$$

- Fission is when a heavy nucleus splits into two smaller nuclei.
  - Chain reactions occur when neutrons emitted in a fission trigger more reactions.
  - Critical mass is the minimum mass needed to sustain chain reactions.
  - A moderator is used to slow the neutrons, so collisions are more likely.
- Fusion is when small nuclei combine to form larger nuclei.
  - Sun's energy comes from fusion reactions.
  - A practical fusion reactor has not yet been built.
  - Radiation damage is measured using dosimetry.
  - Effect of the absorbed dose depends on the type of radiation.

Practice Problems: Atomic and Nuclear Structure

1. In a nuclear reaction, the mass of the products is less than the mass of the reactants. Why is this not observed in a chemical reaction?

A. In chemical reactions, the mass is held constant by the nucleus
B. In chemical reactions, the mass deficit is balanced by a mass surplus
C. The mass deficit in chemical reactions is too small to be observed
D. The mass does not convert to energy in chemical reactions
E. None of the above are true

2. The isotope $^{13}_{7}N$ decays by positron emission to what isotope?

A. $^{14}_{6}C$
B. $^{11}_{7}N$
C. $^{13}_{6}C$
D. $^{12}_{6}C$
E. $^{12}_{7}N$

3. The radioactive gas radon is:

I. more hazardous to smokers than nonsmokers
II. the single greatest source of human radiation exposure
III. a product of the radioactive decay series of uranium

A. I only
B. II only
C. III only
D. I and III only
E. I, II and III

4. Which of the statements best describes the strong nuclear force?

A. The strength of the force increases with distance
B. The force is powerful and is effective over a large range of distances
C. The electrical force is stronger than the nuclear force
D. The force is powerful but is effective only within a short range of distances
E. None of the above are correct statements

5. Natural line broadening can be understood regarding the:

A. Schrodinger wave equation
B. Pauli exclusion principle
C. de Broglie wavelength
D. uncertainty principle
E. quantum numbers

6. A blue photon has a:

A. longer wavelength than a red photon and travels with a higher speed
B. longer wavelength than a red photon and travels with the same speed
C. shorter wavelength than a red photon and travels with a higher speed
D. longer wavelength than a red photon and travels with a lower speed
E. shorter wavelength than a red photon and travels with the same speed

7. Which type of nuclear radiation is powerful light energy that is *not* deflected as it passes between electrically charged plates?

A. Gamma
B. Beta
C. Alpha
D. Nuclide
E. None of the above

8. The main reason that there is a limit to the size of a stable nucleus is the:

A. weakness of the electrostatic force
B. weakness of the gravitational force
C. short-range effect of the strong nuclear force
D. limited range of the gravitational force
E. long-distance effect of the strong nuclear force

9. Elements combine in fixed mass ratios to form compounds. This fixed mass ratio requires the following elements:

A. have unambiguous atomic numbers
B. are always chemically active
C. are composed of continuous matter without subunits
D. are composed of discrete subunits called atoms
E. none of the above are required

10. What is rem?

A. A unit for measuring rapid electron motion
B. The number of radiation particles absorbed per second
C. The number of radiation particles emitted per second
D. The maximum exposure limit for occupational safety
E. A unit for measuring radiation exposure

11. The intensity of X-rays, gamma rays, or any other radiation is:

A. inversely proportional to the square of the distance from the source
B. inversely proportional to the distance from the source
C. directly proportional to the square of the distance from the source
D. directly proportional to the distance from the source
E. inversely proportional to the cube of the distance from the source

12. Ionizing radiation is:

A. a neutron that has acquired a charge, resulting in the formation of an ion
B. high-energy radiation that removes electrons from atoms or molecules
C. radiation that only interacts with ions
D. equivalent to a proton
E. released by ions and reacts with nuclei

13. What nucleus results when ^{55}Ni decays by positron emission?

A. ^{55}Ca
B. ^{55}Ni
C. ^{55}Co
D. ^{55}Fe
E. None of the above

14. Which of the following types of radiation has the highest energy?

A. γ rays
B. Visible light rays
C. α particles
D. β particles
E. All have the same energy

15. What is its temperature if a star has a peak intensity at 580 nm? (Use Wien's displacement constant $b = 2.9 \times 10^{-3}$ K·m)

A. 5,000 °C
B. 2,000 °C
C. 5,000 °F
D. 2,000 K
E. 5,000 K

Solutions: Atomic and Nuclear Structure

1. D is correct.

Mass is conserved in a chemical reaction; no particles are created or destroyed. Atoms are rearranged to form products from reactants.

In nuclear reactions, the mass difference is due to mass converting into energy.

2. C is correct.

Positron emission occurs during β^+.

The decay equation:

$$^{13}_{7}\text{N} \rightarrow ^{13}_{6}\text{C} + e^+ + \nu_e$$

A proton converts into a neutron, and a positron is ejected with an electron neutrino. The daughter nuclide is ^{13}C.

3. E is correct.

Radon gas is a natural decay product of uranium that accounts for the greatest source of yearly radiation exposure in humans. Radon is more hazardous to smokers due to the combined carcinogenic effects of smoking and radiation exposure.

4. D is correct.

The strong nuclear force is the strongest of the four fundamental forces. However, it only acts within a small distance range (about the nucleus's diameter).

5. D is correct.

Natural line broadening extends a spectral line over a range of frequencies. This occurs partly due to uncertainty principle, which relates the time an atom is excited to emitted photon energy.

$$\Delta E \Delta t > \hbar / 2$$

where $\hbar$ is the symbol for the reduced Planck's constant

Because energy is related to frequency by:

$$E = hf$$

where h is Planck's constant

Range of frequencies observed (broadening) is due to the uncertainty in energy outlined by the uncertainty principle.

6. E is correct.

Speed is constant for all electromagnetic waves and only changes due to the transmission medium, not the frequency or wavelength of the wave.

Blue photons have higher energy and thus higher frequencies than red light due to:

$$E = hf$$

However, the frequency is inversely proportional to wavelength:

$$\lambda = c / f$$

Thus, blue photons have shorter wavelengths than red photons due to their higher frequencies.

7. A is correct.

Gamma radiation is a high-energy electromagnetic wave, and as such, it has no charge and will not deflect within an electric field.

8. C is correct.

Nuclei with atomic numbers greater than 83 are inherently unstable and thus radioactive. This limit in size is because the strong nuclear force has a short range, and as the nucleus gets larger, the strong nuclear force cannot overcome Coulomb repulsion from protons within the nucleus.

9. D is correct.

Its molar mass can measure the mass of an element or compound.

Molar mass relates the mass of the element or compound to a discrete number of subunits (atoms for elements, molecules for compounds).

10. E is correct.

The rem is short for the Roentgen equivalent in man and is designed to measure the biological damage of ionizing radiation. It does not measure the number of particles absorbed or emitted, nor is it the maximum occupational safety exposure limit for radiation.

11. A is correct.

Intensity of electromagnetic radiation concerning distance from the point source:

$$I = S / 4\pi r^2$$

where I = intensity, S = point source strength, r = radial distance from the point source

Radiation intensity from the point source is inversely proportional to the square of the distance from the source.

12. B is correct.

Ionizing radiation can be high-energy charged particles (alpha, beta) or electromagnetic waves (X-rays, gamma rays).

All are ionizing because they possess enough energy to remove and ionize electrons from atoms (or molecules).

13. C is correct.

Nuclear reaction:

$$^{55}_{28}\text{Ni} \rightarrow {}^{55}_{27}\text{Co} + e^+ + \nu_e$$

where Co = product, e^+ = positron and ν_e = electron neutrino

In positron emission (β^+ decay), a proton in the nucleus converts to a neutron while releasing a positron and an electron neutrino.

The atomic number decreases by one, but the mass number stays constant.

14. A is correct.

Gamma rays are high-energy electromagnetic waves.

Gamma rays have the highest energy radiation (e.g., alpha, beta, electromagnetic spectrum).

15. E is correct.

Use Wien's Displacement Law:

$$\lambda_{max} = b / T$$

$$T = (2.9 \times 10^{-3}\ \text{K·m}) / (580 \times 10^{-9}\ \text{m})$$

$$T = 5{,}000\ \text{K}$$

Notes for active learning

Notes for active learning

APPENDIX

Common Equations and Conversions

Constants and Conversion Factors

1 unified atomic mass unit	$1\ u = 1.66 \times 10^{-27}$ kg $1\ u = 931$ MeV/c^2
Proton mass	$m_p = 1.67 \times 10^{-27}$ kg
Neutron mass	$m_n = 1.67 \times 10^{-27}$ kg
Electron mass	$m_e = 9.11 \times 10^{-31}$ kg
Electron charge magnitude	$e = 1.60 \times 10^{-19}$ C
Avogadro's number	$N_0 = 6.02 \times 10^{23}$ mol^{-1}
Universal gas constant	$R = 8.31$ J/(mol·K)
Boltzmann's constant	$k_B = 1.38 \times 10^{-23}$ J/K
Speed of light	$c = 3.00 \times 10^8$ m/s
Planck's constant	$h = 6.63 \times 10^{-34}$ J·s $h = 4.14 \times 10^{-15}$ eV·s $hc = 1.99 \times 10^{-25}$ J·m $hc = 1.24 \times 10^3$ eV·nm
Vacuum permittivity	$\varepsilon_0 = 8.85 \times 10^{-12}$ C^2/N·m^2
Coulomb's law constant	$k = 1/4\pi\varepsilon_0 = 9.0 \times 10^9$ N·m^2/C^2
Vacuum permeability	$\mu_0 = 4\pi \times 10^{-7}$ (T·m)/A
Magnetic constant	$k' = \mu_0/4\pi = 10^{-7}$ (T·m)/A
Universal gravitational constant	$G = 6.67 \times 10^{-11}$ m^3/kg·s^2
Acceleration due to gravity at Earth's surface	$g = 9.8$ m/s^2
1 atmosphere pressure	1 atm $= 1.0 \times 10^5$ N/m^2 1 atm $= 1.0 \times 10^5$ Pa
1 electron volt	1 eV $= 1.60 \times 10^{-19}$ J
Balmer constant	$B = 3.645 \times 10^{-7}$ m
Rydberg constant	$R = 1.097 \times 10^7$ m^{-1}
Stefan constant	$\sigma = 5.67 \times 10^{-8}$ W/m^{2}K^4

Units

Name	Symbol
meter	m
kilogram	kg
second	s
ampere	A
kelvin	K
mole	mol
hertz	Hz
newton	N
pascal	Pa
joule	J
watt	W
coulomb	C
volt	V
ohm	Ω
henry	H
farad	F
tesla	T
degree Celsius	°C
electronvolt	eV

Prefixes

Factor	Prefix	Symbol
10^{12}	tera	T
10^{9}	giga	G
10^{6}	mega	M
10^{3}	kilo	k
10^{-2}	centi	c
10^{-3}	mili	m
10^{-6}	micro	μ
10^{-9}	nano	n
10^{-12}	pico	p

Newtonian Mechanics

Translational Motion	$v = v_0 + a\Delta t$
	$x = x_0 + v_0\Delta t + \frac{1}{2}a\Delta t^2$
	$v^2 = v_0^2 + 2a\Delta x$
	$\vec{a} = \frac{\sum \vec{F}}{m} = \frac{\vec{F}_{net}}{m}$
Rotational Motion	$\omega = \omega_0 + \alpha t$
	$\theta = \theta_0 + \omega_0 t + \frac{1}{2}\alpha t^2$
	$\omega^2 = \omega_0^2 + 2\alpha\Delta\theta$
	$\vec{\alpha} = \frac{\sum \vec{\tau}}{I} = \frac{\vec{\tau}_{net}}{I}$
Force of Friction	$\lvert\vec{F}_f\rvert \leq \mu\lvert\vec{F}_n\rvert$
Centripetal Acceleration	$a_c = \frac{v^2}{r}$
Torque	$\tau = r_{\perp}F = rF\sin\vartheta$
Momentum	$\vec{p} = m\vec{v}$
Impulse	$\vec{J} = \Delta\vec{p} = \vec{F}\Delta t$
Kinetic Energy	$K = \frac{1}{2}mv^2$
Potential Energy	$\Delta U_g = mg\Delta y$
Work	$\Delta E = W = F_{\parallel}d = Fd\cos\theta$
Power	$P = \frac{\Delta E}{\Delta t} = \frac{\Delta W}{\Delta t}$
Simple Harmonic Motion	$x = A\cos(\omega t)$
	$x = A\cos(2\pi f t)$

a = acceleration
A = amplitude
E = energy
F = force
f = frequency
h = height
I = rotational inertia
J = impulse
K = kinetic energy
k = spring constant
ℓ = length
m = mass
N = normal force
P = power
p = momentum
L = angular momentum
r = radius of distance
T = period
t = time
U = potential energy
v = velocity or speed
W = work done on system
x = position
y = height
α = angular acceleration

Newtonian Mechanics (*continued*)

Center of Mass	$x_{cm} = \frac{\sum m_i x_i}{\sum m_i}$	μ = coefficient of friction
		θ = angle
Angular Momentum	$L = I\omega$	τ = torque
Angular Impulse	$\Delta L = \tau \Delta t$	ω = angular speed
Angular Kinetic Energy	$K = \frac{1}{2} I\omega^2$	
Work	$W = F\Delta r \cos\theta$	
Power	$P = Fv \cos\theta$	
Spring Force	$\lvert\vec{F}_s\rvert = k\lvert\vec{x}\rvert$	
Spring Potential Energy	$U_s = \frac{1}{2} kx^2$	
Period of Spring Oscillator	$T_s = 2\pi\sqrt{m/k}$	
Period of Simple Pendulum	$T_p = 2\pi\sqrt{\ell/g}$	
Period	$T = \frac{2\pi}{\omega} = \frac{1}{f}$	
Gravitational Body Force	$\lvert\vec{F}_g\rvert = G\frac{m_1 m_2}{r^2}$	
Gravitational Potential Energy of Two Masses	$U_G = -\frac{Gm_1 m_2}{r}$	

Optics

Wavelength to Frequency	$\lambda = \frac{v}{f}$	d = separation
		f = frequency or focal length
Index of Refraction	$n = \frac{c}{v}$	h = height
		L = distance
Snell's Law	$n_1 \sin\theta_1 = n_2 \sin\theta_2$	M = magnification
Thin Lens Equation	$\frac{1}{s_i} + \frac{1}{s_0} = \frac{1}{f}$	m = an integer
		n = index of refraction
Magnification Equation	$\lvert M\rvert = \left\lvert\frac{h_i}{h_o}\right\rvert = \left\lvert\frac{s_i}{s_o}\right\rvert$	R = radius of curvature
		s = distance
Double Slit Diffraction	$d \sin\theta = m\lambda$	v = speed
	$\Delta L = m\lambda$	x = position
Critical Angle	$\sin\theta_c = \frac{n_2}{n_1}$	λ = wavelength
		θ = angle
Focal Length of Spherical Mirror	$f = \frac{R}{2}$	

Sound

Standing Wave/ Open Pipe Harmonics	$\lambda = \frac{2L}{n}$	f = frequency
Closed Pipe Harmonics	$\lambda = \frac{4L}{n}$	L = length
Harmonic Frequencies	$f_n = nf_1$	m = mass
Speed of Sound in Ideal Gas	$v_{sound} = \sqrt{\frac{yRT}{M}}$	M = molecular mass
Speed of Wave Through Wire	$v = \sqrt{\frac{T}{m/L}}$	n = harmonic number
Doppler Effect (approaching stationary observer)	$f_{observed} = (\frac{v}{v - v_{source}})f_{source}$	R = gas constant
Doppler Effect (receding stationary observer)	$f_{observed} = (\frac{v}{v + v_{source}})f_{source}$	T = tension
Doppler Effect (observer moving towards source)	$f_{observed} = (1 + \frac{v_{observer}}{v})f_{source}$	v = velocity
Doppler Effect (observer moving away from source)	$f_{observed} = (1 - \frac{v_{observer}}{v})f_{source}$	y = adiabatic constant
		λ = wavelength

Electrostatics, Magnetism, Circuits

Electric Field	$\vec{E} = \frac{\vec{F}_E}{q}$	
Electric Field Strength	$\lvert\vec{E}\rvert = \frac{1}{4\pi\varepsilon_0}\frac{\lvert q\rvert}{r^2}$	
Electric Field Strength	$\lvert\vec{E}\rvert = \frac{\lvert\Delta V\rvert}{\lvert\Delta r\rvert}$	
Electrostatic Force Between Charged Particles	$\lvert\vec{F}_E\rvert = \frac{1}{4\pi\varepsilon_0}\frac{\lvert q_1 q_2\rvert}{r^2}$	
Electric Potential Energy	$\Delta U_E = q\Delta V$	
Electrostatic Potential due to a Charge	$V = \frac{1}{4\pi\varepsilon_0}\frac{q}{r}$	
Capacitor Voltage	$V = \frac{Q}{C}$	
Capacitance of Parallel Plate Capacitor	$C = \kappa\varepsilon_0\frac{A}{d}$	
Electric Field Inside a Parallel Plate Capacitor	$E = \frac{Q}{\varepsilon_0 A}$	
Capacitor Potential Energy	$U_C = \frac{1}{2}Q\Delta V = \frac{1}{2}C(\Delta V)^2$	
Current	$I = \frac{\Delta Q}{\Delta t}$	
Resistance	$R = \frac{\rho l}{A}$	
Power	$P = I\Delta V$	
Current	$I = \frac{\Delta V}{R}$	
Resistors in Series	$R_s = \sum_i R_i$	
Resistors in Parallel	$\frac{1}{R_p} = \sum_i \frac{1}{R_i}$	

A = area

B = magnetic field

C = capacitance

d = distance

E = electric field

ϵ = emf

F = force

I = current

l = length

P = power

Q = charge

q = point charge

R = resistance

r = separation

t = time

U = potential energy

V = electric potential

v = speed

κ = dielectric constant

ρ = resistivity

θ = angle

Φ = flux

Electrostatics, Magnetism, Circuits (*continued*)

Capacitors in Parallel	$C_p = \sum_i C_i$
Capacitors in Series	$\frac{1}{C_s} = \sum_i \frac{1}{C_i}$
Magnetic Field Strength (from a long straight current-carrying wire)	$B = \frac{\mu_0 I}{2\pi r}$
Magnetic Force	$\vec{F}_M = q\vec{v} \times \vec{B}$ $\vec{F}_M = \lvert q\vec{v}\rvert \lvert\sin\theta\rvert \lvert\vec{B}\rvert$ $\vec{F}_M = I\vec{l} \times \vec{B}$ $\vec{F}_M = \lvert I\vec{l}\rvert \lvert\sin\theta\rvert \lvert\vec{B}\rvert$
Magnetic Flux	$\Phi_B = \vec{B} \cdot \vec{A}$ $\Phi_B = \lvert\vec{B}\rvert \cos\theta \, \lvert\vec{A}\rvert$
Electromagnetic Induction	$\epsilon = \frac{-\Delta\Phi_B}{\Delta t}$ $\epsilon = Blv$

Fluid Mechanics and Thermal Physics

Density	$\rho = \frac{m}{V}$
Pressure	$P = \frac{F}{A}$
Absolute Pressure	$P = P_0 + \rho g h$
Buoyant Force	$F_b = \rho V g$
Fluid Continuity Equation	$A_1 v_1 = A_2 v_2$
Bernoulli's Equation	$P_1 + \rho g y_1 + \frac{1}{2}\rho v_1^2 = P_2 + \rho g y_2 + \frac{1}{2}\rho v_2^2$
Heat Conduction	$\frac{Q}{\Delta t} = \frac{kA\Delta T}{d}$
Thermal Radiation	$P = e\sigma A(T^4 - T_C^4)$
Ideal Gas Law	$PV = nRT = Nk_B T$
Average Energy	$K = \frac{3}{2}k_B T$
Work	$W = -P\Delta V$
Conservation of Energy	$\Delta E = Q + W$
Linear Expansion	$\Delta l = \alpha l_o \Delta T$
Heat Engine Efficiency	$n_c = \lvert W/Q_H \rvert$
Carnot Heat Engine Efficiency	$n_c = \frac{T_H - T_C}{T_H}$
Energy of Temperature Change	$Q = mc\Delta T$
Energy of Phase Change	$Q = mL$

A = area

c = specific heat

d = thickness

e = emissivity

F = force

h = depth

k = thermal conductivity

K = kinetic energy

l = length

L = latent heat

m = mass

n = number of moles

n_c = efficiency

N = number of molecules

P = pressure or power

Q = energy transferred to system by heating

T = temperature

t = time

E = internal energy

V = volume

v = speed

W = work done on a system

y = height

σ = Stefan constant

α = coefficient of linear expansion

ρ = density

Modern Physics

Photon Energy	$E = hf$
Photoelectric Electron Energy	$K_{max} = hf - \phi$
Electron Wavelength	$\lambda = \frac{h}{p}$
Energy Mass Relationship	$E = mc^2$
Rydberg Formula	$\frac{1}{\lambda} = R(\frac{1}{n_f^2} - \frac{1}{n_i^2})$
Balmer Formula	$\lambda = B(\frac{n^2}{n^2 - 2^2})$
Lorentz Factor	$\gamma = \frac{1}{\sqrt{1 - \frac{v^2}{c^2}}}$

B = Balmer constant

c = speed of light

E = energy

f = frequency

K = kinetic energy

m = mass

p = momentum

R = Rydberg constant

v = velocity

λ = wavelength

ϕ = work function

γ = Lorentz factor

Geometry and Trigonometry

Rectangle	$A = bh$	A = area
		C = circumference
Triangle	$A = \frac{1}{2}bh$	V = volume
		S = surface area
Circle	$A = \pi r^2$	b = base
	$C = 2\pi r$	h = height
Rectangular Solid	$V = lwh$	l = length
		w = width
Cylinder	$V = \pi r^2 l$	r = radius
	$S = 2\pi rl + 2\pi r^2$	θ = angle
Sphere	$V = \frac{4}{3}\pi r^3$	
	$S = 4\pi r^2$	
Right Triangle	$a^2 + b^2 = c^2$	
	$\sin\theta = \frac{a}{c}$	
	$\cos\theta = \frac{b}{c}$	
	$\tan\theta = \frac{a}{b}$	

Trigonometric Functions for Common Angles

θ	$\sin\theta$	$\cos\theta$	$\tan\theta$
0°	0	1	0
30°	1/2	$\sqrt{3}/2$	$\sqrt{3}/3$
37°	3/5	4/5	3/4
45°	$\sqrt{2}/2$	$\sqrt{2}/2$	1
53°	4/5	3/5	4/3
60°	$\sqrt{3}/2$	1/2	$\sqrt{3}$
90°	1	0	∞

Notes for active learning

Isotopic Decay Half-Life Reference Table

Isotope	Mode of decay	Half-life
Tritium	Beta	12.26 years
Carbon-14	Beta	5,930 years
Argon-41	Beta, gamma	1.83 hours
Iron-55	Electron capture	2.7 years
Colbalt-58	Beta, gamma	71 days
Colbalt-60	Beta, gamma	5.26 years
Nickel-63	Beta	92 years
Krypton-85	Beta, gamma	10.76 years
Strontium-89	Beta	5.4 days
Strontium-90	Beta	28 years
Yttrium-91	Beta	59 days
Zirconium-93	Beta	9.5×10^5 years
Zirconium-95	Beta, gamma	65 days
Niobium-95	Beta, gamma	35 days
Technetium-99	Beta	2.1×10^5 years
Ruthenium-106	Beta	1 year
Iodine-129	Beta	1.6×10^7 years
Iodine-131	Beta, gamma	8 days
Xenon-133	Beta, gamma	5.27 days
Cesium-134	Beta, gamma	2.1 years
Cesium-135	Beta	2×10^6 years
Cesium-137	Beta	30 years
Cerium-141	Beta	32.5 days
Cerium-144	Beta, gamma	285 days

Isotope	Mode of decay	Half-life
Promethium-147	Beta	2.6 years
Samarium-151	Beta	90 years
Europium-154	Beta, gamma	16 years
Lead-210	Beta	22 years
Radon-222	Alpha	3.8 days
Radium-226	Alpha, gamma	1,620 years
Thorium-229	Alpha	7,300 years
Thorium-230	Alpha	26,000 years
Uranium-234	Alpha	2.48×10^5 years
Uranium-235	Alpha, gamma	7.13×10^8 years
Uranium-238	Alpha	4.51×10^9 years
Neptunium-237	Alpha	2.14×10^6 years
Plutonium-238	Alpha	89 years
Plutonium-239	Alpha	24,360 years
Plutonium-240	Alpha	6,760 years
Plutonium-241	Beta	13 years
Plutonium-242	Alpha	3.79×10^5 years
Americium-241	Alpha	458 years
Americium-243	Alpha	7,650 years
Curium-242	Alpha	163 days
Curium-244	Alpha	18 years
Thorium-229	Alpha	7,300 years

Glossary of Physics Terms

A

Ab initio – a mathematical model to describe atomic nuclei by solving the non-relativistic Schrödinger equation for constituent nucleons and forces existing between them. *Such methods yield precise results for very light nuclei but become more approximate for heavier nuclei.*

Abbe number (*V-number* and *constringence*) (*optics and lens design*) – a measure of a transparent material's dispersion (a variation of refractive index versus wavelength). *High values of V indicate low dispersion.*

Absolute electrode potential (*electrochemistry*) – the electrode potential of a metal measured with respect to a universal reference system (without any additional metal–solution interface).

Absolute humidity (or *saturation value*) – the ratio of water vapor in a sample of air to the volume of the sample. Maximum amount of water vapor in 1 m^3 of air at a temperature.

Absolute magnitude – a classification scheme that compensates for the differences in the distance to stars; calculates the brightness that stars would appear to have if they were at a defined, standard distance of 10 parsecs.

Absolute pressure – is zero-referenced against a perfect vacuum, using an absolute scale, so it is equal to gauge pressure plus atmospheric pressure.

Absolute scale – 1. temperature scale set so that zero is the theoretically lowest temperature possible (this would occur when the random motion of molecules has ceased). 2. any system of measurement that begins at a minimum (or zero point) and progresses in only one direction. Zero point of an absolute scale is a natural minimum, leaving only one direction in which to progress. *In contrast, an arbitrary or "relative" scale begins at some point selected by a person and can progress in both directions.*

Absolute zero – the theoretically lowest temperature possible, at which the molecular motion vanishes; –273.16 °C, 0 K. or –459.67 °F. The theoretical lower limit of the thermodynamic temperature scale, at which enthalpy and entropy of a cooled ideal gas reach minimum values and fundamental particles of nature have minimal vibrational motion.

Absorptance – the ratio of the total absorbed radiation to the total incident radiation.

Absorption spectroscopy – radiation absorption as a function of frequency (or wavelength) due to interaction with a sample which a absorbs energy (i.e., photons) from radiating field. Absorption intensity varies as a function of frequency, and this variation is the absorption spectrum. *Absorption spectroscopy is performed across the electromagnetic spectrum.*

Acceleration – the rate of change of velocity of a moving object concerning time; the SI units are m/s^2. This change in velocity can result from a change in speed, direction, or a combination of changes in speed and direction.

Acceleration due to gravity – the rate of change in velocity produced in a body due to the Earth's attraction; denoted by the letter *g* (SI unit is m/s^2); on the surface of the Earth, its average value is 9.8 m/s^2; increases when going towards the poles from the equator; decreases with altitude and with depth inside the Earth; the value of g at the center of the Earth is zero.

Accelerometer – measures the proper acceleration of a body irrespective of other forces.

Achromatic – capable of transmitting light without decomposing it into its constituent colors.

Acid rain – rainwater with a pH of less than 5.7 is acid rain. It is caused by the gases NO_2 (e.g., car exhaust fumes) and SO_2 (e.g., burning fossil fuels) dissolving in the rain. *Acid rain kills fish, kills trees, and destroys buildings and lakes.*

Acoustics – the science of sound production, transmission, and effects.

Acoustic shielding – a sound barrier that prevents the transmission of acoustic energy.

Adhesion – the tendency of dissimilar particles or surfaces to cling. Contrast *cohesion.*

Adiabatic – any change in which there is no gain or loss of heat.

Adiabatic cooling – the decrease in temperature of an expanding gas that involves no additional heat flowing out of the gas; it is the cooling from the energy lost by expansion.

Adiabatic heating – the increase in temperature of the compressed gas that involves no additional heat flowing into the gas; the heating from the energy gained by compression.

Adiabatic process – occurs without transfer of heat or mass of substances between a thermodynamic system and surroundings. In an adiabatic process, energy is transferred to the surroundings only as work. *Adiabatic process provides a rigorous conceptual basis for the first law of thermodynamics, and as such, it is a crucial concept in thermodynamics.*

Aerodynamics – the study of the motion of air, particularly its interaction with a solid object (e.g., an airplane wing). *It is a sub-field of fluid and gas dynamics, and many aspects of aerodynamics theory are common to these fields.*

Afocal lens – a lens of zero convergent power whose focal points are infinitely distant.

Air mass – 1. a large, uniform body of air with nearly identical temperature and moisture conditions throughout. 2. (*meteorology*) an air volume defined by temperature and water vapor content. Air masses may cover hundreds or thousands of square miles and adapt to the characteristics of the surface below them. *They are often classified according to their latitude and their source regions*. 3. (*astronomy*) the "amount of air that one is looking through" when observing a star or other celestial source from a vantage point within Earth's atmosphere. *It is formulated as the integral of air density along the light ray.*

Air mass coefficient – defines the direct optical path length through the Earth's atmosphere, expressed as a ratio relative to the path length vertically upwards (i.e., at the zenith). *The air mass coefficient can be used to characterize the solar spectrum after solar radiation has traveled through the atmosphere.*

Albedo – the fraction of the total light incident on a reflecting surface, especially a celestial body, which is reflected in all directions.

Allotropic forms – elements with structures having different physical properties (e.g., graphite and diamond).

Alloy – a chemical mixture of a metal with one or more other metals or elements.

Alpha decay (*α-decay*) – radioactive decay in which an atomic nucleus emits an alpha particle and thereby transforms or "decays" into a different atomic nucleus, with a mass number reduced by four and an atomic number reduced by two.

Alpha particle (*α, $α^{2+}$, He^{2+}*) – the nucleus of a helium atom (i.e., 2 protons and 2 neutrons) emitted as radiation from a decaying heavy nucleus (α-decay). A subatomic particle with two protons and two neutrons bound into a particle identical to the nucleus of a helium-4 ion. It has a charge of +2 *e* and a mass of 4 u. *Alpha particles are produced in radioactive alpha decay but may be produced in other ways and given the same name.*

Alternating current (AC) – an electric current in which the movement of electric charge periodically reverses direction. The flow of charge (i.e., current) traveling in one direction for one-hundredth of a second but the opposite direction for the next hundredth. An electric current that first moves in one direction, then in the opposite direction with a regular frequency. Contrast *direct current.*

Ammeter – an instrument used to measure electric current.

Amorphous – describes solids that have neither definite form nor structure.

Amorphous solid – a solid which does not have a definite geometrical shape.

Amp (A or *ampere*) – a unit of electric current equivalent to Coulomb/second.

Ampere (A or *amp*) – the unit amp; the SI unit of electric current; one ampere is the flow of one coulomb of charge per second.

Amplifier (*electronic amplifier or amp*) – an electronic device that can increase the power of a signal (a time-varying voltage or current). It is a two-port electronic circuit that uses electric power from a power supply to increase the amplitude of a signal applied to its input terminals, producing a proportionally greater amplitude signal at its output. *An amplifier's amplification is measured by its gain: the ratio of output voltage, current, or power to input. An amplifier is a circuit that has a power gain greater than one.*

Amplitude (*of an oscillation*) – maximum absolute value attained by disturbances of a wave or any quantity that varies periodically. *Height of the wave crest above the average position.* The maximum displacement of a body from its mean position during an oscillatory motion.

Amplitude (*of waves*) – the maximum displacement of particles of the medium from their mean positions during wave propagation.

Angle of contact – the angle between tangents to the liquid surface and the solid surface inside the liquid; both tangents are drawn at the point of contact.

Angle of incidence (*optics*) – the angle between a ray incident on a surface and the line perpendicular to the surface at the point of incidence, called the normal. *The angle of an incident (arriving) ray or particle to a surface, measured from a line perpendicular to the surface (the normal). Any wave can form a ray (e.g., optical, acoustic, microwave, or X-ray).*

Angle of reflection (*optics*) – the change in the direction of a wavefront at an interface between two media so that it returns to the medium from which it originated. The angle of a reflected ray or particle from a surface is measured from a line perpendicular to the surface (the normal). *The law of reflection states that for specular reflection, the angle at which the wave is incident on the surface equals the angle at which it is reflected. Mirrors exhibit specular reflection.*

Angle of refraction – the angle between the refracted ray and the normal.

Angle of repose – the degree of inclination of a plane with the horizontal such that a body placed on the plane is on the verge of sliding but does not.

Angstrom (Å) – a unit of length primarily used to measure subatomic particles equal to 10^{-10} meters (one ten-billionth of a meter) or 0.1 nanometers.

Angular acceleration – the rate of change of angular velocity of a body moving along a circular path. The time rate of change of angular velocity. In three dimensions, it is a pseudovector. In SI units, it is measured in radians per second squared (rad/s^2) and is usually denoted by Greek letter alpha (α). Like angular velocity, there are two types of angular acceleration: 1) *spin angular acceleration* and 2) *orbital angular acceleration*, representing the time rate of change of spin angular velocity and orbital angular velocity. Unlike linear acceleration, angular acceleration need not be caused by a net external torque. *For example, a figure skater can speed up her rotation (thereby obtaining an angular acceleration) simply by contracting her arms inwards, which involves no external torque.*

Angular displacement – the angle described at the center of a circle by a moving body along a circular path. It is measured in *radians*.

Angular frequency (ω) (*angular speed, radial frequency, circular frequency, orbital frequency, radian frequency,* and *pulsatance*) – a scalar measure of rotation rate. It refers to the angular displacement per unit time (e.g., in rotation), the rate of change of the phase of a sinusoidal waveform (e.g., in oscillations and waves), or the rate of change of the argument of the sine function. Angular frequency (or angular speed) is the magnitude of the vector quantity, which is angular velocity. *Angular frequency vector* is sometimes a synonym for the vector quantity angular velocity. One revolution equals 2π radians, where ω is the angular frequency or angular speed (measured in radians per second), T is the period (measured in seconds), and f is the ordinary frequency (measured in hertz) (symbolized with v).

Angular momentum (*moment of momentum* or *rotational momentum*) – the cross-product of position vector and momentum. The rotational equivalent of linear momentum. *It is an important quantity in physics because it is a conserved quantity; the total angular momentum of a closed system remains constant.*

Angular momentum quantum number (*quantum mechanics*) – one of four descriptions of the energy state of an electron wave; describes the energy sublevels of electrons within the principal energy levels of an atom.

Angular velocity (*ω*) – how fast an object rotates or revolves relative to another point (i.e., how fast the angular position or orientation of an object changes with time). *Spin angular velocity* is how fast a rigid body rotates with respect to its center of rotation. *Orbital angular velocity* is how fast a rigid body's center of rotation revolves around a fixed origin (i.e., the time rate of change of its angular position relative to the origin). Angular velocity is measured in angle per unit time (e.g., *radians per second*). SI unit of angular velocity is expressed as radians/sec, with the radian having a dimensionless value of unity; thus, the SI units of angular velocity are listed as *1/sec. Angular velocity is usually represented by the Greek* letter *omega (ω, sometimes Ω). By convention, positive angular velocity indicates counterclockwise rotation, while negative is clockwise.*

Anion – a negatively charged ion. Contrast *cation.*

Annihilation (*particle physics*) – when a particle and an antiparticle combine and release their rest energies in other particles. Occurs when a subatomic particle collides with its respective antiparticle to produce other particles, such as an electron colliding with a positron to produce two photons. The total energy and momentum of the initial pair are conserved in the process and distributed among a set of other particles in the final state. *Antiparticles have opposite additive quantum numbers from particles, so the sum of all quantum numbers of such an original pair is zero. Hence, any set of particles may be produced with a total quantum number of zero as long as energy conservation and momentum conservation are obeyed.*

Anode – electrode through which a conventional electric current flows into a polarized electrical device; the direction of current flow is, by convention, opposite to the direction of electron flow, so electrons flow *out of* the anode. In a galvanic cell, the anode is the negative terminal or pole that emits electrons toward the external part of an electrical circuit. In an electrolytic cell, the anode is the wire or plate having an excess positive charge, so named because negatively charged anions tend to move towards it. Contrast cathode.

Anti-gravity – a theory of creating a place or object free from the force of gravity. It does not refer to the lack of weight under gravity experienced in free fall or orbit or to balancing the force of gravity with some other force, such as electromagnetism or aerodynamic lift.

Antimatter – matter composed of the antiparticles (or "partners") of the corresponding particles in "ordinary" matter and can be thought of as matter with reversed charge, parity, and time, known as CPT reversal. Antimatter occurs in natural processes like cosmic ray collisions and radioactive decay. However, only a tiny fraction of these have successfully been bound together in experiments to form antiatoms. Minuscule numbers of antiparticles can be generated at particle accelerators; however, only a few nanograms have been produced. *No macroscopic amount of antimatter has ever been assembled due to the extreme cost and difficulty of production and handling Nonetheless, antimatter is an essential component of widely available applications related to beta decay (e.g., positron emission tomography, radiation therapy, and industrial imaging).*

Antineutron (n) – the antiparticle of the neutron; has zero mass and spin ½. It differs from the neutron only in that some properties have equal magnitude but opposite signs. It has the same mass as the neutron and no net electric charge but has the opposite baryon number (+1 for the neutron, −1 for the antineutron). This is because the antineutron is composed of antiquarks, while neutrons are composed of quarks. The antineutron consists of one up antiquark and two down antiquarks.

Antiparticle (*particle physics*) – every particle has an associated antiparticle with the same mass but opposite physical charges such as electric charge. For example, the electron's antiparticle is the antielectron (i.e., *positron*). *While the electron has a negative electric charge, the positron has a positive one and is produced naturally, particularly in radioactive decay. Some particles, such as the photon, are their antiparticle. Otherwise, for each pair of antiparticle partners, one is designated as "normal" matter (the kind comprising all matter with which humans interact), and the other (usually given the prefix "anti-") as antimatter.*

Antiproton – the antiparticle of the proton, which is stable but typically short-lived since any collision with a proton will cause both particles to be annihilated in a burst of energy. The existence of the antiproton with an electric charge of $-1\ e$, opposite to the electric charge of $+1\ e$ of the proton, was predicted by Paul Dirac in his 1933 Nobel Prize lecture. Dirac received the Nobel Prize for his 1928 publication of his Dirac equation that predicted the existence of positive and negative solutions to Einstein's energy equation ($E = mc^2$) and the existence of the positron (i.e., the antimatter analog of the electron, with opposite charge and spin).

Antiquark (*particle physics*) – for every quark flavor, there is a corresponding antiparticle known as an antiquark that differs from the quark only in that some of its properties (e.g., electric charge) have equal magnitude but opposite signs.

Arc length – the distance between two points along a section of a curve.

Archimedes principle – a body immersed in a fluid experiences an apparent weight loss equal to the weight of the fluid displaced. *Upward buoyant force exerted on a body immersed in a fluid, whether wholly or partially submerged, equals the weight of the fluid that the body displaces and acts in the upward direction at the center of mass of the displaced fluid.*

Area – the amount of surface enclosed within the boundary lines.

Area moment of inertia – a geometrical property of an area that reflects how its points are distributed about an arbitrary axis.

Astronomical unit (A.U.) – the radius of the Earth's orbit defined as one astronomical unit.

Astrophysics – the branch of astronomy focused on the physics of the Universe, especially the compositional nature of celestial bodies rather than their positions or motions in space.

Atom – the smallest unit of an element that can exist alone or in combination with other elements. A basic unit of matter consisting of a dense central nucleus surrounded by a cloud of negatively charged electrons. *The atomic nucleus contains a mix of positively charged protons and electrically neutral neutrons.*

Atomic mass unit (AMU) – relative mass unit (amu) of an isotope based on a standard ^{12}carbon isotopes; one atomic mass unit (1 amu) = 1/12 mass of a ^{12}C atom = 1.66×10^{-27} Kg.

Atomic number (Z) (*atomic mass number* or *nucleon number*) – the number of protons in the nucleus of an atom. *It is most often used to classify elements within the periodic table.*

Atomic orbital – a function describing an electron's location and wave-like behavior in an atom. This function describes the electron's charge distribution around the atom's nucleus and can be used to calculate the probability of finding an electron in a specific region around the nucleus.

Atomic physics – atoms as isolated systems of electrons and atomic nuclei.

Atomic structure – the atom is composed of various subatomic particles. The constituent particles of an atom are the electron, the proton, and the neutron. In the Standard Model of physics, electrons are truly elementary particles with no internal structure, whereas protons and neutrons are composite particles composed of elementary particles called quarks. Atoms have two types of quarks, each having a fractional electric charge. Protons comprise two up quarks (each with a charge of +2/3) and one down quark (with a charge of −1/3).

Neutrons consist of one up quark and two down quarks. This distinction accounts for the difference in mass and charge between the two particles. Quarks are held together by the strong interaction (or strong force), which is mediated by gluons. The protons and neutrons, in turn, are held to each other in the nucleus by the nuclear force, which is a residuum of the strong force with somewhat different range properties. The gluon is a member of the family of gauge bosons, which are elementary particles that mediate physical forces.

Atomic weight (A) – the total of protons (or electrons) and neutrons within an atom. The weighted average of the masses of stable isotopes of an element as they occur in nature based on the abundance of each and the atomic mass of the isotope compared to ^{12}carbon.

Avogadro constant – the ratio of the number of constituent particles in a substance, usually atoms or molecules, to the amount of substance, of which the SI unit is the mole. It is defined as precisely 6.02214076×10^{23} mol^{-1}.

Avogadro's number – the number of ^{12}carbon atoms in exactly 12.00 g of C is 6.02×10^{23} atoms or other chemical units: the number of chemical units in one mole of a substance. The number of molecules in 1 mole of a substance equals 6.02214076×10^{23}.

Avogadro's Law – under the same temperature and pressure conditions, equal volumes of gases contain an equal number of molecules. *The volume of gases is equal at the same temperature and pressure and contains equal numbers of molecules.*

Axion (*particle physics*) – a hypothetical subatomic particle postulated to account for the rarity of processes that break charge-parity symmetry. It is exceptionally light, electrically neutral, and pseudoscalar.

Axis – the imaginary line about which a planet or other object rotates.

Azimuthal quantum number (*l*) – a quantum number for an atomic orbital that determines its orbital angular momentum and describes aspects of the angular shape of the orbital. The azimuthal quantum number is the second of a set of quantum numbers that describe the unique quantum state of an electron (the others being the principal quantum number n, the magnetic quantum number m_ℓ, and the spin quantum number m_s). For a given value of the principal quantum number n (*electron shell*), the possible values of ℓ are the integers from 0 to $n - 1$.

B

Babinet's principle (*optics*) – the diffraction pattern from an opaque body is identical to that from a hole of the same size and shape except for the overall forward beam intensity.

Background radiation – ionizing radiation (e.g., alpha, beta, gamma rays) from natural sources. *The ubiquitous ionizing radiation to which Earth is exposed.*

Balanced forces – when some forces act on a body, and the resultant force is zero. See *resultant forces.*

Balmer series (or *Balmer lines*) (*atomic physics*) – one of six named series describing the spectral line emissions of the hydrogen atom. Lines in the spectrum of the hydrogen atom in the visible range are produced by the transition between a high energy level to $n = 2$, with n being the principal quantum number. *The Balmer series is calculated using the Balmer formula, an empirical equation discovered by Johann Balmer in 1885.*

Bar – a unit of pressure equal to 10^5 Pascal.

Barometer – an instrument measuring atmospheric pressure used in weather forecasting and determining elevation above sea level. *Pressure forecasts short-term weather changes.*

Baryon (*particle physics*) – subatomic particle composed of three quarks. A subatomic particle such as a proton or a neutron is made of (usually) three quarks. Nearly all matter humans are likely to encounter is baryonic matter.

Battery – a combination of two or more electrical cells which produce electricity.

Beam – a structural element capable of withstanding load primarily by resisting bending. Beams are traditionally descriptions of building or civil engineering structural elements. *However, smaller structures such as truck or automobile frames, machine frames, and mechanical or structural systems contain beam structures designed and analyzed similarly.*

Beat (*acoustics*) – the periodic variation in sound intensity due to the superposition of waves differing slightly in frequency; rhythmic increases and decreases of volume from constructive and destructive interference between two sound waves of slightly different frequencies.

Bending (or *flexure*) – the behavior of a slender structural element subjected to an external load applied perpendicularly to a longitudinal axis of the element.

Bending moment – the reaction induced in a structural element when an external force or moment is applied to the element, causing the element to bend. *The simplest structural element subjected to bending moments is the beam.*

Bernoulli's principle (*fluid dynamics*) – an increase in the speed of a fluid coincides with a decrease in pressure or a decrease in the fluid's potential energy.

Bernoulli's theorem (*fluid dynamics*) – the total energy per unit volume of a non-viscous, incompressible fluid in a streamlined flow will remain constant.

Bessel function – a canonical solution $y(x)$ of Friedrich Bessel's differential equation for an arbitrary complex number α, the *order* of the Bessel function Although α and $-\alpha$ produce the same differential equation, it is conventional to define different Bessel functions for these two values so that the Bessel functions are mostly smooth functions of α. *The most critical cases are when α is an integer or half-integer. Bessel functions for integer α are cylinder functions or cylindrical harmonics because they appear in the solution to Laplace's equation in cylindrical coordinates. Spherical Bessel functions with half-integer α are obtained when the Helmholtz equation is solved in spherical coordinates.*

Beta decay (or *β-decay*) (*nuclear physics*) – radioactive decay in which a beta particle is emitted from an atomic nucleus, transforming the original nuclide into its isobar.

Beta particle (*nuclear physics*) – a high-energy, high-speed electron or positron emitted by certain radioactive atomic nuclei.

Beta (β) particle (or *beta ray*) – a high-energy electron emitted as ionizing radiation from a decaying nucleus (β-decay).

Big Bang theory – the current model of galactic evolution in which the universe is assumed to have been created by an intense and brilliant explosion from a primeval fireball.

Binding energy – the net energy required to break a nucleus into its constituent protons and neutrons; the energy equivalent released when a nucleus is formed. *A bound system typically has a lower potential energy than the sum of its constituent parts.*

Biomass – the chemical energy stored in fast growing plants.

Biophysics – an interdisciplinary science using methods and theories from physics to study biological systems.

Black body – an ideal body that would absorb all incident radiation and reflect none. A hypothetical idealized physical body that completely absorbs all incident electromagnetic radiation, regardless of frequency or angle of incidence. *In many theoretical discussions of thermodynamics, perfect black bodies are imagined as substitutes for actual physical bodies, and the construction of nearly perfect black bodies in the real world remains a topic of interest for materials engineers.* Contrast *white body.*

Black-body radiation – electromagnetic radiation within or surrounding a body in thermodynamic equilibrium with its environment or emitted by a black body (an opaque and non-reflective body) held at constant, uniform temperature. The radiation has a specific spectrum and intensity that depends only on the temperature of the body.

Block and tackle – a system of two or more pulleys with a rope or cable threaded between them, usually used to lift, or pull heavy loads.

Black body radiation – electromagnetic radiation emitted by an ideal material (the black body) perfectly absorbs and perfectly emits radiation.

Black hole – the remaining theoretical core of a supernova so dense that light cannot escape.

Bohr model – the structure of the atom that attempted to correct the deficiencies of the solar system model and account for the Balmer series.

Boiling point – the temperature at which a phase change of liquid to gas occurs through boiling, the same temperature as the *condensation point.*

Boiling point elevation – the boiling point of a liquid (e.g., solvent) increases when another compound is added, meaning that the resulting solution has a higher boiling point than the pure solvent. *This happens when a non-volatile solute (e.g., salt) is added to a pure solvent (e.g., water). Boiling point can be measured accurately using an ebullioscope.*

Boltzmann constant – a physical constant that relates the average kinetic energy of the particles in a gas with the temperature of the gas. It is the gas constant R divided by the *Avogadro constant.*

Boson (*particle physics*) – a subatomic particle that behaves according to Bose-Einstein statistics and possesses integer spin. Bosons include elementary particles such as photons, gluons, W and Z bosons, Higgs bosons, and the hypothetical graviton, as well as certain composite particles such as mesons and stable nuclides of even mass number. Bosons constitute one of two main classes of particles, the other being fermions. *Unlike fermions, there is no limit to the number of bosons occupying the same quantum state.*

Boundary – the division between two regions of differing physical properties.

Boyle's Law – for a given mass of a gas at a constant temperature, the volume of the gas is inversely proportional to the pressure. *The volume of a given mass of a gas at constant temperature is inversely proportional to its pressure.*

Bremsstrahlung – radiation emitted by the acceleration of unbound charged particles.

Brewster's Law – states that the refractive index of a material is equal to the tangent of the polarizing angle for the material.

British thermal unit (BTU) – the amount of energy or heat needed to increase the temperature of one pound of water one-degree Fahrenheit. An imperial unit of energy as the amount needed to heat 1 pound of water by 1 °F; 1 btu equals about 1,055 joules. *In scientific contexts, the BTU has been replaced by the SI unit of energy, the joule.*

Brittleness (*materials*) – the tendency to break without significant plastic deformation when subjected to stress. Brittle materials absorb relatively little energy prior to fracture, even those of high strength. A snapping sound often accompanies breaking.

Brownian motion (or *pedesis*) – the continuous random motion of solid microscopic particles suspended in a fluid (i.e., gas or liquid) medium due to their ongoing bombardment by atoms and molecules.

Bulk modulus – a measure of a substance's resistance to uniform compression defined as the ratio of the infinitesimal pressure increase to the resulting relative volume decrease. Its base unit is the *Pascal.*

Bulk modulus of elasticity – a ratio of normal stress to volumetric strain produced in a body.

Buoyancy (or *buoyant force*) – upward force exerted by a fluid that opposes the weight of an immersed object.

C

Calculus – a branch of mathematics that studies change and has two major sub-fields: differential calculus (i.e., concerning rates of change and slopes of curves) and integral calculus (i.e., concerning accumulation of quantities and the areas under and between curves). These two branches are related to each other by the fundamental theorem of calculus.

Calorie – a unit of heat; 1 Calorie = 4.186 joule.

Candela – the SI unit of luminous intensity defined as the intensity in each direction of a source that emits monochromatic photons of frequency 540×10^{12} Hz and has a radiant intensity in that direction of 1/683 W/sr.

Capacitance – the ratio of the charge stored per increase in potential difference. The ratio of the change in the electric charge of a system to the corresponding change in its electric potential. Two closely related notions are *self-capacitance* and *mutual capacitance*. Any object that can be electrically charged exhibits *self-capacitance*. A material with a large self-capacitance holds more electric charge at a given voltage than one with low capacitance. *Mutual capacitance* is particularly important for understanding the operations of the capacitor, one of the three elementary linear electronic components (along with resistors and inductors).

Capacitive reactance – a capacitor consists of two conductors separated by an insulator, known as a dielectric. *Capacitive reactance is an opposition to the change of voltage across an element. Capacitive reactance is inversely proportional to the signal frequency (or angular frequency, ω) and the capacitance.*

Capacitor – an electrical device that stores charge and energy in the electrical field.

Capillarity – the rise or fall of a liquid in a fine bore tube.

Carnot cycle – a theoretical ideal thermodynamic cycle proposed by French physicist Nicolas Léonard Sadi Carnot in 1824 and expanded upon by others in the 1830s and 1840s. It provides an upper limit on the efficiency that any classical thermodynamic engine can achieve during the conversion of heat into work or, conversely, the efficiency of a refrigeration system in creating a temperature difference by applying work to the system. *It is not an actual thermodynamic cycle but is a theoretical construct.*

Carnot's theorem – no engine operating between two temperatures can be more efficient than a reversible engine working between the same two temperatures.

Cartesian coordinate system – a coordinate system that specifies each point uniquely in a plane by a set of numerical coordinates, which are the signed distances to the point from two fixed perpendicular oriented lines, measured in the same unit of length. Each reference line is a *coordinate axis* or *axis* (plural *axes*) of the system, and the point where they meet is the *origin* at the ordered pair (0, 0). *Coordinates are also the positions of the perpendicular projections of the point onto the two axes, expressed as signed distances from the origin.*

Cathode – the electrode through which a conventional electric current flows out of a polarized electrical device; the direction of current flow is, by convention, opposite to the direction of electron flow, so electrons flow *into* the cathode. *In a galvanic cell, the cathode is the positive terminal or pole which accepts electrons flowing from the external part of an electrical circuit. However, in an electrolytic cell, the cathode is the wire or plate having an excess negative charge, so named because positively charged cations tend to move towards it.* Contrast *anode.*

Cathode rays – negatively charged particles (i.e., electrons) emitted from a negative terminal in an evacuated glass tube.

Cation – a positively charged ion. Contrast *anion.*

Celsius scale of temperature (or *centigrade scale*) – the ice-point is taken as the lower fixed point (0 °C), and the steam-point is taken as the upper fixed point (100 °C); the interval between the ice-point and the steam-point is divided into 100 equal divisions; the unit division on this scale is 1 °C; previously called the centigrade scale; the relationship relates the temperatures on the Celsius scale and the Fahrenheit scale, $C/100 = (F - 32) / 180$; the temperature of a healthy person is 37 °C or 98.6 °F.

Center of gravity – the point through which all the weight of an object appears to act. The point in a body around which the resultant torque due to gravity forces vanishes. *Near the Earth's surface, where gravity acts downward as a parallel force field, the center of gravity and the center of mass are the same.*

Center of mass – within a given distribution of mass, the unique point in space at which the weighted relative position of the distributed mass sums to zero.

Central-force problem – a classic problem in potential theory involves the determination of the motion of a particle in a single central potential field. *The solutions to such problems are important in classical mechanics since many naturally occurring forces, such as gravity and electromagnetism, are central forces.*

Centrifugal force – an apparent outward force on an object in circular motion; a consequence of the Third Law of Motion. *The apparent outward force that draws a rotating body away from the center of rotation. Inertia of the body causes it as the path is continually redirected.*

Centigrade – See *Celsius scale.*

Centripetal force – the radial force required to keep an object moving in a circular path. *A body moves uniformly along a circular path directed along a radius towards the center.*

Chain reaction – a self-sustaining reaction where some of the products can produce more reactions of the same kind (e.g., in a nuclear chain reaction, neutrons are the products that produce more nuclear reactions in a self-sustaining series).

Charles' Law – volume is directly proportional to temperature for a gas at constant pressure.

Chemical physics – studies chemical processes by investigating physicochemical phenomena using techniques from atomic and molecular physics and condensed matter physics.

Chromatic aberration (*optics*) – an optical lens defect causing color fringes due to the lens bringing different light colors to focus at different points.

Circular motion – the motion of a body along a circular path.

Classical mechanics (or *Newtonian mechanics*) – a sub-field of mechanics concerned with the set of physical laws describing the motion of bodies under the collective actions of a system of forces.

Closed system – the system which cannot exchange heat or matter with the surroundings.

Coefficient of areal expansion – the fractional change in surface area per degree of temperature change; see *coefficient of thermal expansion.*

Coefficient of linear expansion – the fractional change in length per degree of temperature change; see *coefficient of thermal expansion.*

Coefficient of thermal expansion – the fractional change in the size of an object per degree of change in temperature at constant pressure; the SI unit is K^{-1}.

Coefficient of volumetric expansion – the fractional change in volume per degree of temperature change; see *coefficient of thermal expansion.*

Coherent source – when there is a constant phase difference between waves emitted from distinct parts of the source.

Cohesion – the tendency of similar particles or surfaces to cling to one another. Contrast *adhesion.*

Combustion (or *burning*) – the combining of a substance with oxygen.

Compass – a magnet that is free to rotate and indicate direction.

Complementary colors – two colors which, when mixed, give white. For example, blue + yellow = white, red + cyan = white, and green + magenta = white.

Compound – a substance made up of two or more elements chemically combined.

Compression – a part of a longitudinal wave in which the density of the medium particles is higher than the typical density.

Compressive stress – a force that tends to compress the surface as the Earth's plates move into each other.

Concave lens – spreads out light rays.

Condensation – the changing of a gas to a liquid state.

Condensation (*sound)* – a compression of gas molecules; a pulse of increased density and pressure that moves through the air at the speed of sound.

Condensation (*water vapor*) – more vapor or gas molecules return to liquid than evaporate.

Condensation nuclei – tiny particles (e.g., dust, soot, or salt crystals) suspended in the air on which water condenses.

Condensation point – temperature when a gas or vapor changes to liquid; see *boiling point.*

Condensed matter physics – studies the physical properties of condensed phases of matter.

Conduction – the transfer of heat from a region of higher temperature to a region of lower temperature by increased kinetic energy moving from molecule to molecule without the movement of the solid.

Constringence (*V-number* and *Abbe number*) (*optics and lens design*) – a measure of a transparent material's dispersion (a variation of refractive index versus wavelength). *High values of V indicate low dispersion.*

Constructive interference – the condition in which two waves arrive at the same place simultaneously and in phase, adding amplitudes to create a new wave.

Control rods – material inserted between fuel rods in a nuclear reactor to absorb neutrons and control the rate of the nuclear chain reaction.

Convection – the transfer of heat by the actual transfer of matter. Heat transfers from a higher to a lower temperature region by the displacement of high-energy molecules. *For example, warmer, less dense (higher KE) air is displaced by cooler, denser (lower KE energy) air.*

Conventional current – the opposite of electron current; an electric current consists of a drift of positive charges that flow from the positive terminal to the negative terminal of a battery.

Convex lens – a lens that brings light rays together.

Corrosion – an undesired process when metal converts to another compound (e.g., rusting).

Coulomb – a unit used to measure the quantity of electric charge, equivalent to the charge resulting from the transfer of 6.24 billion particles such as the electron.

Coulomb's Law – the relationship between charge, distance, and magnitude of the electrical force between two bodies; the force between two charges is directly proportional to the product of charges and inversely proportional to the square of the distance between charges.

Covalent bond – a chemical bond formed by sharing a pair of electrons.

Covalent compound – chemical compound held by covalent bonds.

Crest – the point of maximum positive displacement on a transverse wave. *The point on a wave with the maximum value or upward displacement within a cycle.*

Critical angle – the limit to the angle of incidence when all light rays are reflected internally.

Critical mass – the mass of fissionable material needed to sustain a chain reaction. *The smallest amount of fissile material needed for a sustained nuclear chain reaction.*

Current – a flow of charge. The unit is Ampere (A).

Curvilinear motion – the trajectory of a body along a curved path. *A moving particle or object that conforms to a known or fixed curve. Such motion is studied with two coordinate systems: planar and cylindrical.*

Cycle – a complete vibration.

Cyclotron – a device used to accelerate charged particles. *A particle accelerator in which charged particles accelerate outwards from the center along a spiral path.*

D

Damping – any influence upon or within an oscillatory system that has the effect of reducing, restricting, or preventing its oscillations. *Damping results from processes that dissipate the energy stored in the oscillation.*

DC motor – a mechanically commutated electric motor powered by direct current.

De-acceleration – negative acceleration when the velocity of a body decreases with time.

Decibel – a unit of sound level; if P_1 & P_2 are two amounts of power, the first is said to be n decibels greater, where n = 10 log10 (P_1/P_2).

Decibel scale – a nonlinear scale of loudness based on the ratio of the intensity level of a sound to the intensity at the hearing threshold.

Deflection – the degree to which a structural element is displaced under a load. It may refer to an angle or a distance.

Deformation – 1. (mechanics) 2. (engineering)

Density (or *mass density*) – the mass of a substance per unit volume. *A physical property of a substance defined as its mass per unit volume.*

Derivative – for a mathematical function of a real variable, a measurement of the sensitivity to change of the function value (output) with respect to a change in its argument (input). *For example, the derivative of the position of a moving object with respect to time is the object's velocity, which measures how quickly the position of the object changes as time changes. Derivatives are a fundamental tool of calculus.*

Destructive interference – two waves arriving at the same point simultaneously out of phase add amplitudes that cancel to create zero total disturbance; see *constructive interference.*

Dewpoint temperature – the temperature at which condensation begins.

Dew – condensation of water vapor into droplets of liquid on surfaces.

Diffraction – the bending of light around the edge of an opaque object.

Diffuse reflection – light rays reflected in many random directions, as opposed to the parallel rays reflected from a perfectly smooth surface such as a mirror.

Diopter – a unit of measure of the refractive power of a lens.

Direct current – an electrical current that always flows in one direction only (i.e., from the positive terminal to the negative terminal).

Direct proportion – two variables increasing or decreasing in the same ratio (at same rate).

Dispersion – the splitting of white light into its component colors of the spectrum.

Displacement – a vector quantity for the change in the position of an object as it moves in a particular direction, the shortest distance between the initial position and the final position of a moving body.
1. (*fluids*) occurs when an object is immersed in a fluid, pushing it out of the way and taking its place. The volume of the immersed object will be exactly equal to the volume of the displaced fluid so that the volume of the immersed object can be deduced if the volume of the displaced fluid is measured.
2. (*vectors*) the shortest distance from the initial to the final position of a point. Thus, it is the length of an imaginary straight path, typically distinct from the path traveled.

Distance – a scalar quantity for the length of the path traveled by a body irrespective of its direction. *A numerical description of how far apart objects are.*

Distillation – vaporization of a liquid by heating and vapor condensation by cooling.

Doppler effect – an apparent change in the frequency of sound or light due to the relative motion between the source of the sound or light and the observer.

Doppler effect – the change in frequency of a wave (or periodic event) for an observer moving relative to its source. *Compared to the emitted frequency, the received frequency is higher during the approach, identical at the instant passing by, and lower during recession.*

Drag – forces on a solid object toward the relative fluid flow velocity. Unlike other resistive forces (e.g., dry friction) independent of velocity, drag forces depend on velocity.

Ductility – a solid material's ability to deform under tensile stress; is often characterized by the material's ability to be stretched into a wire.

Dynamics – classical mechanics that studies forces and torques and their effects on motion, as opposed to kinematics, which studies motion without reference to these forces.

Dyne – a unit of force in the centimeter-gram-second system of physical units, equal to the force for a mass of one gram an acceleration of one centimeter per second per second. One dyne equals 0.00001 Newton.

E

Echo – a reflected sound distinguished from the original sound, usually arriving 0.1 s or more after the original sound.

Einstein mass-energy relation – $E = mc^2$; E is the energy released, m is the mass defect, and c is the speed of light.

Elastic potential energy – the potential energy of a body by its configuration (i.e., shape).

Elastic strain – an adjustment to stress in which materials recover their original shape after stress is released.

Elasticity – the tendency of a material to return to its original shape after deforming.

Electric charge – a physical property of matter that causes it to experience a force near other electrically charged matter. There are two types of electric charges: positive and negative.

Electric circuit – an electrical network consisting of a closed loop giving a return path for the current. Consists of a voltage source that maintains an electrical potential, a continuous conducting path for a current to follow, and a device where the electrical potential does work; a switch completes or interrupts the conducting path.

Electric current – the flow of electric charge; the electric force field produced by an electrical charge. *The flow of electric charge through a conductive medium.*

Electric field – the region of space surrounding electrically charged particles and time-varying magnetic fields. *The electric field depicts the force exerted on other electrically charged objects by the electrically charged particle the field is surrounding.*

Electric field line – an imaginary curve tangent to which at any given point gives the direction of the electric field at that point.

Electric field lines – a map of an electric field representing the direction of the force that a test charge would experience; lines of force show the direction of an electric field.

Electric force – a fundamental force that results from the interaction of electrical charges; it is the most powerful force in the universe.

Electric generator – a mechanical device that uses wire loops rotating in a magnetic field to produce electromagnetic induction to generate electricity.

Electric potential energy – due to the position of a charge near other charges.

Electric power – the rate at which an electric circuit transfers electric energy.

Electrical conductor – any material that contains movable electric charges and, therefore, can conduct an electric current under the influence of an electric field. *Materials with free electrons that move throughout the material (e.g., metals) allow electric current to flow through the material.*

Electrical energy – energy from electromagnetic interactions.

Electrical insulator (*electrical nonconductor*) – any material whose internal electric charges do not flow freely and, therefore, do not conduct an electric current under the influence of an electric field. *Electrical nonconductors or materials that obstruct the flow of electric current.*

Electrical network – an interconnection of electrical elements such as resistors, inductors, capacitors, voltage sources, current sources, and switches.

Electrical nonconductor (*electrical insulator*) – materials with electrons that do not move easily within the material (e.g., rubber).

Electrical resistance – the property of opposing or reducing electric current. *The opposition to the passage of an electric current through an electrical element.*

Electricity – the physical phenomena associated with the flow of electric charge.

Electrode – a conductor that dips into an electrolyte and allows the electrons to flow to and from the electrolyte.

Electrolysis – the production of a chemical change using electricity. Electrolysis can be used to split up water into hydrogen and oxygen.

Electrolyte – water solution of ionic substances that will conduct an electric current.

Electromagnet – a magnet formed by a solenoid turned on and off by regulating current.

Electromagnetic force (***EM field** or **EM***) – a physical field produced by moving electrically charged objects. *One of four fundamental forces: the force of attraction or repulsion between two charged particles.*

Electromagnetic induction – the process in which current is induced in a coil whenever there is a change in the magnetic flux linked with the coil.

Electromagnetic radiation (***EM radiation** or **EMR***) – energy emitted and absorbed by charged particles, which exhibits wave-like behavior as it travels through space.

Electromagnetic waves – the waves due to oscillating electrical and magnetic fields and do not need any material medium for their propagation; can travel through a material medium (e.g., light waves and radio waves); travel in a vacuum with speed 3.0×10^8 m/s.

Electromotive force (EMF) – the electrical intensity or "pressure" developed by a source of electrical energy (e.g., battery or generator) and measured in volts. *Any device that converts other forms of energy into electrical energy provides electromotive force as its output.*

Electron – a subatomic particle with the smallest negative charge possible; usually found in an orbital of an atom but is gained or lost when atoms become ions.

Electron configuration – arrangement of electrons in orbits and sub-orbits about the atom's nucleus.

Electron current – the opposite of conventional current; electric current consists of a drift of negative charges that flows from the negative terminal to the positive terminal of a battery.

Electron pair – a pair of electrons with different spin quantum numbers in an orbital.

Electron paramagnetic resonance (*electron spin resonance (ESR) and electron magnetic resonance (EMR)* – a method for studying materials with unpaired electrons that makes use of the Zeeman effect. It shares some basic principles of *nuclear magnetic resonance* (NMR).

Electron volt (eV) – the energy gained by an electron moving across a potential difference of 1 V equals 1.60×10^{-19} Joules. *It is the amount of energy gained by the charge of a single electron moved across an electric potential difference of one volt.*

Electronegativity – a chemical property that describes the tendency of an atom or a functional group to attract electrons (or electron density) towards itself. *The comparative ability of atoms of an element to attract bonding electrons.*

Electroplating – where metal is covered with another metal layer using electricity.

Electrostatic charge – an accumulated electric charge on an object from a surplus or a deficiency of electrons.

Electronics – a field focused on electrical circuits involving active electrical components such as vacuum tubes, transistors, diode integrated circuits, and associated passive interconnection technologies.

Element – a pure chemical substance that cannot be broken down into anything simpler by chemical or physical means; there are over 100 known elements, the fundamental materials of which matter is made.

Endothermic process – the process in which heat is absorbed. A process or reaction in which a system absorbs energy from its surroundings, usually in the form of heat but also as light, electricity, or sound. Contrast *exothermic*.

Energy – the capacity to do work; it is a scalar quantity; the SI unit is the Joule. There are five forms of energy: mechanical, chemical, radiant, electrical, and nuclear.

Entropy (S) – a quantity that describes the randomness of a substance or system.

Equilibrium – the condition of a system when neither its motion nor its internal energy tends to change with time; a balanced object is in equilibrium.

Escape velocity – the minimum velocity with which an object must be thrown upward to overcome the gravitational pull and escape into space; the escape velocity depends on the mass and radius of the planet/star but not on the mass of the body thrown upward.

Evaporation – a process of more molecules leaving a liquid for the gaseous state than returning from the gas to the liquid; can occur at any given temperature from the surface of a liquid; takes place from the surface of the liquid; causes cooling; faster if the surface of the liquid is large, the temperature is higher, and the surrounding atmosphere does not contain a large amount of vapor of the liquid.

Exothermic process – the process in which heat is evolved; *a process or reaction that releases energy from a system, usually in the form of heat but also as light, electricity, or sound.* Contrast *endothermic*.

F

Fahrenheit scale of temperature – the ice-point (lower fixed point) is taken as 32 °F, and the steam-point (upper fixed point) is taken as 212 °F; the interval between these two points is divided into 180 equal divisions; the unit division on the Fahrenheit scale is 1 °F; the relationship relates the temperatures on the Celsius scale and the Fahrenheit scale, C/100 = (°F – 32) / 180; the temperature of a healthy person is 37 °C or 98.6 °F.

Falling bodies – objects moving towards a body with greater gravitational influence, such as a planet.

Farad – the SI unit of capacitance; the capacitance of a capacitor that, if charged to 1 C, has a potential difference of 1 V.

Faraday – the electric charge required to liberate a gram equivalent of a substance; 1 Faraday = 9,6485 coulomb/mole.

Fermat's principle – an electromagnetic wave takes a path involving the least time when propagating between two points.

Fermion (*particle physics*) – a particle that behaves according to Fermi–Dirac statistics, obeys the Pauli exclusion principle, and possesses half-integer spin. Fermions include all quarks, leptons, and composite particles made of an odd number of these (such as all baryons and many atoms and nuclei). *Fermions constitute one of two main classes of particles, the other being bosons.*

First Law of Motion – *every object remains at rest or in uniform straight-line motion unless acted on by an unbalanced force*.

Fission – a nuclear reaction or a radioactive decay process in which the nucleus of an atom splits into smaller parts (lighter nuclei), often producing free neutrons and photons (as gamma rays) and releasing a considerable amount of energy.

Flexure (or *bending*) – the behavior of a slender structural element subjected to an external load applied perpendicularly to a longitudinal axis of the element.

Fluid – matter flows or pours; individual fluid molecules move, rolling by another.

Focus (*optics*) – the point to which rays initially parallel to the axis of a lens or mirror converge or from which they appear to diverge.

Force (F) – any interaction that, when unopposed, changes the motion of a physical body. A push or pull which tends to change the state of rest or uniform motion, the direction of motion or the shape and size of a body; a vector quantity; SI unit is a Newton, denoted by N; one N is the force which when acting on a body of mass 1 kg produces an acceleration of 1 m/s^2.

Force of gravitation – when two objects are attracted by their masses and act even if they are not connected; an action-at-a-distance force.

Fossil fuels – formed from the remains of plants and animals that lived millions of years ago.

Fracture strain – an adjustment to stress in which materials crack or break from stress.

Fraunhofer lines – the dark lines in the Sun or star spectrum.

Freefall (*kinematics*) – the motion of a body falling to Earth with no other force except the force of gravity acting on it; free-falling bodies are weightless.

Freezing – the changing of a liquid to a solid state.

Freezing point – the temperature at which a liquid to solid phase changes; the same temperature as the melting point for a given substance.

Frequency – the number of oscillations completed in 1 second by an oscillating body.

Frequency (of oscillations) – the number of oscillations by an oscillating body per second.

Frequency (of waves) – the number of waves produced per second.

Friction – the force that resists the motion of one surface relative to another with which it is in contact; *caused by the humps and crests of surfaces, even those on a microscopic scale; the area of contact is small, and the consequent high pressure leads to local pressure welding of the surface; in motion, the welds are broken and remade continually.*

Fuel – any substance that burns oxygen to produce heat.

Fuel rod – long zirconium alloy tubes with fissionable material for nuclear reactors.

Fundamental charge – the smallest common charge known; the magnitude of the charge of an electron and a proton, which is 1.60×10^{-19} coulombs.

Fundamental frequency – the lowest frequency (longest wavelength) at which a system vibrates freely and can set up standing waves in an air column or on a string.

Fundamental properties – a property that cannot be defined in terms other than to describe how it is measured; the fundamental properties are length, mass, time, and charge.

Fuse – a safety device in an electric circuit. If the current (i.e., the flow of charge) gets too high, the fuse wire melts, breaking the circuit and switching off the current.

Fusion – 1. (or *melting*) a physical process that results in the phase transition of a substance from a solid to a liquid. *The changing of a solid to a liquid state.* 2. nuclear reaction when atomic nuclei join (or *fuse*) to form a heavier nucleus.

G

g – a symbol representing the acceleration of an object in free fall due to the force of gravity; its magnitude is 9.80 m/s^2.

Gamma (γ) ray – a high energy photon of short wavelength electromagnetic radiation emitted by decaying nuclei (γ-decay). *A form of electromagnetic radiation of high frequency and, therefore, high energy.*

Gases – a phase of matter composed of molecules relatively far apart, moving freely in constant, random motion and have weak cohesive forces acting between them, resulting in the characteristic indefinite shape and indefinite volume of a gas.

Graham's Law of Diffusion – the diffusion rate of a gas is inversely proportional to the square root of its density.

Gram-atomic weight – the mass in grams of one mole of an element numerically equal to its atomic weight.

Gram-formula weight – the mass in grams of one mole of a compound numerically equal to its formula weight.

Gram-molecular weight – the gram-formula weight of a molecular compound.

Gravitation (or *gravity*) – a natural phenomenon by which physical bodies attract each other with a force proportional to their masses.

Gravitational constant (G) (*G; universal gravitational constant, and Newton's constant*) – a physical constant calculating gravitational force between two bodies. A term in the equation for Newton's Law of Gravitation, equal to the force of gravitation, which acts between two bodies with a mass of 1 kg, each separated by 1 m; the value of G is 6.67×10^{-11} Nm^2/kg^2.

Gravitational energy – the potential energy associated with the gravitational field.

Gravitational field – a model used to explain the influence that a massive body extends into the space around itself, producing a force on another massive body. Thus, a gravitational field is used to explain gravitational phenomena. *It is measured in newtons per kilogram* (N/kg).

Gravitational potential – equals the work (energy transferred) per unit mass done by the force of gravity to move an object to a fixed reference location.

Gravitational potential at a point – the amount of work done against the gravitational forces to move a particle of unit mass from infinity to that point.

Gravitational potential energy (PE) – the energy a body possesses by its height from the ground; equals *mgh*.

Gravitational wave – a ripple in the curvature of spacetime propagates as a wave and is generated in specific gravitational interactions, traveling outward from their source.

Gravity – the gravitational attraction at the surface of a planet or other celestial body.

Greenhouse effect – the process of increasing the temperature of the lower parts of the atmosphere through redirecting energy back toward the surface; the absorption and re-emission of infrared radiation by carbon dioxide (CO_2), water vapor, and a few other gases in the atmosphere.

Ground state – the energy of an atom when its electrons are at the lowest energy state.

H

Half-life – the time required for a quantity to fall to half its value as measured at the beginning of the period. In physics, half-life typically refers to a property of radioactive decay but may refer to any quantity that follows an exponential decay. *Time required for one-half of the unstable nuclei in a radioactive substance to decay into a new element.*

Halogens – the elements in group seven in the periodic table

Hard water – water that is difficult to form a lather with soap.

Heat – energy that makes a body hot or cold; measured by the temperature effect, the SI unit is the Joule (J).

Heisenberg uncertainty principle – fundamental limit to the precision with which specific pairs of physical properties of a particle (i.e., complementary variables) can be known simultaneously (e.g., one cannot measure the exact momentum and position of a subatomic particle at the same time – the more certain is one, the less certain is the other).

Hertz (Hz) – a unit of frequency equivalent to one cycle per second.

Homokinetic – the physics of complex, self-organizing systems.

Hooke's Law – stress is directly proportional to strain within the elastic limit.

Horsepower – a unit of power; 1 hp = 746 Watts.

Humidity – the ratio of water vapor in a sample of air to the volume of the sample.

Huygens' principle – each point on a light wavefront can be regarded as a source of secondary waves, the envelope of these secondary waves determining the position of the wavefront later.

Hypothesis – a tentative explanation of a phenomenon compatible with the data and provides a framework for understanding and describing that phenomenon.

I

Ice-point – a physical process that results in the phase transition of a substance from a liquid to a solid. The melting point of ice under 1 atm pressure equals 0 °C or 32 °F.

Ideal gas equation – PV = nRT.

Immiscible liquids – do not mix to form a solution (e.g., oil and water).

Impedance – measures opposition a circuit presents to a current when a voltage is applied.

Impulse – equal to the product of the force acting on a body and the time for which it acts; if the force is variable, the impulse is the integral of Fd_t from t_0 to t_1; the impulse of a force acting for a given time interval is equal to change in momentum produced over that interval; $J = m(v - u)$, assuming that the mass m remains constant while the velocity changes from v to u; the SI units are kg m/s.

Impulsive force – acts on a body for a brief time but produces a large change in momentum.

Incandescent – matter emitting visible light because of elevated temperature (e.g., a light bulb, a flame from any burning source, the Sun).

Incident ray – representing the direction of motion of incoming light to a boundary.

Index of refraction – the ratio of the speed of light in a vacuum to light in a material.

Indicator – a substance that shows a color change if a substance is acidic or basic.

Inertia – the property of matter that causes it to resist any change in its state of rest or of uniform motion; there are three kinds of inertia: the inertia of rest, the inertia of motion, and the inertia of direction; the mass of a body is a measure of its inertia.

Infrasonic – sound waves at a frequency below the range of human hearing (less than 20 Hz).

Insulators – materials that are poor conductors of heat or electricity (e.g., wood or glass); air pockets slow the movement of heat because the air molecules are far apart.

Intensity – a measure of the energy carried by a wave.

Interference – the redistribution of energy due to the superposition of waves with a phase difference from coherent sources, resulting in alternate light and dark bands.

Intermolecular forces – interaction between molecules.

Internal energy – the sum of the kinetic energy (KE) and potential energy (PE) of the molecules of an object.

International System of Units (SI) – the modern metric system. *A system of units of measurement devised around seven base units and the convenience of the number ten.*

Inverse proportion – the relationship in which the value of one variable increase while the value of a second variable decreases at the same rate (in the same ratio).

Ion exchange – a method of removing hardness from water, replacing the positive ions that cause the hardness with H^+ ions.

Ion – a charged atom or group of atoms (e.g., Na^+, OH^- or NH_4^+). An atom or molecule in which the total number of electrons is not equal to the number of protons, giving the atom a net positive or negative electric charge.

Ionic bond – a force of attraction between oppositely charged ions in a compound, resulting from a transfer of electrons.

Ionization – a process of forming ions from molecules. Converting an atom or molecule into an ion by adding or removing charged particles such as electrons or other ions.

Ionized – an atom or a particle with a net charge because it has gained or lost electrons.

Isobaric process – pressure remains constant.

Isochoric process – volume remains constant.

Isostasy – a balance or equilibrium between adjacent blocks of Earth's crust.

Isothermal process – in which temperature remains constant.

Isotope – atoms of the same element with the same atomic number (i.e., number of protons) but with a different mass number (i.e., number of neutrons). *While all isotopes of a given element share the same number of protons, each isotope differs in its number of neutrons.*

J

Joule (J) – a derived unit of energy, work, or amount of heat in the International System of Units. The unit used to measure work and energy can measure heat: $1\ J = 1 N \cdot m$.

Joule's Law of Heating – the heat produced when current (I) flows through a resistor (R) for a given time (t) is given by $Q = I^2Rt$.

K

Kelvin – a scale and unit used to measure temperature. Kelvin scale is an absolute thermodynamic temperature scale with zero as its null point.

Kelvin scale of temperature (K) – the lower fixed point is 273.15 K (0 °C), and the steam-point (the upper fixed point) is 373.15 K (100 °C); the interval between these two points is divided into 100 equal parts; each division is equal to 1 K.

Kelvin's statement of Second Law of Thermodynamics – it is impossible that heat has been extracted from a reservoir at the end of a cycle of changes, and an equal amount of work has been produced without producing some other effect.

Kepler's Laws of Planetary Motion – the three laws describing the motion of the planets.

Kepler's First Law – in planetary motion, each planet moves in an elliptical orbit, with the Sun at one focus.

Kepler's Second Law – a radius vector between the Sun and a planet moves over equal areas of the ellipse during equal time intervals.

Kepler's Third Law – the square of the period of an orbit is directly proportional to the cube of the radius of the orbit's major axis.

Kilocalorie (Kcal) – the energy required to raise the temperature of 1 kg of water by 1 °C; 1 Kcal = 1,000 calories.

Kilogram – the fundamental unit of mass in the metric system of measurement.

Kinematics – the branch of classical mechanics that describes the motion of points, bodies (objects), and systems of bodies (i.e., groups of objects) without consideration of the causes of motion. Kinematics is the "*geometry of motion.*"

Kinetic energy (KE) – possessed by a body due to its motion; $KE = \frac{1}{2}mv^2$, where m is mass and v is velocity. The energy a physical body possesses due to its motion; the work needed to accelerate a body of a given mass from rest to its stated velocity. *The body continues to maintain this kinetic energy unless its velocity changes*. Contrast *potential energy*.

Kirchhoff's circuit laws (*Kirchhoff's rules* or *Kirchhoff's Laws*) – two approximate equalities that address the current and voltage in electrical circuits.

Kirchhoff's equations (*fluid dynamics*) – the motion of a rigid body in an ideal fluid.

L

Laminar flow (or *streamline flow*) – when a fluid flows in parallel layers with no disruption between the layers.

Laplace–Runge–Lenz vector (*LRL vector*) – describes the shape and orientation of the orbit of one astronomical body around another (e.g., a planet revolving around a star). *For two bodies interacting by Newtonian gravity, the LRL vector is a constant of motion; it is the same no matter where it is calculated on the orbit; equivalently, the LRL vector is conserved.*

Laser (light amplification by stimulated emission of radiation) – a device that produces a coherent stream of light through stimulated emission of radiation.

Latent heat – energy released (or absorbed) during a constant-temperature phase change.

Latent heat of vaporization – the heat absorbed when one gram of a substance changes from the liquid phase to the gaseous phase; *the heat released when one gram of gas changes from the gaseous phase to the liquid phase.*

Latent heat of fusion – the quantity of heat required to convert one unit mass of a substance from a solid to a liquid at its melting point without a temperature change; the SI unit is J kg^{-1}.

Latent heat of sublimation – the quantity of heat required to convert one unit of mass of a substance from a solid to a gas *without* a change in its temperature.

Law of Conservation of Energy – energy cannot be created or destroyed but transformed.

Law of Conservation of Mass – mass (including single atoms) can neither be created nor destroyed in a chemical reaction.

Law of Conservation of Matter – matter cannot be created or destroyed in chemical rxns.

Law of Conservation of Momentum – the total momentum of a group of interacting objects remains constant without external forces.

Law of the lever – when a lever is balanced, the sum of the clockwise moments equals the sum of the anti-clockwise moments.

Lenz's Law – induced current flows in such a direction that it opposes the cause producing it.

Lepton (*particle physics*) – an elementary particle that does not undergo strong interactions but is subject to the Pauli exclusion principle. *Two main leptons are charged leptons (electron-like leptons) and neutral leptons (neutrinos).*

Lever – a rigid body that can turn about a fixed point called the fulcrum. *A machine with a beam or rigid rod pivoted at a fixed hinge or fulcrum, one of six classical simple machines.*

Light – electromagnetic radiation that occupies a specific range of wavelengths within the electromagnetic spectrum. In physics, the term sometimes refers to electromagnetic radiation of any wavelength, in which light includes gamma rays, X-rays, microwaves, and radio waves. *In common usage, light refers to visible light.*

Light-year – distance that light travels in a vacuum in one year (365.25 days); 9.46×10^{15} m.

Line spectrum – an emission (e.g., light, sound, or other radiation) spectrum consisting of separate isolated lines (discrete frequencies or energies); can be used to identify the elements in a matter of unknown composition.

Linear actuator – a motor that generates a linear movement directly.

Linear algebra – the branch of mathematics concerning vector spaces, often finite or countably infinite dimensional, and linear mappings between such spaces.

Linear elasticity – the mathematical study of how solid objects deform and become internally stressed due to prescribed loading conditions. Linear elasticity simplifies the more general *nonlinear theory of elasticity* and is a branch of *continuum mechanics*.

Lines of force – lines drawn to make an electric field strength map, with each line originating on a positive charge and ending on a negative charge; each line represents a path on which a charge would experience a constant force; *having the lines closer indicates a more energetic electric field.*

Liquids – one of four classical states of matter with a definite volume but no fixed shape. *A phase of matter composed of molecules that have interactions stronger than those found in gas but not strong enough to keep the molecules near the equilibrium positions of a solid, resulting in the characteristic definite volume but the indefinite shape of a liquid.*

Liouville's theorem – phase space volume is conserved.

Liquid crystal (LC) – matter in a state with properties between a conventional liquid and a solid crystal. *An LC may flow like a liquid, but its molecules may be oriented like a crystal.*

Liter – a metric system unit of volume, usually used for liquids.

Longitudinal strain – the ratio of change in the length of a body to its initial length.

Longitudinal waves – particles of the medium oscillate along a wave's propagation direction (e.g., sound waves).

Loudness – a subjective interpretation of a sound related to the energy of the vibrating source, the condition of the transmitting medium, and the distance involved.

Lubricant – a substance that reduces friction (i.e., force opposing the direction of motion).

Luminosity – the total energy radiated into space each second from the surface of a star.

Luminous – objects that produce visible light (e.g., Sun, stars, light bulbs, burning materials).

Lunar eclipse – when the Earth passes between the sun and the moon.

Lyman series – a group of lines in the ultraviolet region in the hydrogen spectrum.

M

M-theory – an extension of string theory that attempts to unify contradictory mathematical formulations and identifies 11 dimensions.

Mach number – a dimensionless quantity representing the ratio of an object's speed moving through a fluid to the local speed of sound.

Machine – any powered tool consisting of one or more parts is constructed to achieve a particular goal. Machines are usually powered by mechanical, chemical, thermal, or electrical means and are frequently motorized.

Machine element – an elementary component of a machine. Three basic types are: 1) structural components, 2) mechanisms, and 3) control components.

Maclaurin series – a representation of a function as an infinite sum of terms calculated from the values of the function's derivatives at a single point.

Magnetic domain – tiny physical regions in permanent magnets, 0.01 to 1 mm, have magnetically aligned atoms, giving the domain an overall polarity.

Magnetic field – a mathematical description of the magnetic influence of electric currents and magnetic materials. The magnetic field at any given point is specified by both a direction and a magnitude (or strength); as such, it is a vector field. *The region around a magnet where other magnetic objects experience its magnetic force; a model used to describe how magnetic forces on moving charges act at a distance.*

Magnetic poles – the ends (or sides) of a magnet about which the force of magnetic attraction seems to be concentrated.

Magnetic quantum number (*quantum mechanics*) – one of four descriptions of the energy state of an electron wave; describes the energy of an electron orbital as the orbital is oriented in space by an external magnetic field, a kind of energy sub-sublevel.

Magnetic reversal – the changing of the polarity of the Earth's magnetic field as the north magnetic pole and the south magnetic pole exchange positions.

Magnetic wave – the spread of magnetization from a small portion of a substance from an abrupt change in the magnetic field.

Magnification – the ratio of the size of the image to the size of the object.

Magnetism – a property of materials that respond to an applied magnetic field.

Magnitude – measurement of a vector; scalar quantities consist of a number and unit only.

Malleable – ability to be hammered into sheets (e.g., metals).

Malus Law – the intensity of light transmitted from the analyzer varies directly as the square of the cosine of the angle between the plane of transmission of the analyzer and the polarizer.

Maser – microwave amplification by stimulated emission of radiation.

Mass (m) – the quantity of matter in a body; the SI unit is the kg; remains the same everywhere; a measure of inertia, which means *resistance to a change of motion.*

Mass balance (or *material balance*) – an application of the law of conservation of mass to analyze physical systems.

Mass defect – the difference between the sum of the masses of the individual nucleons of a nucleus and the mass of the nucleus.

Mass density – see *density*.

Mass flux – the rate of mass flow per unit area. Common symbols are j, J, φ, or Φ, sometimes with subscript m to indicate mass is the flowing quantity. Its SI units are kg s^{-1} m^{-2}.

Mass moment of inertia – a property of a distribution of mass in space that measures its resistance to rotational acceleration about an axis.

Mass number (*atomic mass number* or *nucleon number*) – the number of protons and neutrons (together known as nucleons) in an atomic nucleus.

Mass number – the sum of the number of protons and neutrons in a nucleus; used to identify isotopes (e.g., Uranium-238).

Materials science – incorporates elements of physics, chemistry, and engineering concerned with the design and discovery of new materials, particularly solids.

Mathematical physics – applying mathematics in physics, developing mathematical methods suitable for such applications, and formulating physical theories.

Mathematics – the abstract study of topics encompassing quantity, structure, space, change, and other properties.

Matrix – a rectangular array of numbers, symbols, or expressions arranged in rows and columns. The individual items in a matrix are its *elements* or *entries*.

Matter – anything that occupies space and has mass. Any substance (often a particle) that has rest mass and (usually) also volume.

Maxwell's equations – a set of partial differential equations that, with the Lorentz force law, are the foundation of classical electrodynamics, classical optics, and electric circuits. *Maxwell's equations describe how electric and magnetic fields are generated and altered by each other, as well as charges and currents.*

Mean life – average time during which a system (e.g., atom, nucleus) is in a specified form.

Measure of central tendency – how quantitative data clusters around some value; any of several ways of specifying this "*central value*."

Mechanical energy – the sum of the potential energy (PE) and the kinetic energy (KE) of a body; energy associated with the position of a body.

Mechanical waves (*elastic waves*) – require a material medium for propagation (e.g., sound waves and water waves).

Mechanics – concerned with the behavior of physical bodies when subjected to forces or displacements and the subsequent effects of the bodies on their environment.

Megahertz (MHz) – a unit of frequency equal to 106 Hertz.

Melting (or *fusion*) – a physical process that results in the phase transition of a substance from a solid to a liquid. *The changing of a solid to a liquid state.*

Melting point – the temperature at which a solid-to-liquid phase change occurs.

Meson (*particle physics*) – a hadronic subatomic particle composed of one quark and one antiquark bound by strong interaction. *All mesons are unstable, with the longest-lived lasting only a few hundredths of a microsecond.*

Metal – matter having physical properties of conductivity, malleability, ductility, and luster.

Meter – the fundamental metric unit of length.

MeV – a unit of energy equal to 1.6×10^{-13} joules.

Millibar – a measure of atmospheric pressure equivalent to 1,000 dynes per cm^2.

Miscible fluids – fluids that can mix in any proportion.

Mixture – matter made of unlike parts with variable composition and separated into their parts by physical means.

Model – a mental or physical representation of something that cannot be observed directly.

Modulus of elasticity – the ratio of stress to strain produced in a body. The mathematical description of an object or substance's tendency to be deformed elastically (i.e., non-permanently) when a force is applied. The elastic modulus of an object is defined as the slope of its stress-strain curve in the elastic deformation region. As such, a stiffer material will have a higher elastic modulus.

Modulus of rigidity – the ratio of tangential stress to the shear strain produced in a body.

Molar mass – a physical property of matter defined as the mass of a given substance divided by the amount of substance and expressed in grams per mole.

Mole – the amount of a substance that contains Avogadro's number of atoms, ions, molecules, or any other chemical unit; 6.02×10^{23} atoms, ions, or other chemical units.

Molecular physics – studies the physical properties of molecules, the chemical bonds between atoms, and molecular dynamics. *It is closely related to atomic physics and overlaps with theoretical chemistry, physical chemistry, and chemical physics.*

Molecule – two or more atoms chemically combined. An electrically neutral group of two or more atoms held together by covalent chemical bonds. *Molecules are distinguished from ions by their lack of electrical charge.*

Moment – a measure of the turning effect of a force. Moment of a force = force × perpendicular distance from the fulcrum.

Moment of force (*torque or moment)* – the tendency of a force to rotate an object about an axis, fulcrum, or pivot. *As force is push or pull, torque can be considered a twist to an object.*

Moment of inertia – a property of a distribution of mass in space that measures its resistance to rotational acceleration about an axis.

Momentum – a measure of the quantity of motion, the product of the mass and velocity of a body; SI units are kg·m /s.

Monochromatic light – consists of a single wavelength.

Motion – any change in the position of an object over time. Motion can be mathematically described in terms of displacement, distance, velocity, speed, acceleration, and momentum. It is observed by attaching a frame of reference to an observer and measuring the change in an object's position relative to that frame. An object's motion cannot change unless a force acts upon it.

Muon (*particle physics*) – an elementary particle, technically classified as a lepton like an electron, with a unitary negative electric charge (−1) and a spin of 1⁄2. *Muons are not believed to have any sub-structure.*

N

Nanoengineering – the practice of engineering on the nanoscale. Nanoengineering is largely a synonym for nanotechnology but emphasizes the applied rather than the pure science aspects of the field.

Nanotechnology (or *nanotech*) – the manipulation of matter on an atomic and molecular scale. A more generalized description of nanotechnology is "*the manipulation of matter with at least one dimension sized from 1 to 100 nanometers*".

Natural frequency – the frequency of oscillation of an elastic object in the absence of external forces; depends on the size, composition, and shape of the object.

Negative electric charge – one of two types of electric charge; repels other negative charges and attracts positive charges.

Negative ion – atom or particle with a surplus of electrons and a negative charge.

Net force – the resulting force after vector forces are summed; if a net force is zero, the vector forces have canceled, and there is no unbalanced force.

Neutralization – the reaction between an acid and a base to give salt and water.

Neutrino (*particle physics*) – an electrically neutral subatomic particle denoted by the Greek letter ν (nu). All evidence suggests that neutrinos have mass but that their mass is tiny, even by the standards of subatomic particles. Their mass has never been measured accurately.

Newton (N) – a unit of force defined as kg·m/s^2; 1 Newton accelerates a 1 kg mass by 1 m/s^2.

Newton's First Law of Motion – a body continues in a state of rest or uniform motion in a straight line unless an external (unbalanced) force acts upon it.

Newton's Law of Gravitation – the gravitational force of attraction acting between two particles is directly proportional to the product of their masses and inversely proportional to the square of the distance between them; the force of attraction acts along the line joining the two particles; real bodies having spherical symmetry act as point masses with their mass assumed to be concentrated at their center of mass.

Newton's Laws of Motion – a set of three physical laws describing the relationship between the forces acting on a body and its motion due to those forces. *Together, they form the basis for classical or Newtonian mechanics.*

Newton's Second Law of Motion – the rate of change of momentum is equal to the force applied; the force acting on a body is directly proportional to the product of its mass and acceleration produced by force in the body.

Newton's Third Law of Motion – for every action, there is an equal and opposite reaction; the action and the reaction act on two different bodies simultaneously.

Newtonian mechanics (or *classical mechanics*) – a sub-field of mechanics concerned with the set of physical laws describing the motion of bodies under the collective actions of a system of forces.

Noise – sounds made up of groups of waves of random frequency and intensity.

Non-uniform acceleration – when the velocity of a body increases by unequal amounts in equal intervals of time.

Non-uniform speed – when a body travels unequal distances in equal intervals.

Non-uniform velocity – when a body covers unequal distances in equal intervals of time in a direction or when it covers equal distances in equal intervals but changes its direction.

Normal (*N*) – a line perpendicular to the surface of a boundary.

Nuclear energy – energy from reactions involving the nucleus.

Nuclear fission – the splitting of a heavy nucleus into more stable, lighter nuclei with an accompanying release of energy.

Nuclear force – one of four fundamental forces; a strong force of attraction that operates over short distances between subatomic particles; *overcomes the electric repulsion of protons in a nucleus and binds the nucleus.*

Nuclear fusion – a nuclear reaction of low-mass nuclei fusing to form a more stable and massive nucleus with an accompanying release of energy.

Nuclear physics – studies the constituents and interactions of atomic nuclei.

Nuclear reactor – a steel vessel in which a controlled chain reaction of fissionable materials releases energy.

Nucleon number (*atomic mass number* or *mass number*) – the number of protons and neutrons (together known as nucleons) in an atomic nucleus.

Nucleons – either a proton or a neutron in its role as a component of an atomic nucleus. *A collective name for protons and neutrons in the nucleus of an atom.*

Nucleus – central, positively charged, dense portion of an atom with protons and neutrons.

Nuclide – an atomic species characterized by the specific composition of its nucleus, i.e., by its number of protons, its number of neutrons, and its nuclear energy state.

O

Octet rule – during bonding, atoms tend to reach an electron arrangement with eight electrons in the outermost shell.

Ohm – the SI derived unit of resistance; 1 ohm = 1volt/ampere.

Ohm's Law – the current flowing through a conductor is directly proportional to the potential difference across the conductor ends. At a constant temperature, voltage across a conductor is proportional to current flowing through it. Voltage = current × resistance ($V = IR$).

Open system – a system across whose boundaries matter and energy can pass.

Optical fiber – a thin thread of fused silica that transmits light by total internal reflection.

Optical tweezers – an optomechanical device used to capture, analyze, and manipulate dielectric objects or particles, which operate via force from the electric field of light.

Optically detected magnetic resonance – an optical technique for initializing and reading quantum spin in some crystal defects.

Optics – involves the behavior and properties of light, including its interactions with matter and the construction of instruments that use or detect it. *Optics usually describes the behavior of visible, ultraviolet, and infrared light; however, because light is one of several forms of electromagnetic radiation, other forms, such as X-rays, microwaves, and radio waves, exhibit comparable properties.*

Orbital – region around the nucleus of an atom where an electron is likely to be.

Origin – point on a graph where the x and y variables have a value of zero simultaneously.

Oscillatory motion (or *vibratory motion*) – periodic motion of a body about its mean position.

Oxidation – the addition of oxygen or loss of electrons.

P

Particle physics – studies the nature of particles, which are the constituents of what is usually referred to as matter and radiation.

Pascal – a unit of pressure equal to a force of 1 N acting uniformly over an area of 1 m^2.

Pascal's Law – a principle in fluid mechanics that pressure exerted in a confined incompressible fluid is transmitted equally throughout the fluid such that initial pressure variations remain the same. *Pressure exerted in liquid is transmitted equally in all directions.*

Paschen series – a group of lines in the infrared region in the hydrogen spectrum.

Pauli exclusion principle (*quantum mechanics*) – no electrons in an atom can have the same four quantum numbers; a maximum of two electrons can occupy a given orbital.

Pedesis (or *Brownian motion*) – the continuous random motion of solid microscopic particles suspended in a fluid (i.e., gas, liquid) due to ongoing bombardment by atoms and molecules.

Peltier effect – the evolution or absorption of heat at the junction of two dissimilar metals carrying current.

Period (of a wave) – the time for a wave to travel through a distance equal to its wavelength; denoted by T; period of a wave = 1/frequency of the wave.

Period (of an oscillation) – the time to complete one oscillation; does not depend upon the mass of the bob and amplitude of oscillation; directly proportional to the square root of the length and inversely proportional to the square root of the acceleration due to gravity.

Periodic motion – a motion that repeats at regular time intervals.

Periodic table of the elements – a tabular display of the chemical elements organized based on their atomic numbers, electron configurations, and recurring chemical properties. *Elements are presented with increased atomic numbers (number of protons).*

Periodic wave – when the particles of the medium oscillate continuously about their mean positions regularly at fixed intervals.

Permeability – the ability to transmit fluids through openings, tiny passageways, or gaps.

pH scale – a scale from 0 to 14. If the pH of a solution is 7 it is neutral; if the pH of a solution is less than 7 it is acidic; if the pH of a solution is greater than 7 it is basic.

Permanent hardness – water that cannot be removed by boiling; caused by calcium sulfate.

Phase – when particles in a wave are in the same state of vibration (i.e., in the same position and the direction of motion).

Phase change – a substance changing from one state of matter to another; it absorbs or releases internal potential energy not associated with a temperature change.

Photons (*quantum physics*) – quanta of energy in the light wave; the particle associated with light. An elementary particle is the quantum of light and all other forms of electromagnetic radiation, and it is the force carrier for the electromagnetic force.

Photoelectric effect – the emission of electrons in some materials when the light of a suitable frequency falls on them.

Physical change – a change in the state of a substance but not in the identity of the substance.

Physical chemistry – the study of macroscopic, atomic, subatomic, and particulate phenomena in chemical systems in terms of laws and concepts of physics.

Physics – the natural science that studies matter and its motion through space and time, along with related concepts such as energy and force. *More broadly, it is the general analysis of nature conducted to understand how the universe behaves.*

Pitch (sound) – depends on the frequency of the wave.

Planck's constant (*quantum mechanics*) – the ratio of the energy of vibrating molecules to their frequency of vibration; a value of 6.63×10^{-34} J·s. *A physical constant that is the quantum of action in quantum mechanics.*

Plasma – a phase of matter; a hot highly ionized gas consisting of electrons and atoms stripped of their electrons because of high kinetic energies.

Plasticity – the property of a solid whereby it undergoes a permanent change in shape or size when subjected to stress.

Plastic strain – an adjustment to stress in which materials become molded or bent out of shape under stress and do not return to their original shape after the stress is released.

Pneumatics – the study and control of mechanical force and movement generated by applying compressed gas.

Polarized light – light whose constituent transverse waves vibrate in the same plane.

Polaroid – a film that transmits only polarized light.

Polaroid or polarizer – a device that produces polarized light.

Positive electric charge – one of the two types of electric charge; repels positive and attracts negative charges.

Positive ion – atom or particle with a net positive charge due to an electron or electrons being torn away.

Positron (*particle physics*) – an elementary particle having the same mass as an electron but an equal and positive charge.

Potential difference (or *voltage*) – the force that moves the electrons around the circuit. The unit is Volt (V).

Potential energy (PE) – possessed by a body by its position or configuration; see gravitational and *elastic potential energy*.

Power – scalar quantity for the rate of doing work; SI unit is Watt; 1 W = 1 J/s. The rate energy is converted between forms. Unit is Watts (W). Power = voltage × current (P = VI).

Pressure – force per unit area (e.g., kilograms per square meter (kg/m^2). Unit is Pascal (Pa). *The ratio of force to the area over which that force is distributed.*

Primary coil – part of a transformer; a coil of wire connected to an alternating current source.

Primary colors – three colors (red, yellow, and blue) combined in various proportions to produce any other color. When red, green, and blue are combined, it results in white. *For example, red + green + blue = white.*

Principal quantum number (*quantum mechanics*) – one of four descriptions of the energy state of an electron wave; describes the main energy level of an electron regarding its most probable distance from the nucleus.

Principle of calorimetry – if two bodies of different temperatures are in thermal contact, and no heat enters or exits the system, heat lost by the body with higher temperature equals heat gained by the body of lower temperature (i.e., the heat lost = the heat gained).

Probability – the expectation that an event will occur, or a statement is true. Probabilities are given a value between 0 (will not occur) and 1 (will occur). *The higher the probability of an event, the more certain one can be that the event will occur.*

Products – chemicals produced in a chemical reaction.

Progressive wave – transfers energy from one part of a medium to another.

Projectile – an object thrown into space horizontally or at an acute angle and under the action of gravity; the path followed by a projectile is its trajectory; the horizontal distance traveled by a projectile is its range; the time is taken from the moment it is thrown until when it strikes the ground is its time of flight.

Proof – a measure of ethanol concentration of an alcoholic beverage; double the concentration by volume (e.g., 50% by volume is 100 proof).

Properties – qualities or attributes that, taken together, are usually unique to an object (e.g., color, texture, size).

Proportionality constant – a value applied to a proportionality statement that transforms the statement into an equation.

Pulley – a wheel on an axle designed to support the movement of a cable or belt along its circumference; one of six classical simple machines. *Pulleys lift loads, apply force, and transmit power.*

Pulse (or *wave pulse*) – a wave of short duration confined to a small portion of the medium at any given time.

Q

Quanta (*quantum mechanics*) – fixed amounts, usually referring to fixed amounts of energy absorbed or emitted by matter.

Quantum limit – the shortest wavelength present in a continuous X-ray spectrum.

Quantum electrodynamics (QED) (*quantum mechanics*) – the relativistic quantum field theory of electrodynamics. It describes how light, and matter interact and is the first theory where complete agreement between quantum mechanics and special relativity is achieved. *QED mathematically describes all phenomena involving electrically charged particles interacting using the exchange of photons and represents the quantum counterpart of classical electromagnetism, giving a complete account of matter and light interaction.*

Quantum field theory – a theoretical framework for constructing quantum mechanical models of subatomic particles in particle physics and quasiparticles in condensed matter physics.

Quantum mechanics (or *wave mechanics*) – physical phenomena at microscopic scales, where the action is on the order of the Planck constant. Quantum mechanics departs from classical mechanics primarily at the quantum realm of atomic and subatomic length scales. *Quantum mechanics provides a mathematical description of much of the dual particle-like and wave-like behavior and interactions of energy and matter. Studies model of the atom based on the wave nature of subatomic particles and the mechanics of electron waves.*

Quantum numbers (*quantum mechanics*) – describe the energy states of an electron; in the Bohr model of the atom, the orbit quantum numbers could be any whole number (e.g., 1, 2, 3, etc.); in the quantum mechanics model of the atom, four quantum numbers describe the energy state of an electron wave (n, m, l, and s).

Quark (*particle physics*) – a fundamental particle with a charge with magnitudes of one-third or two-thirds of an electron. *Quarks combine to form composite particles called hadrons, the most stable of which are protons and neutrons, the components of atomic nuclei.*

R

Rad – a measure of radiation received by a material (radiation-absorbed dose).

Radiant energy – travels through space (e.g., visible light; electromagnetic spectrum).

Radiation (*nuclear physics*) – the emission and propagation of waves transmitting energy through space or medium. Heat transfers through invisible rays, which travel outwards from the hot object without a medium.

Radioactive decay – the natural, spontaneous disintegration or decomposition of a nucleus.

Radioactive decay constant (*nuclear physics*) – the ratio of the rate of nuclear isotopic disintegrations per unit of time to the total number of radioactive nuclei.

Radioactive Decay Law (*nuclear physics*) – the radioactive substance's disintegration rate is directly proportional to the number of undecayed nuclei.

Radioactive decay series – begins with one radioactive nucleus that decays into a second nucleus, decays into a third nucleus, and so on until a stable nucleus is reached.

Radioactivity – emission of particles or energy from an atomic nucleus as it disintegrates.

Radionuclide (*radioactive nuclide, radioisotope,* or *radioactive isotope*) (*nuclear physics*) – any nuclide possessing excess nuclear energy to the point that it is unstable. Such excess energy is emitted through several radioactive decay processes, resulting in a stable nuclide or sometimes another unstable radionuclide, which can then undergo further decay. *Specific radionuclides can occur naturally; many others can be produced artificially in nuclear reactors, cyclotrons, particle accelerators, or radionuclide generators.*

Rarefaction – a part of a longitudinal wave where the density of the medium particles is less than the typical density.

Real image (*optics*) – generated by a lens or mirror that can be projected onto a screen.

Reactants – chemicals that react together in a chemical reaction.

Rectifier – converts alternating current to direct current.

Rectilinear motion – a body in a straight line.

Redshift (*optics*) – occurs when light seen from an object moving away is proportionally increased in wavelength or "shifted" to the red end of the visible light spectrum.

Reduction – the removal of oxygen or the gaining of electrons.

Reflected ray – a line representing the direction of motion of light reflected from a boundary.

Reflection (*optics*) – the bouncing back of light from a surface.

Refraction – the bending of a light wave, a sound wave, or another wave from its straight-line path as it travels from one medium to another. The change in the direction of a wave as it passes from one transmission medium to another or because of a gradual change in the medium. *Though most used in the context of refraction of light, other waves, such as sound and fluid waves, also experience refraction.*

Refractive index – the ratio of the speed of light in a vacuum to that in the medium.

Relative density (*specific gravity*) – the ratio of the density (mass of a unit volume) of a substance to the density of a reference material. *Specific gravity* usually means relative density concerning water. *Relative density* is more common in modern scientific usage.

Relative humidity – the percentage of the amount of water vapor present in a specific air volume to the amount needed to saturate it.

Resistance (R) – the opposition of a conductor to current (i.e., the flow of charge). A good conductor has a low resistance, and a bad conductor has a high resistance.

Resolving power – the ability of an optical instrument to produce separable images of different points of an object.

Resonance – the frequency of an external force matches the natural frequency of the body.

Restoring force – tends to bring an oscillating body back to its mean position whenever it is displaced from the mean position.

Resultant force – a single force that acts on a body to produce the same effect on it as other forces collectively; see *balanced forces*.

Reverberation – apparent increase in the volume of sound caused by reflections from the boundary surfaces, usually arriving within 0.1 seconds after the original sound.

Rigid body – an idealized extended body whose size and shape are fixed and unaltered when forces are applied.

Rigid body – an idealization of a solid body in which deformation is neglected. In other words, the distance between any two given points of a rigid body remains constant in time regardless of the external forces exerted on it. *Even though such an object cannot physically exist due to relativity, objects can normally be assumed to be perfectly rigid if they are not moving near the speed of light.*

Rotational energy (or *angular kinetic energy*) – the kinetic energy due to the rotation of an object that forms part of its total kinetic energy.

Rotational speed (or *speed of revolution*) – the number of complete rotations or revolutions a rotating body makes per unit of time.

Rydberg formula (*atomic physics*) – describes the wavelengths of spectral lines of many chemical elements.

S

Salt – when a metal replaces the hydrogen of an acid.

Saturated air – air in which an equilibrium exists between evaporation and condensation; the relative humidity is 100 percent.

Saturated solution – the apparent limit to dissolving a given solid in a specified amount of water at a given temperature; a state of equilibrium exists between dissolving solute and solute coming out of solution.

Saturation value (or *absolute humidity*) – the ratio of water vapor in a sample of air to the volume of the sample. Maximum amount of water vapor in 1 m^3 of air at a temperature.

Scalar – any simple physical quantity that can be described by a single number (as opposed to vectors, tensors) which are described by several numbers such as magnitude and direction) and is unchanged by coordinate system rotations or translations (in Newtonian mechanics) or by Lorentz transformations or central-time translations (in relativity).

Scalar quantity – a physical quantity described entirely by its magnitude.

Scattering – the general physical process by which some forms of radiation, such as light, sound, or moving particles, are forced to deviate from a straight trajectory by one or more localized non-uniformities in the medium through which they pass.

Schrödinger equation (*quantum mechanics*) – a mathematical expression that describes the time evolution of wave functions in quantum mechanics.

Science – a systematic enterprise that builds and organizes knowledge through testable explanations and predictions about the universe.

Scientific law – a relationship between quantities (usually an equation in physical sciences) that describes a broader range of phenomena and is more important than a scientific principle.

Scientific principle – a relationship between quantities within a specific range of observations and behavior.

Screw – a mechanism that converts rotational motion to linear motion and a torque (rotational force) to a linear force; one of six classical simple machines.

Second – the standard unit of time in the metric and English measurement system.

Second Law of Motion – the acceleration of an object is directly proportional to the net force acting on that object and inversely proportional to the object's mass.

Secondary coil – part of a transformer; a coil of wire in which the voltage of the original alternating current in the primary coil is stepped up or down by electromagnetic induction.

Secondary colors – formed when two primary colors (i.e., red, green, blue) are mixed. The three secondary colors are yellow, magenta, and cyan. *For example, red + green = yellow; red + blue = magenta; and blue + green = cyan.*

Second's pendulum – a simple pendulum whose period on the surface of Earth is 2 seconds.

Semiconductors – elements whose electrical conductivity is intermediate between a conductor and an insulator.

Shear strain – ratio of relative displacements of one plane to its distance from a fixed plane.

Shear stress – restoring force developed per unit area when deforming force acts tangentially to the body's surface, producing a change in the body's shape without volume change.

Siemens – the derived SI unit of electrical conductance; equal to the conductance of an element with a resistance of 1 ohm; also written as ohm^{-1}.

Simple harmonic motion – the vibratory motion when the restoring force is proportional to the displacement from the mean position and is directed opposite to the displacement.

Simple machine – a mechanical device that changes the direction or magnitude of a force. *In general, a set of six simple classical machines identified by Renaissance scientists, drawing from Greek texts on technology, are collectively defined as the simplest mechanisms that can provide mechanical advantage* (also called *leverage*).

Simple pendulum – a heavy point mass (e.g., small metallic ball) suspended by a light inextensible string from frictionless rigid support; *simple machine based on gravity.*

Siphon – a tube in an inverted U shape that causes a liquid to flow uphill without pumps, powered by the fall of the liquid as it flows down the tube by the pull of gravity. *It may more generally refer to a wide variety of devices involving the flow of liquids through tubes.*

Snell's Law – the ratio of *sin* i to *sin* r is a constant equal to the refractive index of the second medium concerning the first.

Solar eclipse – happens when the moon passes between the sun and the Earth.

Solenoid – a cylindrical coil of wire that becomes electromagnetic when current passes.

Solids – a phase of matter with molecules that remain close to fixed equilibrium positions due to strong interactions between the molecules, resulting in the characteristic definite shape and definite volume of a solid.

Solubility – the tendency of a solid, liquid, or gaseous chemical substance (i.e., *solute*) to dissolve in another solid, liquid, or gaseous substance (i.e., *solvent*) to form a homogeneous solution of the solute in the solvent. *The solubility of a solute fundamentally depends on the specific solvent as well as on temperature and pressure.*

Solution – a mixture of a solute (usually a solid) and a solvent (usually a liquid).

Sonic boom – sound waves that pile up into a shock wave when a source travels faster than the speed of sound.

Sound – a form of energy; *a mechanical wave is an oscillation of pressure transmitted through a solid, liquid, or gas composed of frequencies within the range of human hearing.*

Specific gravity (or *relative density*) – the ratio of the density (mass of a unit volume) of a substance to the density of a reference material. *Specific gravity* usually means relative density concerning water. *Relative density* is more common in modern scientific usage.

Specific heat – the amount of heat energy required to increase the temperature of 1 g of a substance by 1 °C; *each substance has its specific heat value.*

Speed – a scalar quantity for the distance traveled by a body per unit of time; if a body covers the distance in time, its speed is given by distance/time; SI units are m/s.

Speed of light (*c*) – a universal physical constant defined as 299,792,458 m/s precisely because the length of the meter is defined from this constant and the international standard for time. *When not otherwise qualified, "speed of light" usually refers to the speed of light in a vacuum instead of through some physical medium.*

Speed of revolution (or *rotational speed*) – the number of complete rotations or revolutions a rotating body makes per unit of time.

Spin quantum number (*quantum mechanics*) – one of four descriptions of the energy state of an electron wave; describes the spin orientation of an electron relative to an external magnetic field.

Stable equilibrium – a body is stable if its center of gravity rises when it is slightly moved.

Stable nuclide – any nuclide not radioactive and does not spontaneously undergo radioactive decay, as opposed to a radionuclide. *When such nuclides are referred to concerning specific elements, they are usually termed stable isotopes.*

Standard Model (*particle physics*) – the theory of particle physics describes three of the four known fundamental forces (the electromagnetic force, the weak force, and the strong force, but not the gravitational force) and classifies all known elementary particles.

Standing waves – the condition where two waves of equal frequency traveling in opposite directions meet and form stationary regions of maximum displacement due to constructive interference and stationary regions of zero displacement due to destructive interference.

State of motion – when a body changes its position concerning a fixed point in its surroundings, the states of rest and motion are relative to the frame of reference.

State of rest – when a body does not change its position concerning a fixed point in its surroundings; the states of rest and motion are relative to the frame of reference.

Statics – the branch of mechanics concerned with analyzing loads (force, torque, or moment) on physical systems in static equilibrium, where the relative positions of subsystems do not vary over time or where components and structures are at a constant velocity.

Steam-point – the temperature of steam over pure boiling water under 1 atm pressure; taken as the upper fixed point (100 °C or 212 °F) for temperature scales.

Stefan-Boltzmann Law – the amount of energy radiated per second per unit area of a perfectly black body is directly proportional to the fourth power of the absolute temperature of the surface of the body.

Stiffness – the rigidity of an object, the extent to which it resists deformation in response to an applied force.

Strain – transforming a body from a reference configuration to a current configuration.

Streamline flow (or *laminar flow*) – when a fluid flows in parallel layers with no disruption between the layers.

Stress – 1. an applied force or system of forces that tends to strain or deform a physical body. 2. a measure of the internal forces acting within a deformable body. 3. a quantitative measure of the average force per unit area of a surface within a body on which internal forces act.

Subatomic particle (*particle physics*) – any particle smaller than an atom.

Sublimation – the transformation process directly from the solid phase to the gas phase without passing through an intermediate liquid phase. Sublimation is an endothermic phase transition that occurs at temperatures and pressures below a substance's triple point in its phase diagram. *The changing of a solid directly to a gas* (e.g., iodine).

Superconductors – a phenomenon of zero electrical resistance and expulsion of magnetic fields occurring in certain materials when cooled below a characteristic critical temperature. *Materials for which, under certain conditions, the electrical resistance approaches zero.*

Super-cooled – water in the liquid phase when the temperature is below the freezing point.

Supersaturated – greater than average saturation amount of a solute at a given temperature.

Surface tension – property of a liquid when its surface behaves like a stretched membrane.

Suspension – a mixture of a liquid and a finely divided insoluble solid.

T

Temperature – a measure of the hotness or coldness of a body; according to the molecular model, it measures the average kinetic energy of the molecules; *heat flows from a body at a higher temperature to a body at a lower temperature.*

Temporary hardness – hardness in water that can be removed by boiling. It is caused by calcium hydrogen carbonate.

Tensional stress – the opposite of compressional stress; when one part of a plate moves away from another that does not move.

Tesla (T) – SI unit of magnetic flux density; the magnetic flux density of a magnetic flux of 1 Wb through an area of 1 m^2.

Thermal capacity – the quantity of heat required to raise the temperature of the body one degree (1 K or 1 °C).

Thermal equilibrium – when two bodies in contact are at the same temperature with no heat flow between them: the average temperature of the bodies in thermal equilibrium. A state in which there is no net flow of thermal energy between two physical systems when the systems are connected by a path permeable to heat. A system may be in thermal equilibrium with itself if the temperature within the system is spatially and temporally uniform. *Systems in thermodynamic equilibrium are in thermal equilibrium, but the converse is not always true.*

Thermal expansion – the increase in the size of an object when heated.

Thermometer – a device used for the measurement of temperature; the mercury thermometer is commonly used.

Third Law of Motion – when two objects interact, the force exerted on one object is equal in size and opposite in direction to the force exerted on the other object; forces always occur in matched pairs that are equal and opposite.

Titration – the process of adding one solution from a burette to a measured amount of another solution to determine how much of each is required to react.

Torque (*moment* or *moment of force*) – the tendency of a force to rotate an object about an axis, fulcrum, or pivot. *As force is push or pull, torque can be considered a twist to an object.*

Total internal reflection – a condition where all light is reflected from a boundary between materials; occurs when light travels from a denser to a rarer medium, and the angle of incidence is greater than the critical angle.

Toughness – the ability of a material to absorb energy and plastically deform without fracturing. Material toughness is the energy per unit volume a material can absorb before rupturing. It is also defined as the resistance to fracture of a material when stressed.

Trajectory – the path that a moving object follows through space as a function of time.

Transformation of energy – converting one form of energy into another (e.g., when a body falls, its potential energy is converted to kinetic energy).

Transverse wave – medium particles oscillate perpendicular to the wave's propagation direction (e.g., water, light, radio waves).

Trigonometry – studies triangles and the relationships between their sides and the angles between these sides.

Triple point – the temperature and pressure at which a given substance's three phases (gas, liquid, and solid) coexist in thermodynamic equilibrium.

Trough – the point of maximum negative displacement on a transverse wave.

U

Ultrasonic – sound waves too high in frequency (> 20,000 Hz) to be heard by the human ear.

Unbalanced forces – when some forces act on a body, and the resultant force is not zero.

Uncertainty principle – any of various mathematical inequalities asserting a fundamental limit to the precision with which specific pairs of physical properties of a particle, such as position x and momentum p, can be known simultaneously.

Uniform acceleration – velocity of a body increases by equal amounts in equal time intervals.

Uniform circular motion – an object in a circular path with uniform speed.

Uniform speed – when a body travels equal distances in equal intervals.

Uniform velocity – when a body travels along a straight line in a direction with equal distance in equal time intervals.

Universal Law of Gravitation – every object is attracted to every other object with force directly proportional to the product of their masses and inversely proportional to the square of the distance between the centers of the two masses.

Unpolarized light – consists of transverse waves vibrating in random directions.

Unstable equilibrium – a body is unstable if its center of gravity falls when slightly moved.

Utility frequency – oscillations of alternating current (AC) in an electric power grid transmitted from a power plant to the end-user.

V

V-number (*Abbe number* and *constringence*) (*optics and lens design*) – a measure of a transparent material's dispersion (a variation of refractive index versus wavelength). *High values of V indicate low dispersion.*

Vacuum – an area of space that contains no matter.

Van der Waals force – a general term for weak attractive intermolecular forces.

Valence electron – associated with an atom participating in a chemical bond.

Valence shell – the outermost electron shell of an atom.

Valency – the number of electrons an atom gains, loses, or shares to have a full outer shell.

Vapor – the gaseous state of a substance generally in a liquid state.

Vector – any quantity that has both magnitude and direction.

Vector quantity – a quantity that needs magnitude and direction to describe it.

Vector space – a mathematical structure formed by a collection of elements called vectors, which may be added together and multiplied (*scaled*) by numbers called *scalars*.

Velocity (v) – distance traveled by a body in a direction per unit time; the displacement of the body per unit time; a vector quantity; the SI units are m/s.

Vibration – a back-and-forth motion that repeats.

Vibratory motion (or *oscillatory motion*) – periodic motion of a body about its mean position.

Virtual image (*optics*) – formed when the reflected or refracted light rays appear to meet; this image *cannot be projected on a screen.*

Visible light – a form of electromagnetic radiation generally defined as the range of wavelengths visible to the average human eye.

Volt (V) – the SI-derived unit for electric potential, electric potential difference, and electromotive force, the difference in electric potential between two points of a conducting wire when an electric current of one ampere dissipates one watt of power between those two points. *A unit of potential difference equivalent to joules/coulomb.*

Voltage (or *potential difference*) – the force that moves the electrons around the circuit. The unit is Volt (V).

Voltage drop – the difference in electric potential across a resistor or other part of a circuit that consumes power.

Voltmeter – an instrument used for measuring the difference in electrical potential between two points in an electric circuit. *Analog voltmeters move a pointer across a scale proportionately to the circuit's voltage.*

Volume – the amount of space an object occupies.

W

Watt (W) – a derived unit of power in the International System of Units (SI) defined as one joule per second. The watt measures the rate of energy conversion or transfer.

Wave – a disturbance or oscillation through a medium accompanied by energy transfer.

Wavelength (λ) – the distance between the two nearest points on a wave in the same phase; the distance between two adjacent crests or troughs.

Wave (mechanical) – a periodic disturbance produced in a material medium due to the vibratory motion of the medium particles.

Wave mechanics (*quantum mechanics*) – physical phenomena at microscopic scales, where the action is on the order of the Planck constant. Quantum mechanics departs from classical mechanics primarily at the quantum realm of atomic and subatomic length scales. *Quantum mechanics provides a mathematical description of much of the dual particle-like and wave-like behavior and interactions of energy and matter. Studies model of the atom based on the wave nature of subatomic particles and the mechanics of electron waves.* An alternate name for quantum mechanics derived from the wavelike properties of subatomic particles.

Wave motion – the movement of a disturbance from one part of a medium to another involving the transfer of energy but not the transfer of matter.

Wave period – the time required for two successive crests (or successive regions) of the wave to pass a given point.

Wave velocity – the distance a wave travels in one second; depends on the medium it passes.

Wavelength – a measure of the distance traversed by a single spatial period of a sinusoidal wave, i.e., the distance over which the wave's shape repeats.

Weak interaction (*weak force* or *weak nuclear force*) (*quantum mechanics*) – one of the four fundamental forces of nature, along with the strong nuclear force, electromagnetism, and gravitation. It is responsible for the radioactive decay of subatomic particles and initiates the process known as hydrogen fusion in stars.

Wedge – a triangular, round tool in the form of a compound and portable inclined plane; one of six classical simple machines.

Weight – the force with which a body is attracted towards the center of the Earth; the SI unit is N; the gravitational units are kg·wt and g·wt; the weight of a body is given by *mg*. Weight = mass × acceleration due to gravity.

Weightlessness – when the apparent weight of a body becomes zero; objects falling freely under the action of gravity are seemingly weightless.

Wheel and axle – a wheel attached to an axle so that these two parts rotate together in which a force is transferred from one to the other: one of six classical simple machines.

White body – a hypothetical idealized physical body that reflects all incident electromagnetic radiation entirely and uniformly in all directions, the opposite of a *black body*.

Wien's Displacement Law – for a black body, the product of the wavelength corresponds to its maximum radiance, and its absolute temperature is constant.

Wind – the flow of gases on a large scale.

Wind shear (or *wind gradient*) – a difference in wind speed and direction over a relatively short distance in the atmosphere. Wind shear can be broken down into vertical and horizontal components, with horizontal wind shear seen across fronts and near the coast and vertical shear typically near the surface, though also at higher levels in the atmosphere.

Work – when a force acting on a body displaces it; work = force × displacement (W = Fd) in the direction of the force; work is a scalar quantity; the SI unit is *Joule*.

X

X-ray – a high-energy photon (between 100 eV and 100 keV) with a wavelength shorter than that of ultraviolet radiation and longer than that of gamma radiation.

Y

Young's modulus of elasticity – a measure of the stiffness of a solid material that defines the relationship between mechanical stress and strain. *The ratio of normal stress to the longitudinal strain produced in a body.*

Z

Zeeman effect – splitting a spectral line into several components in the presence of a static magnetic field by lifting degeneracy in electronic states.

Zeroth Law of Thermodynamics – if body A is in thermal equilibrium with body B, and B is also in thermal equilibrium with C, then A is necessarily in thermal equilibrium with C.

Notes for active learning

Frank J. Addivinola, Ph.D.

This study guide's lead author and chief editor is Dr. Frank Addivinola. With his outstanding education, laboratory research, and decades of university science teaching, Dr. Addivinola lent his expertise to oversee the development of this series.

During his extensive career, Dr. Addivinola held faculty positions at colleges and universities, including Harvard University, Johns Hopkins University, University of Maryland, and Northeastern University, and taught undergraduate and graduate-level courses in biology, biochemistry, organic chemistry, inorganic chemistry, physics, anatomy and physiology, medical terminology, nutrition, and medical ethics. He received several awards for his research and presentations.

Dr. Frank Addivinola conducted original research in developmental biology as a doctoral candidate and pre-IRTA fellow in Molecular and Cell Biology at the National Institutes of Health (NIH). His dissertation advisor was Nobel laureate Marshall W. Nirenberg, Chief of the Biochemical Genetics Laboratory at the National Heart, Lung, and Blood Institute (NHLBI). Before NIH, Dr. Addivinola researched prostate cancer in the Cell Growth and Regulation Laboratory of Dr. Arthur Pardee at the Dana Farber Cancer Institute of Harvard Medical School.

Dr. Addivinola holds an undergraduate degree in biology from Williams College. He completed his Masters at Harvard University, Masters in Biotechnology at Johns Hopkins University, and five other graduate degrees at the University of Maryland University College, Suffolk University, and Northeastern University.

www.ingramcontent.com/pod-product-compliance
Ingram Content Group UK Ltd.
Pitfield, Milton Keynes, MK11 3LW, UK
UKHW051240300926
15471UKWH00015B/1127

9 798885 573030